Radek Silhavy · Petr Silhavy
Editors

Artificial Intelligence for System Oriented Design

Proceedings of 8th Computational Methods in Systems and Software 2024
Volume 1

Springer

Editors
Radek Silhavy
Faculty of Applied Informatics
Tomas Bata University in Zlín
Zlín, Czech Republic

Petr Silhavy
Faculty of Applied Informatics
Tomas Bata University in Zlín
Zlín, Czech Republic

ISSN 2367-3370 ISSN 2367-3389 (electronic)
Lecture Notes in Networks and Systems
ISBN 978-3-031-96797-9 ISBN 978-3-031-96798-6 (eBook)
https://doi.org/10.1007/978-3-031-96798-6

This Springer imprint is published by the registered company Springer Nature Switzerland AG
The registered company address is: Gewerbestrasse 11, 6330 Cham, Switzerland

Lecture Notes in Networks and Systems 1489

The series "Lecture Notes in Networks and Systems" publishes the latest developments in Networks and Systems—quickly, informally and with high quality. Original research reported in proceedings and post-proceedings represents the core of LNNS.

Volumes published in LNNS embrace all aspects and subfields of, as well as new challenges in, Networks and Systems.

The series contains proceedings and edited volumes in systems and networks, spanning the areas of Cyber-Physical Systems, Autonomous Systems, Sensor Networks, Control Systems, Energy Systems, Automotive Systems, Biological Systems, Vehicular Networking and Connected Vehicles, Aerospace Systems, Automation, Manufacturing, Smart Grids, Nonlinear Systems, Power Systems, Robotics, Social Systems, Economic Systems and other. Of particular value to both the contributors and the readership are the short publication timeframe and the world-wide distribution and exposure which enable both a wide and rapid dissemination of research output.

The series covers the theory, applications, and perspectives on the state of the art and future developments relevant to systems and networks, decision making, control, complex processes and related areas, as embedded in the fields of interdisciplinary and applied sciences, engineering, computer science, physics, economics, social, and life sciences, as well as the paradigms and methodologies behind them.

Indexed by SCOPUS, EI Compendex, INSPEC, WTI Frankfurt eG, zbMATH, SCImago.

All books published in the series are submitted for consideration in Web of Science.

For proposals from Asia please contact Aninda Bose (aninda.bose@springer.com).

Preface

We are pleased to introduce this year's five-volume proceedings of the 2024 International Conference on Computational Methods in Systems and Software (CoMeSySo 2024). These volumes represent the latest achievements in our field, showcasing a variety of theoretical insights, practical applications, and interdisciplinary collaborations. Compiled through a rigorous peer-review process, the collection underscores the international community's dedication to advancing research on complex systems and software.

The first volume features contributions that revisit foundational theories while proposing new computational approaches. These papers offer refined methods that address persistent challenges in systems engineering and provide fertile ground for further innovation. With an emphasis on reliability and precision, the research presented here demonstrates the continual evolution of core techniques.

In the second volume, authors examine emerging topics spanning simulation technologies, model-based solutions, and state-of-the-art optimisation strategies. Their collective work speaks to the vitality of our discipline, illustrating how novel frameworks and flexible architectures can overcome obstacles posed by real-world constraints.

The third volume shifts the focus to cross-disciplinary research, highlighting how computational methods intersect with fields such as data science, artificial intelligence, and human-computer interaction. This section is particularly noteworthy for illustrating how the fusion of diverse perspectives sparks inventive applications and enhances the functionality of contemporary systems.

Within the fourth volume, readers will find papers that delve into software architectures, testing methodologies, and quality assurance practices. These contributions underscore the importance of robust processes in delivering reliable products, discussing how best to balance theoretical insights with the demands of industry and society.

Finally, the fifth volume addresses broader aspects of systems design and deployment. It features studies that tackle large-scale implementation, project governance, and the integration of complex infrastructures. In this section, the authors propose forward-looking strategies that can support sustainable development and maintain the resilience of future systems.

We hope the research's breadth and depth will inspire you as you explore each volume. We thank the authors, reviewers, and the organising committee for their invaluable roles in making this possible.

Radek Silhavy
Petr Silhavy

Organization

Program Committee

Program Committee Chairs

Program Committee Members

Elarbi Badidi	College of Information Technology, United Arab Emirates University, Al Ain, United Arab Emirates
Luis Alberto Morales Rosales	Head of the Master Program in Computer Science, Superior Technological Institute of Misantla, Mexico
Mariana Lobato Baes	Research-Professor, Superior Technological of Libres, Mexico
Abdessattar Chaâri	Laboratory of Sciences and Techniques of Automatic Control & Computer Engineering, University of Sfax, Tunisian Republic
Gopal Sakarkar	Shri. Ramdeobaba College of Engineering and Management, Republic of India
V. V. Krishna Maddinala	GD Rungta College of Engineering & Technology, Republic of India
Anand N. Khobragade	Scientist, Maharashtra Remote Sensing Applications Centre, Republic of India
Abdallah Handoura	Computer and Communication Laboratory, Telecom Bretagne, France
Almaz Mobil Mehdiyeva	Department of Electronics and Automation, Azerbaijan State Oil and Industry University, Azerbaijan

Technical Program Committee Members

Ivo Bukovsky, Czech Republic
Maciej Majewski, Poland
Miroslaw Ochodek, Poland
Bronislav Chramcov, Czech Republic
Eric Afful Dazie, Ghana
Michal Bliznak, Czech Republic
Donald Davendra, Czech Republic
Radim Farana, Czech Republic
Martin Kotyrba, Czech Republic
Erik Kral, Czech Republic
David Malanik, Czech Republic
Michal Pluhacek, Czech Republic
Zdenka Prokopova, Czech Republic
Martin Sysel, Czech Republic
Roman Senkerik, Czech Republic
Petr Silhavy, Czech Republic
Radek Silhavy, Czech Republic
Eva Volna, Czech Republic

Janez Brest, Slovenia
Ales Zamuda, Slovenia
Boguslaw Cyganek, Poland
Krzysztof Okarma, Poland
Monika Bakosova, Slovak Republic
Pavel Vaclavek, Czech Republic
Olga Brovkina, Czech Republic
Elarbi Badidi, United Arab Emirates

Organizing Committee Chair

Radek Silhavy — Tomas Bata University in Zlin, Faculty of Applied Informatics, email: comesyso@openpublish.eu

Conference Organizer (Production)

Silhavy s.r.o.
Web: https://comesyso.openpublish.eu
Email: comesyso@openpublish.eu

Conference Website, Call for Papers

https://comesyso.openpublish.eu

Contents

Analysis of the Psychophysiological Characteristics of Athletes

Fedor Kulikov[1], Zulfiya Kamaldinova[1], Anton Ivaschenko[2(✉)], and Denis Zheikov[2]

[1] Samara State Technical University, 244, Molodogvardeyskaya, Samara, Russia
[2] Samara State Medical University, 89, Chapaevskaya, Samara, Russia
anton.ivashenko@gmail.com

Abstract. The paper discusses the problem of analysing the psychophysiological indicators of athletes using modern information technologies for decision-making support. There is proposed a solution using the example of tennis, used to predict athletes' victories, cluster tennis players by level of play, and deter-mine the playing style using a heat map of shots. An original mathematical model is used in the tournament recommendation module to calculate predicted tournament ranking points. The modal was probated using a dataset describing 1500 matches held in the Samara region in 2021. Clustering of tennis players allows evaluating changes in the level of play of different regions. The cluster-ing method uses three parameters to determine groups of players: age, rating and number of matches played per year. Based on the heat map, which shows the places where the player most often encounters the ball, the algorithm draws conclusions about the tennis player's playing styles. Combination of the pro-posed solutions supports automated coach decision-making to improve the effectiveness of athletes' training. #COMESYSO1120.

Keywords: Psychological Testing · Predicting · Clustering · Decision-making Support

1 Introduction

Analysis of the psychophysiological state of athletes is an important component of assessing and ensuring their performance. To solve this task, information technologies for statistical analysis of the performance of individual athletes and groups are widely used, which makes it possible to predict and manage their development. Modern Big Data analysis technologies make it possible to build fully functional analytics and decision-making support systems for coaches and doctors involved in training sessions and competitions.

Modern research notes the dependence of the performance of athletes not only on their physical form, but also on the emotional state and individual psychological attitude and state of mind [1, 2]. Sports psychology is deeply studied in competitive sports [3, 4]. The dependence of the athlete's physical condition on his mental state is noted as well

R. Silhavy and P. Silhavy (Eds.): CoMeSySo 2024, LNNS 1489, pp. 1–8, 2025.
https://doi.org/10.1007/978-3-031-96798-6_1

[5, 6]. Additionally, the influence of the athlete's psycho-emotional state on his burnout and possible injury is being studied [7, 8].

As part of the algorithms and software development for decision-making support systems in this area, the challenging problem is rational choice and visualization of the most critical indicators [9–11]. The system should adequately display the most complete description of the psychophysiological parameters of the athlete's condition. However, this description should be concise to avoid information overload and con-fusing calculations.

2 Predicting an Athlete's Victory

The main purpose of developing a mathematical model for predicting an athlete's victory in a match is to provide statistical information to players. For example, a tennis player can look at the tournament grid to see which opponent he will most likely play with in 1, 2 or more tournament rounds in order to begin preparing for him in advance. Additionally, a mathematical model is used in the tournament recommendation module to calculate predicted tournament ranking points.

On the basis of system analysis of the problem domain there were formulated the following indicators:

$f_{ste} = \sum_i score_i \cdot time_i / \sum_i time_i$ - efficiency (score) in matches, taking into account the strength of opponents and how long ago the matches were played;
$f_{count} = 1/(1 + e^{-N/10})$ - number of matches per year;
$f_{div} = n / \sum_i unique\_matches_i$ - adaptation to the current opponent's technique;
$f_{per} = \sum_i score_i \cdot time_i / \sum_i time_i$ - match performance against current opponent;
$f_{dt} = \sqrt{N_{max}^2 / (\max\{N_{max}\} \cdot N_i)}$ - adaptability to the game at the current time of day, where N_{win} is the number of wins at this time of day, N_i is the number of wins in the i-th time interval);
$f_{fatigue} = \sum f_{duration} \bullet f_{hours}$ - fatigue from previous matches;
$f_{moral} = f_{test}^{self} / f_{test}^{enemy} \cdot winrate_{self} / winrate_{enemy}$ - moral qualities (concentration, motivation, staying calm);
$f_{style} = k_{style}^{self} / k_{style}^{enemy}$ - effectiveness of play style against current opponent;
$f_{rating} = (p_{self}/p_{enemy})^{0.25}$ - player rating.

An athlete training model can be described as a number of factors:

$$Y = F(X),$$

$$X = \{f_{ste}, f_{count}, f_{div}, f_{per}, f_{dt}, f_{fatigue}, f_{moral}, f_{style}, f_{rating} \cdot\} \tag{1}$$

The target variable y is the calculated probability of the tennis player winning. To obtain a training sample, you cannot use the classic binary classification "1" – victory and "0" – defeat, since the accuracy of the model will be extremely low. Instead, the y value is the score (2–3) of the match, which reflects the approximate strength ratio.

$$y = s_{self} / (s_{self} + s_{enemy}), \tag{2}$$

where $s_x = 1 \cdot Game + 3 \cdot Set + 4 \cdot Match$.

This means that a tennis match is played up to two sets won (Set), and at major international tournaments – up to three. To win a set you need to win 6 games, or with a score of 7–5 or 7–6 if the game score was 5–5 (6–6). According to the formula, 1 point is given for each winning game, 3 points for a set, and 4 points for winning a match. This formula is subjective, but it relatively good at representing the balance of power of the players. As a result, the model should give an approximate probability of victory.

The training dataset of match results comprised 25% of the total number of matches. The data was collected manually from 1500 matches held in the Samara region in 2021. All calculations of factors were carried out in C #. Then the obtained data is transferred to the MySql DBMS (since, in addition to the machine learning problem being solved, they are also used in statistical information in the application itself). And the processing of the calculated values is implemented in Python. Correlations between factors, as well as the correlation of factors with the target function, were assessed (see Fig. 1).

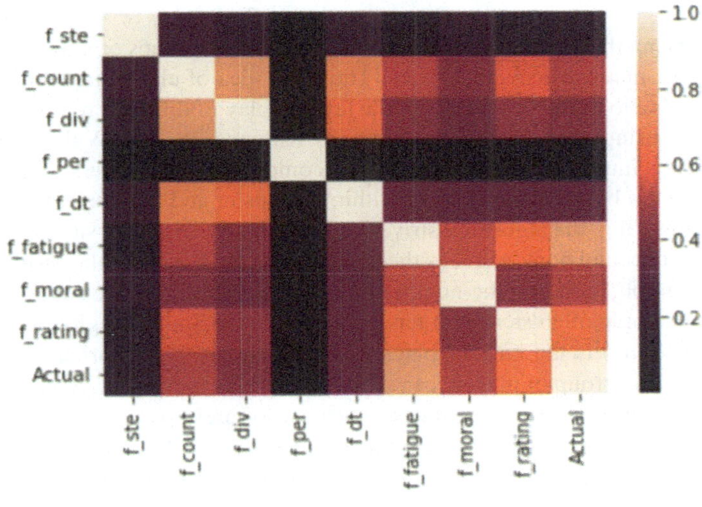

Fig. 1. Heat map of sporting activity factor correlations.

Using the heat map, you can process the f personal factor, the correlation value of which is close to zero with all other factors. This is due to the fact that the sample of matches is currently small: the players, with rare exceptions, have met each other no more than once. This problem will be solved with an increase in the sample.

The correlation values of the remaining factors generally range from 30 to 70% with each other, which indicates that there are no interdependent factors. Linear regression coefficients are captured in the following model:

$$Y = 0.032 \cdot f_{ste} + 0.036 \cdot f_{count} - 0.029 \cdot f_{div} - 0.032 \cdot f_{per} + 0.061 \cdot f_{dt} + +0.031 \cdot f_{fatigue} +0.083 \cdot f_{moral} - 0.1 \cdot f_{style} + 0.394 \cdot f_{rating}. \quad (3)$$

The accuracy of this model was 68.3%. Application of this model in practice proves its adequacy. There are still remaining a few ways to improve it.

The range of values cannot fit from 0 to 100%. Both negative values and values greater than 100% appear in the model, which does not correspond to the probability of victory. Alternatively, you can "cut" such values and bring them to 0 or 100%, respectively, but this is not acceptable.

Besides the approach should have a function for excluding certain factors from the model at the user's request. For example, a user may decide to exclude a rating factor, which, although it reflects the overall level of a tennis player's play, is essentially just a number and should not affect the result. To get rid of a factor or add a new factor to a regression equation, you have to completely recalculate the coefficients. This means that you will have to store the coefficients for every possible combination of factors or calculate the coefficients in real time, but the recalculation will take about a minute.

3 Clustering Tennis Players by Level of Play

The clustering method uses three parameters to determine groups of players: age, rating and number of matches played per year. The main idea of clustering tennis players is that it is possible to evaluate changes in the level of play of different regions. Clustering is carried out using nearest neighbor methods KMeans. WCSS (Within-Cluster Sum of Squares) is a measure used to evaluate the compactness of clusters. It is the sum of squared distances between each object within the cluster and its centroid.

When clustering, the algorithm strives to minimize WCSS. The smaller the WCSS, the more compact and homogeneous the clusters will be. Based on the resulting WCSS dependency graph Based on the number of clusters 4 clusters were selected. Figure 2 presents an example of clustering of tennis players in the Samara region.

Cluster 0 (blue) is the bulk of players, they regularly play tournaments and have some results. This group of players can serve as an indicator of the overall level of play in the region. Cluster 1 (red) - includes players who rarely participate in tournaments and play more for fun. Cluster 2 (green) are young players. This cluster is an indicator of the development of the level of play in the region. Cluster 3 (yellow) - includes the best players in the region who have fewer matches played in tournaments in Russia, but more matches in tournaments in other countries, which allowed them to earn a lot of rating points.

The proposed solution was used to provide decision-making support for tournament recommendation for players. Russian tennis tournament holds more than 1500 tournaments a year; and players choose which tournaments are best to participate in. Currently, tennis players learn about tournaments through the official website and look for them among the general list.

Sometimes a player needs to choose which of several tournaments taking place at the same time is better for him to play in order to get in shape or increase his rating, which is more important to him at the moment. In this case, the player will face a problem, as soon as the list of tournament participants is constantly changing. This may affect the maximum number of rating points that can be earned in a tournament and the player's expectations for the tournament.

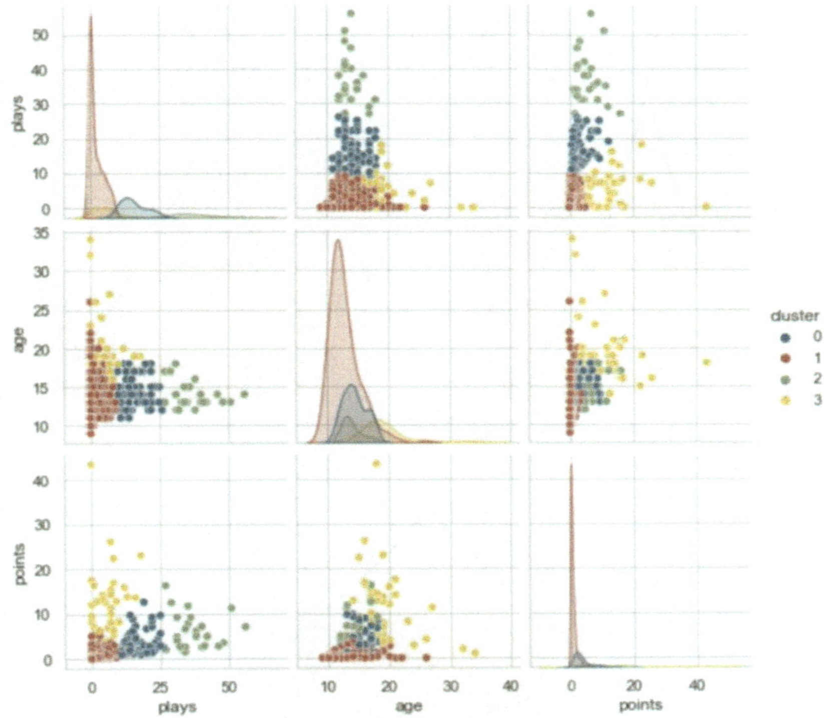

Fig. 2. Clustering of players in the Samara region.

Thus, a tool is required that will track important tournaments for the player so that he does not miss them due to the notification system in the application, offer suitable tournaments and maintain up-to-date information about the composition of the participants in each tournament. The criteria by which the algorithm sorts the priority of tournaments are as follows: distance to the city; predicted range of rating points; aver-age rating of participants in the previous tournament; percentage of players with un-familiar equipment.

4 Building a Heat Map of Sporting Activity

The heat map (see Fig. 3) shows the places where the player most often encounters the ball. Based on the heat map, the algorithm must draw conclusions about the tennis player's playing styles. Most of the points should be close to the black curves.

Each playing style has its own curves. Adaptability to style is calculated by the sum of squared distances from each point to the curve. The active style involves constant pressure, high tempo of play and an active position on the court. Mostly flat shots and half-volley shots or shots at the net are used. The main goal is to independently create a winning situation by knocking your opponent out of the court and complete the rally. The

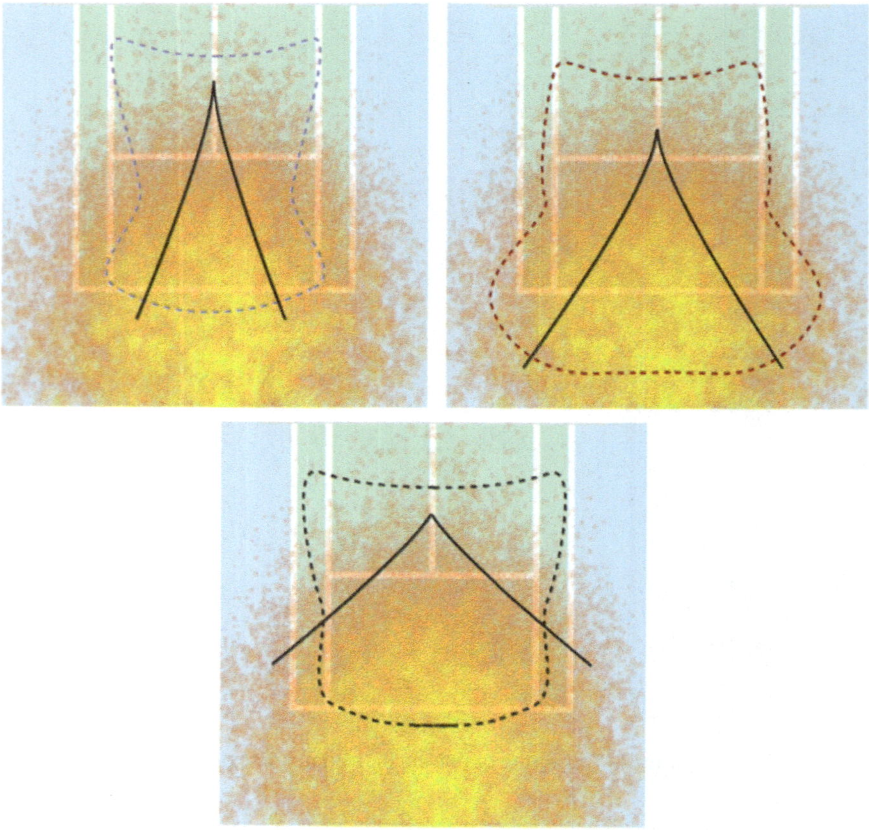

Fig. 3. Playing style curves: Active (top left), Defensive (top right), Reactive (bottom).

high pace of the game forces the opponent to make mistakes, which are immediately exploited. This style requires a complete understanding of pressure and position and is characterized by low reliability. Suitable for confident and morally charged players, since any doubt in the shot will most likely lead to a mistake.

The defensive style consists of using tight and twisting blows, positioning the player 2–3 meters from the back line. The main goal is to physically and mentally exhaust a strong opponent, and to outplay against weak opponents due to reliability and endurance. It is suitable for mentally tough players with good physical fitness.

The reactive style is about fully understanding the game as a whole, constantly moving around the court, changing the pace of play and disrupting the opponent's play. The main goal is to play as inconveniently as possible for the opponent, concentrating on his movements, and use any mistake by the opponent as an advantage in the play. The entire arsenal of strikes and serves is used. It requires a high level of concentration (Fig. 4).

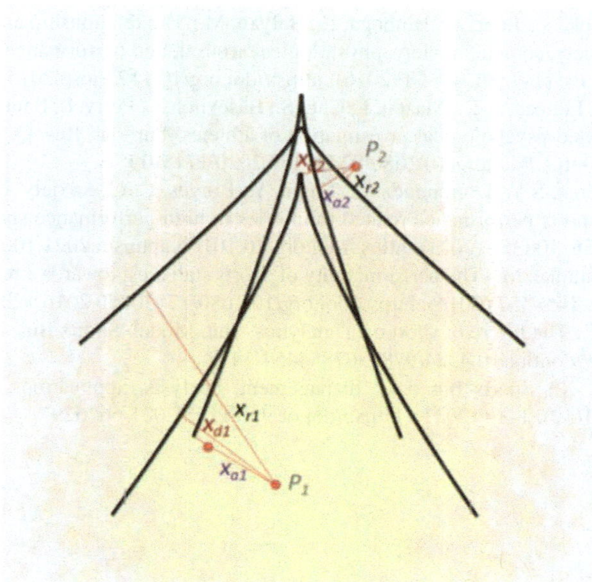

Fig. 4. Adaptation to a playing style.

5 Conclusion

The proposed solution was implemented as a part of information and analytical sys-tem being developed in integration with a web application to automate tennis tournaments. The system is currently used for athlete's decision-making support as a tournament rec-ommendation module. Next steps are concerned with improving the pre-diction accuracy and implementation of Artificial Intelligence.

References

1. Palazzolo, J.: Anxiety and performance. Encéphale. **46**(2), 158–161 (2020). https://doi.org/10.1016/j.encep.2019.07.008
2. Fadare, S., Isong, L.M., Lambaco, E., Montalban, K., Ken, P.: Athletes' health and well-being: a review of psychology's state of mind. Am. J. Multidis. Res. Innov. **1**, 44–50 (2022). https://doi.org/10.54536/ajmri.v1i4.551
3. He, G., Dong, M.: Research and development of sports psychology in the field of competitive sports. J. Psychol. Res. 1(1), 1 (2018). https://doi.org/10.30564/jpr.v1i1.196
4. Sushko, R., Vysochina, N., Vorobiova, A., Doroshenko, E., Pastuhova, V., Vysochin, F.: Psychological selection in game sports on the basketball example. J. Phys. Educ. Sport. **19**, 1708–1714 (2019). https://doi.org/10.7752/jpes.2019.03250
5. Slimani, M., Taylor, L., Baker, J., Elleuch, A., Ayadi, F., Chamari, K., Cheour, F.: Effects of mental training on muscular force, hormonal and physiological changes in kickboxers. J. Sports Med. Phys. Fitness. **57**, 1069–1079 (2017)

6. Balyan, K., Tok, S., Tatar, A., Binboğa, E., Balyan, M.: The relationship among personality, cognitive anxiety, somatic anxiety, physiological arousal, and performance in male athletes. J. Clin. Sport Psychol. **10**, 48–58 (2016). https://doi.org/10.1123/jcsp.2015-0013
7. Monfared, S., Lebeau, J.-C., Mason, J., Cho, S., Basevitch, I., Perry, I., Baur, D., Tenenbaum, G.: A bio-physio-psychological investigation of athletes' burnout. Res. Q. Exerc. Sport. **92**, 1–10 (2020). https://doi.org/10.1080/02701367.2020.1715911
8. Ríos, G.J., Pérez, S.Y., Domínguez, E., Moya, Y., Borges, C.R.: Anxiety and psychological variables of sports performance related to injuries in high-performance sportsmen. Apunts. Sports Med. **56**, 100358 (2021). https://doi.org/10.1016/j.apunsm.2021.100358
9. Yarrow, D., Kranke, M.: The performativity of sports statistics: towards a research agenda. J. Cult. Econ. **9**, 445–457 (2016). https://doi.org/10.1080/17530350.2016.1202856
10. Hayduk III, T.: The future of sport data analytics. Stat. Model. Sports Bus. Anal. **1**, 159–173 (2020). https://doi.org/10.4324/9780367854454-14
11. Bai, Z., Bai, X.: Sports big data: management, analysis, applications, and challenges. Complexity. **1**(10), 1–11 (2021). https://doi.org/10.1155/2021/6676297

Using the Fuzzy Logic Models for Analysis of Time Phases of Crude Oil Prices: Brent-Europe

Seyit Kerimkhulle[1](✉), Shakharzat Kuttykozhayeva[2], Gulnara Turtkarayeva[2], Toty Seitova[3], Nazgul Ospanova[2], and Daniyar Toleubay[1]

[1] Information Systems Department, L.N. Gumilyov Eurasian National University, 2, Satpayev Street, 010008 Astana, Kazakhstan
{kerimkul_sye,toleubai_dm}@enu.kz

[2] Physics, Mathematics and Informatics Sciences Department, Sh. Ualikhanov Kokshetau University, 76, Abay Street, 020000 Kokshetau, Kazakhstan
{skyttykojaeva,gtyrtkaraeva,nospanova}@shokan.edu.kz

[3] Information and Communication Technologies Department, Sh. Ualikhanov Kokshetau University, 76, Abay Street, 020000 Kokshetau, Kazakhstan
tseitova@shokan.edu.kz

Abstract. In this work we note that using the Fuzzy Logic theory, in particular fuzzification of terms: Increased, Balanced and Decreased and defuzzification of fuzzy numbers based on the membership function we obtain of forecasting data for the audit, analyze and diagnostic of crude oil prices: Brent – Europe according to criteria the Over / Undervalue, Excess / Deficit and Equilibrium. Further, these studies also allow us to obtain the following classifications: – the time phase of the Increased that covers the periods from 1987 to 2005, in which the crude oil prices moved from 18.53 to 56.38 dollars per barrel; – the time phase of the Balanced that covers the periods from 1994 to 2016, in which the crude oil prices moved from 15.86 to 43.64 dollars per barrel; – the time phase of the Decreased that covers the periods from 2005 to 2023, in which the crude oil prices moved from 54.57 to 82.49 dollars per barrel.

Keywords: Fuzzy Logic Models · Increased · Balanced · Decreased · Analyze and Diagnostic · Crude Oil Prices: Brent-Europe

1 Introduction

The applied problems for the crude oil prices: Brent – Europe have been studied quite a lot, however, some areas require more in-depth research in areas such as the development of a methodology for applying the theory of fuzzy logic, which is the subject of this work. In particular, time-frequency analysis of the interaction mechanism between European carbon and crude oil markets; time series impact on crude oil prices; assessment of the impact of time lags and of geopolitical risk and of volatility markets and portfolio risk and of statistical and financial performance metrics and sentiment scores on crude oil prices; the effects of two benchmarks on Russian crude oil prices and others.

R. Silhavy and P. Silhavy (Eds.): CoMeSySo 2024, LNNS 1489, pp. 9–17, 2025.
https://doi.org/10.1007/978-3-031-96798-6_2

So in [1] reveals a complex interplay between carbon and oil prices, showcasing non-linear and multi-frequency dynamics. Through a novel time-frequency approach, the research identifies a stable in-phase relationship between the two markets, with oil leading at medium and lower frequencies, yet sensitivity to external factors during abnormal events and economic downturns. Thus in [1] findings offer crucial insights for regulators to develop adaptable risk management strategies in carbon markets and aid investors in hedging risks across various timeframes.

Further, in work [2] suggest that the pass-through of global Brent oil prices to fuel prices in Germany's retail market is influenced by the number of time lags considered in the analysis, with longer lag lengths resulting in extended pass-through durations. Additionally, the uniformity of impulse response functions across different brands implies a significant level of competition among retailers, indicating a shared market dynamic in responding to changes in oil prices.

So, in [3] investigate the impact of geopolitical risk on various financial markets and portfolio diversification during the COVID-19 pandemic and Russia-Ukraine conflict. Here, in [3] utilize wavelet-based approaches and empirical data spanning from 2012 to 2022 to demonstrate the fluctuating relationships among global asset markets and geopolitical risk indicators, shedding light on both leading and lagging positions of different assets during distinct crisis periods.

Analyzing the impact of price shocks on Russian crude oil prices, the author in work [4] finds subtle asymmetries in world markets. Thus, taking into account the complexity of the dynamics of the global oil market, we find that as the scale of shocks increases, the asymmetric nature of these effects increases.

In examining the dynamics of volatility spillovers between Brent crude oil and various stock markets, in [5] find that these spillover effects are not only time-varying but also crisis-sensitive and frequency-dependent. Notably, during significant financial crises such as the Global Financial Crisis of 2008–2009, the oil price plunge starting in mid-2014, and the COVID-19 pandemic, these effects intensified.

In work [6] emphasize the complexity of predicting crude oil prices and advocate for a systematic comparison of various prediction models and variables, considering both statistical and financial aspects. Thus in [6] highlight the significance of evaluating prediction models not only based on statistical measures but also on their practical financial implications, demonstrating that incorporating qualitative information like sentiment analysis can enhance both statistical accuracy and financial performance.

Also in the works [7] investigates European crude oil options pricing, considering nonlinear dynamics, stochastic behavior, and jump risks, while accounting for convenience yield and extreme price fluctuations. Here through an equilibrium method, it derives a closed-form solution for option prices in an incomplete market, incorporating risk-neutral dynamics. Additionally, it extends to mean-reverting stochastic volatility, providing analytical formulas for option prices, consistent with recent literature findings.

The decline in crude oil prices from 2014 to 2016, increased U.S. shale production, is challenged in the analysis in work [8], which suggests a broader explanation rooted in global economic conditions and demand dynamics. The evidence presented suggests that weakening oil demand, rather than shale oil expansion, was the primary driver behind the significant drop in oil prices during this period.

In the study [9] assessment of the interdependence between carbon futures and crude oil futures prices, it was discovered that the impact of crude oil prices on carbon prices is asymmetric across various market conditions.

Further research [10] a novel approach to forecasting real prices of crude oil is introduced, employing real-time forecast combinations using time-varying parameter (TVP) models with a single predictor. Thus in work [10] reveals substantial predictability across all horizons up to 24 months, with significant improvements in mean squared predictive error reduction and directional accuracy compared to benchmark models.

It should also be noted the works of the following authors [11–13], who made significant efforts to develop the theory, methodology and practice of modern features of the crude oil market.

The purpose of this research is to develop a methodology for the application of fuzzy logic theory to solve applied problems of analysis and diagnostics of time phases, Over / Undervalue and Excess / Deficit for the of crude oil prices: Brent – Europe. To achieve the purpose received:

- Creating Membership functions for terms: Increased, Balanced and Decreased (Sect. 3.1),
- Analysis and diagnostics of time phases of crude oil prices: Brent – Europe (Sect. 3.2).

2 Methods

Let $f(t), t = \overline{1987, 2023}$, are as the flows the observation data for the Crude Oil Prices: Brent – Europe, Dollars per Barrel, Annual, Not Seasonally Adjusted (see diagram with round marker darker shade by 25%, Fig. 1), which we obtained from source of the Federal Reserve Economic Data of the Federal Reserve Bank of St. Louis [14].

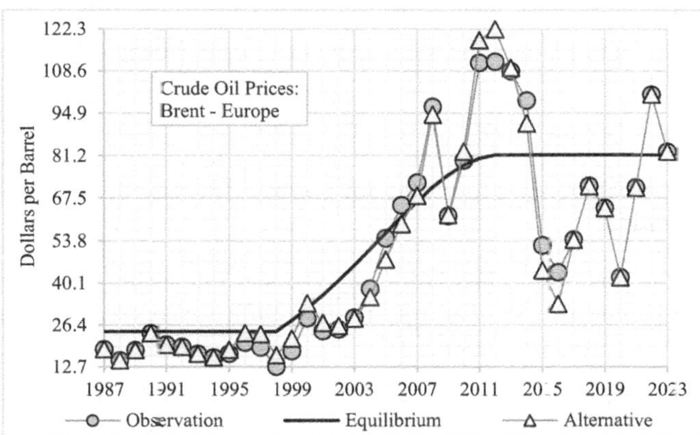

Fig. 1. Database for analyze and diagnostic of crude oil prices: Observation data, Equilibrium trajectory and Alternative strategy. Compiled by the author based on Source [14]).

Then the algorithm for solving the problem of analysis and diagnostics of time phases are:

- Created the data for the stocks $s(t)$, $t = \overline{1987, 2023}$ as the sum of flows:

$$s(t) = \sum_{j=1987}^{t} f(j), \quad t = \overline{1987, 2023}, \tag{1}$$

- Build trajectory for the Equilibrium growth of the Crude Oil Prices: Brent – Europe of world market statistics, here we use the first order quasilinear Verhulst logistic growth difference equation in the form of a logorathmic function [15]:

$$\begin{cases} 0, & t = \overline{1987, 1997}, \\ \ln \frac{1-s(t)/k}{s(t)} = \ln \frac{1-s(1998)/k}{s(1998)} + r \bullet (t - 1997), & t = \overline{1998, 2012}, \\ \Delta^2 s(2012) = 0, & t = 2012, \\ 0, & t = \overline{2013, 2023}, \end{cases} \tag{2}$$

where $t = \overline{1987, 2023}$ are dates for forecasting; k is the capacity of stock data; $r \geq 0$ is the logistic growth rate; $\Delta^2(2012) \stackrel{\text{def}}{=} s(2011) - 2s(2012) + s(2013)$ is the second difference operator; $s(1998)$ is an initial stock data of transactions.

- Used the least squares for get the solution to regression model of the logistic growth Eq. (2):

$$\hat{s}(t) = \begin{cases} \frac{s(1998) \bullet k \bullet e^{\hat{r} \bullet (1998-1997)}}{k - s(1998) + s(1998) \bullet e^{\hat{r} \bullet (1998-1997)}} & t = \overline{1987, 1997}, \\ \frac{s(1998) \bullet k \bullet e^{\hat{r} \bullet (t-1997)}}{k - s(1998) + s(1998) \bullet e^{\hat{r} \bullet (t-1997)}}, & t = \overline{1998, 2012}, \\ \frac{s(1998) \bullet k \bullet e^{\hat{r} \bullet (2012-1997)}}{k - s(1998) + s(1998) \bullet e^{\hat{r} \bullet (2012-1997)}} & t = \overline{2013, 2023}, \end{cases} \tag{3}$$

where $\hat{r}$ is the estimate of slope parameter of logistic growth rate; $\hat{s}(t), t = \overline{1987, 2023}$ are a calculate value of stock data of transactions.

The results of implementing algorithm (1)–(3) are shown in Fig. 1.

3 Results

3.1 Creating Membership Functions for Terms: Increased, Balanced and Decreased

To implement the algorithm (1)–(3) for analyze and diagnostic of Crude Oil Prices: Brent – Europe, Dollars per Barrel, Annual, $\overline{1987, 2023}$, we are creating a result-oriented data set by following fuzzy numbers [15]:

- for the term of Increased (or *inc*) we have a fuzzy triangular number

$$T_{inc} = (1987{,}1998{,}2005), \tag{4}$$

and respectively the membership function of this term is of the form:

$$\mu_{inc}(t) = \begin{cases} 1 & \text{for } 1987 \leq t \leq 1998, \\ \frac{2005-t}{7} & \text{for } 1998 \leq t \leq 2005, \end{cases} \tag{5}$$

– for the term of Balanced (or *bln*) we have a fuzzy triangular number

$$T_{bln} = (1994{,}2005{,}2016), \tag{6}$$

and respectively the membership function of this term is of the form:

$$\mu_{bln}(t) = \begin{cases} \frac{t-1994}{11} & \text{for } 1994 \le t \le 2005, \\ \frac{2016-t}{11} & \text{for } 2005 \le t \le 2016, \end{cases} \tag{7}$$

– for the term of Decreased (or *dcr*) we have a fuzzy triangular number

$$T_{dcr} = (2005{,}2012{,}2023), \tag{8}$$

and respectively the membership function of this term is of the form:

$$\mu_{dcr}(t) = \begin{cases} \frac{t-2005}{7} & \text{for } 2005 \le t \le 2012, \\ 1 & \text{for } 2012 \le t \le 2023. \end{cases} \tag{9}$$

3.2 Analysis and Diagnostics of Time Phases of Crude Oil Prices: Brent – Europe

Indeed, using the Fuzzy Logic theory [15] for defuzzification of fuzzy numbers (4), (6), (8) and the membership function (5), (7), (9), we obtain of forecasting data for the analysis and diagnostics of time phases of crude oil prices: Brent – Europe (see diagram with triangular marker darker shade by 5%, Fig. 1):

$$\begin{aligned} \hat{f}(t) &= (\mu_{inc}(t) + \mu_{bln}(t) + \mu_{dcr}(t)) \\ &\bullet f(t) \bullet D^{t-1987},\ t = \overline{1987, 2023}. \end{aligned} \tag{10}$$

where $\mu_{inc}(t), \mu_{bln}(t), \mu_{dcr}(t)$ are membership functions, respectively for terms: Increased, Balanced and Decreased; $f(t)$ are observation data for the crude oil prices: Brent – Europe, Dollars per Barrel, Annual, $\overline{1987, 2023}$; D is deflator for the crude oil prices.

To forecasting of time phases for the analysis and diagnostics on Increased, Balanced and Decreased terms, we will be use the first concept of Over / Undervalue as the difference between Observation and Alternative strategies, further the second concept of Excess / Deficit as the difference between Observation and Equilibrium to crude oil prices: Brent – Europe:

– the time phase of the Increased that covers the periods from 1987 to 2005, in which the crude oil prices moved from 18.53 to 56.38 dollars per barrel. Indeed, the reason for the change in the time phase of the Increased may be, firstly, the Undervalue of crude oil prices on the world market at 15.84 dollars per barrel (see scatter plot with rectangular marker, white background darker shade 5%, Fig. 2), second – Deficit supply pressure of the crude oil prices on the world market at 148.68 dollars per barrel during from 1987 to 2005 (see scatter plot with diamond marker, white background darker than 25%, Fig. 2). In total, market pressure was exerted of the crude oil prices by – 15.84 – 148.68 = – 164.52 dollars per barrel, which means to crude oil prices has Increased 56.38–18.53 = 37.85 dollars per barrel during from 1987 to 2005;

- the time phase of the Balanced that covers the periods from 1994 to 2016, in which the crude oil prices moved from 15.86 to 43.64 dollars per barrel. Indeed, the reason for the change in the time phase of the Balanced may be, firstly, the Neutral value of crude oil prices on the world market (see scatter plot with rectangular marker, white background darker shade 5%, Fig. 2), second – Deficit supply pressure of the crude oil prices on the world market at 47.23 dollars per barrel during from 1994 to 2016 (see scatter plot with diamond marker, white background darker than 25%, Fig. 2). In total, market pressure was exerted of the crude oil prices by 00.00 + 47.23 = – 47.23 dollars per barrel, which means to crude oil prices has Increased +43.64–15.86 = 27.78 dollars per barrel during from 1994 to 2016;
- the time phase of the Decreased that covers the periods from 2005 to 2023, in which the crude oil prices moved from 54.57 to 82.49 dollars per barrel. Indeed, the reason for the change in the time phase of the Decreased may be, firstly, the Overvalue of crude oil prices on the world market at 22.68 dollars per barrel (see scatter plot with rectangular marker, white background darker shade 5%, Fig. 2), second – Deficit supply pressure of the crude oil prices on the world market at 23.43 dollars per barrel during from 1987 to 2005 (see scatter plot with diamond marker, white background darker than 25%, Fig. 2). In total, market pressure was exerted of the crude oil prices by +22.68–148.68 = – 164.52 dollars per barrel, which means to crude oil prices has Increased 56.38–18.53 = 37.85 dollars per barrel during from 1987 to 2005.

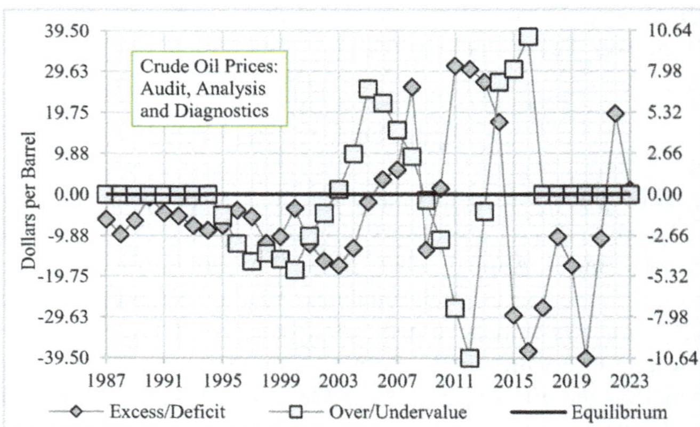

Fig. 2. Database for analyze and diagnostic of crude oil prices: Observation data, Equilibrium trajectory and Alternative strategy. Compiled by the author based on Source [14]).

4 Discussion

Here we note that the audit, analysis and diagnostics of time phases, Over / Undervalue and Excess / Deficit for the of crude oil prices: Brent – Europe may have some developments in decision making. Indeed, in assessing the effect of oil prices on the economic growth of oil exporting countries bordering the Caspian Sea; the impact of oil price shocks on stock exchanges in Caspian Basin countries; the impact of oil prices on the Stock Market and Real Exchange Rate and the impact of Russia-Ukraine war on crude oil prices.

This work [16] highlights the substantial impact of oil price volatility on the economic growth and energy security of Caspian Sea-bordering countries, emphasizing the urgent need for strategic measures to address this challenge.

The study [17] explores how oil price shocks affect stock exchanges in Iran, Kazakhstan, and Russia, revealing a stronger impact from negative shocks. To mitigate this, the countries should focus on maintaining macroeconomic stability and prioritizing industrial production.

The research [18] examines the relationship between KASE stock prices and oil prices from 2016 to 2021, concluding that changes in oil prices impact stock prices, with Brent crude oil prices positively impacting prices closure of KASE while the real exchange rate has a negative impact.

Results of the work [19] evaluates the predictive accuracy of linear and nonlinear models in forecasting daily crude oil prices. So the results demonstrate the superior performance of certain linear models over the random walk method, with implications for investment strategies favouring the use of the moving average convergence divergence indicator.

The authors of the work [20] it uncovers a significant correlation between the conflict of the Russia-Ukraine war and crude oil price fluctuations, underscoring the imperative for strategic responses and risk mitigation measures by relevant stakeholders.

5 Conclusion

In conclusion, we note that using the Fuzzy Logic theory, in particular fuzzification of terms: Increased, Balanced and Decreased and defuzzification of fuzzy numbers based on the membership function we obtain of forecasting data for the audit, analyze and diagnostic of crude oil prices: Brent – Europe according to criteria the Over / Undervalue, Excess / Deficit and Equilibrium. Further, these studies also allow us to obtain the following classifications: – the time phase of the Increased that covers the periods from 1987 to 2005, in which the crude oil prices moved from 18.53 to 56.38 dollars per barrel; – the time phase of the Balanced that covers the periods from 1994 to 2016, in which the crude oil prices moved from 15.86 to 43.64 dollars per barrel; – the time phase of the Decreased that covers the periods from 2005 to 2023, in which the crude oil prices moved from 54.57 to 82.49 dollars per barrel.

Acknowledgments. This research was funded by the Science Committee of the Ministry of Science and of Higher Education the Republic of Kazakhstan (Grant No. AP26196337).

References

1. Wu, Y., Zhang, C., Yun, P., Cao, W., Wagan, Z.A.: Time-frequency analysis of the interaction mechanism between European carbon and crude oil markets. Energy Environ. **32**(7), 1331–1357 (2021). https://doi.org/10.1177/0958305X211002457

2. Frondel, M., Vance, C., Kihm, A.: Time lags in the pass-through of crude oil prices: big data evidence from the German gasoline market. Appl. Econ. Lett. **23**(10), 713–717 (2016). https://doi.org/10.1080/13504851.2015.1102836
3. Raza Rabbani, M., Hassan, M.K., Jamil, S.A., Sahabuddin, M., Shaik, M.: Revisiting the impact of geopolitical risk on Sukuk, stocks, oil and gold markets during the crises period: fresh evidence from wavelet-based approach. Manag. Financ. **50**(3), 514–533 (2023). https://doi.org/10.1108/MF-12-2022-0587
4. Sahin, G., Dogan, N., Berument, M.H.: The effects of two benchmarks on Russian crude oil prices. Econ. Chang. Restruct. **56**(2), 733–748 (2023). https://doi.org/10.1007/s10644-022-09441-0
5. Mensi, W., Hammoudeh, S., Vinh Vo, X., Hoon Kang, S.: Volatility spillovers between oil and equity markets and portfolio risk implications in the US and vulnerable EU countries. J. Int. Financ. Mark. Inst. Money. **75**, 101457 (2021). https://doi.org/10.1016/j.intfin.2021.101457
6. Haas, C., Budin, C., d'Arcy, A.: How to select oil price prediction models – The effect of statistical and financial performance metrics and sentiment scores. Energy Econ. **133**, 107466 (2024). https://doi.org/10.1016/j.eneco.2024.107466
7. Hu, Z., Yang, B.-Z., He, X.-J., Yue, J.: Equilibrium pricing of European crude oil options with stochastic behavior and jump risks. Math. Comput. Simul. **219**, 212–230 (2024). https://doi.org/10.1016/j.matcom.2023.12.020
8. Prest, B.C.: Explanations for the 2014 oil price decline: supply or demand? Energy Econ. **74**, 63–75 (2018). https://doi.org/10.1016/j.eneco.2018.05.029
9. Ren, X., Dou, Y., Dong, K., Li, Y.: Information spillover and market connectedness: multi-scale quantile-on-quantile analysis of the crude oil and carbon markets. Appl. Econ. **54**(38), 4465–4485 (2022). https://doi.org/10.1080/00036846.2022.2030855
10. Kristjanpoller, W.: A hybrid econometrics and machine learning based modeling of realized volatility of natural gas. Financ. Innov. **10**(1), 45 (2024). https://doi.org/10.1186/s40854-023-00577-0
11. Urekeshova, A., Rakhmetulina, Z., Dubina, I., Barykin, S., Mottaeva, A.B., Niyazbekova, S.U.: The impact of digital finance on clean energy and green bonds through the dynamics of spillover. Int. J. Energy Econ. Policy. **13**(2), 441–452 (2023). https://doi.org/10.32479/ijeep.13987
12. Abramova, M., Varzin, V., Niyazbekova, S., Isaeva, E., Chertova, D.: Features of the mechanism for implementing sustainable development through the green economy. E3S Web Conf. **402**, 08030 (2023). https://doi.org/10.1051/e3sconf/202340208030
13. Butova, T., Varzin, V., Niyazbekova, S., Yessymkhanova, Z., Anzorova, S.: Effectiveness of Regulatory Mechanisms of the Oil and Gas Industry in the Context of Environmental Problems. Lecture Notes in Networks and Systems, vol. 575 LNNS, pp. 38–49 (2023). https://doi.org/10.1007/978-3-031-21219-2_4
14. FRED Homepage: Federal Reserve Economic Data of Federal Reserve Bank of St. Louis. [Electronic resource]. https://fred.stlouisfed.org. Accessed 12 Mar 2024
15. Bojadziev, G., Bojadziev, M.: Fuzzy Logic for Business, Finance, and Management, 2nd edn, World Scientific Publishing Co., pp. 1–232 (2007). https://doi.org/10.1142/6451
16. Bolganbayev, A., Myrzabekkyzy, K., Baimaganbetov, S., Kelesbayev, D.: The effect of oil prices on the economic growth of oil exporting countries bordering the Caspian Sea: panel data analysis. Int. J. Energy Econ. Policy. **11**(6), 432–437 (2021). https://doi.org/10.32479/ijeep.11835
17. Köse, N., Ünal, E.: The impact of oil price shocks on stock exchanges in Caspian Basin countries. Energy. **190**, 116383 (2020). https://doi.org/10.1016/j.energy.2019.116383

18. Myrzabekkyzy, K., Bolganbayev, A., Baimaganbetov, S.: The impact of oil prices on the stock market and real exchange rate: the case of kazakhstan. Int. J. Energy Econ. Policy. **12**(1), 163–168 (2022). https://doi.org/10.32479/ijeep.11880
19. Dbouk, W., Jamali, I.: Predicting daily oil prices: linear and non-linear models. Res. Int. Bus. Financ. **46**, 149–165 (2018). https://doi.org/10.1016/j.ribaf.2018.01.003
20. Zhang, Q., Hu, Y., Jiao, J., Wang, S.: The impact of Russia-Ukraine war on crude oil prices: an EMC framework. Hum. Soc. Sci. Commun. **11**(1), 8 (2024). https://doi.org/10.1057/s41599-023-02526-9

Forecasting for Japanese Yen to Dollar Exchange Rate Based on Fuzzy Logic

Seyit Kerimkhulle[1(✉)], Shakharzat Kuttykozhayeva[2], Gulnara Turtkarayeva[2], Toty Seitova[3], Nazgul Ospanova[2], and Yerassyl Iskakov[1]

[1] Information Systems Department, L.N. Gumilyov Eurasian National University, 2, Satpayev Street, 010008 Astana, Kazakhstan
{kerimkul_sye,iskakov_yek}@enu.kz

[2] Physics, Mathematics and Informatics Sciences Department, Sh. Ualikhanov Kokshetau University, 76, Abay Street, 020000 Kokshetau, Kazakhstan
{skyttykojaeva,gtyrtkaraeva,nospanova}@shokan.edu.kz

[3] Information and Communication Technologies Department, Sh. Ualikhanov Kokshetau University, Kokshetau 76, Abay Street, 020000, Kazakhstan
tseitova@shokan.edu.kz

Abstract. In this work we use the Fuzzy Logic theory, in particular fuzzification of terms: Convergence, Divergence, Balanced and defuzzification of fuzzy numbers based on the membership function we obtain of forecasting data for the Japanese Yen to U.S. Dollar Spot Exchange Rate. Further, we use them to solve applied problems of the Over / Undervalue as the difference between Observation and Alternative strategies and in determining and to analyzing of Excess / Deficit volumes as the difference between Observation and Equilibrium to Japanese Yen to U.S. Dollar Spot Exchange Rate. In particular, we got the time phase of Convergence that covers the periods from 1971 to 1985 and from 1996 to 2012, further, the time phase of Divergence that covers the periods from 1986 to 1995 and from 2013 to 2023, finally, the time phase of Balanced that covers the periods from 1971 to 2023.

Keywords: Fuzzy Logic Models · Convergence · Balanced · Divergence · Japanese Yen · Exchange Rate

1 Introduction

This work is devoted to the development of a model for forecasting the Japanese Yen to U.S. Dollar Spot Exchange Rate using fuzzy logic. Which is due to the fact that financial markets constantly attract the attention of both researchers and market participants due to their high degree of complexity, variability and uncertainty. In today's world, where economic processes are becoming increasingly interconnected, integrated and complex, forecasting currency exchange rates plays a key role in making informed decisions not only in the financial sector, but also in the real sector of the economy.

One of the many objects for analysis and forecasting is the exchange rate of the Japanese Yen to U.S. Dollar Spot Exchange Rate. Japan is a major global player with a

R. Silhavy and P. Silhavy (Eds.): CoMeSySo 2024, LNNS 1489, pp. 18–26, 2025.
https://doi.org/10.1007/978-3-031-96798-6_3

leading economy that occupies an influential position, and its currency, the yen, is one of the most traded currencies on the international foreign exchange market. Fluctuations in currency exchange rates have a powerful economic impact, affecting a wide range of market participants, including governments, banks, corporations and individuals. Keeping track of currency exchange rates is critical in many ways [1]. Research [2] highlights that real factors such as supply and demand have a significant impact on the fluctuations of the yen-dollar exchange rate.

Forecasting exchange rates between the Japanese yen and the U.S. dollar over the long term is supported by empirical research. In [3] emphasizes the importance of government debt levels, as well as monetary and financial factors in determining the exchange rate. Thus, emphasizing the complexity of the forecasting task and the need to take into account many variables when analyzing foreign exchange markets.

Thus, studies [4] point to external pressure, especially from the U.S. government, as a factor contributing to the appreciation of the Japanese yen. At the same time, [5] emphasize trade imbalances between Japan and the United States, which can worsen the trade deficit and impede the flow of capital between these countries.

Further, analyzing the impact of portfolio flows on exchange rates, which is confirmed in [6]. The authors note that capital flows from Japan to the United States have a significant impact on the dollar-yen exchange rate, especially in the context of a stronger dollar.

We also note that research [7] shows that during certain periods of crisis, the Japanese yen manifests itself as a “safe haven”, strengthening when uncertainty in the U.S. economy and politics increases. This is due to low interest rates in Japan, making the yen the main funding currency for speculative trading strategies.

Factors related to global events, such as the spread of COVID-19, also play an important role in forecasting. In research [8] reveals a significant relationship between the depreciation of the Japanese yen against the dollar and the increase in Japanese stock returns, especially during the pandemic, indicating changes in the dynamics of the foreign exchange market during the crisis.

The establishment of the euro in 1999 in some countries, as the authors of [9] highlight a significant impact, reducing the utility of the Japanese yen while increasing the utility of the euro. In addition, the study indicates a downward trend in the utility of the Japanese yen over time, especially during the global financial crisis and the Eurozone crisis.

In addition to external factors affecting the yen exchange rate, internal factors such as the policies of the Central Bank of Japan and monetary policy also play an important role. Research [10] highlights the subtle effectiveness of Bank of Japan interventions in influencing the yen exchange rate. However, interventions can sometimes have unintended consequences and lead to increased exchange rate volatility.

In the study [11], the authors examine the effectiveness of the Bank of Japan's interventions in the USD/JPY market and identify changes in the effectiveness of interventions after 1995. They found that the Federal Reserve's actions significantly influenced the Bank of Japan's intervention strategies, highlighting the subtle impact of interventions on exchange rate trends and market dynamics. It should also be noted the works of the following authors [12–14], who made considerable efforts in developing the theory and practice of the currency market.

The purpose of this research is to develop a methodology for the application of fuzzy logic theory to solve applied problems of forecasting time phases, Over/Undervalue and Excess/Deficit for the Japanese Yen to U.S. Dollar Spot Exchange Rate. To achieve the purpose received:

- Creating Membership functions for terms: Convergence, Balanced and Divergence (Sect. 3.1),
- Forecasting of time phases, Over / Undervalue and Excess / Deficit for Japanese yen to U.S. dollar spot exchange rate (Sect. 3.2).

2 Methods

Let $f(t)$, $t = \overline{1971, 2023}$, are as the flows the observation data for the Japanese Yen to U.S. Dollar Spot Exchange Rate, Japanese Yen to One U.S. Dollar, Annual, Not Seasonally Adjusted (see diagram with round marker darker shade by 25%, Fig. 1), which we obtained from source of the Federal Reserve Economic Data of the Federal Reserve Bank of St. Louis [15].

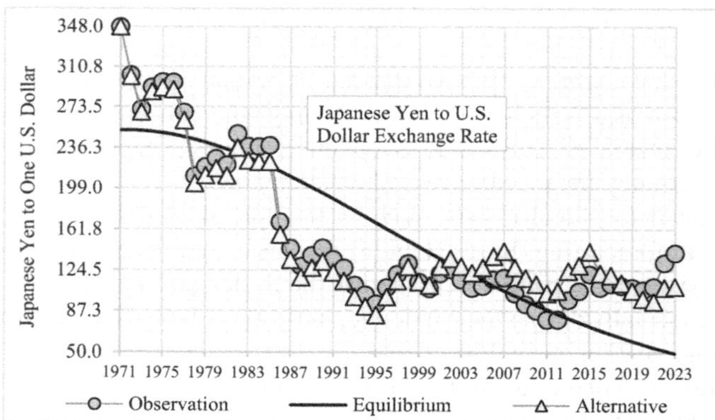

Fig. 1. Forecasting for Japanese yen to U.S. dollar spot exchange rate: Observation data, Equilibrium trajectory and Alternative strategy. Compiled by the author based on Source [15]).

Then the algorithm for solving the problem are:

- Created the data for the stocks $s(t)$, $t = \overline{1971, 2023}$ as the sum of flows:

$$s(t) = \sum_{j=1971}^{t} f(j), \quad t = \overline{1971, 2023}, \tag{1}$$

- Build trajectory for the Equilibrium growth of the Japanese Yen to U.S. Dollar Spot Exchange Rate of world market statistics, here we use the first order quasilinear

Verhulst logistic growth difference equation in the form of a logorathmic function [16]:

$$\begin{gathered}\ln\frac{1-s(t)/k}{s(t)}=\ln\frac{1-s(1971)/k}{s(1971)}+r\bullet(t-1970),\\ \Delta^2 s(1971)=0,\ \ t=\overline{1971,2023},\end{gathered}\tag{2}$$

where $t=\overline{1971,2023}$ are dates for forecasting; k is the capacity of stock data; $r\geq 0$ is the logistic growth rate; $\Delta^2(1971)\overset{\text{def}}{=}s(1972)-2s(1971)+s(1970)$ is the second difference operator; $s(1971)$ is an initial stock data of transactions.

- Used the least squares for get the solution to regression model of the logistic growth Eq. (2):

$$\hat{s}(t)=\frac{s(1971)\bullet k\bullet e^{\hat{r}\bullet(t-1970)}}{k-s(1971)+s(1971)\bullet e^{\hat{r}\bullet(t-1970)}},\ t=\overline{1971,2023},\tag{3}$$

where $\hat{r}$ is the estimate of slope parameter of logistic growth rate; $\hat{s}(t),t=\overline{1971,2023}$ are a calculate value of stock data of transactions.

The results of implementing algorithm (1)–(3) are shown in Fig. 1.

3 Results

3.1 Creating Membership Functions for Terms: Convergence, Balanced and Divergence

To implement the algorithm (1)–(3) for forecasting time phases to Japanese Yen to U.S. Dollar Spot Exchange Rate, we are creating a result-oriented data set by following fuzzy numbers [17]:

- for the term of Convergence (or *cnv*) we have a fuzzy triangular number (see long broken line, Fig. 2)

$$T_{cnv}=(1971{,}1971{,}1995),\tag{4}$$

and respectively the membership function of this term is of the form (see long broken line, Fig. 2):

$$\mu_{cnv}(t)=\begin{cases}1 & \text{for}\quad t=1971,\\ \frac{1995-t}{24} & \text{for } 1971\leq t\leq 1995,\end{cases}\tag{5}$$

- for the term of Balanced (or *bln*) we have a fuzzy trapezoidal number (see long dotted line, Fig. 2)

$$T_{bln}=(1971{,}1995{,}2012{,}2023),\tag{6}$$

and respectively the membership function of this term is of the form (see long dotted line, Fig. 2):

$$\mu_{bln}(t) = \begin{cases} \frac{t-1995}{24} & \text{for } 1971 \le t \le 1995, \\ 1 & \text{for } 1995 \le t \le 2012, \\ \frac{2023-t}{11} & \text{for } 2012 \le t \le 2023, \end{cases} \tag{7}$$

- for the term of Divergence (or *div*) we have a fuzzy triangular number (see short broken line, Fig. 2)

$$T_{div} = (1995{,}2023{,}2023). \tag{8}$$

and respectively the membership function of this term is of the form (see short broken line, Fig. 2):

$$\mu_{div}(t) = \begin{cases} \frac{t-1995}{28} & \text{for } 1995 \le t \le 2023, \\ 1 & \text{for } \quad t = 2023. \end{cases} \tag{9}$$

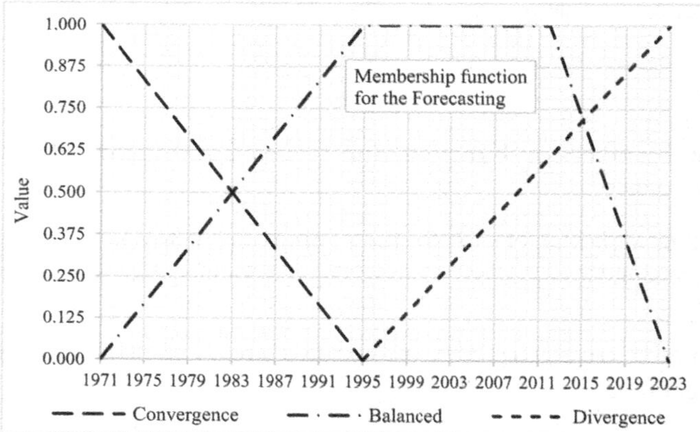

Fig. 2. Membership functions for terms: Convergence, Balanced and Divergence. Compiled by the author based on Source [15].

3.2 Forecasting of Time Phases for Japanese Yen to U.S. Dollar Spot Exchange Rate

Indeed, using the Fuzzy Logic theory [17] for defuzzification of fuzzy numbers (4), (6), (8) and the membership function (5), (7), (9), we obtain of forecasting data for the Japanese Yen to U.S. Dollar Spot Exchange Rate (see diagram with triangular marker darker shade by 5%, Fig. 1):

$$\begin{aligned} \hat{f}(t) &= (\mu_{cnv}(t) + \mu_{bln}(t) + \mu_{div}(t)) \\ &\bullet f(t) \bullet D^{t-1971}, \ t = \overline{1971, 2023}. \end{aligned} \tag{10}$$

where $\mu_{cnv}(t)$, $\mu_{bln}(t)$, $\mu_{div}(t)$ are membership functions, respectively for terms: Convergence, Balanced and Divergence; $f(t)$ are observation data for the Japanese Yen to U.S. Dollar Spot Exchange Rate; D is deflator for exchange rate.

To forecasting of time phases, we will be use the first concept of Over / Undervalue as the difference between Observation and Alternative strategies, further the second concept of Excess / Deficit as the difference between Observation and Equilibrium to Japanese Yen to U.S. Dollar Spot Exchange Rate:

- the first time phase of Convergence that covers the periods from 1971 to 1985, in which the Spot Exchange Rate moved from 347.79 to 238.47 Japanese yen to U.S. dollar. Indeed, the reason for the change in the first time phase of the Convergence of Spot Exchange Rate may be, firstly, the Overvalue Japanese yen over the U.S. dollar on the currency market at 117.66 points, second – Excess pressure of the Japanese yen over the U.S. dollar on the currency market at 302.99 points. In total, market pressure was exerted of Japanese yen to U.S. dollar by 420.65 = 117.66 + 302.99 points, which means to currency has strengthened one U.S. dollar on 347.79–238.47 = 109.32 Japanese yen;
- the first time phase of Divergence that covers the periods from 1986 to 1995, in which the Spot Exchange Rate moved from 168.35 to 93.96 Japanese yen to U.S. dollar. Indeed, the reason for the change in the first time phase of the Divergence of Spot Exchange Rate may be, firstly, the Overvalue Japanese Yen over the U.S. Dollar on the currency market at 111.21 points, second – Deficit pressure of the Japanese yen over the U.S. dollar on the currency market at 622.84 points. In total, market pressure was exerted of Japanese yen to U.S. dollar by 111.21–622.84 = – 511.63 points, which means to currency has strengthened one U.S. dollar on 168.35–93.96 = 74.38 Japanese yen;
- the second time phase of Convergence that covers the periods from 1996 to 2012, in which the Spot Exchange Rate moved from 108.78 to 79.82 Japanese yen to U.S. dollar. Indeed, the reason for the change in the second time phase of the Convergence of Spot Exchange Rate may be, firstly, the Undervalue Japanese yen over the U.S. dollar on the currency market at 211.91 points, second – Deficit pressure of the Japanese yen over the U.S. dollar on the currency market at 223.24 points. In total, market pressure was exerted of Japanese yen to U.S. dollar by – 211.91 – 223.24 = – 435.15 points, which means to currency has strengthened one U.S. dollar on 108.78–79.82 = 28.96 Japanese yen;
- the second time phase of Divergence that covers the periods from 2013 to 2023, in which the Spot Exchange Rate moved from 97.60 to 140.50 Japanese yen to U.S. dollar. Indeed, the reason for the change in the second time phase of the Convergence of Spot Exchange Rate may be, firstly, the Undervalue Japanese yen over the U.S. dollar on the currency market at 16.96 points, second – Excess pressure of the Japanese yen over the U.S. dollar on the currency market at 543.09 points. In total, market pressure was exerted of Japanese yen to U.S. dollar by – 16.96 + 543.09 = – 526.13 points, which means to currency has relaxed one U.S. dollar on 97.60–140.50 = – 42.90 Japanese yen;
- the time phase of Balanced that covers the periods from 1971 to 2023, in which the Spot Exchange Rate moved from 347.79 to 140.50 Japanese yen to U.S. dollar.

Indeed, the reason for the change in the time phase of the Balanced of Spot Exchange Rate may be, firstly, in the absence of Over / Undervalue or equal to zero the difference between Observation and Alternative strategies, in second, in the absence of Excess / Deficit or equal to zero the difference between Observation and Equilibrium to Japanese Yen to U.S. Dollar Spot Exchange Rate.

4 Discussion

Here we note that forecasting of time phases, Over / Undervalue and Excess / Deficit for Japanese yen to U.S. dollar spot exchange rate may have some developments in decision making a new approach to analyzing currency trends in global financial markets; in determining the effectiveness of foreign exchange interventions; in the implementation of new approaches to the dollar/yen exchange rate in the monetary policy model and macroeconomic statistics news.

In particular, the study [18] presents a new approach to the analysis of currency trends using the epsilon-tau procedure to identify up and down trends in exchange rates by analyzing tolerance and patience levels. When applied to the USD/JPY market, this approach identifies characteristic downtrends that often coincide with major events, such as the impact of the 2016 Brexit referendum.

Also, in work [19] emphasizes the effectiveness of foreign exchange interventions by the Japanese authorities in changing the level of the exchange rate and reducing exchange rate volatility. The authors found that "large-in-size" interventions play a significant role in achieving these goals, which is consistent with coordination channel theory.

Further, the study [20] revises the monetary exchange rate models for the dollar/yen exchange rate. Using a new time-varying coefficient approach, it identifies benchmarks associated with significant policy changes or economic events, highlighting the changing impact of macroeconomic fundamentals on the exchange rate over different periods.

Thus, the study [21] demonstrates the effectiveness of interventions in the foreign exchange market through a coordination channel, in which official signals of intervention act as a coordinating mechanism to stabilize speculation, which is especially noticeable in the exchange rate of the yen to the U.S. dollar.

In turn, the study [22] shows that news of Japanese macroeconomic statistics consistently influences the dollar to yen exchange rate within a few minutes after their publication, while the impact on trading activity and price volatility is observed immediately after the news is released.

It is also important to note that in global financial markets, the impact of exchange rate fluctuations may extend to third countries. The study [23] reveals the impact of exchange rate volatility on Japan's trade with third countries, which can have a significant impact on the global economy, which is also of interest to other market players.

5 Conclusion

In conclusion, we note that using the Fuzzy Logic theory, in particular fuzzification of terms: Convergence, Divergence, Balanced and defuzzification of fuzzy numbers based on the membership function we obtain of forecasting data for the Japanese Yen to U.S.

Dollar Spot Exchange Rate. Further, we use them to solve applied problems of the Over / Undervalue as the difference between Observation and Alternative strategies and in determining and to analyzing of Excess / Deficit volumes as the difference between Observation and Equilibrium to Japanese Yen to U.S. Dollar Spot Exchange Rate. In particular, we got the time phase of Convergence that covers the periods from 1971 to 1985 and from 1996 to 2012, further, the time phase of Divergence that covers the periods from 1986 to 1995 and from 2013 to 2023, finally, the time phase of Balanced that covers the periods from 1971 to 2023.

Acknowledgments. This research was funded by the Science Committee of the Ministry of Science and of Higher Education the Republic of Kazakhstan (Grant No. AP26196337).

References

1. Wang, N.: When Econometrics Met Information Theory: A Real-Time Approach to Track Yen Exchange Rate Springer Proceedings in Complexity, pp. 179–194 (2021). https://doi.org/10.1007/978-3-030-67318-5_11
2. Hamori, S.: Nominal and real exchange rate fluctuations of yen-dollar rates. In: Exchange Rates: Policies, Effects and Fluctuations, pp. 131–145 (2021)
3. Litsios, I., Pilbeam, K.: The role of national debts in the determination of the yen-dollar exchange rate. Econ. Inq. **57**(2), 1182–1195 (2019). https://doi.org/10.1111/ecin.12735
4. Aminian, N., Fung, K.C., Garíca-Herrero, A., Lin, C.C.: The political economy of exchange rates: the case of the Japanese yen. Jpn. World Econ. **24**(3), 193–196 (2012). https://doi.org/10.1016/j.japwor.2012.04.001
5. Thorbecke, W.: Global imbalances, triangular trading patterns, and the yen/dollar exchange rate. J. Jpn. Int. Econ. **22**(4), 503–517 (2008). https://doi.org/10.1016/j.jjie.2007.02.005
6. Menla Ali, F., Spagnolo, F., Spagnolo, N.: Portfolio flows and the US dollar – yen exchange rate. Empir. Econ. **52**(1), 179–189 (2017). https://doi.org/10.1007/s00181-016-1075-7
7. Lee, Z.-Z., Su, C.W., No Tao, R.: Longer a safe haven currency? A fresh evidence of Japanese yen under uncertainty. Panoeconomicus. **71**(1), 119–134 (2024). https://doi.org/10.2298/PAN190329021L
8. Narayan, P.K., Devpura, N., Wang, H.: Japanese currency and stock market – What happened during the COVID-19 pandemic? Econ. Anal. Policy. **68**, 191–198 (2020). https://doi.org/10.1016/j.eap.2020.09.014
9. Eiji, O., Makoto, M.: Declining Japanese yen in the changing international monetary system. East Asian Econ. Rev. **21**(4), 317–342 (2017). https://doi.org/10.11644/KIEP.EAER.2017.21.4.333
10. Guyot, O., Montgomery, H., Sheng, D.: The effectiveness of foreign exchange interventions in Japan. J. Infrastruct. Policy Dev. **7**(2), 2171 (2023). https://doi.org/10.24294/jipd.v7i2.2171
11. Kim, S.-J., Sheen, J.: Interventions in the yen-dollar spot market: a story of price, volatility and volume. World Sci. Stud. Int. Econ. **64**, 73–106 (2018). https://doi.org/10.1142/9789813223585_0003
12. Urekeshova, A., Rakhmetulina, Z., Dubina, I., Barykin, S.E., Mottaeva, A.B., Niyazbekova, S.U.: The impact of digital finance on clean energy and green bonds through the dynamics of spillover. Int. J. Energy Econ. Policy. **13**(2), 441–452 (2023). https://doi.org/10.32479/ijeep.13987

13. Bogodukhova, E., Britvina, V., Konyukhov, V., Semenov, A., Niyazbekova, S., Mottaeva, A.: Statistical analysis of the integration of additive and information technologies for the artificial cultivation of plant cells. BIO Web Conf. **65**, 01004 (2023). https://doi.org/10.1051/bioconf/20236501004
14. Stupak, Y., Syzdykova, E., Jazykbayeva, B., Shirshova, L., Issayeva, A., Niyazbekova, S.: Unification of technical solutions used in customs control as a factor in the development of economic systems. E3S Web Conf. **371**, 05053 (2023). https://doi.org/10.1051/e3sconf/202337105053
15. FRED Homepage: Federal Reserve Economic Data of Federal Reserve Bank of St. Louis. [Electronic resource]. https://fred.stlouisfed.org. Accessed 10 March 2024
16. Tsoularis, A., Wallace, J.: Analysis of logistic growth models. Math. Biosci. **179**(1), 21–55 (2002). https://doi.org/10.1016/S0025-5564(02)00096-2
17. Bojadziev, G., Bojadziev, M.: Fuzzy Logic for Business, Finance, and Management, 2nd edn, pp. 1–232. World Scientific Publishing Co. (2007). https://doi.org/10.1142/6451
18. de Sousa, A.M.Y.R., Takayasu, H., Takayasu, M.: Segmentation of time series in up- and downtrends using the epsilon-tau procedure with application to USD/JPY foreign exchange market data. PLoS One. **15**(9), e0239494 (2020). https://doi.org/10.1371/journal.pone.0239494
19. Wan, J.-Y., Kao, C.-W.: Effects of Japanese intervention on yen/dollar exchange rate volatility: a conditional jump dynamics approach. Appl. Econ. Lett. **17**(4), 367–373 (2010). https://doi.org/10.1080/13504850701735799
20. Beckmann, J., Belke, A., Kühl, M.: Cointegration, structural breaks and monetary fundamentals of the dollar/yen exchange. Int. Adv. Econ. Res. **17**(4), 397–412 (2011). https://doi.org/10.1007/s11294-011-9315-2
21. Reitz, S., Taylor, M.P.: FX intervention in the Yen-US dollar market: a coordination channel perspective. IEEP. **9**(2), 111–128 (2012). https://doi.org/10.1007/s10368-012-0208-5
22. Hashimoto, Y., Ito, T.: Effects of Japanese macroeconomic statistic announcements on the dollar/yen exchange rate: high-resolution picture. J. Jpn. Int. Econ. **24**(3), 334–354 (2010). https://doi.org/10.1016/j.jjie.2010.01.003
23. Bahmani-Oskooee, M., Hegerty, S.W., Xi, D.: Third-country exchange rate volatility and Japanese–US trade: evidence from industry-level data. Appl. Econ. **48**(16), 1452–1462 (2016). https://doi.org/10.1080/00036846.2015.1100264

Development of an IoT-Based Smart Watering System for Cucumber Plantations

Nur Amlya Abd Majid[1], Mohd Fahmi Mohamad Amran[2(✉)], Muhammad Fairuz Abd Rauf[2], Suziyanti Marjudi[3], and Azlinda Abdul Aziz[4]

[1] Tamhidi Centre, Universiti Sains Islam Malaysia, 71800 Bandar Baru Nilai, Negeri Sembilan, Malaysia

[2] Faculty of Defence Science and Technology, National Defence University Malaysia, 57000 Kuala Lumpur, Malaysia
fahmiamran@upnm.edu.my

[3] Faculty of Computer Science and Information Technology, Universiti Tun Hussein Onn Malaysia, 86400 Johor, Malaysia

[4] College of Computing, Informatics and Mathematic, Universiti Teknologi MARA (UiTM) Melaka Campus Jasin, 77300 Melaka, Malaysia

Abstract. Over the last decade, the farming industry has experienced substantial growth. Previous studies have demonstrated that smart technologies including artificial intelligence (AI) and the Internet of Things (IoT) can be applied across all agricultural processes. The implementation of an IoT system for watering cucumber plantations can streamline the growth and development of plants including aspects such as marketing, farm management and monitoring thereby enhancing the efficiency of nearly all processes. This research aims to develop an IoT-based smart watering device specifically for cucumber plants. This tool incorporates a soil moisture sensor to determine soil moisture levels and automatically control the watering process. Data from the device will be displayed on an Android-based smartphone application. Data on the tool's features, responsiveness and customization were collected through interviews. The analysis aims to assess customer satisfaction based on the prototype's requirements. Results indicate both the functional and non-functional requirements of the system. The data also serve as a benchmark to evaluate the suitability of soil moisture levels for cucumber plants. The amount of water supplied during the watering process affects soil moisture levels. The IoT-based smart watering system can automate watering by using soil moisture percentages as a benchmark. Additionally remote control of watering can be achieved via a WiFi connection between the smartphone client and the microcontroller.

Keywords: Smart plant · Smart farming · agriculture · IoT

1 Introduction

Industrial agriculture refers to the modern, industrialized production of livestock, poultry, fish and crops [1]. This method relies on technological, economic and political innovations including advancements in agricultural machinery and farming techniques,

R. Silhavy and P. Silhavy (Eds.): CoMeSySo 2024, LNNS 1489, pp. 27–42, 2025.
https://doi.org/10.1007/978-3-031-96798-6_4

genetic technology, methods for achieving economies of scale, the creation of new markets, patent protection for genetic information and global trade. These approaches are widely used in the farming industry today, often under the umbrella of smart farming.

This research aims to examine how information technology (IT) innovations can facilitate farmers' work. Smart Watering is an automated system that uses environmental data to optimize plant watering [2]. The Internet of Things (IoT) involves interconnected devices, mechanical and digital machines, objects, animals, or people with unique identifiers (UIDs) that can transfer data over a network without human-to-human or human-to-computer interaction [2]. This project focuses on applying IoT technology to smart watering systems. Users can maintain their crops more efficiently, benefiting the farm environment, accelerating plant growth, stabilizing the economy, and enhancing commercial crop production.

Maintaining both crop production and environmental health promotes a green environment, ensuring healthy plants. The advantages of using IoT Smart Watering include disease and weed prevention specific to each plant, water and time conservation, and improved soil structure and nutrient preservation. Hand watering, especially in the early morning and evening, can be time-consuming. IoT Smart Watering allows users to monitor watering through their mobile phones, which helps prevent root disease associated with compacted soil. Using drip or sprinkler systems produces smaller water droplets, preserving nutrients and reducing soil compaction [12].

The IoT expands our ability to gather and utilize accurate data, making tasks more efficient. The IoT Smart Watering system for cucumber plants is designed for cucumber plantations because Cucumber (Cucumis sativus L.) is an important vegetable crop in the world and the third largest vegetable crop in Malaysia in terms of planting area [3]. Cucumbers are widely available in night markets, wholesale markets, and satay stalls, and can be harvested every two months. This makes cucumber plants an ideal subject for testing the IoT Smart Watering system, given their prominence in Malaysia.

2 Literature Review

2.1 Internet of Things (IoT)

The term "Internet of Things" (IoT) was first coined by Kevin Ashton in 1999 in the context of supply chain management [4]. The IoT refers to a network of interconnected, sensor-equipped electronic devices that collect data, communicate with each other, and can be monitored or controlled remotely over the Internet [4]. The main goal of IoT development is to extend internet connectivity from digital devices to physical objects, enabling communication between digital devices, objects, and other systems. The data collected can be shared between people, between machines and people, or between machines, with data being stored and managed in the cloud. Looking to the future, Cisco IBSG predicts that there will be 50 billion devices connected to the Internet by 2020 [5].

Using the IoT, the smart watering system is designed to be user-friendly and efficient, automating the irrigation process and reducing the need for manual intervention [9]. By utilizing precise sensors to monitor soil moisture levels, the system ensures that plants receive the optimal amount of water, thereby conserving water and preventing over-irrigation. The integration of advanced sensors allows for accurate monitoring of soil conditions, ensuring that plants receive the right amount of water based on real-time data. Once installed, the system requires minimal maintenance, reducing overall costs associated with traditional irrigation methods. The system provides users with real-time updates and notifications about the field conditions, enabling better management and timely interventions. Currently, there is no standardized protocol for tagging and monitoring plants with sensors, which can lead to compatibility issues between different components and systems. The use of IoT devices raises significant privacy and security concerns [10]. Data about plant status and crop quantities must be encrypted to protect sensitive information from unauthorized access. The reliance on software and internet connectivity introduces potential safety risks. The system could be vulnerable to hacking, leading to misuse of personal information, alteration of system settings, or unauthorized access to account details. These risks place the burden of ensuring safety and security on the consumer.

2.2 Devices

Arduino is a small circuit board with immense potential, capable of reading inputs and outputs according to [6]. The Arduino processor, developed by the Atmel Corporation in California, powers the Arduino Uno microcontroller board, which is based on the ATmega328. It features 14 digital input/output pins, of which 6 can be used as PWM outputs, and 6 analog inputs. Additionally, it includes a 16 MHz ceramic resonator, a USB connection, a power jack, an ICSP header, and a reset button. Arduino is designed to help people understand modern technology, facilitating quick task completion and easy connectivity with various components such as relays, motors, and sensors. Furthermore, this device allows these electronics to be programmed using the IDE software, which is user-friendly and compatible with platforms like Windows, Linux, and Mac.

Other devices were Raspberry Pi. It's powerful, small computer about the size of a credit card, created to inspire a new generation of learners to be creative. According to [7], the ESP8266 is a Chinese-manufactured system on a chip (SoC) microprocessor designed by Espressif Systems. It features a built-in TCP/IP protocol stack and a 2.4GHz Wi-Fi module that supports WPA/WPA2, capable of running applications independently. The ESP8266 can work with other modules to build higher-performing devices and is suitable for various applications, such as IoT devices, tiny web servers, and data collection. In this research, Arduino was selected due to its suitability for straightforward, repetitive tasks like operating a watering system or managing related components. Moreover, Arduino offers a wider range of compatible components compared to Raspberry Pi. For the obstacle detection sensor, the soil moisture sensor consists of two probes that measure the volumetric content of water. These probes allow current to pass through the soil, and the sensor measures the resulting resistance to determine moisture levels. When water is present, the sensor conducts electricity with less resistance, resulting in a higher moisture level reading. Conversely, dry soil conducts less electricity, leading to higher resistance and a lower moisture level reading.

While the rain sensor module detects rainfall and can act as a switch when rain drops fall on the board, also calculating the rain intensity. The module consists of a separate rain board and control board for convenience, a power indicator LED, and an adjustable sensitivity meter. The sensor outputs data to detect the amount of rainfall.

When connected to a 5 V power supply, the LED indicator turns on to inform the user of rainfall.

Meanwhile the buzzer produces sound upon contact with an object. For instance, if the product encounters rain, the buzzer will emit a sound to notify the user.

3 Problem Statement

The adoption of industrial agriculture practices has significantly modernized farming, incorporating advanced technology, economic strategies, and political policies to enhance production efficiency. However, traditional farming methods still face significant challenges, particularly in the areas of resource management, crop health, and environmental sustainability.

One of the major issues is the inefficient use of water resources, leading to wastage and increased labor demands. Traditional hand-watering techniques, for instance, are time-consuming and often result in over or under-watering, which can adversely affect plant health and yield. Furthermore, soil compaction and nutrient loss are common problems associated with improper irrigation practices.

In the context of cucumber cultivation, which is one of the most prominent crops in Malaysia, these issues are particularly pressing. Cucumber plants require precise watering schedules to prevent root diseases and optimize growth. Despite its popularity and economic importance, current irrigation methods in cucumber farming are not sufficiently leveraging modern technology to address these challenges.

Based on the farmers' experience, planting cucumbers involves several steps. They must start by planting the seeds manually and then transferring them to polybags. When watering their plants, farmers need to check whether the plants are still in the seed stage or have started growing, as the water requirements vary. Newly planted seeds need at least 100 mL of water every 2 h, while plants that have been growing for a week require at least 300 mL per polybag and need to be watered at least six times a day. The water needs of the plants depend on their age [11].

Additionally, watering depends on the daily weather. During heavy rains, farms can become waterlogged, causing stagnant water in the polybags of newly planted seeds. Farmers must remove this stagnant water because excess water can prevent seeds from absorbing nutrients and performing photosynthesis, leading to their death.

In terms of supply chain management, after planting the cucumber seeds, farmers can start harvesting cucumbers after two months. They then sell the cucumbers to consumers and receive payment. The price of cucumbers is flexible and depends on the current market price, which can be around RM3 per kilogram or less, depending on the grade of the cucumber. However, farmers face challenges in sustaining their operations in the farming industry. Some farmers have more advanced technologies and better experience in planting, giving them a competitive advantage.

The emergence of the Internet of Things (IoT) offers a potential solution through smart farming technologies such as automated watering systems [9]. IoT Smart Watering systems can provide precise and efficient irrigation by monitoring environmental conditions and adjusting water delivery accordingly. This innovation promises to conserve water, save time, and improve crop health and yield.

Therefore, this research aims to explore the implementation of an IoT Smart Watering system for cucumber plantations in Malaysia. By integrating IoT technology, the project seeks to enhance water management, promote sustainable farming practices, and ultimately improve the economic viability of cucumber farming. The focus will be on how information technology can streamline agricultural operations, making farming more efficient and environmentally friendly.

4 Methodology

4.1 Environment and Data Collection

The methodology involves six phases of prototype development which are: requirement gathering and analysis, quick design, building the prototype, customer evaluation, refining the product and engineering the product. These phases are based on the System Development Life Cycle (SDLC) of the Prototype Model as Fig. 1. The prototyping model is an attractive approach for complicated and large systems where no manual process or existing system can help determine the requirements [8]. The prototypes are usually incomplete systems and many details about how the user develops the system are not included in the prototype. The goal of this model is to provide a system with overall functionality.

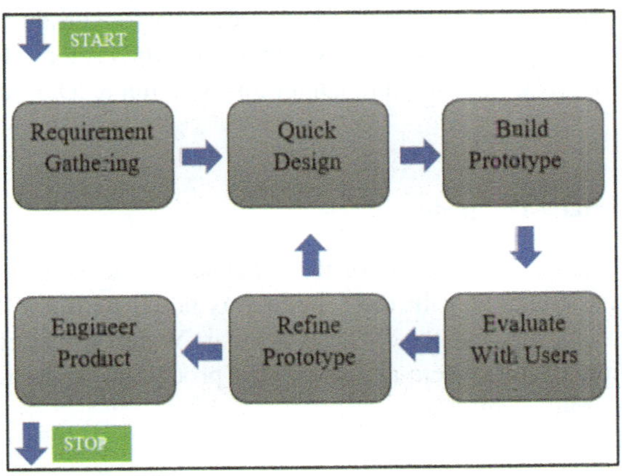

Fig. 1. Prototype Model.

For the Requirement Gathering, the product analysis team identifies the basic requirements. This involves understanding the user's functional needs. Once the team fully understands the user's needs and wants, they prepare a product requirement document. However, detailed internal design and external aspects like performance and security are ignored at this stage.

In Quick Design Phase, the development team creates an initial prototype of the system. The basic requirements are reflected in the user interface. These features may not exactly match the final software but provide a similar look and feel to the final product for the customer.

Building Prototypes Phase involves taking the output from the quick design and building the prototype logically. This prototype gives a look and feel reminiscent of the final product.

After constructing the prototype, the evaluation with users phase is evaluated by the customer. The customer checks the functionality to ensure it works as expected and provides feedback to the development team. Feedback from customers and stakeholders is collected in an organized manner and used for further enhancements.

For the Refine Prototype Phase also known as the revise and enhance phase, this stage involves discussing feedback and reviews with the customer. Any problems identified are addressed, and the prototype is refined to meet customer requirements. This stage may involve negotiations with the customer regarding time, budget constraints, and the technical feasibility of the actual implementation. The cycle repeats until customer satisfaction is achieved.

5 Design

Design is a stage where the product evolves from conceptual ideas into detailed plans. During the design process, various diagrams are created before product development. All requirements are translated into diagrams which include the Project Flowchart Diagram, Block Diagram, Network Architecture Diagram, Schematic Diagram and Prototype Design.

5.1 Project Flowchart Diagram

A flowchart is a visual representation of the sequence of steps and decisions required to perform a process. Each step in the sequence is noted within a diagram shape. In this flowchart diagram, the process of the IoT Smart Watering system is illustrated. It provides an overview of the system and shows the process flow, including how the soil moisture sensor operates (Fig. 2).

5.2 Block Diagram

The IoT Smart Watering system with a soil sensor and rain detector follows a specific procedure to produce optimal results. This research aims to assist farmers by leveraging IoT technology to improve their irrigation practices. The soil sensor measures soil moisture levels to ensure adequate water supply. When the soil sensor detects moisture levels, it sends the information to the microcontroller for processing. The processed information then triggers the output indicators if the conditions are met, reducing the need for farmers to manually check which polybags need water.

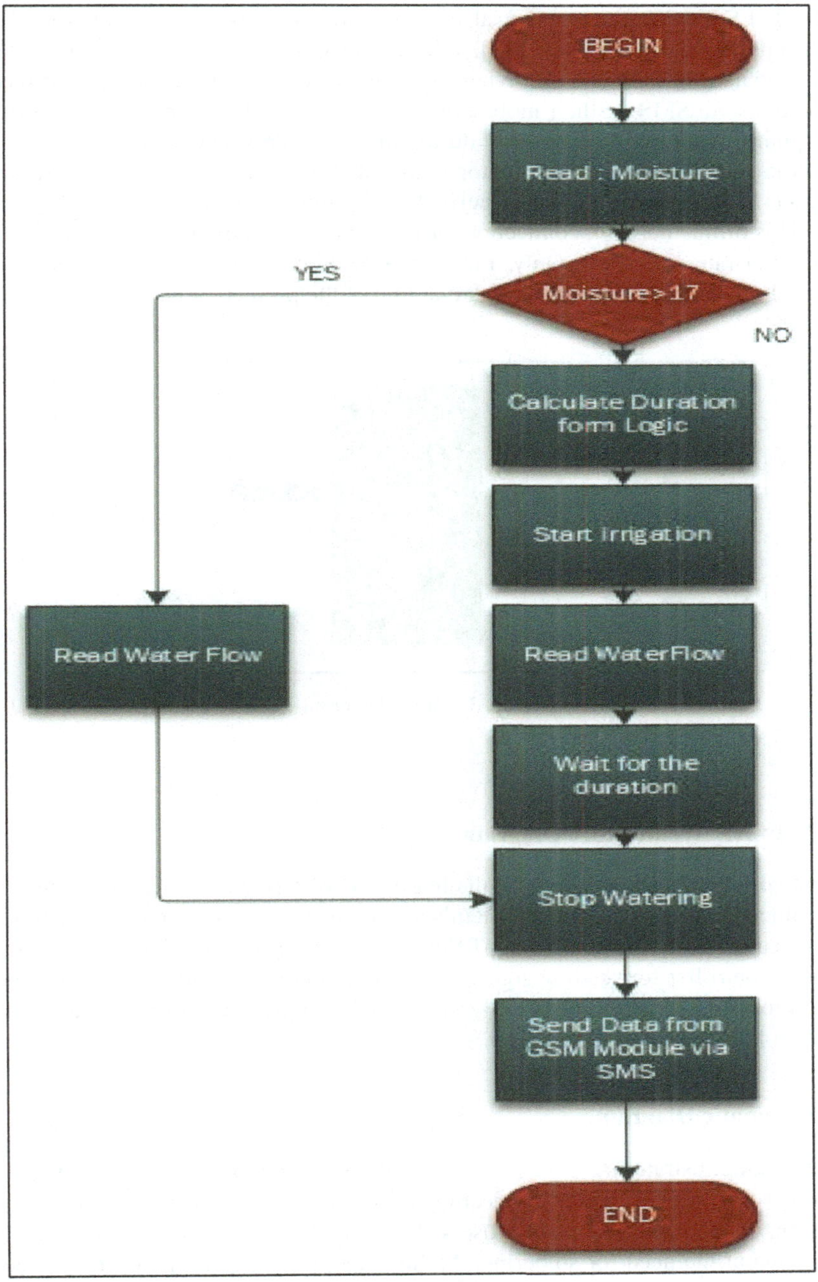

Fig. 2. Flowchart Diagram.

The IoT Smart Watering system also integrates a rain sensor with the microcontroller to detect potential heavy rainfall [13]. The rain detector, installed on the Arduino board, activates when it starts to rain. When the rain sensor detects rain, the user receives notifications via SMS on their mobile phone. This allows the user to easily control the water pump, ensuring it is turned off during rainfall to prevent overwatering.

Additionally, a water pump motor is installed to manage water flow. The system uses a GSM module to send and receive signals between the product and the user via SMS. This offline network connection ensures reliable communication without relying on Wi-Fi or internet access. Only if the GSM module is used, signals will be received clearly [14]. Figure 3 shows the block diagram of the project.

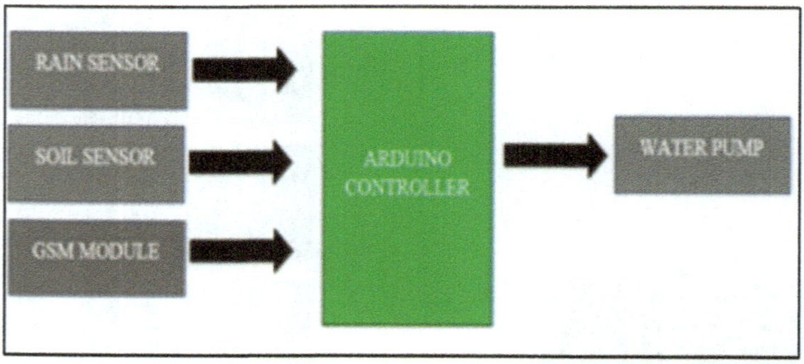

Fig. 3. Block Diagram.

5.3 Network Architecture Diagram

The network architecture design for the IoT Smart Watering encompasses all the essential components required for the system's development process. Figure 4 illustrates the network architecture design for the IoT Smart Watering. The network facilitates access to the microcontroller for sending signals to the smartphone, providing real-time updates on the product's status. Subsequently, the signal is transmitted to the smartphone via SMS, notifying the user accordingly.

5.4 Schematic Diagram

This diagram illustrates the connection from the Arduino to the breadboard. The breadboard serves as a platform for connecting sensors and other devices without requiring soldering, making it convenient for users. They can directly connect components to the board without the need for soldering. Additionally, the diagram delineates the correct connection methods to ensure error-free project implementation. Schematic diagrams play a vital role in describing the high-level functioning of a system or process [15]. They simplify complex relationships between system objects by visually representing them, making the connections more evident (Fig. 5).

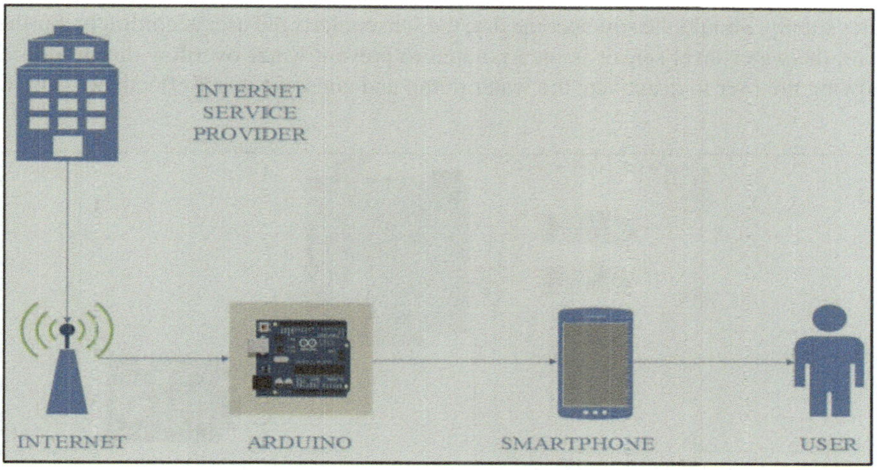

Fig. 4. Network Architecture Diagram.

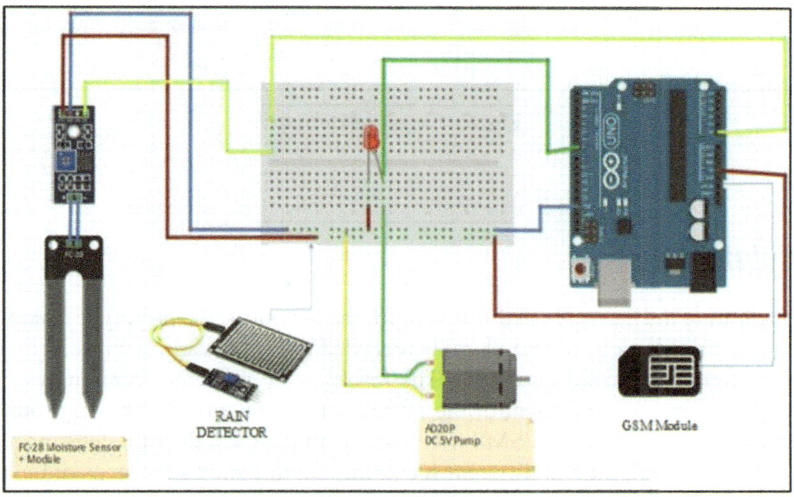

Fig. 5. Schematic Diagram.

5.5 Prototype Design

This overview presents the prototype's functionality, particularly focusing on the Arduino's operational flow for plant management. The inclusion of a water pump serves as a motor to regulate water flow in and out of the system. Additionally, the prototype integrates a GSM Module, enabling user notifications via SMS. The moisture sensor installed in the plant plays a crucial role in determining soil moisture, ensuring optimal water supply. Should the soil become dry, the sensor alerts the user accordingly. Furthermore, the water level sensor is incorporated to prevent water overflow during rainfall, allowing the user to deactivate the water pump and conserve water effectively (Fig. 6).

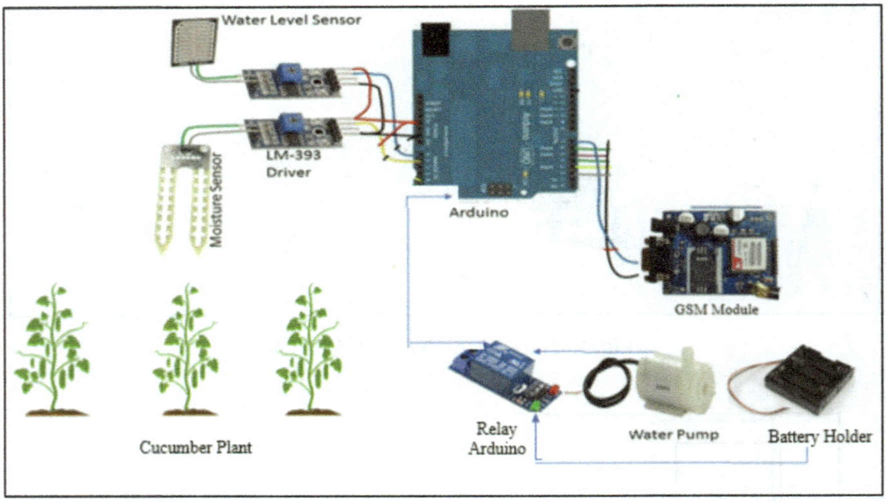

Fig. 6. Prototype Design.

6 Testing

After integrating the hardware and software for the assembly, the prototype is now ready to be tested. Information for this phase is recorded based on various aspects, including expected results and actual outcomes. The results were gathered from analysis based on testing and proper assessment of the project's functionality. The testing focuses on analyzing the response time of SMS notifications to the user, the rain detector, the soil sensor, and the water pump in different scenarios. Table 1 shows the steps of the testing process.

Table 1. Water Sensor Testing.

No	Test step	Test data	Expected result	Actual result	Status (Pass/Fail)
1	Power up the prototype	On and Off	The prototype is power up and operate normally	The result is same as the expected result	Pass
2	Test the detection based on 1 cm water depth	If the reading of the rain sensor detect the water presence on the rain detector more than 10%, the water pump will not watering the plant and give the SMS notification to user	Water pump will not operate due to rain and will operate if there is no rain detected	Water pump will not operate due to rain and will operate if there is no rain detected	Pass
3	Test the detection based on 4 cm water depth	If the reading of the rain sensor detect the water presence on the rain detector more than 10%, the water pump will not watering the plant and give the SMS notification to user	Water pump will not operate due to rain and will operate if there is no rain detected	Water pump will not operate due to rain and will operate if there is no rain detected	Pass
4	Test the detection based on 7 cm water depth	If the reading of the rain sensor detect the water presence on the rain detector more than 10%, the water pump will not watering the plant and give the SMS notification to user	Water pump will not operate due to rain and will operate if there is no rain detected	Water pump will not operate due to rain and will operate if there is no rain detected	Pass
5	Documentation	Record the result and analyse the data	Record the result and analyse the data	Record the result and analyse the data	Pass

After testing the water sensor, the user will be informed about the presence of water, and the water pump will stop watering the plant. Table 2 shows the soil sensor testing.

After completing all the sensor detection tests, the user can easily operate the prototype without any issues. The user will receive alerts from the soil sensor and rain detector, ensuring proper functionality. This system also helps the user confirm that all sensors are working correctly without any problems. Table 3 shows the GSM module testing.

Table 2. Soil Sensor Testing.

No	Test step	Test data	Expected result	Actual result	Status (Pass/Fail)
1	Power up the prototype	On and Off	The prototype is power up and operate normally	The result is same as the expected result	Pass
2	Test the soil sensor in moist plant	If the sensor detects the soil, it will read the moisture of the plant	If the moisture of the plant reading percentage is more than 30% than there will be no watering occur	If the moisture of the plant reading percentage is more than 30% than there will be no watering occur	Pass
3	Test the soil sensor in dry plant	If the sensor detects the soil, it will read the moisture of the plant	If the moisture of the plant reading percentage is below than 10% than it will give a signal to water pump to start watering the plant until it get enough amount of water	If the moisture of the plant reading percentage is below than 10% than it will give a signal to water pump to start watering the plant until it get enough amount of water	Pass
4	Test the soil sensor when the rain is occurred	If the sensor detects the soil, it will read the moisture of the plant	If the moisture of the plant is below thsn 10% suddenly rain occur therefore the automatic watering will pending due to rain	If the moisture of the plant is below thsn 10% suddenly rain occur therefore the automatic watering will pending due to rain	Pass
5	Documentation	Record the result and analyze the data.	Record the result and analyze the data.	Record the result and analyze the data.	Pass

After integrating the system, the results demonstrate the effectiveness of SMS notifications sent to the user. The SMS messages are triggered each time the rain sensor or soil sensor is activated, providing real-time updates on the plant's condition. However, if the prototype is not powered up, the user will not receive any notifications. The example results are based on a forest environment, as this is where the user's cucumber field is located (Table 4).

Table 3. GSM Module Testing.

No	Test step	Test data	Expected result	Actual result	Status (Pass/Fail)
1	Power up the prototype	On and Off	The prototype is power up and operate normally	The result is same as the expected result	Pass
2	Obtaining the network signal	There will be blinking red LED every 3 second when the GSM Module is Ready	The red LED on GSM Module are blinking every 3 seconds	The red LED on GSM Module are blinking every 3 seconds	Pass
3	Sending SMS test	If the prototype is switch on, the user will receive a message	SMS will send to the user when the prototype is ready to be use.	SMS will send to the user when the prototype is ready to be use.	Pass
4	Documentation	Record the result and analyze the data.	Record every result and analyze the data	Record the data and analyze it	Pass

Table 4. SMS (Short Messaging Service) Respond Time According To Signal.

Telco	Telco Signal Bar (1–4)	Condition	Time taken for user getting message (min)
Celcom to Unifi	4	Forest environment	2 min
Digi to Maxis	3	Forest environment	5 min
U Mobile to U Mobile	1	Forest environment	10 min
Maxis to Digi	2	Forest environment	5 min
Digi to Celcom	3	Forest environment	4 min

SMS notifications are sent every time the rain sensor and soil sensor are activated on the plant. These messages inform the user about the plant's current condition. However, if the prototype is not powered on, the user will not receive any notifications from the plant (Fig. 7).

Fig. 7. SMS Notification.

7 Conclusion

The adoption of IoT Smart Watering systems in cucumber cultivation has significant potential to modernize and enhance the efficiency of agricultural practices in Malaysia. This technology can address key challenges faced by traditional farming methods such as inefficient water usage, soil compaction and nutrient loss by providing precise and automated irrigation solutions. The scalability of such systems offers a transformative opportunity for larger agricultural applications, extending the benefits beyond cucumber farming to other crops and farming practices. The scalability of the IoT Smart Watering System can be enhanced water management by precisely monitoring and managing water delivery based on real-time environmental data such as soil moisture, weather conditions and plant growth stages. This precision can be scaled to larger farms and different crop types, ensuring optimal water use across diverse agricultural operations.

IoT Smart Watering improved crop health and yield by customized irrigation schedules. The system can be tailored to the specific needs of various crops, adjusting watering schedules to match the unique requirements of each plant species. This adaptability can enhance crop health and yield across a wide range of agricultural products. It is also can enhance labor efficiency and cost savings by automating the processes. The automated processes automate the labor-intensive task of manual watering, freeing up farmers' time for other critical activities. This automation can reduce labor costs and increase operational efficiency on larger farms. By minimizing the need for manual intervention, larger agricultural operations can manage more extensive areas with fewer workers, leading to significant cost savings.

The scalability of IoT Smart Watering systems holds great promise for revolutionizing agriculture in Malaysia and beyond. By addressing critical challenges in water management, crop health, labor efficiency, environmental sustainability and supply chain optimization these systems can significantly enhance the productivity and sustainability of agricultural operations. The adoption of such scalable technology can lead to a more efficient, competitive and sustainable agricultural sector, benefiting farmers, consumers and the environment alike.

8 Future Works

This research can be enhanced with several features to increase its effectiveness and efficiency. The concern about the power source for the prototype should be addressed. It is recommended that future researchers use rechargeable batteries with solar technology for better mobility and durability. Implementing solar panels in the system allows the prototype to gather solar power from the sun throughout the day, improving the system's usage duration. This also helps reduce the cost of current usage at night, as the solar system can power the prototype all night. By adding a Wi-Fi module would allow the prototype to run more efficiently. Currently, the GSM module used as the network platform is limited, as it only provides notifications and stores data in message history. Using both Wi-Fi and GSM modules simultaneously would ensure continuous network connectivity, offering offline SMS notifications as a backup. Moreover, integrating a Wi-Fi module would enable the addition of a database system to store notifications in a cloud server, allowing easy access and backup of data. Adding more sensors, such as solenoid valves, can enhance the system's functionality. The solenoid valve controls the water flow in and out, which is crucial for watering cucumber plants according to their age. This addition helps regulate water flow and ensures that not all water is used at once. It is recommended that future researchers integrate the prototype functions with third-party applications or custom-built applications. This would improve data collection and storage and allow users to control the prototype's capabilities via a smartphone.

Acknowledgement. We would like to express our gratitude to everyone involved for their invaluable support and guidance throughout this research. Special thanks to Universiti Pertahanan Nasional Malaysia for providing the necessary resources and facilities. We also appreciate the insightful feedback from our colleagues and reviewers, which significantly improved this work.

References

1. https://www.fao.org/home/en/
2. Mitra, A., Pooja, Saini, G.: Automated smart irrigation system (Asis). In: 2019 International Conference on Computing, Communication, and Intelligent Systems (ICCCIS), Greater Noida, India, pp. 327–330 (2019). https://doi.org/10.1109/ICCCIS48478.2019.8974466. keywords: {Irrigation system;Internet of Things;soil moisture sensor;temperature sensor;crop quality fertilizer;NPK sensor;ARIMA}

3. Goundar, S.P., Ramani, S., Mohd, K.S., Aik, C.K., Fatihah, H.N.N.: Evaluation of new cucumber (Cucumis sativus L.) hybrids for agronomic characteristics and begomovirus resistance special issue. Malays. J. Biochem. Mol. Biol. **2**, 30–37 (2022) 5th International Plant Breeding Conference 2022
4. Kramp, T., van Kranenburg, R., Lange, S.: Introduction to the internet of things. In: Bassi, A., et al. (eds.) Enabling Things to Talk. Springer, Berlin, Heidelberg (2013). https://doi.org/10.1007/978-3-642-40403-0_1
5. Abeid, A.S.: Internet of things issues and challenges. Int. J. Acad. Manag. Sci. Res. (IJAMSR). **4**(5), 71–75 (2020)
6. Ismailov, A., Jo'rayev, Z.: Study of arduino microcontroller board. Sci. Educ. **3**(3), 172–179 (2022)
7. Soy, H.: ESP8266 and ESP32 series of SoC microcontrollers. In: Kocer, S., Dundar, O., Butuner, R. (eds.) Programmable Smart Microcontroller Cards, pp. 110–125. ISRES Publishing (2021)
8. Http://www.lumitex.com
9. Obaideen, K., Yousef, B.A.A., AlMallahi, M.N., Tan, Y.C., Mahmoud, M., Jaber, H., Ramadan, M.: An overview of smart irrigation systems using IoT. Energy Nexus. **7**, 100124 (2022). https://doi.org/10.1016/j.nexus.2022.100124
10. Tawalbeh, L., Muheidat, F., Tawalbeh, M., Quwaider, M.: IoT privacy and security: challenges and solutions. Appl. Sci. **10**(12), 4102 (2020). https://doi.org/10.3390/app10124102
11. https://www.haifa-group.com/cucumber-fertilizer/crop-guide-growing-cucumbers
12. Lim, C., Osman, B.B., Yusof, Y., Rahmat, A.R.: Integrating Iot and mobile applications for efficient plant watering: the Iot smart watering system (SWS). J. Digit. Syst. Dev. **2**(1), 124–139 (2024). https://doi.org/10.32890/jdsd2024.2.1.8
13. Kurundkar, S., Dhandar, P., Lahudkar, M., Chaudhari, L., Kandare, U., Boge, R.: IoT-enabled smart irrigation system for precision farming using microcontroller. In: Data Science and Intelligent Computing Techniques, pp. 945–952 (2024). https://doi.org/10.56155/978-81-955020-2-8-80
14. Ratnadewi, R., et al.: Control and notification automatic water pump with arduino and SMS gateway. IOP Conf. Ser. Mater. Sci. Eng. **407**, 012160 (2018)
15. Suffian, D., Ibrahim, S.: Adopting six sigma in formulating a defect prediction model for system testing. Int. J. Digit. Content Technol. Appl. **6**, 60–76 (2012). https://doi.org/10.4156/jdcta.vol6.issue21.8

Construction of a Multidimensional Data Cube for a Factual Database: Feature Extraction Using the Principal Component Analysis Based on the Example of Repression Statistics by Region

Alexander Solovyev[1,2], Anna Bogacheva[2], and Vladimir Tishchenko[1,2(✉)]

[1] Federal Research Center 'Computer Science and Control' of Russian Academy of Sciences, Moscow, Russia
vtischenko@isa.ru
[2] St. Tikhon's Orthodox University, Moscow, Russia

Abstract. Based on statistical processing of data from the database 'for Christ suffered' using the principal component analysis (PCA), factual data related to repressions by region in the period 1929–1933 are analyzed. As a result of applying PCA, the dimension of the problem is reduced from 35 regions to 5 principal components (PC). The principal components by region are interpreted as waves of anti-church arrests, represented by trend regions, during the period of collectivization. Also, based on the first two PC in the dual problem, the regions of arrests are classified by month of the period. The result of the intelligent analysis of the selected multidimensional data is presented in the form of a graph of measure of the number of arrests and a thematic map of arrests, as a geoinformation layer through a mapping service. The graph and layer were generated using the 'NKWSystem' hypertext system for the NIKA OODBMS. As a result of the analysis, trend regions were identified, sorted in order of decreasing significance: Moscow city, Moscow Region, Arkhangelsk Region, Vologda Region, Leningrad city. The main trend regions are Moscow city and the Moscow Region with corresponding peaks of the second wave of arrests in December 1930 and May 1931.

Keywords: statistical processing of historical facts · SOLAP data mining · NIKA DBMS · a measure of the number of arrests · thematic map of arrests

1 Introduction

Extracting knowledge from the data of the factual database of historical facts "for Christ suffered"[1] — book edition of the same name by the editors by Archpriest Vladimir Vorobyov (Vorobyev, 2015) — allows specialists in the subject area to put forward more

[1] Short title. Database "for Christ suffered". Full title: "New Martyrs, Confessors, those who suffered for Christ during the years of persecution of the Russian Orthodox Church in the twentieth century." (c) St. Tikhon Orthodox University for the Humanities (Institute until 2004)

R. Silhavy and P. Silhavy (Eds.): CoMeSySo 2024, LNNS 1489, pp. 43–68, 2025.
https://doi.org/10.1007/978-3-031-96798-6_5

probable hypotheses or confirm existing ones based on data analysis. The multidimensional data cube model used in existing data analysis systems can be naturally used in the case of the NIKA DBMS (Emelyanov et al. 1996), on the basis of which the specified factual database is built. OLTP technology allows to extract reduced knowledge from the database based on regulated SQL queries. Regularity means that the request is predetermined. This leads to low flexibility. To obtain new knowledge, queries need to be re-programmed in the form of SQL code, which creates a high complexity of knowledge retrieval. A more advanced technology is online analytical processing OLAP. It involves working with a multidimensional data cube, the cells of which contain the necessary statistical data on different dimensions of the cube and in their various combinations. The most advanced technology is data mining, which uses methods of mathematical statistics. However, the most promising way to extract knowledge seems to be a combination of the listed technologies, which leads to a combined method of knowledge extraction — OLAP Data mining (Parsaye, 1998, Makarov et al. 2013). Finally, modern analysis involves the use of spatial OLAP analysis based on fuzzy logic (Duraciova and Chalachanova, 2018). In this study, georeferencing of data allows the analysis results to be displayed on thematic maps, which is an element of the spatial data analysis technology SOLAP.

The work (Soloviev et al. 2023) uses this technology to study the process of persecution over the years. The basis was the persecution graph by-year constructed by Professor N. Emelyanov. In this case, trend detection methods were used as statistical methods. To compare the intensity of persecution in the graphs for different groups of believers, the least squares method was used to calculate the statistics of the multiple coefficient of determination. The next stage of studying persecution for individual periods, corresponding to peaks or waves of persecution, requires consideration of graphs of persecution by month. One of the peaks of persecution in 1930 corresponds to the so-called period of collectivization. This wave of persecution covers 5 years from 1929 to 1933. A more detailed look at the months reveals that the peak was 1930 decomposes into many smaller peaks in different months of the period. It is of interest to discover the causes of this phenomenon. A preliminary examination of the statistics for peaks by region yielded different groupings of regions for different peaks. In some cases, the first region in the group by the number of repressions is dominant. The statistical research method in this case is the principal component analysis (PCA), which is at the same time not only a method of dimensionality reduction, but also a classification method. As initial indicators, statistics are considered for various geographical regions of the country, in which persecution could occur to varying degrees and at different times.

2 Methods

2.1 Identification of Component Classes with the Maximum Number of Arrests

Division into Stages of the Entire Period of Persecution by the Soviet Regime

The general graph of persecutions by year (Emelianov, 1998, 2005) requires a more

pstbi@ccas.ru (c) Brotherhood in the Name of the All-Merciful Savior spasbr@ccas.ru.http://martyrs.pstbi.ru.

detailed examination by month for a deeper understanding of the processes that occurred especially at the peak of persecution. The peaks divide the entire period under consideration 1917–1956 into separate sub-periods corresponding to the peaks: 1917–1918 — seizure of power by the Bolsheviks in 1917; 1921–1923 — seizure of church valuables in 1922; 1929–1933 — collectivization in 1930; 1936–1938 — "great terror" in 1937. Collectivization, which took place in 1930, is the third "wave" of persecution. It corresponds to the period 1929–1933.

The Historical Landscape: A Detailed Examination of the Statistics of the Period of Persecution 1929–1933

The repressions that are recorded in the database "for Christ suffered" can be described through the following non-terminal vertices, reflecting individual facts of persecution of various types: Arrests, Convictions, Imprisonments, Deaths. Despite the incompleteness of the data entered into the database, when detailed by month, the sample of data from the general population can be considered representative. This is proven, for example, by the number of arrests of 11 621 in the period 1929–1933 out of the total statistics of the number of arrests of 37 378 in the period 1917–1956. Figure 1 gives graphs that show statistics on the facts of repression by month for 1929–1933.

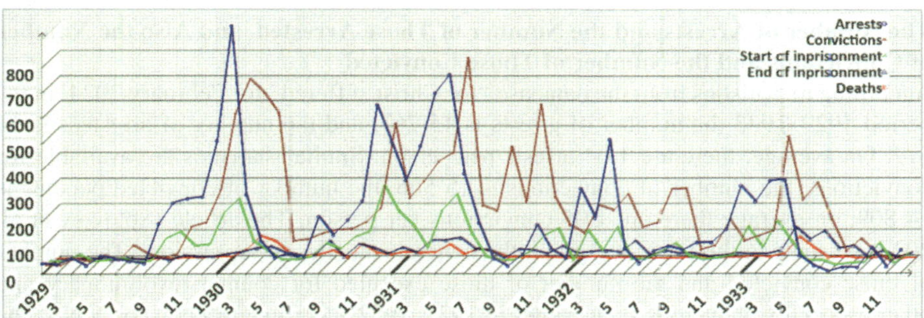

Fig. 1. Statistics for the fields of year-month of Arrest (blue color), Conviction (brown color), Start of imprisonment (green color), End of imprisonment (dark blue color), Death (red color) in the period 1929–1933.

As can be seen from the graphs, the wave of persecutions in 1930, when examined by month, consists of many peaks of different sizes. The supposed reason for this split into separate peaks is the implementation of the instructions of the Soviet authorities about the arrests of Orthodox Christians in different regions in different months of the period 1929–1933. The graphs of arrests and convictions are similar in form, but the graph of convictions is shifted in time by 1–2 months to the right. This is explained by the fact that first there was an arrest, and then, as a result of the investigation, a verdict was passed. The graph of the number of those whose imprisonment began prevails over the graph of the number of those whose imprisonment ended. During this period, the number of prisoners increased. And the deaths graph reflects the fact that there were significantly fewer shootings than arrests. Based on these observations, it can be concluded that a separate examination of the arrests graph will reveal the main features of persecution in a given period of time.

Before considering the statistics of the period under study, one remark needs to be made. Expanding the scope to include other periods of repression is possible by considering the waves of persecution mentioned in the previous paragraph in the form of graphs of persecution by month. However, the PCA *with original features in the form of statistics of arrests by region*, used for the period 1929–1933, may not be effective for a number of reasons when considering similar statistics for other periods. Statistics for other periods require a different approach in each case, because the shape of the persecution graphs of different periods differs in type. For example, the graph of persecutions by month during the period of the "great terror" has one large peak and differs little from the graph of persecutions by year. The same applies to the other two periods of persecution. On the other hand, consideration of statistics for individual regions turns out to be essential in the case of the period 1929–1933, when there is a clear splitting of the peak of 1930–1931 on the graph by year into many comparable peaks on the graph by month. The reason for this split is investigated using the PCA, in which arrest statistics by region are taken as the original features. As for other types of repressions besides arrests, qualitatively, but not quantitatively, in the form of the graphs they approach the shape of the arrest graph.

The Number of Arrests and the Number of Those Arrested, and Also the Number of Convictions and the Number of Those Convicted

According to statistics from the database "for Christ suffered" for February 2024. in the period 1929–1933 the number of arrests is 15 263, and the number of arrestees is 14 672. On average, there are 1.04 arrests per person. Similar statistics are available for convictions: the number of convictions is 15 566, the number of convicted persons is 14 809, on average there are 1.05 convictions per person. The article (Soloviev et al. 2023) presents in the form of a graph the general statistics of repression "The number of those convicted and the number of those executed for counter-revolutionary and other especially dangerous state crimes in 1921–1952 according to Zemskov V.N." The inaccuracy of Zemskov's terminology is that the term "number of those convicted" must be read from him as "number of convictions" (Ilyashenko, 2011). The difference between these two concepts can be easily felt using the example of the total number of convictions in the database "for Christ suffered": the number of convictions is 38 256, the number of those convicted is 31 237, an average of 1,23 convictions per person. A concrete example can be given. Bishop Afanasy Sakharov was sentenced to exile and imprisonment 7 times. Hence it can be concluded that the number of those convicted by Zemskov cannot coincide with the number of convictions. According to Zemskov, the total number of convictions is 4,060,306, which means there were approximately 4,060,306/1,23 $\approx$ 3 301 061 convicted. To study the facts of repression by region, the number of arrests was chosen as statistics, and not the number of convictions, because the facts of arrests have an explicit geographical reference — the terminal vertex of PlaceOfArrest (Tishchenko, 2024), but the facts of convictions do not have such a reference. Arrest statistics will be discussed below. The figures are based on the number of arrests, not the number of arrestees. This consideration is also supported by the fact that the distributions of the number of arrests (Soloviev et al. 2023) and the number of convictions by year have the same peaks, are similar in shape, and their total number differs little from each other. To compare with the above-mentioned numbers of convictions and convicted persons, it can be cited the number of arrests in the database "for Christ suffered" — 38 571 and the number of arrestees — 31 626.

Graph of Arrests by Month of the Period of Persecution 1929–1933

Figure 2 shows a graph of arrests by month with the peak values marked.

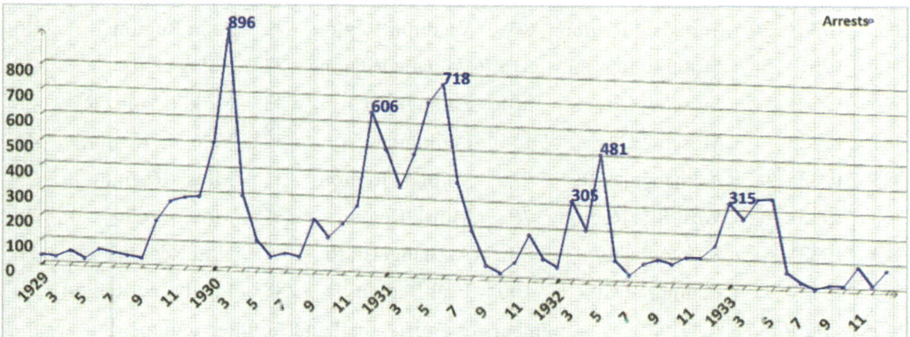

Fig. 2. Statistics for the fieldы of year-month of Arrest in the period 1929–1933.

Table 1 shows the corresponding number of arrests by month over 5 years.

Table 1. Statistics by year-month fields of arrest in the period 1929–1933.

1929^{1}	2^{2}	3^{3}	4^{4}	5^{5}	6^{6}	7^{7}	8^{8}	9^{9}	10^{10}	11^{11}	12^{12}
37	33	55	28	64	53	44	36	178	252	268	272
1930^{13}	2^{14}	3^{15}	4^{16}	5^{17}	6^{18}	7^{19}	8^{20}	9^{21}	10^{22}	11^{23}	12^{24}
474	**897**	282	115	57	72	61	203	136	190	261	**606**
1931^{25}	2^{26}	3^{27}	4^{28}	5^{29}	6^{30}	7^{31}	8^{32}	9^{33}	10^{34}	11^{35}	12^{36}
473	335	457	648	**718**	359	180	53	23	69	174	85
1932^{37}	2^{38}	3^{39}	4^{40}	5^{41}	6^{42}	7^{43}	8^{44}	9^{45}	10^{46}	11^{47}	12^{48}
56	**306**	196	**481**	89	35	80	97	83	110	110	155
1933^{49}	2^{50}	3^{51}	4^{52}	5^{53}	6^{54}	7^{55}	8^{56}	9^{57}	10^{58}	11^{59}	12^{60}
314	257	**333**	**335**	64	27	6	21	20	92	22	83

Notes to the Table. The number of arrests time series contains arrest counts by month of the year in the first column. The superscript indicates the serial number of the month in the period under consideration. The first month of the year is represented by the year, i.e. 1929, 1930, etc.

List of Regions with the Largest Number of Arrests in the Period 1929–1933
To obtain a list of regions with the largest number of arrests, it is necessary to consider the peaks in the graph of arrests by month (see Fig. 2). Table 2 below lists the top three regions by arrest rate for each peak.

Table 2. The largest number of arrests by regions in the period 1929–1933.

A	*Y*	*M*	Region 1	*A*1	Region 2	*A*2	Region 3	*A*3
897	1930	2	Yaroslavl Region	80	Vladimir Region	79	Altai Region	46
718	1931	5	Moscow Region	315	Vologda Region	77	Kaluga City	36
606	1930	12	Moscow City	361	Leningrad City	43	Middle Volga R.	28
481	1932	4	Moscow City	135	Leningrad City	57	Vologda Region	41
335	1933	4	Arkhangelsk R.	73	Vologda Region	31	Minsk Region	25
338	1933	3	West Sib. Region	76	Minsk Region	35	Moscow City	28
314	1933	1	Moscow City	50	Arkhangelsk R.	49	Ural Region	26
306	1932	2	Leningrad City	100	Ural Region	88	Leningrad Region	10

Notes to the Table. $A, A1, A2, A3$ — numbers of arrests: total and for three regions, Y — year, M — month. In the table, peaks and regions are ordered in descending order of the number of arrests. The maximum peak of arrests was in February 1930. represented by different regions with similar numbers of arrests. This is followed by peaks with a dominant first and, in some cases, a second region.

2.2 Dimensionality Reduction in Data Analysis Using PCA

Arrest Statistics by Region
To calculate statistics by region, an additional RegionOfArrest index was built in the database using the PlaceOfArrest field (Tishchenko, 2024; Hutchinson and Veenendaal, 2013). The RegionOfArrest index included the names of those regions in which mass arrests took place. The threshold value is 250 arrests, which is more than half a percent 0.5% of the total number of arrests for the entire period of persecution. In accordance with this rule, 35 regions were selected for the RegionOfArrest index, shown in Table 3 in descending order of the number of arrests.

In Table 3, the total values of arrests A by region, ordered in descending order, are combined into 4 groups. Groups G are formed by ranges of arrest numbers. As can be seen from Table 3, the first group (700–1000) includes 2 regions (1–2), the second (400–700) — 4 regions (3–6), the third (100–400)) — 13 regions (7–19), in the fourth (1–100) — 16 regions (20–35). The regions with the highest number of arrests are the first two in Table 3 — Moscow and the Moscow Region. Statistics are also associated with each region: Mx — mat. Expectation, $<x>$ — average value, s — standard deviation.

Table 3. Statistics of arrests by grouped regions in the period 1929–1933.

N	*G*	Region	*A*	*Mx*	<*x*>	*s*	*N*	*G*	Region	*A*	*Mx*	<*x*>	*s*
1	1	Moscow City	968	181,1	16,1	51,58	19	3	Leningrad Region	101	9,9	1,7	3,73
2	1	Moscow Region	712	153	11,9	40,92	20	4	Arkhangelsk City	95	11,4	1,6	3,95
3	2	Tver Region	538	53,3	9	19,93	21	4	Ivanovo Region	87	17,5	1,5	4,83
4	2	Arkhangelsk R.	523	29,6	8,7	13,48	22	4	Smolensk Region	87	12,3	1,5	3,96
5	2	Vologda Region	508	40,7	8,5	16,53	23	4	Orenburg Region	72	8	1,2	2,85
6	2	Leningrad City	500	48,3	8,3	18,25	24	4	Tomsk	71	25,4	1,2	5,35
7	3	Vladimir Region	372	32,9	6,2	12,87	25	4	Kirov Region	52	4,3	0,9	1,70
8	3	Bashkir Repub.	324	11,7	5,4	5,81	26	4	Ryazan Region	41	4,9	0,7	1,67
9	3	Yaroslavl Region	316	34,7	5,3	12,44	27	4	Voronezh Region	36	2,7	0,6	1,14
10	3	Middle Volga. R.	296	26,4	4,9	10,30	28	4	Krasnoyarsk Region	33	3,3	0,6	1,22
11	3	Altai Region	265	18,2	4,4	7,79	29	4	Tambov Region	33	3,2	0,6	1,22
12	3	Pskov Region	242	35,7	4	11,48	30	4	N.Novgorod R.	25	2	0,4	0,84
13	3	Tatar ASSR	234	22,6	3,9	8,54	31	4	Kuibyshev Region	20	2,7	0,3	0,89
14	3	Saratov Region	221	13,3	3,7	5,94	32	4	Orel Region	18	4,6	0,3	1,14
15	3	Kursk Region	166	14	2,8	5,57	33	4	Novgorod Region	12	1	0,2	0,45
16	3	Vitebsk Region	158	7,8	2,6	3,67	34	4	Lipetsk Region	11	2,3	0,2	0,63
17	3	Kalinin Region	116	9,4	1,9	3,81	35	4	Petrograd	3	1,7	0,1	0,32
18	3	Penza Region	110	5,4	1,8	2,55							

Representation in the Form of a Multidimensional Cube of Data on Repressions and a Section of the Cube on Arrests
The listed fields relating to the facts of repression can be understood as dimensions of a multidimensional data cube: Arrests, Convictions, Start of Imprisonment, End of Imprisonment, Deaths and a time axis of 60 months from 1929–1933. The data elements are the corresponding statistics. As was shown in the previous paragraph 2.1, the facts of arrests reflect the main surges of persecution by month and it makes sense to consider the section of a multidimensional cube associated with arrests. The results of OLAP queries on arrests by month and region form a section of a multidimensional cube.

Table 3 contains data aggregated by month of the period under review as a result of an OLAP query using the RegionOfArrest index. The total distribution of the number of arrests is decomposed into the distribution of the number of arrests for 35 regions. In its turn, each of the regions can be considered as a dimension in the feature space. However, the large number of dimensions makes the interpretation of statistical results difficult. A linear method for reducing the dimension of a feature space is the PCA, the use of which makes it possible to visualize the source data, allows to correctly interpret the statistical conclusions obtained, and also significantly reduce the amount of statistical information without loss of information content (Aivazyan et al. 1989). It must be emphasized that this method is purely statistical. Statistically insignificant data is by no means insignificant in the domain sense. Even one arrest is a personal feat of faith (see Appendix A). In this sense, for example, the Lipetsk Region, with 11 arrests, is the same region that has been persecuted as Moscow, although to a lesser extent.

Dependence of Arrest Statistics Between Regions and Principal Components
Arrest statistics by region are measured in the same units. In this case, in the monograph devoted to dimensionality reduction (Aivazyan et al. 1989), it is recommended to use a covariance matrix (rather than a correlation matrix with dimensionless quantities) to analyze data using PCA. The largest dispersion of values has two regions with the maximum number of arrests — Moscow and the Moscow Region, the dispersion of which is an order of magnitude higher than the dispersion of other regions: D (Moscow) = 2660.90 and D (Moscow Region) = 1674.20. Examples of the largest covariances are Moscow and Vologda regions: 460.50; Moscow and Leningrad: 367.30. Since the sum of all region variances along the main diagonal of the covariance matrix is equal to the sum of the variances across all PC, it can be assumed that the statistics for these two regions will form the basis for the structure of the first two principal components (PC), which will explain most of the variance. In addition, according to the definition of PCA, the variances of the PC are the eigenvalues of the original covariance matrix ordered in descending order and constitute the diagonal covariance matrix of the PC. Thus, unlike the space of original features, there are no mutual correlations in the space of PC.

Principal Components of the Arrest Process
The Essence of the Method: the Process of PCA and Its Application to Data

The original arrest statistics are correlated and duplicate information. In the PCA, through a linear orthonormal transformation, a transition occurs to uncorrelated PC, which compress statistical information on arrests with the least loss. In this case, the PC highlight those of the original features that have *the greatest variance*. To calculate PC, the following steps must be performed.

1. Obtaining the initial set of features X_i, $i = 1, \ldots, n$ in the form of multidimensional statistics of arrests by regions based on the section of the multidimensional cube of repression data through OLAP-queries. 60-dimensional observations were obtained in 35 regions ($n = 35$) in the framework of the five-year period under consideration since 1929 to 1933. Table 3 in the column A shows the total statistics on all measurements for each of the 35 regions.
2. The estimated selective vector of the average values μ and the covariance matrix Σ are taken as estimates of the unknown vector of the average and the covariance matrix for the n-dimensional set of features X (Aivazyan et al. 1989). Table 3 in the column <*x*> shows the average values, in the column *s* shows standard deviations for the statistics of arrests in the regions, and the dispersions of s^2 for all features of the X_i are the diagonal of the selective covariance matrix Σ.
3. The PC of the Y_i are searched as linear orthogonal normalized combinations of the initial features X_i. As the first PC, the component with the greatest dispersion is taken. As a kth PC, a component is taken, which is not correlated with the k-1 previous PC and has the greatest dispersion. From the conditions of the maximum of dispersions, it can be obtained that the maximum dispersions of the PC are equal to their eigenvalues λ_i, which are the roots of the characteristic equation of the covariance matrix Σ: $|\Sigma\text{-}\lambda I| = 0$. From the equations $(\Sigma\text{-}\lambda I)l_i = 0$ are all components of the l_i vector for all eigen vectors $i = 1, \ldots, n$. Then the PC are in the form of $Y = LX$, where the lines of the matrix L make their eigenvectors l_i (Ayvazyan et al. 1989) To calculate λ_i and l_i, the Jacobi method for a symmetrical real matrix was used.
4. Determination of the number of significant PC $p < n$. As a result of component analysis, $p = 5$ PC were obtained, which significantly reduces the dimension of the feature space with dimension $n = 35$. Eigenvalues and eigenvectors for significant PC are given in and below Table 7.

Essential within the framework of the method is the use of *significance criteria* to determine the number of significant PC. There is no single-valued method for determining the number of significant components (Mirkes et al. 2020). It is assumed that all eigenvalues are ordered in descending order.

Graphic Criteria for the Significance of the Number of Principal Components. The First Two Bipolar Principal Components of the Arrest Process: Moscow–Arkhangelsk Region and Moscow Region–Tver Region

Several graphical methods are used below.

Figure 3a, b depict the so-called scree test $\lambda(p)$ and its adaptation $\log_2\lambda(p)$ (Cangelosi and Goriely, 2007). In the case of $\lambda(p)$, the "steep cliff" criterion is used. The point on the graph where $\lambda(p–1)> > \lambda(p)$ gives the result at p = 2, leaving two PC with maximum variance. The dependence $\log_2\lambda(p)$, starting from number $p + 1$, should represent a straight line, where p is the number of significant PC. This graph does not give any clear result regarding the dimension of the PC space. Figure 3c, d show the dependences of the measures of relative distortions $\kappa(p)$ and $\gamma(p)$ (Aivazyan et al. 1989), calculated using the formulas (1) and (2):

$$\kappa = \frac{\lambda_{p+1} + \cdots + \lambda_n}{\lambda_1 + \cdots + \lambda_n}, \tag{1}$$

$$\gamma = \frac{\lambda_{p+1}^2 + \cdots + \lambda_n^2}{\lambda_1^2 + \cdots + \lambda_n^2}, \tag{2}$$

Here n is the dimension of the space of original features, and p is the dimension of the space of significant PC. The mentioned "steep cliff" method gives the dimension $p = 1$ in the case of $\kappa(p)$ and $\gamma(p)$. However, in this case the condition is not met that the amount of distortion is small enough, because $\kappa(1) = 0.57$ and $\gamma(1) = 0.39$. Thus, the graphical method of dependence $\lambda(p)$ gives the minimum possible number of significant components $p = 2$ with acceptable accuracy, when the PC explain the variance by approximately 72%. The result was the following two PC, calculated using formula (3):

$$\begin{aligned} Y_1 &= +0,966{*}X\,(\text{Moscow}) - 0,133{*}\text{X(Arkhangelsk Region)} \\ Y_2 &= +0,915{*}X\,(\text{MoscowRegion}) - 0,283{*}X\,(\text{TverRegion}) \end{aligned} \tag{3}$$

Here $X(<Region_i>) = X_i$ denotes the original feature expressed by the number of arrests in the region $<Region_i>$. The expressions for the PC include only those features for which the correlation between the variable X_i and the PC Y_j exceeds 0,40 in absolute value (Afifi and Azen, 1979) and the coefficients before variables X_i exceed 0,133 in absolute value. Table 4 shows statistics on arrests by month (original features), corresponding to the minimum and maximum values along the two axes of the PC Y_1 and Y_2.

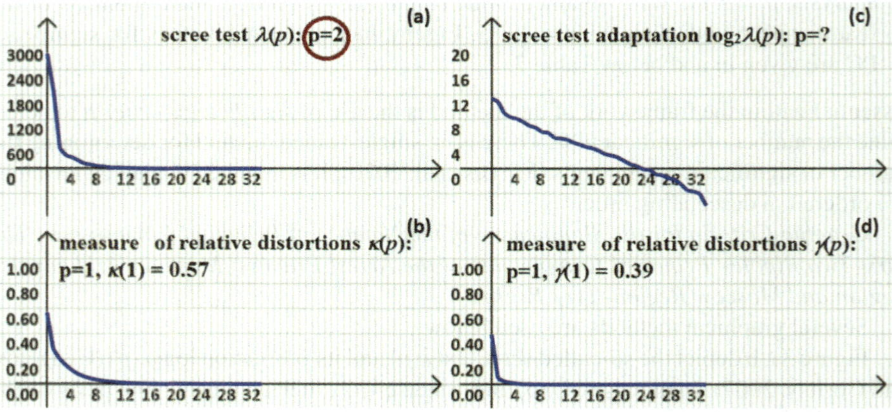

Fig. 3. Graphs: (a) the dependence of the eigenvalue on its number $\lambda(p)$, (b) logarithmic eigenvalues $\log_2\lambda(p)$, (c) measure of relative distortion $\kappa(p)$, (d) measure of relative distortion $\gamma(p)$.

Point $N = 30$ in the III quadrant of the Y_1 and Y_2 plane gives close to minimal values of the number of arrests in Moscow and Moscow Region, and in the features with

Table 4. Arrest statistics for minimum and maximum values of Y_1 and Y_2.

N	*Y*	*M*	AR	M	MR	TR	V
24	1930	12	7	361	11	4	$\max\{Y_1\}$
29	1931	5	15	2	315	35	$\max\{Y_2\}$
30	1931	6	10	0	8	134	$\min\{Y_1,Y_2\}$

Notes to the Table. N — serial number of the month, *Y* — year, *M* — month. Number of arrests in: AR — Arkhangelsk Region, M — Moscow, MR — Moscow Region, TR — Tver Region, V — value of *Y*.

negative coefficients — Arkhangelsk Region, and Tver Region — on the contrary, there has been a slight increase in the number of arrests. From the formulas for the PC Y_1 and Y_2 it follows that they can be considered as bipolar components (Lawley and Maxwell, 1963), because in both components the number of positive and negative coefficients is the same. In this case, it does not matter which half of the coefficients are taken with positive signs. In the Y_1 component, the coefficients have opposite signs in front of the signs of the number of arrests in Arkhangelsk Region and in Moscow reflect some characteristic differences in the processes of persecution in these two regions, i.e. on average, with a significant increase in the number of arrests in Moscow, in the Arkhangelsk Region the number of arrests decreased, or remained at the same level, or increased slightly and vice versa. This is illustrated in Fig. 4.

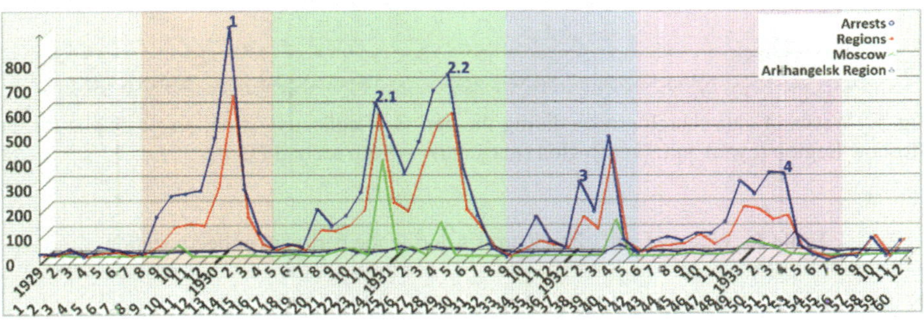

Fig. 4. Waves of anti-church arrests during the period of collectivization: arrest graphs for Arkhangelsk Region (dark blue) and Moscow (green) — from the first PC Y_1, for all regions (blue) and for selected regions (red).

In Fig. 4, on the abscissa axis the months of the period 1929–1933 are numbered from 1 to 60. The ordinate axis indicates the number of arrests. The selected regions include the top 35 regions based on the number of arrests, considered in PCA. The first PC Y_1 explains the variance by approximately 43% and its interpretation in the subject area can be as follows: on the one hand, there was a predominant number of arrests of Orthodox Christians in the city of Moscow, on the other hand, arrests that did not coincide in months with Moscow occurred in the Arkhangelsk Region. The sample correlation between the numbers of arrests in Moscow and the Arkhangelsk Region is

0.14. The graph in Fig. 5 for regions from the second PC Y_2 shows that the peaks of persecution (except for the first — month 14) in Moscow Region and Tver Region were shifted in time relative to each other. The correlation in this case is approximately 0.25. The following properties of the process of arrests during the period of collectivization can be distinguished.

1. What is common to the arrest processes in all regions is that they were initiated from the center, which is Moscow.
2. The processes of arrests in cities could be intensified by the movement of peasants from villages to cities.
3. The anti-religious goal of collectivization is to distract peasants from Orthodox life in the church, which was the spiritual center for peasants in any village or hamlet.

From the graphs of arrests for all regions and for selected regions, four component waves of arrests are distinguished, indicated by different background colors in Fig. 4 (see Table 5).

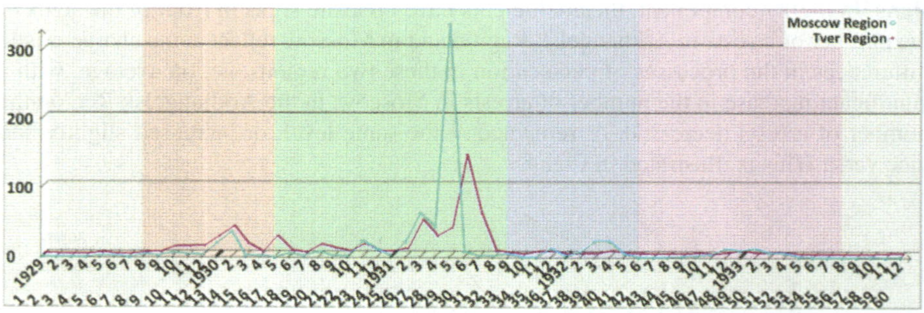

Fig. 5. Waves of anti-church arrests during the period of collectivization: arrest graphs for the Moscow Region (cyan) and Tver Region (magenta) — from second PC Y_2.

Table 5. Waves of anti-church arrests during the period of collectivization.

N	Period		Historical context
	Months	Months /Years	
1	8–17	8/1929–5/1930	The "first wave" of forced collectivization, accompanied by the mass closure of churches, ended with the withdrawal of peasants from collective farms
2	18–33	6/1930–9/1931	The "second wave" of collectivization, during which grain growers accused of sabotage were sent into "kulak exile"
3	34–42	10/1931–6/1932	Deportation of peasants 1930–31. Peasant revolts of 1929–32
4	43–55	7/1932–7/1933	Famine of 1932–33. Resolution on death penalty for theft of collective farm property

The first wave of arrests practically did not affect Moscow (but not the Moscow Region), since there were no peasants who had arrived from the countryside. The second wave of arrests covers both Moscow itself and the Moscow Region (see Figs. 4 and 5). The third wave and partly the fourth also affect Moscow and Moscow Region. As can be seen from Figs. 4 and 5 highest peaks in Moscow (24th month) and Moscow Region (29th month) correspond to the peaks on the general graph of arrests (Fig. 4) in the second wave of persecution. The number of arrests in Moscow and in the Moscow Region make up approximately half in the "Moscow" peaks (see Table 2, lines 2–3). Arkhangelsk Region has its peaks in all four waves of arrest, and Tver Region only in the first two.

Analytical Criteria for the Significance of the Number of Principal Components. The First Five Principal Components of the Arrest Process: Waves of Anti-Church Arrests During the Period of Collectivization

Returning to the analysis of the number of significant PC, in addition to graphical methods, it is necessary to consider analytical methods. The book (Lawley and Maxwell, 1963) describes the Bartlett test, which is designed to test the hypothesis that a given number of smallest eigenvalues is equal. This method practically does not reduce the number of significant PC for arrest processes. In (Harman, 1968; Afifi and Azen, 1979; Aivazyan et al. 1989) it is proposed to consider as a criterion the proportion of total variance explained by significant PC. The method assumes that there is no separation between "signal" and "noise". In this sense, any predefined precision makes sense. The Guttman-Kaiser test of the mean or half-mean eigenvalue corrects this shortcoming. The criterion selects PC with eigenvalues λ_i:

$$\lambda_i \geq \overline{\lambda} = \frac{1}{n} trC, \tag{4}$$

where tr$C$ is the trace of the sample covariance matrix, n is the number of initial features. The criterion is applied in the case when significant PC have $\lambda_i \gg \overline{\lambda}$, and the remaining values have $\lambda_i \ll \overline{\lambda}$,. In other cases, the method may produce many significant PC (Rencher, 1998). In accordance with the *Broken stick model* (Cangelosi and Goriely, 2007), the PC for which the inequality holds should be retained:

$$\frac{\lambda_i}{trC} > l_i = \frac{1}{n} \sum_{j=i}^{n} \frac{1}{j}, \tag{5}$$

The method tends to underestimate the number of significant PC. Both of the mentioned methods are sensitive to irrelevant features. This can be taken into account using the condition number of the covariance matrix, i.e. the ratio of the maximum eigenvalue to the minimum $\kappa = \lambda_1/\lambda p$, which should not be very large. The criterion leaves the PC with eigenvalues (Mirkes et al. 2020):

$$\lambda_i > \frac{\lambda_1}{\kappa_0}, \tag{6}$$

where κ_0 is some threshold value, for example $\kappa_0 = 10$.

Table 6 shows the results of applying the described methods in the form of the number p of significant PC of the arrest process. The third column for comparison shows the proportion of the total variance I_p that is explained by p significant PC.

Table 6. Number of significant PC in ascending order.

Methods	p	I_p
Broken stick model	3	0,8
Condition value	4	0,849
Proportion of total variance explained	5	0,892
Average eigenvalue	6	0,922

Assuming that there is no separation between "signal" and "noise", the method of proportion of total variance explained is general and applicable in the case of the PC of arrest processes with some specified accuracy. For this purpose, in (Aivazyan et al. 1989) a measure of information content is introduced, associated with the transition to a smaller dimension of the space of the PC compared to the space of the original features, which in the case under consideration represent statistics of arrests by region. Let X be the initial arrest statistics for n regions, $Y'(X)$ and $Y(X)$ be the n and p PC of the arrest process, respectively, with $p << n$. In addition, $I_p(Y(X))$ is a measure of the information content of the p-dimensional feature system $Y(X)$. In the general case, to move to a smaller set of features $Y(X)$ with minimal losses in information content, it is necessary to maximize the value of the functional:

$$I_p(Y(X)) = \max_{\mathrm{Y} \in \mathrm{F}} \{ I_p(Y'(X)) \}, \tag{7}$$

where F is the class of admissible transformations of the original features X. In the case of PCA, $I_p(Y(X))$ has the following form:

$$I_p(Y(\mathrm{X})) = \frac{\lambda_1 + \cdots + \lambda_p}{\lambda_1 + \cdots + \lambda_n}, \tag{8}$$

Here $\lambda_1,\ldots,\lambda_p,\ldots,\lambda_n$ are the eigenvalues of the covariance matrix for the vector of multidimensional observations according to the original characteristics, ordered in descending order, i.e. arrest statistics by region ($n = 35$, see Table 3). The introduced measure of information content $I_p(\mathrm{Y}(X)) = I_p$ is the proportion of total variance explained and should be in the range of 70–90%. For the proportion of explained variance of 90%, the number of significant PC is equal to 5: $I_p = 0.892 \approx 0.9$ (see Table 7). Expressions for the PC are given below the table.

Features with the maximum absolute value coefficients are highlighted in expressions (9). Figure 6 shows graphs for significant PC. The peaks highlighted in the graphs correspond to the selected features in the PC Y_i, $i = 1,\ldots,5$. The interpretation of the PC of Y_i can be given based on the features that have the greatest weight in the components. The five selected regions correspond to four waves of arrests (WoA) in Fig. 4: Y_1 — Moscow, first peak of second wave (24th month); Y_2 — Moscow Region, second peak of second wave (29th month); Y_3 — Arkhangelsk Region, fourth wave (52nd month); Y_4 — Vologda Region, first wave (14th month), Y_5 — Leningrad, third wave (38th month).

Identifying Trends in Historical Context Based on Arrest Waves and Trending Regions

Table 7. Eigenvalues and the corresponding measure of information content.

p	1	2	3	4	5	6
λ_i	2788,0	1890,5	490,9	319,9	274,8	194,7
$I(p)$	0,431	0,724	0,800	0,849	0,892	0,922

Y_1=
+0.966 * *X*(Moscow)–0.133 * *X*(Arkhangelsk Region), (9)
Y_2=
+0.915 * *X*(Moscow Region)–0.283 * *X*(Tver Region)–0.130 * *X*(Tatar Autonomous Soviet Socialist Republic)–0.103 * *X*(Middle Volga Region)+0.038 * *X* (Leningrad Region)+0.014 * *X* (Kuibyshev Region)+0.011 * *X*(Nizhny Novgorod Region)–0.005 * *X*(Novgorod Region)–0.003 * *X* (Petrograd)
Y_3=
–0.648 * *X*(Arkhangelsk Region)+0.560 * *X*(Tver Region)+0.410 * *X*(Leningrad)
Y_4=
+0.632 * *X*(Vologda Region)–0.566 * *X*(Pskov Region)–0.270 * *X*(Middle Volga Region)–0.114 * *X*(Smolensk Region)
Y_5=
+0.863 * *X*(Leningrad)+0.413 * *X*(Arkhangelsk Region)–0.058 * *X*(Kirov Region) +0.028**X*(Voronezh Region)–0.023 * *X*(Kuibyshev Region)

In the months indicated in the previous paragraph, the number of arrests in the selected (trend) regions was the largest compared to other months. The above is summarized in Table 8.

Below is a possible interpretation of the waves of arrests. Basically, trend regions determine the trend and historical context in which waves of arrests took place. In chronological order, the first wave of persecution corresponds to the peak in February 1930 with the trend region Vologda Region wave has a maximum peak in the number of arrests, which took place in a large number of regions with fairly close average numbers of arrests. As a result, despite the fact that a trend region is formally identified, Vologda Region should not be considered dominant in the number of arrests. The wave corresponds to the first wave of forced collectivization and dekulakization, which covered many regions of Russia. The defining wave of anti-church arrests should be considered the second wave with two peaks in December 1930 and in May 1931 with trend regions in Moscow and the Moscow Region respectively. Both regions are dominant and qualitatively exceed the rest in the number of arrests After the first wave of collectivization, when the ruling Bolshevik Party condemned the violation of the principle of voluntariness, the majority of peasants left the collective farms. Then the second wave of collectivization was carried out with the use of propaganda teams. In essence, this wave was also violent, because peasants were forced to pay unaffordable taxes and were thus economically forced to join the collective farm. At the same time, a very important detail is that the repressions directed against the peasants were actually directed against the authority of the church among the peasants, against the parish, which was headed by the village priest. The center of the spiritual life of the peasants was the Orthodox Church in the villages, which they visited, coming to services on Sundays and church holidays.

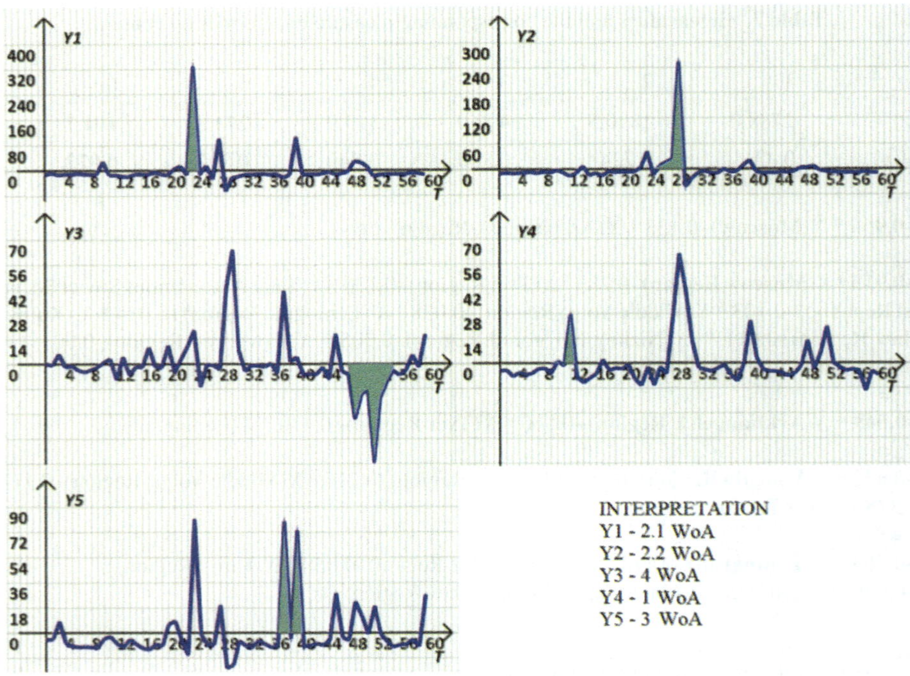

Fig. 6. Significant PC and waves of arrests (WoA).

Table 8. Principal components PC, waves of arrests WoA, and the most significant regions R_a.

PC	WoA	Month	Month/Year	R_a
Y_1	2.1	24	12/1930	Moscow
Y_2	2.2	29	05/1931	Moscow Region
Y_3	4	52	04/1933	Arkhangelsk Region
Y_4	1	14	02/1930	Vologda Region
Y_5	3	38	02/1932	Leningrad

The peasants followed the priest. This did not suit the Bolsheviks, and they decided to change the existing order of things, replacing the parish with a collective farm led by a chairman, who was subordinate to the Bolsheviks. One of the reasons for anti-church persecution in cities during the period of collectivization, for example in the trend region of Moscow, could be the so-called self-dekulakization, when peasants themselves sold or left their farms and moved from villages to cities. The second region in the number of arrests after Moscow is the Moscow region, which is not surprising, since both regions are central. In the third wave, with a peak in February 1932, Leningrad predominated in anti-church arrests. This wave can be explained by the arrests during this period of the so-called Josephites, followers of Metropolitan Joseph Petrov (Mazyrin, 2002). The deportation of peasants from the central regions, in particular to the Arkhangelsk region, during the period of collectivization could have caused a surge in persecution. A history of the Russian Orthodox Church during the period of persecution in the twentieth century in English can be found in the book (Pospielovsky, 1984).

Dual Task: Classification of Arrest Regions

To classify regions of arrests, a dual problem is considered (Aivazyan et al. 1989). In this case, the observations are the regions of arrests, and the features are the months of arrests. In accordance with the geometric extremal properties of the PC (Aivazyan et al. 1989), projecting the original set of observations onto a space of lower dimension $p < n$, which includes the p first PC, "least distorts its geometric configuration". Figure 7 shows the initial observations in the plane of the first two PC by month of arrest. Points on the plane are regions.

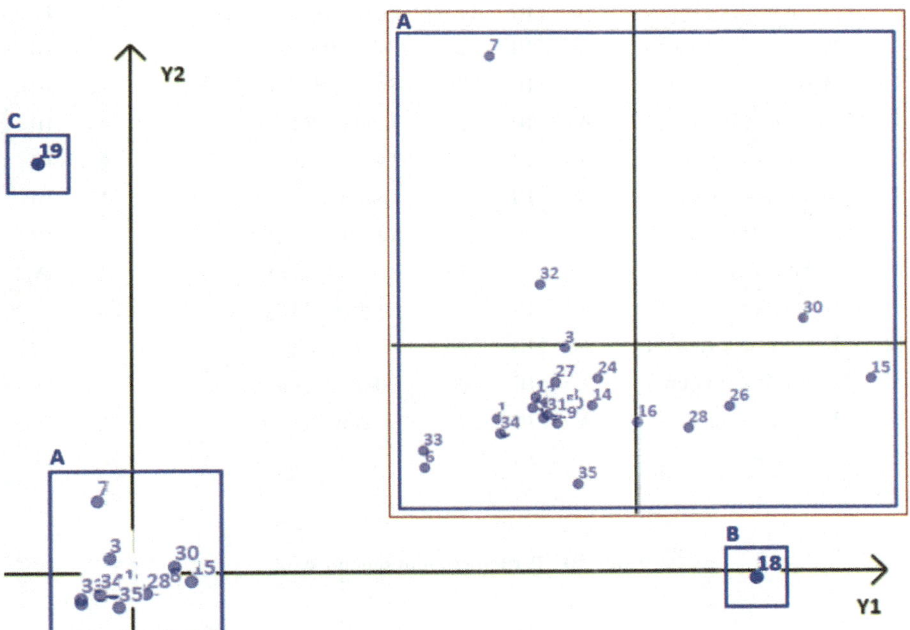

Fig. 7. Clustered observations on the plane of the first two PC by month of arrest. Points are regions of arrests.

Table 9 contains a numbered list of arrest regions in order of increasing quadrant Q on the plane of the first two PC, indicating the class of region C (Fig. 7). The resulting classification identifies three classes of regions: 1) Moscow (class B — maximum in Y_1), 2) Moscow Region (class C — maximum in Y_2), 3) other regions (class A — grouped near the origin of coordinates). PC Y_1 mainly describes peak 2.1 (24th month), and Y_2 — peak 2.2 (29th month). As can be seen from Fig. 4, these peaks correspond to the second wave of arrests, which was the longest from 18 (6/1930) to 33 (9/1931) months.

The classification characterizes the degree of participation of regions in the second wave of arrests, in its first Y_1 and second peak Y_2. Qualitatively, the degree of participation can be described by the location of the region in a certain quadrant Q (see Table 10).

Table 9. Arrest regions list in order of increasing quadrant Q.

N	Region	*C*	*Q*	*N*	Region	*C*	*Q*
30	Middle Volga Region	A	I	20	Nizhny Novgorod Region	A	III
7	Vologda Region	A	II	21	Novgorod Region	A	III
19	Moscow Region	C	II	22	Orenburg Region	A	III
32	Tatar ASSR	A	II	23	Oryol Region	A	III
1	Altai Region	A	III	24	Penza Region	A	III
2	Arkhangelsk city	A	III	25	Petrograd	A	III
3	Arkhangelsk Region	A	III	26	Pskov Region	A	III
4	Bashkiria	A	III	29	Smolensk Region	A	III
5	Vitebsk Region	A	III	31	Tambov Region	A	III
6	Vladimir Region	A	III	33	Tver Region	A	III
8	Voronezh Region	A	III	34	Tomsk city	A	III
9	Ivanovo Region	A	III	35	Yaroslavl Region	A	III
10	Kalinin Region	A	III	15	Leningrad city	A	IV
11	Kirov Region	A	III	16	Leningrad Region	A	IV
12	Krasnoyarsk Region	A	III	18	Moscow city	B	IV
13	Kuibyshev Region	A	III	26	Pskov Region	A	IV
14	Kursk Region	A	III	28	Saratov Region	A	IV
17	Lipetsk Region	A	III	–	–	–	–

Table 10. Participation in second WoA.

WoA *Q*	I	II	III	IV
2.1	+	–	–	+
2.2	+	+	–	–

The "+" sign indicates the participation of the region to one degree or another in the corresponding peak of arrests 2.1 and/or 2.2. The "–" sign means a decline in the number of arrests in the region. Quadrant I highlights the regions that participated in both peaks of the second wave of arrests.

3 Results

As a result of component analysis, it is proposed to introduce some integral characteristic. Let the measure of the number of arrests $M(Y) = \mu_a(t)$ be defined as the distance from the origin in the PC coordinate system to a given point with coordinates $Y_1(t),\ldots,Y_p(t)$.

$$\mu_a(t) = \sqrt{\sum_{i=1}^{p} Y_i^2(t)}, \tag{9}$$

where $t \in T = \{1,\ldots,60\}$ are months, $p = 5$ is the number of significant PC. The elements of the set T are the numbers of the months of the period of persecution under consideration since 1929 to 1933 Then the measure $\mu_a(t)$ defined by formula (9) is some general numerical characteristic of the number of arrests. The measure $\mu_a(t)$ is based on PC and reflects the type of all four waves of arrests. At the same time, the PC that determine the measure of arrests identify regions R_a with maximum coefficients in absolute value, characteristic of the peaks of waves of arrests (see Table 8). Arrest statistics for selected regions shape the main structure of waves of arrests, and statistics for other regions, included in the PC, adjust this structure.

In comparison with the general graph of arrests during the period under review (see Fig. 4 for a graph for all regions), the measure $\mu_a(t)$ reflects waves of arrests through trend regions R_a (see Fig. 8). This leads to the fact that the ratio of peaks in the measure $\mu_a(t)$ will be different from that of the general graph. Each peak of the measure $\mu_a(t)$ will correspond to the peak of arrests for the trend region: first wave of arrests — Vologda Region, first peak of second wave of arrests — Moscow, second peak of second wave of arrests — Moscow Region, third wave of arrests — Leningrad, fourth wave of arrests — Arkhangelsk Region. At the same time, the first peak of the second wave of arrests in Moscow becomes the maximum, while in the general graph this peak is the third largest after the first wave of arrests and the second peak of the second wave of arrests. The second largest is the second peak of the second wave of arrests in the Moscow Region, then the third wave of arrests in Leningrad, the fourth wave of arrests in the Arkhangelsk Region and the first wave of arrests in the Vologda Region, which has the maximum peak on the general graph. This is explained by the fact that the first wave of arrests does not have a region that stands out significantly in terms of the number of arrests, as is the case, for example, in the second wave of arrests, but consists of a larger number of regions with a number of arrests that differs little from the average. Trendy Vologda Region is not even among the top three regions in terms of the number of arrests (see Table 2, line 1). Waves of arrests in trend regions R_a, represented by the measure $\mu_a(t)$, can naturally be displayed in the form of a thematic map, correlating peak values with a certain color transparency (see Table 11 and Fig. 8).

$$\mu_a(\Delta t) = \sum_{t=t_1}^{t=t_2} \mu_a(t), \quad P_a(\Delta t) = \frac{\mu_a(t)}{\sum_{t=1}^{t=60} \mu_a(t)} 100\%, \quad O(t) = \frac{\mu_a(t)}{\mu_{cmax}} 100\%, \quad \mu_{amax} = \mu_a(24)$$

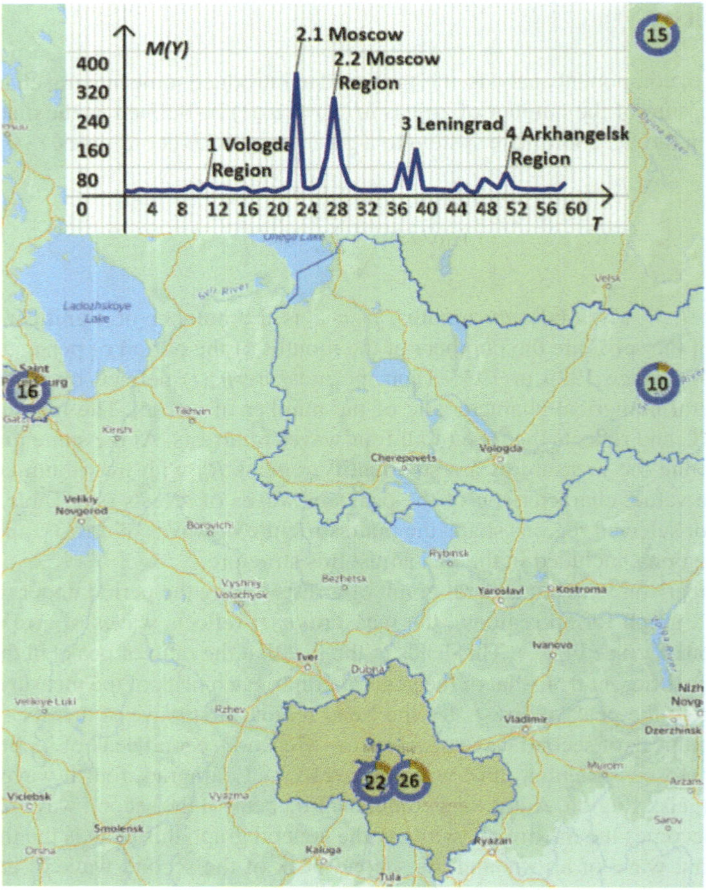

Fig. 8. Graph for a measure of the number of arrests $M(Y) = \mu_a(t)$, thematic map of trending regions color coded with opacity $O(t)$ and percentage Pa of the total arrest measure for each wave of arrests.

The thematic density map also shows the percentage P_a of the total arrest measure for each wave of arrests. The scope of application of SOLAP technology to geo-objects and display of their state on thematic maps is studied in the article (Evsyukov, 2017). The thematic map in Fig. 8 is created based on a geoinformation layer in the form of a KML document (OGC KML, 2015). KML document is generated as a result of applying component analysis to the RegionOfArrest index for the PlaceOfArrest field. The proportion P_a of the total number of arrests arranges the PC in a slightly different order than in the PCA. In the sense of $\mu_a(t)$ P_a shows the degree of influence that a given wave of WoA arrests with trend region R_a had on the entire process of arrests in the period 1929–1933. As can be seen from Table 11 and Fig. 8 Moscow and the Moscow Region which are central regions, had the greatest influence on the process of arrests.

Table 11. Ratio of waves of arrests and trending regions, arranged in a row in descending order by the value of the measure of arrests ($\mu_a t$).

WoA	2.1	2.2	3	4	1
PC	Y_1	Y_2	Y_5	Y_3	Y_4
Months, $t_1 - t_2$	18–26	27–33	34–42	43–55	8–17
Months/Years	6/1930–2/1931	3/1931–9/1931	10/1931–6/1932	7/1932–7/1933	8/1929–5/1930
$\mu_a(\Delta t)$	492	575	346	338	228
Percent, $P_a(\Delta t)$	22,3%	26,1%	15,7%	15,4%	10,4%
R_a	**Moscow**	**Moscow Region**	**Leningrad**	**Arkhangelsk Region**	**Vologda Region**
Month of R_a, t	24	29	38	52	14
Month/Year of R_a	12/1930	5/1931	2/1932	4/1933	2/1930
$\mu_a(t)$	342	274	93	65	24
Opacity, $O(t)$	100%	80%	27%	19%	7%

4 Discussions

It is interesting to compare the methods considered: the method of *maximum arrest regions* (Fig. 2) and the measure $\mu_a(t)$ based on the PCA (Fig. 8) give slightly different results. The main difference is that measure based on the PCA includes the *Vologda Region*, not the Yaroslavl Region; the *ratios of peaks* on the graphs are different in both methods; the differences are due to the fact that measure based on the PCA identifies *trending regions of arrests*, and not just the maximum ones. The graph of the arrest measure function $\mu_a(t)$ distinguishes *arrests by trend regions*.

The initial set of characteristics X, represented by arrest statistics by region, can be expressed through a smaller number of auxiliary indicators Y, which correspond to waves of arrests. If the Y values are specified, then regression analysis is applied. Otherwise, factor analysis or component analysis can be applied. From the point of view of the original problem of the arrest process, it is correct not to make any assumption about the number p of Y indicators. This allows us to do a component analysis, highlighting the most significant components. Factor analysis assumes that the number p of factors is initially known. Also, component analysis allows not to make any assumptions about the type of random distribution of X values. The conditions for using factor analysis require the symmetry of the distribution of the original variables X, and in the maximum likelihood method they are subject to a multivariate normal distribution (Lawley and Maxwell, 1963). Let the dimension $p < n$. Component analysis (Lawley and Maxwell, 1963) states the following

$$\exists Y_1, Y_2, \ldots, Y_n : X_i = \sum_{j=1}^{p} b_{ij} Y_j;\ i = 1, \ldots, n$$

Factor analysis (Lawley and Maxwell, 1963) introduces mutually independent random variables $e_1, e_2, \ldots, e_n$: $X_1 - e_1, X_2 - e_2, \ldots, X_n - e_n$ do not depend on them:

$$\exists Y_1, Y_2, \ldots, Y_n : X_i - e_i = \sum_{j=1}^{p} b_{ij} Y_j; \; i = 1, \ldots, n$$

From the above definitions it follows that component analysis is a special case of factor analysis. Firstly, the above does not mean that, having obtained a factor model, a component model can be derived from it in the form of a special case. Second, PCA is the only mathematically valid factor analysis method (Kim et al. 1986). Finally, PCA can be considered separately from factor analysis. Brown believes that PCA is not a factor analysis method because it does not split the variance of indicators into common and unique (Brown, 2015).

5 Conclusion

In the process of research, a cross-section of a multidimensional data cube on repressions was constructed, associated with statistics on *regions of arrests*. As a result of applying component analysis to statistics on arrests, it can be concluded that anti-church persecution in the period 1929–1933 was most pronounced in the central regions — Moscow and Moscow Region. This reflects the second central and largest wave of persecution, which continued from June 1930 to September 1931. Moscow peaks occurred in the 24th month (December 1930) and in the 29th month (May 1931). This centralization of persecution confirms the fact that persecution in all regions was initiated from a single center. The splitting of the arrest graph into separate waves and peaks means that the order from the center on collectivization, which is inextricably linked with anti-church arrests, was carried out in different regions non-simultaneously, in different months. So, the PC can be interpreted as *waves of arrests* (WoA), which correspond to trend regions (see Table 8). PC have a more complex structure than the statistics of trend regions; most of them are bipolar and reflect the complexity of the arrest processes. The peaks of arrests in Moscow and Arkhangelsk Region (Y_1), in Moscow Region and Tver Region (Y_2) are in antiphase or shifted in time. In Fig. 4, periods of WoA are highlighted with different background colors. Each WoA, due to the overlapping of separate waves by region, has a complex shape with several peaks. The PC Y_i are ordered in terms of the proportion of the total variance explained of arrest statistics by trend regions according to the significance of the influence on the overall process of arrests: Y_1 — Moscow, Y_2 — Moscow Region, Y_3 — Arkhangelsk Region, Y_4 — Vologda Region, Y_5 — Leningrad. Trending regions form the basic structure of the corresponding PC or persecution waves. Regions included in the expressions for PC with weighting coefficients that are smaller in absolute value than trend regions adjust this structure. Consideration of the *dual problem* in the PCA allows us to obtain a classification of *three classes of regions* in the plane of the first two PC (Fig. 7). They are as follows: 1) Moscow (peak in month 24 — maximum in Y_1); 2) Moscow Region (the peak in the 29th month — maximum in Y_2); 3) the remaining regions (grouped in the neighborhood of the origin of coordinates). In accordance with the classification, Tables 9 and 10 define the degree of participation of regions in the second WoA. Simultaneously with Moscow city (IV quadrant in Fig. 7), surges in arrests are observed in Leningrad city, Pskov Region, Saratov Region, Leningrad Region. Together with Moscow Region (II quadrant in Fig. 7) surges in arrests are observed in the Vologda Region, Tatar ASSR. The Middle Volga Region (I quadrant in Fig. 7) belongs to both peaks of the second wave of arrests. Using the PCA, a graph of the measure $\mu_a(t)$ of the number of arrests was obtained with four WoA marked in the period 1929–1933, and a corresponding thematic map of trend regions was constructed (see Table 11 and Fig. 8). The arrest measure represents peaks according to trend regions (Fig. 8) rather than the total number of arrests (Fig. 2): Moscow, Moscow Region, Leningrad, Arkhangelsk Region, Vologda Region. The conducted research is relevant from the point of view of the subject area, because confirms and supplements historical events with statistical conclusions, creating a picture of the persecution of Orthodox Christians in the period 1929–1933.

Further direction of research may be related to a detailed consideration: the waves of persecution during the period of the Bolshevik seizure of power (1917–1918); the waves of persecution during the period of seizure of church valuables (1921–1923); the waves of persecution during the period of "great terror" (1937–1938). Another direction is related to the application of the big data analysis method (Kresse and Danko, 2023) in the context of geospatial data.

Appendix A

Biographical note of Hieroschemamonk John Kevroletin[2] from the database "for Christ suffered".

Reverend Confessor Hieroschemamonk John (Kevroletin).

JOHN (Kevroletin Ivan Afanasyevich), hieroschemamonk.

1875, May 23 - was born in the locality: Tobolsk province. Troitsky county village Kulakovo.

Born into a peasant family, self-taught. In 1893–1894. served as a sailor on a barge in a shipping company in Tyumen.

November 26,1894–1902 - Ministry. Perm province, Irbitsky county, Kyrtomsky Cross Exaltation Monastery.

Worked on the obedience of leather dressing.

October 22,1902–December 20,1907 - Ministry. Perm province, Verkhoturye, Nicholas Monastery.

1902–1919 - Service. Perm province, Verkhoturye, Nicholas Monastery.

Since April 3, 1914 steward of the bishop's country house in the village of Bolshoy Istok, in 1915 again in the monastery. In February 1915 was sent to perform pastoral duties in parishes and monasteries of the Yekaterinburg diocese.

1919–1922 - Service. Perm province., Yekaterinburg county (Sverdlovsk region, Nevyansky District), Serbishino village (Serbishna).

2 See the biographical information of Hieroschemamonk John (Kevroletin Ivan Afanasyevich) in the database "for Christ suffered".http://martyrs.pstbi.ru/bin/nkws.exe/spc_1_prn/ans/newmr/?HYZ9EJxGHoxITYZCF2JMTdG6Xbu6fe0hfGcWeKWs8ufdS8cfG06ce0h661Yc8qUsSmZceXb8E

In 1922 served 9 months. in the Kazan temple of the monastery (?), replacing the priest.

1922–1923 - Service. Perm province, Yekaterinburg county, settlement of Verkh-Isetsky plant.

Replaced the priest for 3 months. Then until 1924 served in St. Sergius Church (?)

1924–1925 - Ministry. Ural Island, Verkhoturye, monastery.

1925–1927 - Service. Ural region, Blagoveshchensky district, Dymkovo village.

1927–1927 - Service. Ural region, Nizhne-Tavdinsky plant settlement.

Served 1 month.

1927–1927 - Service. Ural (Sverdlovsk) region, Irbit.

Served 3 months.

1927–1927 - Service. Ural (Sverdlovsk) region, settlement Karabashsky plant.

Served 3 months.

1928–1928 - Service. Ural (Sverdlovsk) region, Dalmatov.

Served 2 weeks.

1928–1928 - Service. Ural (Sverdlovsk) region, Kislovo village.

Served 6 months.

1928–1929 - Service. Ural (Sverdlovsk) region, Turin mines settlement.

1929–1931 - Service. Ural (Sverdlovsk) region, Fedkovka village.

1931–1931 - Service. Ural (Sverdlovsk) region, Verkhoturye.

1931–1931 - Service. Ural (Sverdlovsk) region, Turinsk.

1931–1932, March 28 - Service. Ural (Sverdlovsk) region, Nevyansky district, Serbishino village (Serbishna).

1932, March 28 - arrested. Ural (Sverdlovsk) region, Nevyansky district, Serbishino village (Serbishna).

1932, September 7 - convicted. Special Meeting at the Collegium of the OGPU of the USSR for 'participation in a priestly-monastic counter-revolutionary organization' (under Articles 58–10,58–11 of the Criminal Code of the RSFSR). Verdict - 3 years of exile in Western Siberia.

1932, March 28–1932 - He was serving a sentence. Ural region, Nizhny Tagil, house of detention.

During the investigation, he was detained in Nizhny Tagil house of detention.

1932–1935 - Served a sentence. Narym Territory.

1935–1937 - Service. Sverdlovsk region, Sosnovka village, Church of the Nativity.

Since 1937 did not serve due to illness (?)

1937 - arrested. Sverdlovsk region, Sosnovka village.

1937 - convicted. The sentence is imprisonment.

1937–1946 - Served a sentence. Forced labor camp (exile?)

1946 - Ural region, Verkhoturye.

In 1946 Father Ignatius returned from the camps. He returned secretly, disabled, and the nuns of the former Pokrovsky and Assumption monasteries, who survived the camps and exiles, hid him in different houses. At night, he changed into women's clothes and went from house to house at the request of the residents. Baptized children, confessed the dying, read prayers.

1948 - Ministry. Ural region, Verkhoturye.

In 1948 went out of staff.

1961 - Ural region, Verkhoturye.

1961, January 27 - date of death. Place of death - Sverdlovsk region, Verkhoturye.

He was buried near the Assumption cemetery church, in 1992. the remains were transferred to the altar of the Church of the Transfiguration of the Verkhotursky Nikolaev Monastery.

** Journal N 7 of the meeting of the Holy Synod of the Russian Orthodox Church of December 27, 2000.

State archive of Yekaterinburg. D.P-41750.

Nechaeva M.Yu. Verkhoturye monasteries in the XX century. Typescript.P.17.

Act of the Jubilee Consecrated Bishops' Council of the Russian Orthodox Church on the conciliar glorification of the New Martyrs and Confessors of Russia of the twentieth century. Moscow, August 12–16, 2000.

Lavrinov V., archpriest. Yekaterinburg diocese. Events. People. Temples. Yekaterinburg: Publishing House Ural. University, 2001. 336P. P.140.

A selection of clergymen, made according to the book: The Book of Memory of the Victims of Political Repressions. Sverdlovsk region. T.1–4. Yekaterinburg: Ural worker, 1999–2003. Typescript.P.15.

http://rus-icons.narod.ru/part13/photo 114.html (website 'World of Russian Icon').

References

Vorobyev, V.N. (archpriest Vladimir Vorobyev): "Suffering for Christ": Persecution of the Russian Orthodox Church 1917-1956. Biographical reference book. Book 1 (A). Publishing house. PSTGU, p. 736. (2015)

Ilyashenko, A.S. (archpriest Alexander Ilyashenko): On the assessment of the number of prisoners and mortality in the GULAG. Economic History: Review. L. I. Borodkin. Moscow State University Publishing House, p. 107–123. (2011)

Mazyrin, A.V. (currently archpriest Alexander Mazyrin): "I follow only Christ…": Metropolitan Joseph (Petrovykh), 1930. Published, introduction and notes by A. Mazyrin. Theological collection. Issue 9. Moscow: PSTBI Publishing House, p. 376–424. (2002)

Afifi, A.A., Azen, S.P.: Statistical Analysis: A Computer-Oriented Approach, 2nd edn, p. 442. Academic, New York (1979)

Aivazyan, S.A., et al.: Applied Statistics: Classifications and Reduction of Dimensionality. Reference edition, p. 607. Finansy i statistika, Moscow (1989)

Brown, T.A.: Confirmatory factor analysis for applied research, vol. 462, 2nd edn. The Guilford Press (2015)

Cangelosi, R., Goriely, A.: Component retention in principal component analysis with application to cDNA microarray data. Biol. Direct. **2**, 2 (2007) https://biologydirect.biomedcentral.com/articles/10.1186/1745-6150-2-2

Duraciova, R., Chalachanova, J.F.: Fuzzy spatio-temporal querying the postgreSQL/postGIS database for multiple criteria decision making. In: Ivan, I., Horak, J., Inspektor, T. (eds.) Dynamics in GIscience Lecture Notes in Geoinformation and Cartography, pp. 81–97. Springer International Publishing AG (2018). https://doi.org/10.1007/978-3-319-61297-3

Emelianov, N.E.: Una storia dei martiri scritta a computer. In: La Nuova Europa, vol. 3, pp. 23–34 (1998)

Emelyanov, N.E.: Studies in history based on the database. In: ‘The 20th Century New Martyrs and confessor of the Russian Orthodox Church’. The Second International Symposium ‘Christianity in Our Life: Past, Present, Future’. Tbilisi 24–26 Nov 2005, pp. 40–41 (2005)

Emelyanov, N.E., Muhanov, I.V., Tishchenko, V.A.: Web server on the basis of NIKA DBMS. In: Proceedings of the Third International Workshop on “Advances in Databases and Information Systems”, vol. 2, pp. 58–59. ACM SIGMOD, Moscow (1996)

Evsyukov, A.A.: Online analytical processing of spatial data for emergency monitoring. In: Ceur Workshop Proceedings, vol. 2033, pp. 180–184 (2017)

Harman, H.H.: Modern Factor Analysis, 2d edn, p. 474. The University of Chicago Press, Chicago and London (1968)

Hutchinson, M.J., Veenendaal, B.: An agent-based framework for intelligent geocoding. Appl. Geomat. **5**(1), 33 (2013)

Kim, J.-O., Mueller, C.W., et al.: Factor Analysis: Statistical Methods and Practical Issues. Sage Publications, Newbury Park. Eleventh Printing (1986)

Kresse, W., Danko, D. Springer Handbook of Geographic Information, 2nd edn (2023)

Lawley, D.N., Maxwell, A.E.: Factor Analysis as a Statistical Method. Butterworth, Kent, England (1963) rev. 1971

Makarov, I.M., et al.: Mathematical Theory of Virtualization of Design Processes and Technology Transfer, p. 316. FIZMATLIT, ISBN: 978-5-9221-1412-7 (2013)

Mirkes, E.M., Allohibi, J., Gorban, A.: Fractional norms and quasinorms do not help to overcome the curse of dimensionality. Entropy. **22**(10), 1105 (2020). https://doi.org/10.3390/e22101105

OGC KML 2.3. Open Geospatial Consortium, Wayland. http://www.opengeospatial.org/standards/kml (2015)

Parsaye, K.A.: Characterization of data mining technologies and processes. J. Data Warehouse. **1**, 12–24 (1998) https://masters.donntu.ru/2007/fvti/kravchenko/library/7.htm?ysclid=m1p4af02xs758391320. Accessed 30 Sept 2024

Pospielovsky, P.: The Russian Church Under the Soviet Regime 1917–1982 St Vladimirs Seminary Pr, 1st edn, p. 535 (1984)

Rencher, A.: Multivariate Statistical Inference and Applications. John Wiley & Sons, Inc., New York (1998)

Soloviev, A.V., Bogacheva, A.N., Tishchenko, V.A.: Elements of analytical data processing for a factual database: statistical processing of historical facts on the example of a database ‘for christ suffered’. In: Silhavy, R., Silhavy, P. (eds.) Networks and Systems in Cybernetics. CSOC 2023 Lecture Notes in Networks and Systems, vol. 723, pp. 23–39. Springer, Cham (2023). https://doi.org/10.1007/978-3-031-35317-8_3

Tishchenko, V.A.: Variant retrospective geocoding using a dictionary of regions, Proceedings of the Institute for System Analysis of Russian Academy of Sciences (ISA RAS), vol.74, issue 1, Moscow, p.73–78. (2024)

Reduction of Uncertainty in the Information Field

Victor Tsvetkov(✉)

Research and Design Institute of Information, Automation and Communications in Railway Transport, Moscow, Russia
cvj7@mail.ru

Abstract. The article investigates the phenomenon of uncertainty. Uncertainty has similarities with the concepts of implicit knowledge, spaces with unknown dimensionality, uncertain set. The paper explores uncertainty and proposes a new approach to reducing uncertainty. Uncertainty is interesting not by itself, but in combination with objects that are characterized by certainty. The antipode of uncertainty is meaningfulness. Uncertainty set the task of its reduction and finding meaningfulness. Dichotomous division and reduction is proposed as a basis for reducing informational uncertainty. Uncertainty can be parametric, temporal and spatial. Interpretability is proposed as a state control. Interpretability can be synonymous with certainty. Finally, a reduction method is proposed to convert uncertainty into certainty. #COMESYSO1120.

Keywords: Uncertainty · Content · Tacit Knowledge · Non-Dichotomous Division · Oppositional Analysis · Information Relations · Content · Composition

1 Introduction

The information field [1] can be considered as an integral model of reality. It contains smaller information models of objects, information models of processes, information models of relations and information models of regularities. A static analog of the information field can serve as a snapshot of reality. Models within the information field can be clear, fuzzy and uncertain.

Uncertainty in the information field is information uncertainty [2–4]. For any phenomenon it is necessary to introduce a description or model. We model uncertainty as an areal or set, the boundaries of which are known, but the elements and their contents are unknown. A physical analog of such a model can be a shadow on a surface.

If the surface is illuminated, objects and details on it are distinguishable. If a dense shadow lies on the surface, the items and elements become invisible. For such a model it is necessary to introduce the concept of "observer". An observer is a subject or an object that is able to perceive or not to perceive elements of the surface.

Uncertainty is always found in scientific research and is often found in the information field. When using shading as an analog of uncertainty, uncertainty can be indicated

R. Silhavy and P. Silhavy (Eds.): CoMeSySo 2024, LNNS 1489, pp. 69–76, 2025.
https://doi.org/10.1007/978-3-031-96798-6_6

in black. Fuzziness or partial uncertainty is denoted in gray. There is a term "gray analysis", which denotes analysis [5] in the presence of partial uncertainty. Uncertainty can characterize parametric space and state space. A close notion of uncertainty is the notion of "uninterpretability". Interpretability [6, 7] can be synonymous with certainty.

Uncertainty can be considered as a state or as a property. It can have qualitative characteristics and less often quantitative characteristics. Informational uncertainty is different from uncertainty in quantum mechanics. The antipode of uncertainty is meaningfulness. Uncertainty poses the problem of its reduction. Reducing uncertainty increases meaningfulness. The existing gaps in the study of uncertainty determine the relevance of this paper.

2 Materials and Methods

The basis of the study is a set-theoretic analysis. The axiomatic approach was used to systematize uncertainties. Qualitative analysis, the method of dichotomous division and the method of reduction in the author's interpretation were used.

3 Results

The result of the research is the introduced model of information uncertainty. Based on this model, a systematization of uncertainty is given. A methodology for reducing uncertainty is proposed.

3.1 Basic Definitions

Let us define the abbreviations used. Uncertainty - UN; tacit knowledge - TK; space of uncertain dimensionality - UDIS; space of certain dimensionality CS; certain set - CS; uncertain set - IS; fuzzy set - FS; definiteness - D; definite set – DS; interpretation - Int.; composition - Comp.; content - Cont.; Reduction - Red.

3.2 Systematics of Uncertainty

There are categorical relations for uncertainty. The first relation is related to union.

$$\mathrm{UN1} \cup \mathrm{UN2} \rightarrow \mathrm{UN3}. \tag{1}$$

Expression (1) and Fig. 1 show that when uncertainties having different bounds are united, a new mixed uncertainty is obtained. The second relation is related to intersection.

$$\mathrm{UN1} \cap \mathrm{UN2} \rightarrow \mathrm{UN4}. \tag{2}$$

Expression (2) and Fig. 2 show that the intersection of uncertainties having different bounds results in a new reduced uncertainty.

$$\mathrm{UN1} \cup \mathrm{UN5} \rightarrow \mathrm{UN1}. \tag{3}$$

Expression (3) shows that when uncertainties are united, of which one (UN5) has boundaries lying inside another uncertainty (UN1), the larger uncertainty absorbs the smaller one. A similar relationship exists for tacit knowledge

$$\text{TK1} \cup \text{TK2} \rightarrow \text{TK3}, \tag{4}$$

$$\text{TK3} > \text{TK1}; \text{TK3} > \text{TK2}. \tag{5}$$

Expressions (4) and (5) show that there is an increase in the tacit knowledge (TK3) when the tacit knowledge (TK) is united, except in a situation like (3). Expressions (1)–(5) occur within the same categories.

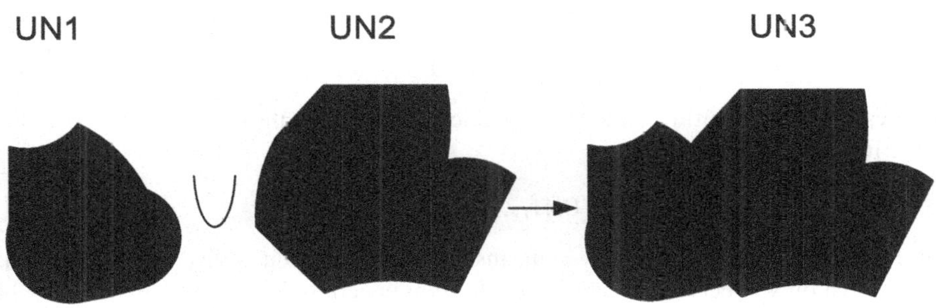

Fig. 1. Union of uncertainties.

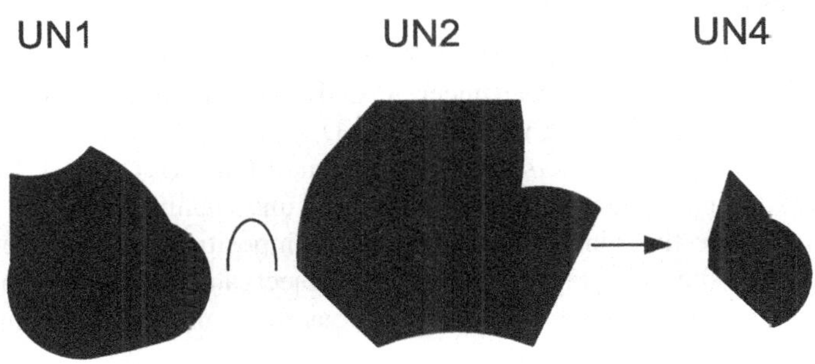

Fig. 2. Intersection of uncertainties.

A similar relationship exists for the union of uncertainty (UN) and definiteness (D)

$$\text{UN1} \cup \text{D1} = \text{UN1}. \tag{6}$$

Expression (6) describes the situation of "definiteness absorption". This situation means that as a result of the interaction between definiteness and uncertainty, the definiteness is absorbed and a new uncertainty is formed.

$$\text{UN1} \cup \text{D2} = \text{Comp}\left(\text{UN}^{*}, \text{D}^{*}\right). \tag{7}$$

Expression (7) describes the situation of "composition of definiteness and uncertainty". This situation means that the interaction of definiteness (D2) and uncertainty (UN1) results in the formation of a composite that includes a changed definiteness (D*) and a changed uncertainty (UN*).

Different situations are possible for expression (7), such as the following relation:

$$\mathrm{UN1} < \mathrm{UN}^*. \tag{8}$$

This relation characterizes the reduction of uncertainty. For another situation, For another situation, the following relation is possible:

$$\mathrm{D}^* < \mathrm{D2}. \tag{9}$$

This relation characterizes the reduction of definiteness. The relation (10) characterizes an increase in definiteness.

$$\mathrm{D2} < \mathrm{D}^*. \tag{10}$$

Expressions (1–10) form a systematics of uncertainty. Using the logical approach allows us to introduce the relation.

$$(\mathbf{UN1}, \ldots, \mathbf{UNn}) \vdash \mathbf{UN}. \tag{11}$$

In expression (11), the symbol $\vdash$ means following or deducibility. Expression (11) means that only uncertainties are deducible from uncertainties. Expression (11) can be interpreted as an "alogism".

3.3 Uncertainty of Interactions

Uncertainty describes situations, relationships, objects and processes. Let us consider a combination of interactions between uncertain and defined objects. Definite objects will be denoted by single capital Latin letters: A, B, D, C, E, G, etc. Undefined objects will be denoted by double Latin letters with the addition of Latin U: UA, UB, UD, UC, UE, UG, etc. The designation UN describes generalized uncertainty.

Let us introduce the concept of uncertainty composition (Comp). Let us call an uncertainty composite an interacting set of certain objects and uncertainty that participate in a common process. There are different situations for an uncertainty composite.

$$\mathrm{Comp}\,(\mathrm{A}, \mathrm{B}, \mathrm{UN}, \mathrm{C}, \mathrm{E}) \rightarrow \mathrm{UN}. \tag{12}$$

Expression (12) describes the absorption of certain objects by uncertainty. This situation exists in discourse or proofs when a counterexample completely refutes a constructed theory based on A, B, C, E.

$$\mathrm{Comp}\,(\mathrm{A}, \mathrm{B}, \mathrm{UN}, \mathrm{C}, \mathrm{E}) \rightarrow \mathrm{Comp}\left(\mathrm{A}, \mathrm{B}, \mathrm{UN}^*, \mathrm{C}, \mathrm{E}\right); \mathrm{UN} > \mathrm{UN}^*. \tag{13}$$

Expression (13) describes the reduction of uncertainty under the influence of arguments A, B, C, E. This situation exists in thesis defense, when the applicant rejects the arguments of the opposing party.

$$\mathrm{Comp}\,(\mathrm{A}, \mathrm{B}, \mathrm{UN}, \mathrm{C}, \mathrm{E}) \rightarrow \mathrm{Comp}\left(\mathrm{A}, \mathrm{B}, \mathrm{UN}'', \mathrm{C}, \mathrm{E}\right); \mathrm{UN} < \mathrm{UN}''. \tag{14}$$

Expression (14) describes the increase of uncertainty when interacting with arguments A, B, C, E. Such a situation exists in thesis defense, when the applicant strengthens the arguments of the opposing side by an incorrect answer.

$$\text{Comp (A, B, UN, C, E)} \rightarrow \text{D}. \tag{15}$$

Expression (15) describes a situation where, under the influence of arguments, uncertainty completely disappears and only definiteness remains. Expression (15) describes a synergistic jump.

$$\text{Comp (A, B, C, E)} \rightarrow \text{UN Comp}. \tag{16}$$

Expression (16) describes a situation where contradictory arguments lead to uncertainty. Such a situation occurs in incorrect conclusions.

$$\text{Comp (A, B, UN, C, E)} \rightarrow \text{Comp (A, B, UN, C, E)}. \tag{17}$$

Expression (17) describes the situation when arguments are ineffective and do not change the uncertainty. Such a situation arises in unreasonable reasoning. Expressions (12)–(17) describe situations of interaction of certain objects with uncertainty.

3.4 Transformations

The state of complete uncertainty is of little interest to the researcher. It is important to describe transitions between uncertainty and definiteness, or transitions from greater uncertainty to lesser uncertainty. Let us consider some content (Cont) that contains uncertainty. Consider the operation of information reduction (Red), which reduces the amount of content.

An example of information reduction is obtaining a two-dimensional photograph of a three-dimensional reality. Simplistically, one of the types of reduction can be considered as a projective transformation reducing the dimensionality of space from four-dimensional to three-dimensional, from three-dimensional to two-dimensional, from n-dimensional to n-1 dimensional.

$$\text{Red [Cont (A, B, UN, C, E)]} \rightarrow \text{Cont (A}*, \text{B}*, \text{UN}*, \text{C}*, \text{E}*). \tag{18}$$

Expression (18) is characterized by the uncertainty reduction UN* < UN. The comparative relation "less" applies to uncertainty, to tacit knowledge, to multidimensional spaces. The reduction operation realizes this relation in conjunction with the succession relation. Recurrence reduction can be used.

$$\text{Red [Cont]} \rightarrow \text{Cont}^*. \tag{19}$$

$$\text{Red} \left[\text{Cont}^*\right] \rightarrow \text{Cont}^{**}. \tag{20}$$

In expression (19) Cont* < Cont. By repeating the procedure (20) many times one can continue to reduce the uncertainty. This procedure can be shown graphically (Fig. 3).

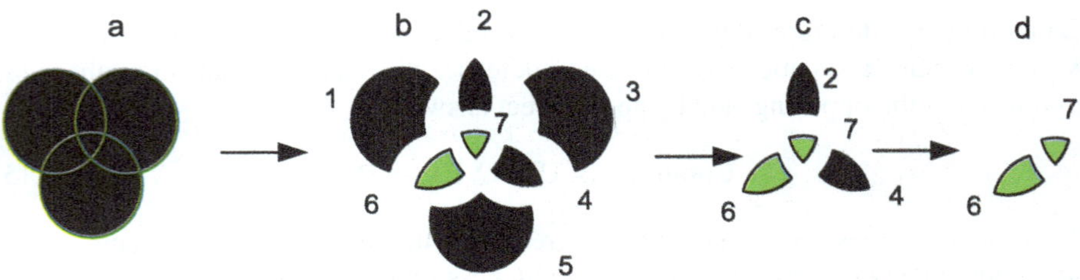

Fig. 3. Reduction that reduces uncertainty.

Figure 3 shows the procedure for reducing total uncertainty (state a). The uncertainty is shown in black (no knowledge). But information about the boundaries exists. The definiteness is shown in green color. Reduction is carried out by the exclusion method. In reality, there may be many reduction steps and transition states. Figure 3 shows the initial state (a), the intermediate states (b, c) and the final state (d). At some stage (b), meaningfulness is revealed. The reduction continues until uncertainty is eliminated and definiteness is revealed (state d).

However, to solve this problem, we need information about the uncertainty bounds (IF) or uncertainty criteria.

$$\text{If } (p1 < p) \rightarrow p1 \in UN, \tag{21}$$

$$\text{If } (p2 > p) \rightarrow p2 \in DIS. \tag{22}$$

Expressions (21) and (22) describe the dichotomous division [8]. Expressions (21) and (22) say that dichotomous analysis or oppositional analysis is applied in reduction [9]. The recursive application of expressions (20)–(22) reduces the uncertainty. A side effect of reduction is that the reduction method reduces the amount of information description (IV).

$$\text{Red}\left[\text{Cont}\left(A, B, UN^{*}, C, E, IV\right)\right] \rightarrow IV^{*} < IV. \tag{23}$$

By repeating the procedure (23) repeatedly one can systematically reduce the information volume. This is a negative factor of reduction.

3.5 Interpretability

Figure 3 can be used to illustrate the relationship between interpretability (Int) and definiteness (D).

$$\text{Int}\,[\text{Cont}(A, B, D, IR)] \rightarrow \text{Sem}\,[\text{Cont}\,(A, B, D, IR)], \tag{24}$$

$$\text{Sem} \neq \varnothing. \tag{25}$$

Expression (24) tells us that applying an interpretation procedure or process to content that includes objects A, B and the information relations IR between them creates the semantic meaningfulness Sem of that content. The information relations create new meaning or new content. The content Cont(A, B, D, IR) can be compared to a sentence in which A, B, D are words with independent dictionary meaning. Combining them into a sentence with the help of information relations (IR) creates a new meaning and a new interpretation of the combination of these words. Expression (25) shows that this meaningfulness is not empty or meaningless. Expressions (24) and (25) can be seen as a condition of interpretability.

When considering Fig. 3, we can consider the individual situations Sit(a), Sit(b), Sin (c), Sit(d). Applying the interpretability procedure to the initial condition leads to uncertainty.

$$\text{Int}\,[\text{Sit}(a)] \rightarrow \text{UN}, \tag{26}$$

$$\text{Sem}(\text{Sit}(a)) = \varnothing. \tag{27}$$

Expression (26) shows that the interpretation of the initial state described by Sit(a) entails uncertainty. Expression (27) says that the meaningfulness of the first state is absent. This is the condition of uninterpretability (27), which is equivalent to uncertainty (26).

Application of the procedure of dichotomous division by the boundaries (which are known) allows us to identify several sets (objects). Their number is equal to 7 (Fig. 3). For each object it is possible to apply the interpretation procedure.

$$\text{Int}\,[\text{Sit}(b1)] \rightarrow \text{UN}, \tag{28}$$

$$\text{Int}\,[\text{Sit}(b2)] \rightarrow \text{UN}, \tag{29}$$

$$\text{Int}\,[\text{Sit}(b3)] \rightarrow \text{UN}, . \tag{30}$$

$$\text{Int}\,[\text{Sit}(b4)] \rightarrow \text{UN}, \tag{31}$$

$$\text{Int}\,[\text{Sit}(b5)] \rightarrow \text{UN}, \tag{32}$$

$$\text{Int}\,[\text{Sit}(b6)] \rightarrow \text{Sem}(b6), \tag{33}$$

$$\text{Int}\,[\text{Sit}(b7)] \rightarrow \text{Sem}(b7). \tag{34}$$

For the first five objects of expression (28)–(32) there is no interpretability and there is uncertainty. For two objects (b6, b7) interpretability and the appearance of meaning are possible. In general, the whole system of objects in this situation is not interpretable. Applying reduction to state (b) and further reduction to state (c) leads to state (d). There is complete interpretability for state (d).

$$\text{Int}\,[\text{Sit}(d6)] \rightarrow \text{Sem}(b6), \qquad (33)$$

$$\text{Int}\,[\text{Sit}(d7)] \rightarrow \text{Sem}(b7). \qquad (34)$$

State (d) corresponds to complete definiteness. Thus the application of dichotomous division and reduction brings the state of complete uncertainty (a) into the state of complete definiteness (d).

4 Conclusion

There are methods of reducing uncertainty. The main methods are dichotomy as division and reduction as uncertainty reduction. Uncertainty is a set that contains no elements. Uncertainty violates the relations of equivalence, correspondence or tautology. To reduce and eliminate uncertainty, we need information about its boundaries and dichotomous criteria for the certainty of uncertainty. The notion of uncertainty composition carries procedural characteristics. The notion of uncertainty systematics reflects patterns. The systematics of uncertainty has features of similarity with the theory of infinitely large numbers. Practical applications of uncertainty theory and the composition model can be applied in solving the problem of big data and in exploring the universe. Finding and reducing uncertainty belongs to the field of comparative analysis. This procedure can be expert and heuristic. There are different types of uncertainty for different kinds of fields and sets. In essence, expressions (22) and (23) are a dichotomous division method. The theory of uncertainty includes procedural and substantive uncertainty as well as elements of alogism. It is of interest to investigate uncertainty by means of interpretability methods.

References

1. Tsvetkov, V.Y.: Information field. Life Sci. J. **11**(5), 551–554 (2014)
2. Zhang, X.F.: Information uncertainty and stock returns. J. Financ. **61**(1), 105–137 (2006)
3. Ferracuti, E.: Information uncertainty and organizational design. J. Account. Econ. **74**(1), 101493 (2022)
4. Keller, A.M., Taylor, H.A., Brunyé, T.T.: Uncertainty promotes information-seeking actions, but what information? Cogn. Res. Princ. Implic. **5**(1), 42 (2020)
5. Zhu, L., Zhao, C., Dai, J.: Prediction of compressive strength of recycled aggregate concrete based on gray correlation analysis. Construct. Build Mater. **273**, 121750 (2021)
6. Zhang, Y., et al.: A survey on neural network interpretability. IEEE Trans. Emerg. Top. Comput. Intell. **5**(5), 726–742 (2021)
7. Erasmus, A., Brunet, T.D.P., Fisher, E.: What is interpretability? Philos. Technol. **34**(4), 833–862 (2021)
8. Deshko, I., Tsvetkov, V.: Dichotomous division logic. AIP Conf. Proc. **2700**(1), 040020 (2023)
9. Deshko, I.P., Tsvetkov, V.Y.: Oppositional division in the information field. In: Informatics and Cybernetics in Intelligent Systems. CSOC 2021. Lecture Notes in Networks and Systems, vol. 228, pp. 599–606. Springer, Cham (2021)

Comparison of Tracking Systems for Monitoring Human Hands and Body in Virtual Reality

Artem Obukhov(✉), Dmitry Pobedinsky, Yaroslav Merkuryev, and Dmitry Kurakin

Tambov State Technical University, 392000 Tambov, Russian Federation
obuhov.art@gmail.com

Abstract. The development of high-quality virtual simulators is impossible without accurate tracking of the position of the human body and hands. When integrating virtual reality systems into various omnidirectional running platforms, one of the challenges is determining the direction of movement, which may not coincide with the rotation of the head. In addition, when implementing interaction, precise positioning of the user's hands is necessary. The article discusses three different approaches to solving these problems: virtual reality trackers, motion capture systems, and computer vision systems with neural networks for human recognition. The study compared these three systems in solving two problems: determining the angle of rotation of the body and the positioning of the hand relative to the virtual reality helmet. The limitations and measurement uncertainty of each tracking system are determined. The obtained results can be used in the implementation of virtual simulators integrated into omnidirectional platforms to provide the best user experience.

Keywords: virtual reality · human tracking systems · omnidirectional platform · computer vision · human body tracking

1 Introduction

When developing high-quality simulators built on the basis of virtual reality (VR), it is necessary to ensure a high level of comfort when implementing the processes of movement and interaction [1, 2]. To solve the first problem, various running platforms are used: active ones, built on the basis of treadmills [3–5], and passive ones, based on the principle of sliding and the use of holding safety mechanisms [6]. The first type is more difficult to manufacture and omnidirectional treadmills have not yet appeared in widespread use [5]. On the other hand, omnidirectional sliding platforms are quite popular in VR clubs [7]. When integrating virtual reality systems into omnidirectional platforms, one of the main tasks is determining the direction of movement, which may not coincide with the rotation of the head [8].

When implementing user interaction, precise positioning of the user's hands is critical [9]. Currently, there are several available methods for tracking human movement, including motion capture systems, computer vision-based systems, and virtual reality

R. Silhavy and P. Silhavy (Eds.): CoMeSySo 2024, LNNS 1489, pp. 77–85, 2025.
https://doi.org/10.1007/978-3-031-96798-6_7

trackers [10–12]. Despite their advantages, each of these methods has its limitations and disadvantages. For example, the range and accuracy of motion capture systems may be limited, while computer vision-based systems may be affected by lighting conditions and clutter. On the other hand, virtual reality trackers may be susceptible to errors caused by environmental factors [13].

This article will review current tracking methods, their limitations and advantages, and provide a fair comparison in body rotation detection and hand tracking tests. A comparative study will be conducted on three different systems: virtual reality trackers, motion capture systems, and computer vision-based systems with neural networks for human recognition. The goal is to gain a clear understanding of the strengths and weaknesses of each method based on accuracy information in the tests above. The results of the study can be used to develop virtual training systems that can accurately track the user's movements and location, which will ultimately lead to an improved user experience and improved training quality.

2 Methods

Let us formulate a scheme of experimental research aimed at solving two problems: determining the angle of rotation of the human body and the position of his hands in the process of interaction with virtual reality. The general diagram is presented in Fig. 1.

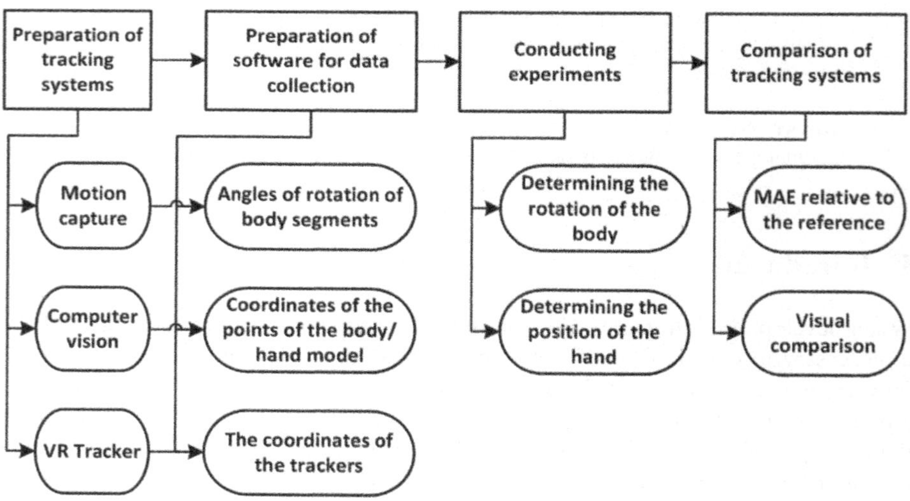

Fig. 1. Scheme of experimental studies.

The first stage of research is related to the preparation of various tracking systems. Within the framework of this work, three systems are considered:

- inertial navigation systems based on gyroscopes and accelerometers, for example, motion capture suits [14];

- virtual reality tracking systems based on infrared sensors, for example, HTC Vive Tracker [15];
- computer vision systems with the integration of neural networks for segmentation of the human body, for example, MediaPipe Pose [16].

Let's look at each of them in detail. The motion capture suit is based on microelectromechanical inertial measurement modules, which are compact, lightweight, low consumption, speed and accuracy of operation, and high frequency of data collection. When operating, this system relies on data from inertial measurement modules. This leads to some problems when using this class of systems: magnetometers used indoors can give inaccurate readings due to distortion of the magnetic field created by various electrical appliances, metal structures and even the building itself. All this leads to data distortion, increased inertia of movement of individual parts of the body, or gross distortions in positioning, which must be taken into account during use. In addition, the use of single sensors may not be as effective as compared to complexes of multiple sensors, since the accumulation of errors on them cannot be compensated by analyzing data from other sensors or other methods of information verification [17].

Virtual reality trackers in their operation rely on analysis of position relative to base stations, so it is necessary to configure the base stations so that there are no foreign objects in their field of view. Also, trackers or controllers need to calibrate the location relative to the virtual reality helmet to correctly collect data. When collecting data from trackers, it is necessary to take into account their location on a person. To be recognized by base stations, trackers should not be covered by clothing or body parts [18]. Trackers are also limited by their size, which makes it impossible to place a large number of sensors on the body.

The computer vision system in the first experiment operates based on processing data received from the camera. The information is input to a neural network for recognizing a human body model, for example, MediaPipe Pose, which forms an array of body points [19, 20]. The limitation of this method is the external conditions (low illumination), as well as the close or too distant location of objects, which makes their detection difficult. The second experiment used a camera with a depth sensor (e.g., Intel RealSense). The information is input to a neural network for recognizing a human body model, for example, MediaPipe Hands, which generates an array of body points.

The next important stage of the research is preparing software for collecting data in the required format. This stage requires, in addition to the use of proprietary software or ready-made libraries, the development and testing of program code that unifies the data into a unified format. This makes it possible to compare and use different tracking systems without changing the architecture. Let's consider the specifics of organizing software for collecting data for each of the tracking systems.

The Perception Neuron 2.0 suit was used as a motion capture system, which required Axis Neuron software (PN32 - Legacy), which provides the necessary functionality for capturing movements: data recording; calculation of orientation and direction, speed and position; data export to various formats, such as: bvh, calc, txt, fbx. The data was recorded in raw format, after which it was exported to fbx binary or calc, the latter could transmit information about body position and rotation angles. The data for the first experiment was obtained by measuring the rotation of the exported fbx model around the Z axis.

The Axis Neuron software did not provide synchronized recording time, so the data was distributed evenly in time according to the timer data. To obtain data on the position of the hand relative to the head, it was necessary to export the model to calc format, and then extract and change the origin of the reference point. For comparison, attitude and heading data were represented as Euler angles.

Data collected using a virtual reality helmet and HTC Vive trackers have the same structure and include seven parameters: time, coordinates along the X, Y, Z axes, as well as rotation angles around the x, y, z axes. Data from devices can be written to CSV, TXT or JSON files at speeds of 60 to 120 times per second. The main limitation with this type of tracking system is sensor overlap, resulting in signal distortion or loss. Also, to obtain not absolute, but relative values relative to the head, it is necessary to subtract their coordinates from the helmet coordinate trackers.

For a computer vision system based on MediaPipe Pose, an array of 33 points is formed, each of which corresponds to an X, Y, Z coordinate. X, Y depend on the position of the object in the frame and vary from 0 to 1, and Z is calculated based on the overlap from the fragments but have no physical meaning. For the experiment, X, Y coordinates are sufficient.

For MediaPipe Hands, an array of 22 points is formed. Since the Z coordinate is not suitable for use in calculations, it is not taken from the MediaPipe model, but directly from the Intel RealSense camera depth sensor using the pyrealsense2 library. The camera is attached to a virtual reality helmet and must constantly capture the object being observed in its field of view. It is also necessary to take into account that the collected data must be transformed (the Y-axis of the camera is directed downward), and also scaled (since the coordinates are calculated relative to the frame, but not relative to the actual size of the objects).

Next, after all the preparatory measures have been carried out, it is necessary to carry out direct experimental research, which will include two experiments:

- determination of the angle of rotation of the body: during this experiment, the user sequentially turns from 0 to 270 degrees (Fig. 2a), which makes it possible to understand how accurately each of the systems records the direction of movement; then the user, within a separate data recording session, rotates 360 degrees (full rotation), the trajectory of the angle change is recorded for all three systems;
- determining the position of the user's hand: in this experiment, the user moves his hand bent at 90 degrees in three key positions (Fig. 2b), the position is determined relative to the head, which allows the collected data to be subsequently used to reconstruct the position of the palm in virtual reality; then the entire trajectory of the hand movement from the first to the last state is recorded.

Finally, the study concludes by comparing the collected data. Firstly, for both experiments, reference values are fixed (angles from 0 to 270 in the first test and the distance between the head and palm in the second), relative to which the mean absolute error (MAE) is calculated. Next, a visual comparison of the trajectories of changes in angles and hand positions is carried out, which makes it possible to evaluate the quality of the tracking systems over time. Based on these assessments, recommendations can be formulated for the use of systems in virtual simulators.

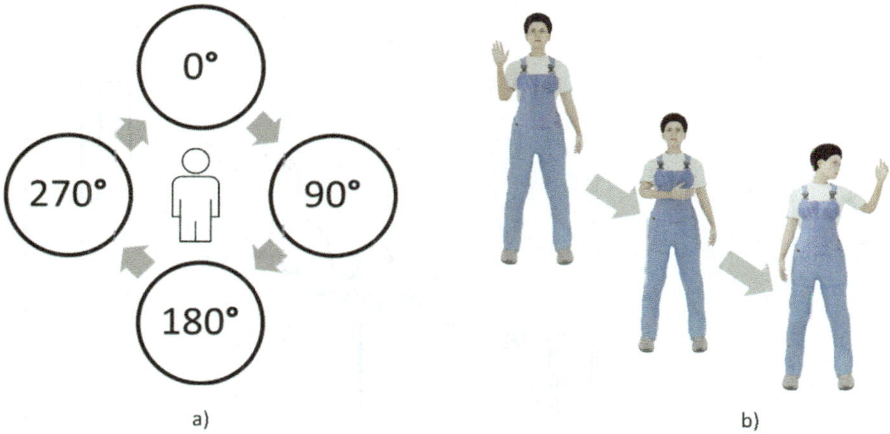

Fig. 2. Diagram of basic human positions for both experiments.

3 Results

In accordance with the previously stated experimental scheme, two experiments were carried out to automatically determine the angle of rotation of the body and the position of the user's hands in space using various tracking systems.

Based on the results of the first experiment, a comparative Table 1 was generated, in which 4 main rotation angles are recorded. The table also shows the average absolute error (MAE) for all measurements.

Table 1. Comparing the accuracy of tracking systems in the first experiment.

The actual value	Tracker	Computer vision	Motion capture
0	11.8	5.9	−0.2
90	84.2	59.2	90.9
180	163.8	180.7	180.9
270	260.7	312.4	270.8
MAE	9.512	19.96	0.712

As a result of the MAE analysis, it was found that the Motion capture suit provides the greatest accuracy, followed by the tracker. The worst results were obtained using computer vision technologies.

We will separately consider a comparison of tracking systems with a person completely turning around its axis (Fig. 3).

Figure 3 shows that in general the trackers and the Motion capture suit are extremely close in return values. On the other hand, a computer vision system, due to fluctuations in

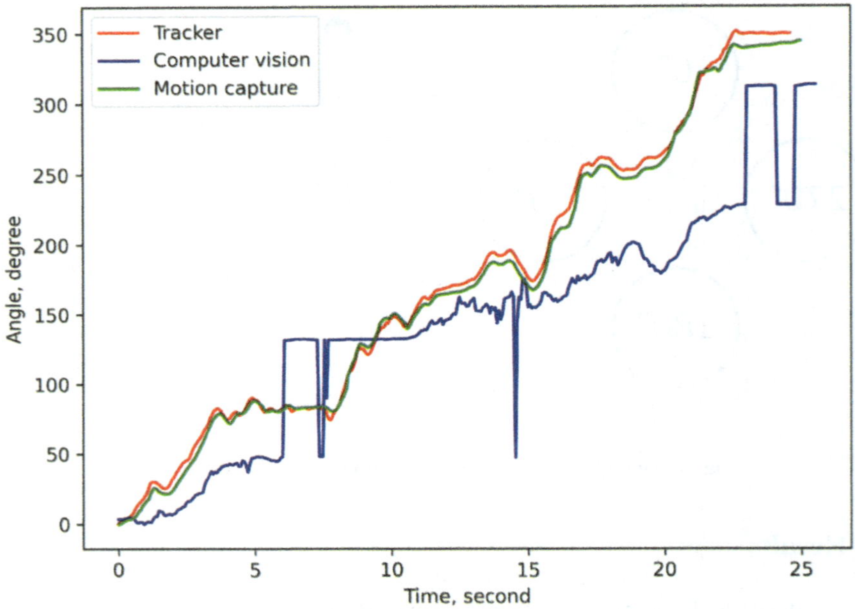

Fig. 3. Comparison of systems with full rotation in the first experiment.

recognized segments, quite often produces contradictory results (sharp jumps in angles are observed in the graph).

Next, let's look at the results of the second experiment. Table 2 shows the calculated position of the user's hand in the format (X; Y; Z). For a clearer comparison of the obtained values, the following assumption is introduced: the coordinates of the hand are normalized relative to the starting position of the hand (the first position in Fig. 2b), thus, the trajectories of the hand begin to move from the point (0;0;0). Then Table 2 shows the corresponding values in positions 2 and 3 (Fig. 2b).

Table 2. Comparing the accuracy of tracking systems in the second experiment.

Position	Tracker	Computer vision	Motion capture
№2	(0.24; −0.04; −0.35)	(0.46; 0.01; −0.31)	(0.39; −0.11; −0.37)
№3	(−0.11; 0.19; −0.54)	(0.02; 0.32; −0.42)	(−0.08; 0.23; −0.56)

A visual comparison of continuous motion trajectories was also carried out, starting from the first to the third (Fig. 4).

Analysis of the movement trajectory along the XY and XZ axes shows that the trackers and the motion capture suit as a whole have similar behavior, differing in the amplitude of movement. The stability of the trajectory when tracking by a computer vision system is low, its applicability for solving this problem is limited, since it is

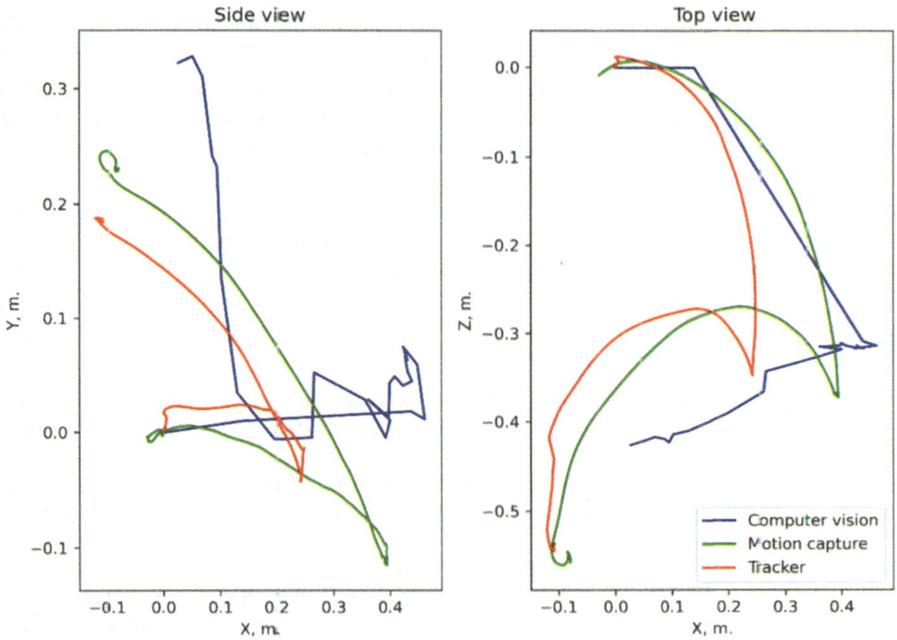

Fig. 4. Comparison of systems on a full trajectory in the second experiment.

obvious that the observed fluctuations will lead to shaking of the virtual hand in the virtual scene. Based on the testing results, it was also revealed that the camera quite often does not detect the hand when leaving the visibility zone, and the motion capture suit, when approaching or touching metal objects or sources of electromagnetic interference, the suit transmits distorted values.

4 Conclusion

The paper examines the problems and prospects for using modern motion tracking systems when integrating them into virtual simulators in solving two problems: determining the angle of rotation of the body and the position of the user's hand.

The specifics of the functioning of three positioning systems (motion capture suit, computer vision system and virtual reality trackers) are considered. An experimental scheme is formulated within which these systems are compared. A set of software and data processing procedures have been defined to ensure the collection of the necessary information about the position and angle of rotation.

In the course of experimental studies, it was found that the task of determining the angle of rotation is best solved by a motion capture suit and virtual reality trackers, and, as a tracker, it is possible to use specialized sensors based on a similar principle, for example, kat loco s or a separate sensor with a gyroscope and an accelerometer. In the second experiment, a virtual reality tracker is used as the optimal positioning method, as it provides accurate absolute positioning. Thus, to track movements of 1–3 points, the use of trackers is recommended; for full body tracking, the use of a suit is preferable. On the other hand, it must be taken into account that sensors and suits are sensitive to overlap (in the first case) and electromagnetic interference (in the second case). In the presence of such complex conditions, it remains to use systems based on computer vision.

The article was carried out with the financial support of the Ministry of Science and Higher Education of the Russian Federation within the framework of the project "Development of an immersive virtual reality interaction system for professional training based on an omnidirectional platform" (FEMU-2024-0005).

References

1. Vlahovic, S., Suznjevic, M., Skorin-Kapov, L.: A survey of challenges and methods for quality of experience assessment of interactive VR applications. J. Multimodal User Interfaces. **16**(3), 257–291 (2022)
2. Obukhov, A.D., Krasnyanskiy, M.N., Dedov, D.L., Nazarova, A.O.: The study of virtual reality influence on the process of professional training of miners. Virtual Real. **27**(2), 735–759 (2023)
3. Krasnyanskiy, M.N., Obukhov, A.D., Dedov, D.L.: Control system for an adaptive running platform for moving in virtual reality. Autom. Remote. Control. **83**(3), 355–366 (2022)
4. Keung, C.C.W., Kim, J.I., Ong, Q.M.: Developing a BIM-based MUVR treadmill system for architectural design review and collaboration. Appl. Sci. **11**(15), 6881 (2021)
5. Pyo, S., Lee, H., Yoon, J.: Development of a novel omnidirectional treadmill-based locomotion interface device with running capability. Appl. Sci. **11**(9), 4223 (2021)
6. Cherni, H., Nicolas, S., Métayer, N.: Using virtual reality treadmill as a locomotion technique in a navigation task: impact on user experience–case of the KatWalk. Int. J. Virtual Real. **21**(1), 1–14 (2021)
7. Polechoński, J., Kostorz, K., Polechoński, P.: Using ankle weights as an effective way to increase the intensity of physical activity while playing immersive virtual reality games on an omnidirectional treadmill. Appl. Sci. **13**(20), 11536 (2023)
8. Hooks, K., Ferguson, W., Morillo, P., Cruz-Neira, C.: Evaluating the user experience of omnidirectional VR walking simulators. Entertain. Comput. **34**, 100352 (2020)
9. Voigt-Antons, J.N., Kojic, T., Ali, D., Möller, S.: Influence of hand tracking as a way of interaction in virtual reality on user experience. In: 2020 Twelfth International Conference on Quality of Multimedia Experience (QoMEX), pp. 1–4. IEEE (2020)
10. Menolotto, M., Komaris, D.S., Tedesco, S., O'Flynn, B., Walsh, M.: Motion capture technology in industrial applications: a systematic review. Sensors. **20**(19), 5687 (2020)
11. Al-Faris, M., Chiverton, J., Ndzi, D., Ahmed, A.I.: A review on computer vision-based methods for human action recognition. J. Imaging. **6**(6), 46 (2020)
12. Obukhov, A., Dedov, D., Volkov, A., Teselkin, D.: Modeling of nonlinear dynamic processes of human movement in virtual reality based on digital shadows. Computation. **11**(5), 85 (2023)
13. Obukhov, A.D., Krasnyanskiy, M.N., Dedov, D.L., Vostrikova, V.V., Teselkin, D.V., Surkova, E.O.: Control of adaptive running platform based on machine vision technologies and neural networks. Neural Comput. Applic. **34**(15), 12919–12946 (2022)
14. Hindle, B.R., Keogh, J.W., Lorimer, A.V.: Inertial-based human motion capture: a technical summary of current processing methodologies for spatiotemporal and kinematic measures. Appl. Bionics Biomech. **2021**(1), 6628320 (2021)
15. Bauer, P., Lienhart, W., Jost, S.: Accuracy investigation of the pose determination of a VR system. Sensors. **21**(5), 1622 (2021)

16. Kim, J.W., Choi, J.Y., Ha, E.J., Choi, J.H.: Human pose estimation using mediapipe pose and optimization method based on a humanoid model. Appl. Sci. **13**(4), 2700 (2023)
17. Zheng, T., Xu, A., Xu, X., Liu, M.: Modeling and compensation of inertial sensor errors in measurement systems. Electronics. **12**(11), 2458 (2023)
18. Arkhipov, A., Lopatko, A., Prokudina, N., Averin, Y., Dedov, D.: Comparison of approaches to human motion tracking in virtual reality. AIP Conf. Proc. **2948**(1), 020011. AIP Publishing (2023)
19. Tu, L.F., Peng, Q.: Method of using real sense camera to estimate the depth map of any monocular camera. J. Electr. Comput. Eng. **2021**(1), 9152035 (2021)
20. Tadic, V., Toth, A., Vizvari, Z., Klincsik, M., Sari, Z., Sarcevic, P., Biro, I.: Perspectives of real sense and zed depth sensors for robotic vision applications. Machines. **10**(3), 183 (2022)

Transparency and Traceability of Agriculture Product Supply Chain with the Use of Blockchain Technology - Desktop Review Paper

N. G. P. Selamolela(✉), N. Sonhera, and T. Zuva

Department of Information and Communication Technology, Vaal University of Technology, Vaal, South Africa
217198864@edu.vut.ac.za

Abstract. The use of blockchain technology in the agriculture sector has merely gradually started to be used in the agricultural sector. In the agri-foods sector, blockchain is acknowledged as an emerging technology that can improve food traceability and offer a transparent, trustworthy, and efficient means of confirming the sustainability, safety, and quality of agri-foods. Very little has been done in the literature to review how blockchain technology has been used in agriculture. Therefore, the purpose of this review paper is to explore how blockchain technology has been used in agriculture, the food supply chain, and the provision of transparency and traceability using blockchain technology models to assist consumers in choosing the food they buy and consume. Research papers searched from conference proceedings, and journals were utilized for this review. Some of the databases used are Google Scholar, ScienceDirect, and IEEE Xplore. The outcome of this review highlighted the barriers to the adoption of blockchain technology, such as data security, preventing discrimination, digital infrastructure, human resources, and handling data that is already in existence. There is still a need for more research to provide solutions to the challenges and open issues. The utilization of blockchain technology presents a viable strategy for promoting a more dependable, safe, healthy, and sustainable agri-food system in the future.

Keywords: Blockchain Technology · agriculture · agri-foods sector · transparency · traceability

1 Introduction

Suppose a customer bought something from a store and the expiration date showed that the food was still good to eat, but the customer experienced diarrhoea after eating it, what the customer would do first is to get in touch with the store. The retailer may then turn to the local regulator for assistance in identifying the root of the problem, because the regulator serves as a central controller with connections to various parties involved in the supply chain, including manufacturers, retailers, farmers, processors, and restaurants.

R. Silhavy and P. Silhavy (Eds.): CoMeSySo 2024, LNNS 1489, pp. 86–94, 2025.
https://doi.org/10.1007/978-3-031-96798-6_8

Nevertheless, there are not many regulated players in the market, and the distributor, packer, trader, or other non-regulated organization could be the root of the problem. Additionally, traditional supply chain management is unidirectional (one-way), therefore it is typically challenging to track down the source of a problem since multiple parties have access to different pieces of information, and some information is hidden due to trade [10]. Additionally, [10] continued to mention that the term supply chain management has no universally accepted definition, it typically refers to the full spectrum of goods and services needed for an agricultural product from the farm to the end-user client.

As a result, tracking down the source of a problem in today's supply chains can be labour-intensive, time-consuming, ineffective, and occasionally an inconclusive procedure. With the aid of an evolving technology such as blockchain, such problems have the potential to be tackled by utilizing distributed, decentralized peer-to-peer communication.

The application of blockchain technology, particularly in the agricultural sector, was explored and encouraged in a review by [22]. The authors of the review put forth a working theory for blockchain technology in data recording and tracking, and they defined blockchain as a network of computer systems that use blockchain technology—also known as distributed ledger technology—as a public bookkeeping mechanism to track transactions in blocks. The authors further explained that every block has three values: the hash of the previous block, the hash of the current block, and data.

In contrast to traditional systems that rely on information supervision centers to transfer and exchange data, [22] further hinted that all the data uploaded into the blockchain system comes directly from primary sources, including farms, markets, warehouse facilities, transit vehicles, customers, etc. Data added to a blockchain must also be consistent, immutable, and free of inconsistencies with previously collected data. A blockchain can only contain data, and every computer system needs to recognize the current data status on the blockchain. Because a single party does not own a public blockchain, the authors noted that it prevents unilateral change. This technology offers traceable transaction histories and immutable data records, which can enhance efficiency, transparency, and traceability.

The demand for locally grown fruit, sustainably sourced meat, and organic goods has grown significantly in recent years, according to an article by [13] that explains the history, origins, and current state of agriculture. The authors also noted that changes in consumer preferences have led to an increase in food fraud. Customers are interested in the origins and components of the food, the writers hinted.

Therefore, with the assistance of blockchain technology, when a consumer walks into a store and decides to purchase a fruit, the consumer can scan the fruit in the store and verify that it is fresh from the tree or if it is chemically induced and how long has it been in the store before purchasing.

Blockchain technology in supply chain management can assist with transparency and traceability of agriculture products. It could give customers the comfort of buying any food they want and the assurance that in the event of a sickness, they could easily track down the root cause, for instance by scanning a QR Code and reading about the lifespan of the product and the ingredients on a blockchain technology system.

Very little has been done in the literature to review how blockchain technology has been used in agriculture. This review paper explores the food supply chain and the provision of transparency and traceability using blockchain technology models to assist consumers in choosing the food they buy and consume.

2 Related Work

2.1 Data Transparency

As elaborated in the introduction, consumers may face health issues due to agricultural products purchased from stores or the negligence of a certain company while producing a product. The traditional systems struggle to track the root cause of a problem or its sources due to multiple parties accessing or having different information or being liable for a certain ingredient in that product's information and trade secrets. As a result, finding the root cause of the problem can be a time-consuming, and tedious process that essentially results in an inconclusive procedure.

According to [21] the authors introduced the concept of blockchain technology, putting forward the application of blockchain technology information security of the food supply chain and comparing it with the traditional supply chain system. The authors mentioned that the current management and relationship with the food supply chain are typically unidirectional (one-way).

The upstream agents can communicate with the downstream agents, for example, the authors mentioned that a manufacturer can communicate with a restaurant, but the restaurant cannot contact the manufacturer, because most supply chain players can only link with directly downstream organizations. Therefore, this potentially results in a communication gap, a lack of data transparency, and information about travel history and the origin of the product could be lost in transit or translation between the upstream and downstream agents including the consumer.

In an article published by [22], the authors presented a working principle of blockchain for data recording and tracking. The authors further discussed four aspects: enhancing data transparency, realizing data traceability, improving food safety and quality monitoring, and finally cost of financial transactions. To demonstrate how blockchain may be used to enhance food traceability, the authors used a case of Walmart's pork traceability system. The authors alluded that, the management of the agriculture food supply can become more dependable and effective with the use of blockchain. The authors further elaborated that blockchain technology can create an unchangeable, traceable data system that connects every link in the food supply chain. The authors allude that with the use of blockchain technology, it is evident that everyone involved in the blockchain system is linked to one another, and that there is a bidirectional and interactive communication pattern.

Each member can be thought of as a node, and nodes have access to all data stored in the blockchain. In other words, using the same example above, the customer can trace all the information about the acquired goods through the full supply chain without the assistance of an intermediary (regulator), as they have direct contact with all the players. This is a clear illustration of how blockchain may be used for data transparency.

2.2 Data Traceability

When a consumer purchases coffee, for instance, and he or she wants to know if it is caffeine-free or not, with the traditional supply chain management systems, the consumer might have to depend on information on the product, and if it is not clear enough, then depend on information monitoring centers such as google for information sharing and transfer.

According to [22] in the study of a working principle of blockchain for data recording and tracking, mentioned that, unlike the conventional approaches that depend on information monitoring centers for information transfer and sharing, every piece of data put into the blockchain system is supplied by primary sources, including farms, marketplaces, storage facilities, transit vehicles, customers, etc. The authors elaborated that the first stage in achieving data traceability is data capture.

There are various methods for data acquisition, including cameras, sensor- Radio frequency identification technology (RFID), and others. The following characteristics can be tracked down from various blockchain users: farmer, packer, processor, shipper, retailer, and customer. [21], made use of RFID to track the movement of cattle using data from farms.

In an article published by [16], the study explored the integration of traceability technology with geographic information systems (GIS) for tracing and retracing the safety of supply chain processes (SCP) in real-world supply chains. A pilot study demonstrated the system's fundamental features, and creative suggestions were made for creating an effective, reliable, and practical SCP tracing system. The authors elaborated that data traceability in the traditional supply chain system, the farm's location details about the soil and fertilizer, and other data can be manually recorded and submitted.

The packer has the option to manually upload data while packing, including the number of things packed, the state of the box, and the packing time and date. Cameras can be used to monitor the entire abattoir production line for data in the manufacturing facilities. In the paper by [16], the authors further mention that all processing data, including temperature, time, and relative humidity, is automatically collected by sensors. The employee must manually enter the product details as well as any other pertinent data, including the manufacturing date, time, location, and packaging details, as well as data about food safety, such as temperature, an inspector's grade, or any further test results. In addition, every vehicle has sensors attached to monitor temperature and relative humidity during shipment; if circumstances rise above predetermined thresholds, the alarm system assists in correcting the situation. Global position systems (GPS) have tracked location shifts and shipping information. Likewise, the application of (GIS) for crop tracing has been fruitful [16].

Additionally, to enable data traceability, food goods might be tracked using smart-tagged bar codes at every stage of the supply chain. Products that are packaged employ bar codes. The food kind is given a unique ID by the system. Since there is no information in the system linked to this unique ID, it first links to a null record. After that, the ID is linked to a special bar code and provided to the employees in the form of a tag. The product bar codes generated by the system are appended to the blockchain upon cutting and packaging. Applying tags to the goods and scanning the barcode to add the ID to the blockchain are the responsibilities of the employees [22].

3 The Use of Blockchain Technology in the Agriculture Supply Chain for Transparency and Traceability

Tracking the provenance of commodities via complex supply networks requires an open, impermeable information structure that is trustworthy for all parties and adaptable enough to change with changing circumstances and legal requirements, as the study [11] explained. In the paper, the authors described how they created the origin Chain system, which makes use of cutting-edge blockchain technology. The authors define transparency and traceability as a system's ability to continuously monitor a product's origin, components, or supply chain location.

Research done by, [17] introduced a Blockchain-Based Soybean Traceability in the Agriculture Supply Chain, where the authors proposed a method for tracking soybeans in the agricultural supply chain using the Ethereum blockchain and smart contracts. It improves efficiency, safety, and integrity by eliminating intermediaries and centralized authority. The approach uses smart contracts for regulation and storage, ensuring transparency and safety in the supply chain ecosystem. The authors further mentioned that transparency and traceability involve communicating between entities and recollecting details about a food product's provenance.

In addition, blockchain technology ensures transparency and immutability in transactions, providing a framework for monitoring the food supply chain, tracing the origin of agricultural products, and ensuring their validity. Considering the above, a successful agriculture traceability and transparency system should include information on ingredients, origins, processing, transportation, and storage processes.

In a study conducted by [22], the authors introduced the concept of blockchain as a new safeguard for agricultural foods and used Walmart's pork traceability system to demonstrate how blockchain may be used to enhance food traceability. The authors explained that the relationship between a food firm and a blockchain corporation serves as the foundation for the current cooperation model. The blockchain organization oversees providing technical support, such as platform setup and database maintenance, while the food company oversees giving information about agri-foods. The authors further stipulated that there is one company that has been acknowledged as the top enterprise blockchain supplier, which is International Business Machines (IBM).

IBM's Trace program uses blockchain technology to secure, share, and permissioned digital records of agricultural food data. This program enhances food safety and improves the food system by allowing users to track food goods from upstream origins to downstream destinations, collaborating with food firms.

In a study conducted by [14], the authors wanted to explore and analyse the supply chain of rice. The authors traced rice using a multi-blockchain, and they developed a Multi-blockchain Rice Refined Supervision Model (MBRRSM) based on a mix of blockchain theory, architecture, information encryption transmission, information security, and the Supervision Proof of Peers algorithm (SPOP). The researchers then collected data both automatically and manually using Radio Frequency Identification (RFI) and Internet of Things (IoT) devices. The authors constructed a prototype system and tested it using various usage scenarios in their article.

In the review paper, [1] also did similar research by introducing a framework that is blockchain-based traceability for the supply chain with a case study of the textile and clothing supply chain. The researchers pointed out that customers struggle to obtain details of the item that can aid in ethical purchasing behaviours or ensure item legitimacy. The study's goal was to research and propose a multi-tiered textile and apparel supply chain traceability and transparency architecture based on blockchain technology. The authors referred to a system that uses the Proof of Work consensus algorithm and the mass balancing validation mechanism to practically provide organic cotton traceability. According to the authors, the Account and Balance method and Rivest-Shamir-Adleman (RSA) algorithm are applied for validation and performance.

In the study conducted by [19] the research aimed to examine the need for a blockchain-based traceability system (BBTS) that allows users to improve the transparency and security of the supply chain. The authors conducted a two-part architecture based on a public permissioned blockchain and a performance-preserving byzantine fault tolerance consensus algorithm for the traceability model which in the paper seems promising for implementing a blockchain network in a solution for food tracking business. The authors utilized six criteria that are derived from the specification framework: differentiation, visibility, transparency, accessibility, authenticity, suitability, forgery protection, and standards that every traceability software solution must meet.

A study by [4] aimed to build a disturbed practical framework for a decentralized and autonomous food supply chain using smart contracts and blockchain technology for traceability. The authors maintained the information gathered in a table of content (TOC) using IoT technology. As a result, records for each member of the food supply chain were saved locally through decentralized storage methods such as the InterPlanetary File System. The suggested model's performance is measured using a set of Key Performance Indicators (KPIs) such as efficiency, responsiveness, required trust assumptions, context requirements, required consensus mechanisms, and food quality, which is also compared to traditional food supply chain traceability mechanisms.

4 Discussion

Blockchain is a technology that offers the potential to improve information flow among stakeholders in the supply chain. In addition, the technology allows for monitoring product information in supply chain management. To help consumers make informed decisions about the food they purchase and eat, this research paper examines the food supply chain and how blockchain technology models can provide transparency and traceability.

Therefore, as evidenced by the literature, blockchain provides traceability and transparency. By contrasting conventional supply chain systems with blockchain technology for information security in the food supply chain, [21] study hinted at this. Data transparency between upstream and downstream agents was judged to be lacking by the authors. Similar remarks were made by [22], whose study examined four ways to improve data openness with Walmart's traceability system. Blockchain technology, according to the authors, can link every party involved in the supply chain, facilitating interactive and two-way communication.

Another application of blockchain technology that can boost effectiveness, security, and integrity is data traceability. Farmers and consumers may profit from the tracking and capturing of data via Radio Frequency Identification (RFI) sensors and cameras [22]. A study by [16] investigated the use of bar codes and unique IDs in geographic information systems (GIS), a different kind of sensor, for traceability systems in actual supply chain operations.

As previously noted, there is not a great deal of literature that reviews the applications of blockchain technology in agriculture; yet numerous studies have put forth models. By doing away with middlemen and centralized authority, in a study conducted by [17], the authors alluded that the approach for tracking soybeans in the agricultural supply chain makes use of smart contracts and the Ethereum blockchain to improve efficiency, safety, and integrity. A system for agriculture traceability and transparency needs to incorporate details about ingredients, origins, processing, transportation, and storage to be effective. Furthermore [22] talked about the IBM Food Trust Program, known as Trace, which creates a digital record of agricultural food data that is secure, shareable, and authorized.

Moreover, a blockchain-based traceability framework for the textile and apparel supply chains was introduced by [1], in addition to other supply chains with a similar focus but that did not center on agriculture. According to the authors, customers find it difficult to find information about products, which could support moral shopping practices or guarantee the legality of the item, according to the authors. A study conducted in 2021 by [19] looked at the necessity of a blockchain-based traceability system (BBTS) to strengthen the supply chain's security and transparency.

All the models, in essence, offer benefits for the use of blockchain technology in agriculture and ensure the system is robust, safe, secure, transparent, and traceable.

5 Challenges and Open Issues

A substantial amount of research backs up the idea that blockchain technology holds immense potential for improving the agriculture food supply chain, nonetheless, its adoption and scalability are still in the early stages. The blockchain system has challenges and open issues that require researchers to continue perfecting the system and modifying the system therefore the next section mentions the challenges and open issues. The key challenges include privacy and data security, limiting prejudice, digital infrastructure, human resources, and managing existing data.

Blockchain technology in some cases lacks trust and information can be easily manipulated, therefore transparency can open doors for potential targets. Most data are manually uploaded, making it more of a tool for suppliers to self-inspect. To increase accessibility, user-friendly scanning systems are needed, and manufacturers must ensure transparency. However, blockchain system administration lacks oversight or regulatory bodies, leading to issues with capital currency transaction security [22].

Consumers may be prejudiced by blockchain technology, especially when a company produces children's food with artificial ingredients and false labels. This could lead to consumers not purchasing or consuming products from underdeveloped or disadvantaged areas.

In a study conducted by [2], the authors mentioned that to address digital infrastructure challenges, initiatives are being made to provide access through inexpensive technology and software for farmers and consumers to collect data on blockchain systems. However, there is a general lack of knowledge about the technical components of blockchain, and there is a need for technical education, skill development, and general knowledge of blockchain.

Blockchain technology can bridge the gap between local farms and manufacturers by uploading pre-existing data to the blockchain or cloud and transferring the gathered data to government regulatory platforms for better tracking and auditing of the food supply chain. Building systems to ensure manufacturers report cases to governmental or regulatory systems is challenging [16].

6 Conclusion

Blockchain technology is gaining momentum in both the public and private sectors, particularly in the agri-food industry. It offers improved traceability and transparency, allowing consumers to keep an eye on products from farm to consumer and improving the quality and safety of food. The advantages of this technology have led to the development of numerous blockchain-based platforms and applications across a wide range of fields, including agriculture.

A new strategy for the agriculture sector is made possible by blockchain technology, which increases consumer options for transparency, traceability, security, effectiveness, and a reduction in counterfeiting events, as well as increased product visibility. Additionally, this technology shields users against dangerous, perhaps expired, products that could lead to a variety of health problems. That kind of important information becomes truly independent like its senders with decentralized access, enabling open and secure access.

Blockchain technology enhances transparency, traceability, globalization, and scalability in agriculture, providing efficient product verification in stores and eliminating issues related to transparency and traceability in the supply chain.

The review aims to improve the transparency and traceability of products to consumers, allowing the agriculture sector to use this model alongside their current processes. By detecting and preventing fraud, complex transparency and traceability processes can be avoided. This article opens the possibility of exploring the potential of blockchain technology in other areas of the agriculture sector.

References

1. Agrawal, T.K., Kumar, V., Pal, R., Wang, L., Chen, Y.: Blockchain-based framework for supply chain traceability: a case example of textile and clothing industry. Comput. Ind. Eng. **154**(107130), 1–13 (2021)
2. Akella, G.K., Wibowo, S., Grandhi, S., Mubarak, S.: A systematic review of blockchain technology adoption barriers and enablers for smart and sustainable agriculture. Big Data Cogn. Comput. **7**(2), 17–19 (2023)
3. Armonk, N.Y.: IBM Announces Major Blockchain Collaboration with Dole, Driscoll's, Golden State Foods, Kroger, McCormick, and Company, McLane Company, Nestlé, Tyson Foods, Unilever, and Walmart to Address Food Safety Worldwide. IBM (2017)

4. Casino, F., Kanakaris, V., Dasaklis, T.K., Moschuris, S., Rachaniotis, N.P.: Modeling food supply chain traceability based on blockchain technology. IFAC-PapersOnline **52**(13), 2728–2733 (2019)
5. Chan, Y.K., Abdullah, J., Khan, S.A.: A Framework for traceable and transparent supply chain management for agri-food sector in Malaysia using blockchain technology. Int. J. Adv. Comput. Sci. Appl. **10**(11), 149–155 (2019)
6. Egels-Zandén, N., Sörum, N.: Supply chain transparency as a consumer or corporate tool: the case of Nudie Jeans Co. J. Consum. Policy **39**(4), 377–395 (2015)
7. Galvez, J.F., Mejuto, J.C., Simal-Gandara, J.: Future challenges on the use of blockchain for food traceability analysis. TrAC Trends Anal. Chem. **107**, 222–232 (2018)
8. Iansiti, M., Lakhani, K.R.: The truth about blockchain. Harv. Bus. Rev. **95**(1), 119–127 (2017)
9. Kaye, L.: Responding to food safety concerns. Walmart invests $25 million in China. In: Food Safety (2016)
10. Khandelwal, C., Singhal, M., Gaurav, G., Dangayach, G.S., Meena, M.L.: Agriculture supply chain management: a review (2010–2020). Mater. Today Proc. **47**(11), 3144–3153 (2021)
11. Lu, Q., Xu, X.: Adaptable blockchain-based systems: a case study for product traceability. IEEE Softw. **34**(6), 21–27 (2017)
12. Matzembacher, D.E., Stangherlin, I., Slongo, L.A.: An integration of traceability elements and their impact on consumer's trust. Food Control **92**, 420–429 (2018)
13. History. https://www.history.com/topics/pre-history/neolithic-revolution. Accessed 18 Jun 2024
14. Peng, X., Zhang, X., Wang, X., Xu, J., Li, H., Zhao, Z., Qi, Z.: A refined supervision model of rice supply chain based on multi-blockchain. Foods **11**(18), 2785 (2022)
15. Peters, A.: In China, You can Track Your Chicken on-You Guessed it-the Blockchain, Fast Company (2018)
16. Qu, X.-H., Zhaung, D.-F., Qui, D.-S.: Studies on GIS based tracing and traceability of safe crop products in China. Agric. Sci. China **6**(6), 724–731 (2007)
17. Salah, K., Nizamuddin, N., Jayaraman, R., Omar, M.: Blockchain-based soybean traceability in agriculture supply chain. IEEE Access **7**, 73295–73305 (2019)
18. Steiner, J., Baker, J., Wood, G., Meiklejohn, S.: Blockchain: The Solution for Transparency in Product Supply Chains. The Provenance Team (2015)
19. Thume, M., Lange, J., Unkel, M.P., Prange, A., Schürmeyer, M.: Blockchain-based Traceability in the Food Industry: Requirements Analysis Along the Food Supply Chain, pp. 1–27 (2021)
20. Tripoli, M., Schmidhuber, J.: Emerging Opportunities for the Application of Blockchain in the Agri-Food Industry. FAO and ICTSD, Rome and Geneva (2018)
21. Tse, D., Zhang, B., Yang, Y., Cheng, C., Mu, H.: Blockchain application in food supply information security. In: IEEE International Conference on Industrial Engineering and Engineering Management (IEEM), pp. 1357–1361. IEEM, Singapore (2017)
22. Xu, J., Guo, S., Xie, D., Yan, Y.: Blockchain: a new safeguard for agri-foods. Artif. Intell. Agric. **4**, 153–161 (2020)

Analysis of Forecasting Accuracy Depending on Various Hyperparameters of Neural Network Training

Vladislav Kukartsev[1,2,3] and Ksenia Degtyareva[1,2(✉)]

[1] Reshetnev Siberian State University of Science and Technology, 660037 Krasnoyarsk, Russia
sofaglu2000@mail.ru

[2] Bauman Moscow State Technical University, 105005 Moscow, Russia

[3] RSAU-MAA named after K.A. Timiryazev, 127434 Moscow, Russia

Abstract. In this paper, a study was conducted on the accuracy of forecasting real estate prices using various neural network architectures. The results showed that the two-layer model with 32 neurons in each layer achieved the highest prediction accuracy (0.98), demonstrating the ability to capture complex data dependencies most effectively and avoid overfitting. This study highlights the importance of choosing the optimal neural network architecture and configuring hyperparameters to achieve high prediction accuracy. The optimal configuration must balance the complexity of the model and its ability to generalize.

Keywords: Neural Networks · Hyperparameters · Model Architecture · Prediction Accuracy · Machine Learning · Data Standardization · Retraining

1 Introduction

Modern machine learning methods and neural networks are actively used to solve a wide range of tasks, including forecasting real estate prices. Using neural networks for forecasting tasks is an effective approach because they are able to model complex nonlinear dependencies in data [1, 2].

This work is devoted to the analysis of the accuracy of predicting real estate prices depending on various hyperparameters of neural network training. The main purpose of the work is to determine the optimal neural network architecture and configure hyperparameters to achieve the best prediction accuracy [3]. The study examined three different models of neural networks with different configurations of layers and the number of neurons: a model with one hidden layer of 32 neurons, a model with two hidden layers of 32 neurons and a model with two hidden layers of 64 neurons.

Each of the models was trained and tested on the same dataset, including various characteristics of houses and their prices. For each model, the accuracy of the prediction was evaluated in order to identify the most effective architecture [4].

R. Silhavy and P. Silhavy (Eds.): CoMeSySo 2024, LNNS 1489, pp. 95–102, 2025.
https://doi.org/10.1007/978-3-031-96798-6_9

2 Research Methods

In this study, machine learning methods, namely artificial neural networks (ANNs), were used to predict real estate prices. The main components of a neural network are layers consisting of neurons, activation functions that determine the output of each neuron, and optimizers that are used to minimize the loss function [5, 6].

Artificial neural networks are computational models inspired by biological neural networks. The main purpose of the INS is to model complex dependencies and interactions between input data and target variables [7]. Neural networks consist of the following components:

1. Input layer: receives the source data.
2. Hidden Layers: Perform calculations to reveal hidden patterns in the data. The number of layers and neurons in each layer is determined by the network architecture.
3. Output layer: Generates predictions based on the processed data.

An important component of a neural network is the activation function, which adds non-linearity to the model and allows the network to train complex patterns. When implementing the models, the ReLU (Rectified Linear Unit) activation function was used. This function promotes fast and efficient learning due to the simplicity of calculations and solves the problem of vanishing gradients faced by other activation functions such as sigmoid or hyperbolic tangent [8–10].

To minimize the loss function and adjust the weights of the neural network in the learning process, the Adam (Adaptive Moment Estimation) optimizer was used. Adam is one of the most popular and effective optimizers in deep learning, as there is an adaptive learning rate for each parameter based on the first and second moments of the gradients, as well as stability and high performance on various tasks and datasets [11–14].

Thus, modern machine learning methods and neural networks were used in this study, which made it possible to conduct a comprehensive analysis of the accuracy of forecasting real estate prices depending on various hyperparameters of learning [15–17].

In this study, a dataset consisting of 50,000 lines was used. Each row represents a separate house with different characteristics that are used to predict its price. The dataset contains the following attributes:

- ID: A unique identifier for each house. This attribute was removed before training the models because it does not carry useful information for price forecasting.
- SquareFeet: The area of the house in square feet. This is a quantitative feature that has a direct impact on the value of the house.
- Bedrooms: The number of bedrooms in the house. This quantitative feature also significantly affects the price of real estate.
- Bathrooms: The number of bathrooms in the house. Like the number of bedrooms, this feature is important for estimating the value of a house.
- Neighborhood: The area where the house is located. Since this is a categorical feature, it has been converted to a numeric format using one-hot encoding. Different neighborhoods can have a significant impact on the price of real estate due to differences in infrastructure, school accessibility, transportation networks, and other factors.

- YearBuilt: The year the house was built. This feature can affect the price, as newer houses are often priced higher due to modern construction standards and new communications.
- Price: The price of the house. This is the target variable that we aim to predict based on the rest of the features.

The description of the dataset is shown in Fig. 1.

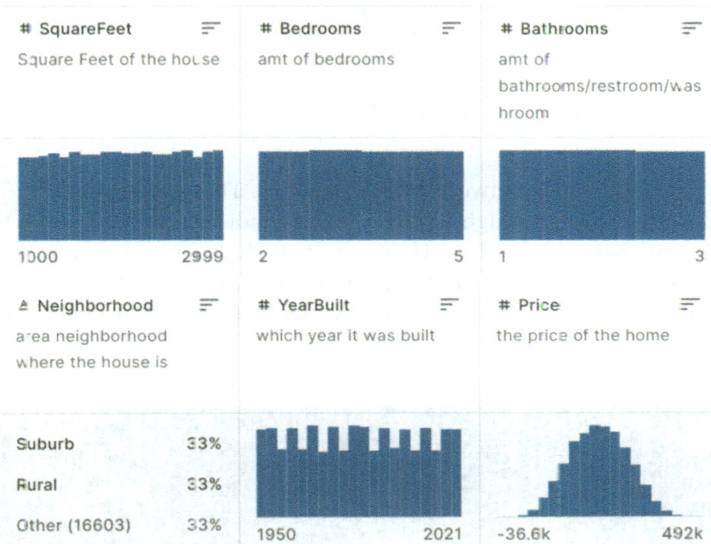

Fig. 1. Description of attributes of the used dataset.

The data was divided into features (X) and a target variable (y). The attributes include all attributes, and the target variable is Price.

All quantitative features (Square Feet, Bedrooms, Bathrooms, Year Built) have been standardized using StandardScaler to bring them to a single scale. This helps to improve the performance of the model, as many machine learning algorithms work better with data that has a standard normal distribution. The dataset was divided into training and test samples in a ratio of 80:20. This allows you to evaluate the accuracy of the model on data that was not used in the training process [18–20]. The use of this dataset made it possible to conduct a comprehensive analysis of various neural network architectures and their ability to predict real estate prices. The next section will present the results of the experiments and their discussion.

3 Results and Discussion

Three different neural network models were considered, each of which was trained and tested on the same dataset. The architectures of each model are described below, their accuracy and the results achieved. Figure 2 shows the accuracy of each of the models.

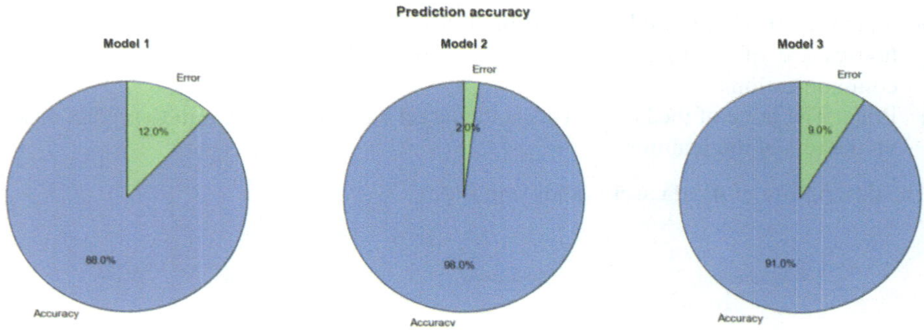

Fig. 2. The accuracy of the implemented models.

Model 1: A single-layer network with 32 neurons. The accuracy is 0.88. This basic model demonstrates a fairly high accuracy, which indicates that it is able to capture the main dependencies in the data. However, a single-layer architecture may not be sophisticated enough to identify all the complex patterns in the data, which limits its ability to make more accurate predictions. The graph of the neural network is shown in Fig. 3.

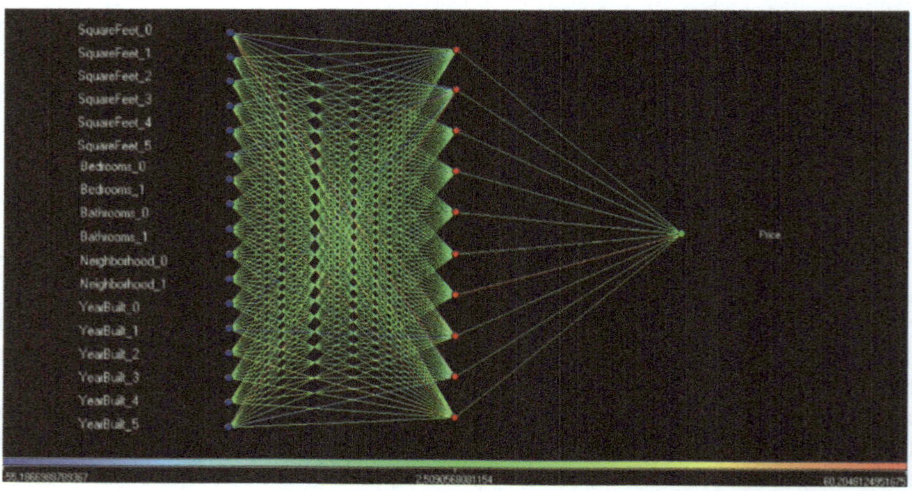

Fig. 3. Neural Network Graph – Model 1.

Model 2: A two-layer network with 32 neurons in each layer. The accuracy is 0.98. Adding a second layer with 32 neurons significantly improved the accuracy of the model. This indicates that a two-layer architecture is better able to capture complex dependencies in data. This model has found the optimal balance between complexity and generalization, which has led to a high level of prediction accuracy. The graph of the neural network is shown in Fig. 4.

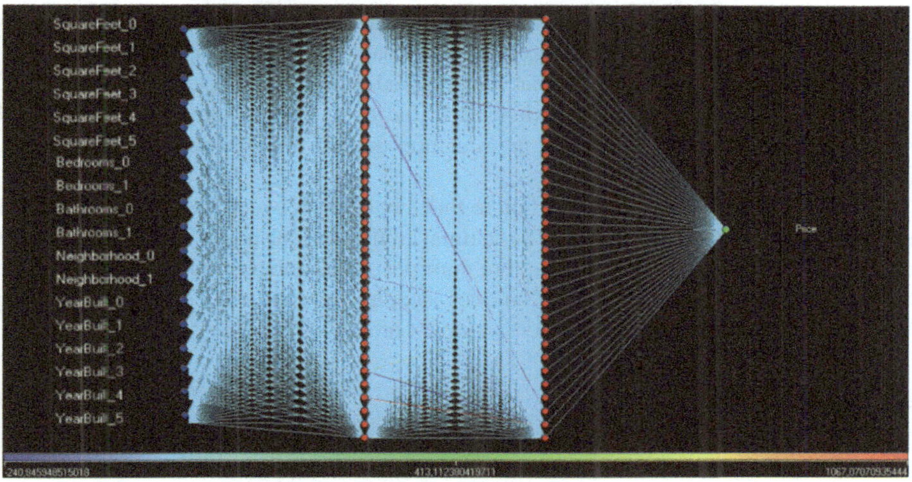

Fig. 4. Neural Network Graph – Model 2.

Model 3: A two-layer network with 64 neurons in each layer. The accuracy is 0.91. An increase in the number of neurons in each layer led to a decrease in accuracy compared to the two-layer model with 32 neurons. Possible causes include overfitting when the model is too complex for the available data, resulting in poor generalization on the test dataset. Also, a model with a large number of neurons may be more sensitive to noise in the data, which may also reduce its accuracy. The graph of the neural network is shown in Fig. 5.

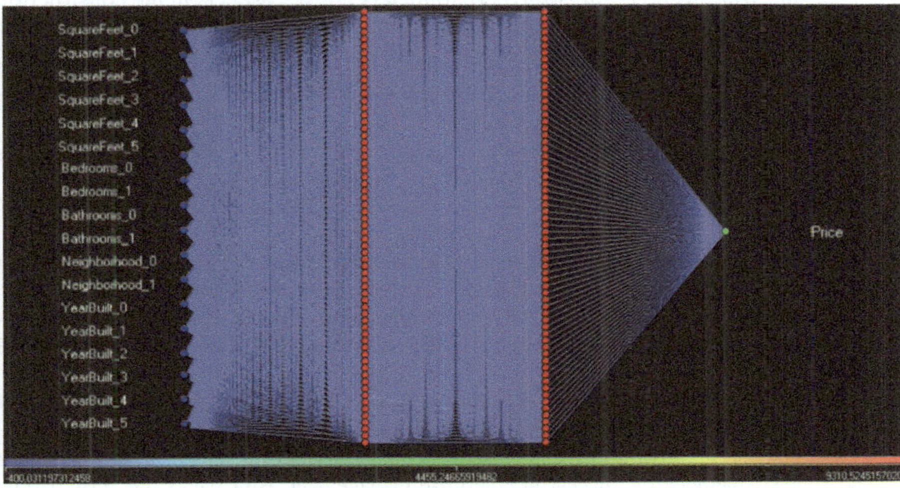

Fig. 5. Neural Network Graph – Model 3.

Model 2 demonstrated the highest prediction accuracy (0.98), which indicates its ability to effectively capture complex patterns in data, while avoiding overfitting.

The single-layer model with 32 neurons showed sufficient accuracy (0.88), but could not identify all the complex dependencies in the data. The two-layer model with 64 neurons, despite the more complex architecture, demonstrated lower accuracy (0.91), probably due to overfitting.

For forecasting tasks, it is important to find a balance between the complexity of the model and its ability to generalize. The optimal configuration should include a sufficient number of layers and neurons to capture complex dependencies, but at the same time avoid overfitting. Regularization techniques such as Dropout or L2 regularization can help improve the performance of more complex models [21, 22].

4 Conclusion

This study was aimed at analyzing the accuracy of forecasting using neural networks of various architectures and hyperparameters. Three models of neural networks with different numbers of layers and neurons were considered [23–25]. The experimental results showed that the two-layer model with 32 neurons in each layer achieved the highest prediction accuracy of 0.98. This architecture proved to be the most optimal, effectively identifying complex patterns in the data and avoiding overfitting [26, 27].

These results highlight the importance of choosing the optimal neural network architecture and configuring hyperparameters to achieve high prediction accuracy.

The architecture of a neural network significantly affects the accuracy of forecasting. Choosing the right number of layers and neurons allows you to improve the results. Complex models do not always provide the best accuracy. An increase in the number of neurons and layers can lead to overfitting, especially with insufficient data or high noise levels. Regularization and tuning of hyperparameters are key aspects in the construction of neural networks [28, 29].

As a result of this study, it was revealed that the architecture of a neural network significantly affects the accuracy of forecasting, and the correct setting of hyperparameters is a key factor for achieving the best results.

References

1. Tynchenko, Y., Kukartsev, V., Gladkov, A.: Assessment of technical water quality in mining based on machine learning methods. Sustain. Dev. Mt. Territ. **16**(1), 56–69 (2024)
2. Kukartsev, V., et al.: Using machine learning techniques to simulate network intrusion detection. In: 2024 International Conference on Intelligent Systems for Cybersecurity (ISCS). IEEE (2024)
3. Alena, R., et al.: Designing a system of step-by-step quality control. In: 2024 12th International Conference on Smart Grid (icSmartGrid). IEEE (2024)
4. Fedorova, N., et al.: Cost-effectiveness of development strategy implementation: key metrics and analysis methods for successful enterprise management. BIO Web Conf. **116**, 05003. EDP Sciences (2024)
5. Kukartsev, V.A., et al.: Study of the influence of the thermal capacity of the lining of acid melting furnaces on their efficiency. Metals. **13**(2), 337 (2023)

6. Fedorova, N., et al.: Analytical methods and tools for business process optimization. BIO Web Conf. **113**, 05009. EDP Sciences (2024)
7. Panfilov, I., et al.: Increasing competitiveness of enterprises by optimizing business processes as a factor of sustainable development of industrial region. E3S Web Conf. **531**, 05019. EDP Sciences (2024)
8. Klyuev, R.V., et al.: Analysis of geological information toward sustainable performance of geotechnical systems. Min. Inf. Anal. Bull. **5**, 144–157 (2024)
9. Makhambetov, M., et al.: Ecological-geological assessment of technogenically disturbed territories of oil fields of the Atyrau region. In: News of the National Academy of Sciences of the Republic of Kazakhstan, Series of Geology and Technical Sciences, vol. 2024, pp. 143–156 (2024)
10. Kukartsev, V., et al.: Optimization of maintenance work by implementing an automated information system at a repair facility. E3S Web Conf. **549**, 09011. EDP Sciences (2024)
11. Rozhkova, A., et al.: Application of statistical data analysis algorithms and determination of the most significant diagnostic factors. BIO Web Conf. **116**, 03015. EDP Sciences (2024)
12. Strateichuk, D.M., et al.: Morphological features of polycrystalline $CdS_{1-x}Se_x$ films obtained by screen-printing method. Crystals. **13**(5), 825 (2023)
13. Stepanova, E., et al.: Development of an automated information system for a logging company. BIO Web Conf. **116**, 03014. EDP Sciences (2024)
14. Rozhkova, A., et al.: Improving the efficiency of water use in agriculture by modelling the classification of groundwater quality. BIO Web Conf. **116**, 03020. EDP Sciences (2024)
15. Olentsova, J., et al.: Predictive modelling of post-monsoon groundwater quality in Telangana using machine learning techniques. BIO Web Conf. **116**, 03021. EDP Sciences (2024)
16. Stepanova, E., et al.: Machine learning approaches for water potability prediction: addressing class imbalance with SMOTE. BIO Web Conf. **116**, 03024. EDP Sciences (2024)
17. Malozyomov, B.V., et al.: Study of supercapacitors built in the start-up system of the main diesel locomotive. Energies. **16**(9), 3909 (2023)
18. Panfilova, T., et al.: Flood susceptibility assessment in urban areas via deep neural network approach. Sustainability. **16**(17), 7489 (2024)
19. Tynchenko, Y., et al.: Landslide assessment classification using deep neural networks based on climate and geospatial data. Sustainability. **16**(16), 7063 (2024)
20. Kukartsev, V., et al.: Application of non-parametric learning method in soil suitability assessment in present day economy. J. Infras. Policy. Dev. **8**(7), 4074 (2024)
21. Bosikov, I.I., et al.: Modeling and complex analysis of the topology parameters of ventilation networks when ensuring fire safety while developing coal and gas deposits. Fire. **6**(3), 95 (2023)
22. Kukartsev, V.V., et al.: Use of analytical methods in variant forecasting tasks. In: International Congress on Information and Communication Technology. Springer Nature Singapore, Singapore (2024)
23. Borodulin, A.S., et al.: Analyzing data by applying neural networks to identify patterns in the data. In: Proceedings of the Computational Methods in Systems and Software, pp. 99–108. Springer Nature, Cham
24. Shutaleva, A., et al.: Sustainability of inclusive education in schools and higher education: teachers and students with special educational needs. Sustainability. **15**(4), 3011 (2023)
25. Kukartsev, V., et al.: Intelligent data analysis as a method of determining the influence of various factors on the level of customer satisfaction of the company. In: Proceedings of the Computational Methods in Systems and Software, pp. 109–128. Springer Nature Switzerland, Cham (2023)
26. Tynchenko, Y., et al.: Soil properties classification in sustainable agriculture using genetic algorithm-optimized and deep neural networks. Sustainability. **16**(19), 8598 (2024)

27. Eremeev, D., et al.: Selection of sources of financing for an emerging networked systems based on mathematical analysis. E3S Web Conf. **458**, 05003. EDP Sciences (2023)
28. Bagaeva, A., et al.: Application software for business process automation. E3S Web Conf. **431**, 05029. EDP Sciences (2023)
29. Pokushko, M., et al.: Algorithm for application of a basic model for the data envelopment analysis method in technical systems. Algorithms. **16**(10), 460 (2023)

Classification of Motor Activity of the Human Upper Limbs Based on the Analysis of Electromyograms

Artem Obukhov(✉), Dmitry Pobedinsky, and Ivan Fedorchuk

Tambov State Technical University, 392000 Tambov, Russian Federation
obuhov.art@gmail.com

Abstract. The development of complex and highly realistic virtual simulators involves the collection and analysis of a large amount of data about the process of human interaction with the virtual environment. One of the key sources of data in the field of professional training is the physical, including muscular activity of the user, on the basis of which conclusions can be drawn about the person's condition and his preparation. To generate a set of such data, various electromyographs are used. The work discusses an approach to collecting and processing electromyography data for the subsequent solution of the problem of classifying user movements. Classification is performed using various machine learning methods, including dense and convolutional neural networks. As a result of the experiment on classification of 6 movements, an accuracy of 95% was obtained on a convolutional neural network. The experiment also analyzed the influence (correlation) of different electromyography channels on a specific movement. The results obtained can be used in professional training complexes as part of biofeedback or for collecting and processing information about the user's muscle activity.

Keywords: Virtual Reality · Electromyography · Muscle Activity · Machine Learning

1 Introduction

Electromyography (EMG) is a widely used method for measuring muscle activity by detecting the electrical signals generated by muscle contractions [1, 2]. EMG sensors can be attached to the surface of the skin or embedded in muscle tissue. Surface EMG sensors are non-invasive and easy to use, making them a popular choice for monitoring muscle activity in a variety of applications.

The study of motor and muscular activity of muscles is of great interest for the medical field, where non-invasive methods for assessing a person's physical condition are in demand, for example, in the process of rehabilitation after various injuries and diseases [3]. In addition, analysis of muscle activity can be useful when organizing professional training, when it is necessary to identify the presence or magnitude of muscle tension in the process of performing any exercises [4, 5]. A systematic review of the issue of using EMG in various rehabilitation tasks as a source of biofeedback was

R. Silhavy and P. Silhavy (Eds.): CoMeSySo 2024, LNNS 1489, pp. 103–114, 2025.
https://doi.org/10.1007/978-3-031-96798-6_10

carried out in [6]. It was revealed that EMG registration at key points of the body can evaluate the effect of exposure to virtual or augmented reality.

This study examines the subject area of professional virtual simulators. Analyzing the performance of EMG sensors attached to users' arms can be useful in determining the type of user movements and estimating the magnitude of muscle tension. This will allow for more functional biofeedback with the user and increase the adaptability of the virtual simulator.

EMG data find practical application in various tasks and areas. A whole area is devoted to the use of EMG to control robotic devices, exoskeletons or prostheses [7, 8]. Thus, in work [9], information about muscle activity was used to classify 15 different finger movements, which can be used both in prosthetics, rehabilitation, and in the development of new human-machine interfaces. This study is also interesting about the mechanisms of fixing electrodes in a circle on the forearm, which makes it possible to record the tension of various muscles. Machine learning and information about muscle activity make it possible to classify movements of the lower extremities, which is presented in [10]. The authors of this work successfully classified gait and leg movements. In [11], EMG information was used to assess muscle tone and implement an intelligent machine learning module to monitor patient progress. Also of interest is the study [12], in which the amount of muscle tension in the hand was analyzed to determine the force of compression of a virtual object, since existing consumer virtual reality controllers do not have such functionality.

The analysis carried out gives an unambiguous idea of the high prospects of combining virtual reality technologies and EMG data for the implementation of more advanced professional training and medical complexes. In this work, preliminary research will be carried out to analyze the possibility of classifying several types of human upper limb movements to develop an algorithm for collecting and processing information, as well as comparing various machine learning models.

Thus, the purpose of the study is to solve the problem of classifying user movements, as well as analyzing muscle activity based on indicators from EMG sensors. To achieve it, it is necessary to collect data from an experimental setup, prepare, train and compare various machine learning algorithms. This will allow us to identify the optimal machine learning algorithm, as well as determine the priority of input features (sensor placement points) when performing various movements.

2 Methods

The research scheme is presented in Fig. 1. Let's consider its main stages.

At the first stage of research, it is necessary to determine a list of typical actions that are planned for classification. A complete list of numbers (identifiers) and descriptions of actions is presented in Table 1. To simplify the understanding of the actions performed, some basic positions are presented in Fig. 2.

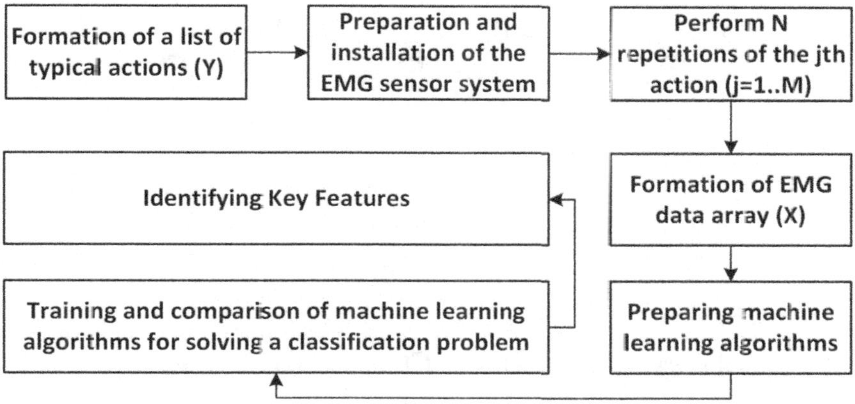

Fig. 1. Experimental research scheme.

Table 1. Description of actions.

ID	Description
1	Raising the arm from position 1 to position 2
2	Lowering the arm from position 2 to position 1
3	Moving the hand to the left from position 3 to position 4
4	Moving the hand to the right from position 4 to position 3
5	Bend the arm at the elbow from position 5 to position 6
6	Arm extension from position 6 to position 5

The presented set of actions is aimed at identifying some specific movement patterns, for example, moving up or down, as well as left and right. Such patterns may be of interest in problems of integrating virtual reality systems and controlled exoskeletons, for example, to determine the direction of movement of the user's hands and create the necessary resistance to this movement [13].

Fig. 2. Basic body movements.

Next, it is necessary to install a data collection system based on EMG sensors. Electrophotography can be carried out using various types of electrodes [14]: gel, dry, isolated - the transfer of biopotential is carried out through a dielectric; non-contact - an insulated electrode, during the transfer of which a layer of tissue is added. The conductive gel electrode has been studied and described in detail for many decades. Its properties have been extensively studied and there is sufficient empirical data on its performance, including low-frequency noise and drift. With proper application and preparation, this electrode is capable of providing an excellent signal. However, the use of this type of electrode may be difficult for multi-channel recording and may also cause discomfort to the patient.

Dry electrodes function by contacting the skin through a layer of moisture that naturally forms on it. Their advantage is that there is no need for additional products such as gel or paste, which simplifies the procedure for the patient. However, when using dry electrodes, signal accuracy may be reduced due to the influence of ambient electrical noise and the potential for skin irritation due to friction. In addition, when dealing with weak signals, dry electrodes may be less effective.

During preliminary studies, several options for ready-made data collection systems were analyzed [15–17]:

- single EMG sensor based on gel electrodes;
- single sensor based on dry electrodes;
- a set of gel electrodes for multi-channel recording;
- a set of dry electrodes in a bracelet form factor for multi-channel recording.

Using a single sensor is incorrect due to the insufficient amount of data (only one area of the arm muscles is monitored), and installing a set of gel electrodes is a rather labor-intensive and inconvenient process. Therefore, during a preliminary comparison based on a set of characteristics (comfort of wearing, connection stability, data volume), a set of dry electrodes in the form factor of a bracelet with 6 sensors was selected, which carries out wireless 12-channel data transmission, one of them is a reference one. The appearance of the bracelet when worn is shown in Fig. 3.

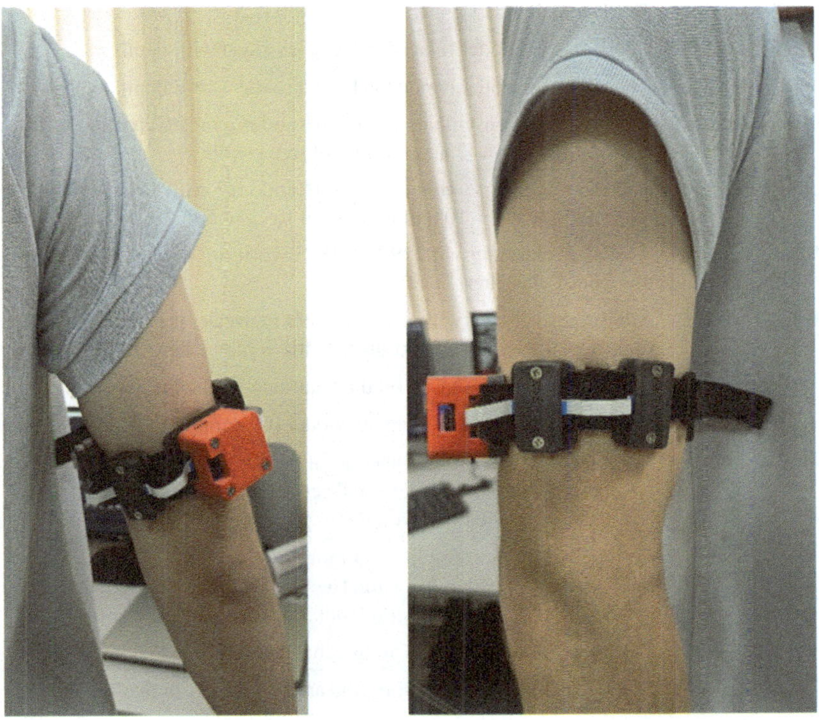

Fig. 3. Scheme of attaching the bracelet to the user's hand.

After installing the data collection system on a person, it is necessary to perform a series of experimental experiments, during which the subject repeatedly performs the actions indicated in Table 1. The recording is made from the beginning to the end of the movement. As a result, a data array X of N*6 records is formed, where N is the number of repetitions. When collecting data, it is necessary to take into account that part of the dataset will be used for testing and validating machine learning models.

The next stage involves preparing and training various machine learning models and algorithms. For comparison, several widely used and fairly effective algorithms are taken, as well as two neural network architectures designed by the authors. A description of machine learning algorithms is presented in Table 2, and the structures of neural networks are presented in Fig. 4 [18–20].

Table 2. Description of machine learning algorithms.

Model	Description
Logistic Regression	Linear model Logistic regression uses the logistic function to estimate the probability of an object belonging to a certain class.
Nearest Neighbors	K-Nearest Neighbors method (KNeighborsClassifier), k = 3
Linear SVM	SVM model with a linear kernel, C = 0.025
Decision Tree (5)	Decision tree model (DecisionTreeClassifier) with depth = 5
Decision Tree (10)	Decision tree model (DecisionTreeClassifier) with depth = 15
Random Forest (3)	Random forest ensemble classifier (RandomForestClassifier) with maximum depth = 3 and number of trees = 50
Random Forest (5)	Random forest ensemble classifier (RandomForestClassifier) with maximum depth = 10 and number of trees = 50
AdaBoost	AdaBoost classifier with "SAMME" algorithm as weak classifier, number of classifiers = 50
AdaBoost (tree)	AdaBoost classifier with decision trees as weak classifiers (AdaBoostClassifier), maximum depth = 5, number of trees = 50
Naive Bayes	Naive Bayes model based on the Gaussian method (GaussianNB)
QDA	Quadratic discriminant analysis model (QuadraticDiscriminantAnalysis)
Stacking Classifier	StackingClassifier ensemble combining LogisticRegression, KNeighborsClassifier, DecisionTreeClassifier with depth 10, RandomForestClassifier with depth 5, final classifier: LogisticRegression
Voting Classifier	VotingClassifier ensemble combining LogisticRegression, KNeighborsClassifier, DecisionTreeClassifier with depth 10, RandomForestClassifier with depth 5, voting method: "soft"
NN	Dense neural network, the architecture is shown in Fig. 4a
CNN	Convolutional neural network, the architecture is shown in Fig. 4b

The final stage of the research consists of comparing the developed models on a test data set, which will determine the best architecture. As an additional stage of research, the analysis of input features and the correlation between them and the output feature - the category of movement - are considered. Since it is statistically impossible to directly compare multiple input values with a category, it is proposed to calculate the average

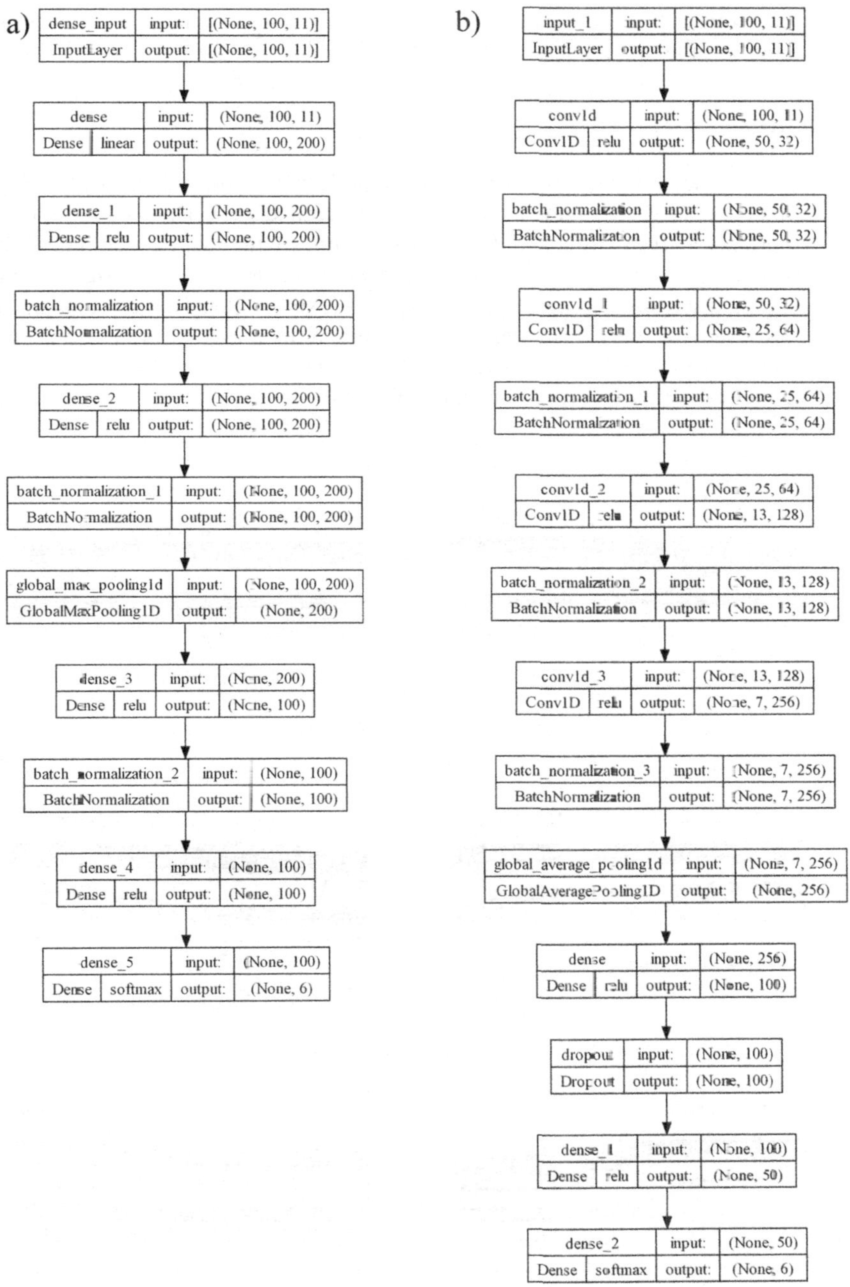

Fig. 4. Architecture of dense (a) and convolutional (b) neural networks.

voltage value for each sensor or the interval between the initial and final values (delta) and use them as input parameters.

3 Results

The duration of one movement was about 4 s. Of the 12 EMG channels, 11 were selected (the seventh channel was excluded since it is the reference one). Thus, 84 records with 11 input parameters were collected, the number of lines with data in each record from 88 to 100. To align the input data, all sequences less than 100 lines were supplemented by copying the last terms. As part of the study, it is assumed that 10 measurements (60 records) are used for training models, and the remaining 4 (24 records) are used for testing. First, let's look at the general view of the collected data for all 84 measurements, indicating the corresponding EMG channels (Fig. 5).

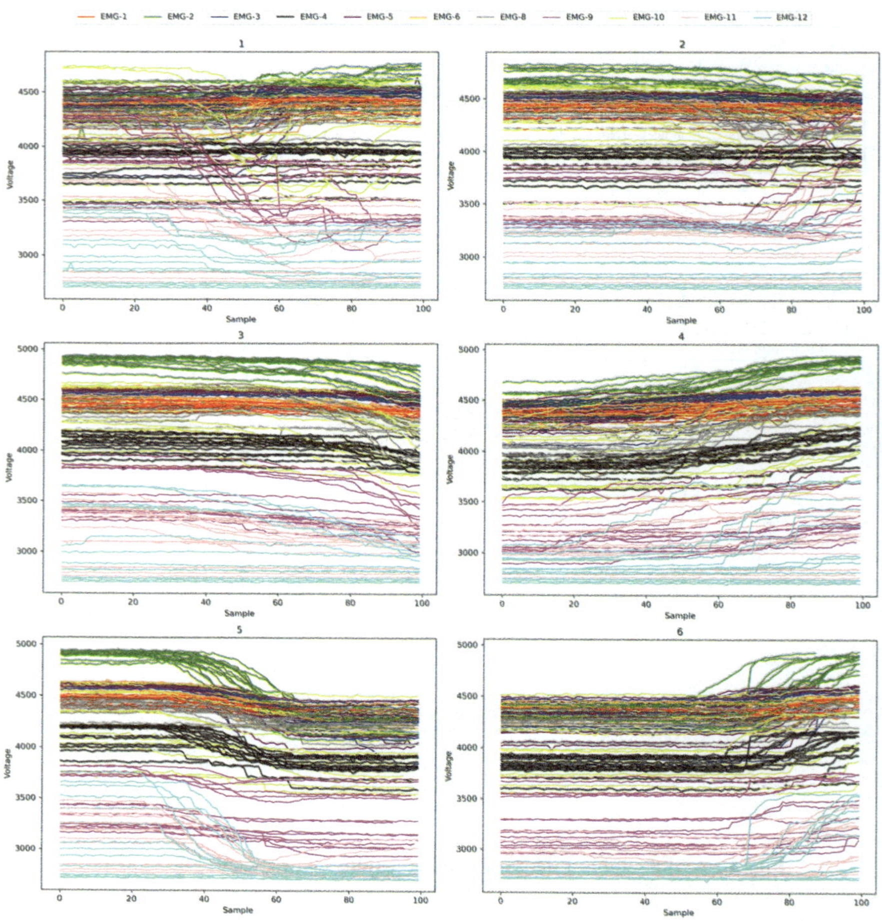

Fig. 5. General view of the collected data.

From Fig. 5 we can conclude that there are significant differences in voltage values at different leads of the EMG systems when performing various movements.

Next, a Pearson correlation matrix was constructed between the numbers of tests (movements) and input features: delta (the difference between the first and last values) and the average voltage values for each lead. For convenience, Fig. 6 does not display the entire matrix, but only the correlations between inputs and outputs.

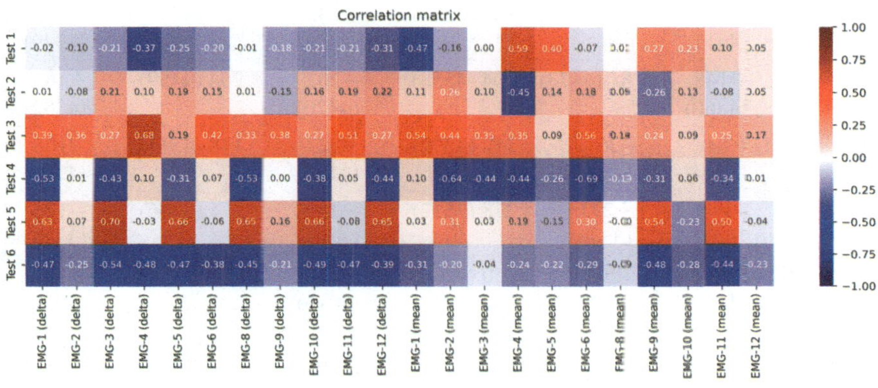

Fig. 6. Pearson correlation matrix between tests and input features.

Analysis of the correlation values showed that each test depends on the voltage of different EMG leads. A pairwise comparison using the delta metric shows that the correlation in the key leads for movement is opposite: in a pair of tests 1–2 these are leads 3–6 and 9–12, in a pair of tests 3–4 - almost all leads, in a pair 5–6 it is also observed significant difference between the correlation. All this suggests that trained machine learning algorithms will be able to identify and classify various movements.

Table 3 shows the delta and average voltage values across all leads for each test.

Table 3. Delta and average voltage values in each test.

EMG Channel / Test	1	2	3	4	5	6
EMG-1	−8 (4384)	−2 (4387)	63 (4437)	−96 (4397)	105 (4403)	−85 (4367)
EMG-2	−209 (4486)	174 (4640)	238 (4834)	−420 (4640)	631 (4600)	−517 (4451)
EMG-3	−175 (4388)	82 (4488)	82 (4563)	−211 (4464)	356 (4429)	−299 (4339)

(continued)

Table 3. (*continued*)

EMG Channel / Test	1	2	3	4	5	6
EMG-4	−13 (3873)	−4 (3884)	141 (4031)	−255 (3925)	288 (3970)	−216 (3865)
EMG-5	−149 (4411)	58 (4506)	122 (4583)	−245 (4473)	338 (4442)	−307 (4349)
EMG-6	−148 (4360)	81 (4473)	104 (4558)	−202 (4470)	264 (4459)	−181 (4392)
EMG-8	−84 (4304)	126 (4328)	212 (4392)	−327 (4193)	148 (4311)	−108 (4293)
EMG-9	449 (3707)	−271 (3546)	278 (3512)	−259 (3293)	171 (3366)	−126 (3324)
EMG-10	−14 (4161)	129 (4213)	350 (4280)	−377 (4016)	200 (4148)	−146 (4090)
EMG-11	156 (3193)	−103 (3136)	143 (3115)	−129 (3103)	292 (2949)	−211 (2919)
EMG-12	72 (2976)	−64 (2975)	190 (3033)	−265 (2955)	383 (2930)	−345 (2835)

At the next stage of the experiment, we will move on to training neural networks and machine learning algorithms.

A comparison of the classification accuracy of all models is presented in Table 4. As a result, the best performance is obtained using a convolutional neural network with the architecture presented in Fig. 4b.

Table 4. Comparison of machine learning models for solving classification problems.

Model/Metric	Accuracy	F1-score
Logistic Regression	0.875	0.877
Nearest Neighbors	0.833	0.843
Linear SVM	0.708	0.706
Decision Tree (5)	0.5	0.446
Decision Tree (10)	0.5	0.446
Random Forest (3)	0.583	0.56
Random Forest (5)	0.833	0.834

(*continued*)

Table 4. (*continued*)

Model/Metric	Accuracy	F1-score
AdaBoost	0.292	0.244
AdaBoost (tree)	0.375	0.38
Naive Bayes	0.708	0.709
QDA	0.375	0.292
Stacking Classifier	0.875	0.877
Voting Classifier	0.875	0.877
NN	0.875	0.869
CNN	**0.958**	**0.958**

Thus, the advantage is given to the convolutional neural network model, which, by summarizing information and extracting features by convolutional layers, provides the highest classification accuracy. Further, models of multilayer neural networks, Voting Classifier and Logistic Regression show lower accuracy.

4 Conclusion

This paper examines the problem of classifying electromyography data using machine learning technologies. Its solution is highly promising in the development of software and hardware training complexes, rehabilitation systems and medical simulators.

Based on the formulated research scheme, data was collected from a bracelet consisting of 6 dry EMG sensors, recording data on 12 channels. 15 machine learning algorithms were prepared and trained to classify 6 types of movements based on muscle activity. An accuracy of 95% was obtained using a convolutional neural network. The correlation of different electromyography channels to a specific type of movement was analyzed. The results obtained can be used in professional training complexes as part of biofeedback or to collect and process information about the user's muscle activity.

The article was carried out with the financial support of the Ministry of Science and Higher Education of the Russian Federation within the framework of the project "Development of an immersive virtual reality interaction system for professional training based on an omnidirectional platform" (FEMU-2024-0005).

References

1. Bengacemi, H., Abed-Meraim, K., Buttelli, O., Ouldali, A., Mesloub, A.: A new detection method for EMG activity monitoring. Med. Biol. Eng. Comput. **58**, 319–334 (2020)
2. Eliseichev, E.A., Mikhailov, V.V., Borovitskiy, I.V., Zhilin, R.M., Senatorova, E.O.: A review of devices for detection of muscle activity by surface electromyography. Biomed. Eng. **56**(1), 69–74 (2022)

3. Al-Ayyad, M., Owida, H.A., De Fazio, R., Al-Naami, B., Visconti, P.: Electromyography monitoring systems in rehabilitation: a review of clinical applications, wearable devices and signal acquisition methodologies. Electronics. **12**(7), 1520 (2023)
4. Jarque-Bou, N.J., Sancho-Bru, J.L., Vergara, M.: A systematic review of EMG applications for the characterization of forearm and hand muscle activity during activities of daily living: results, challenges, and open issues. Sensors. **21**(9), 3035 (2021)
5. Taborri, J., Keogh, J., Kos, A., Santuz, A., Umek, A., Urbanczyk, C., Rossi, S.: Sport biomechanics applications using inertial, force, and EMG sensors: a literature overview. Appl. Bionics Biomech. **2020**(1), 2041549 (2020)
6. Toledo-Peral, C.L., Vega-Martínez, G., Mercado-Gutiérrez, J.A., Rodríguez-Reyes, G., Vera-Hernández, A., Leija-Salas, L., Gutiérrez-Martínez, J.: Virtual/augmented reality for rehabilitation applications using electromyography as control/biofeedback: systematic literature review. Electronics. **11**(14), 2271 (2022)
7. Tang, Z., Zhang, K., Sun, S., Gao, Z., Zhang, L., Yang, Z.: An upper-limb power-assist exoskeleton using proportional myoelectric control. Sensors. **14**(4), 6677–6694 (2014)
8. Kotov-Smolenskiy, A.M., Khizhnikova, A.E., Klochkov, A.S., Suponeva, N.A., Piradov, M.A.: Surface EMG: applicability in the motion analysis and opportunities for practical rehabilitation. Hum. Physiol. **47**(2), 237–247 (2021)
9. Amezquita-Garcia, J., Bravo-Zanoguera, M., Gonzalez-Navarro, F.F., Lopez-Avitia, R., Reyna, M.A.: Applying machine learning to finger movements using electromyography and visualization in opensim. Sensors. **22**(10), 3737 (2022)
10. Wang, J., Dai, Y., Si, X.: Analysis and recognition of human lower limb motions based on electromyography (EMG) signals. Electronics. **10**(20), 2473 (2021)
11. Covaciu, F., Pisla, A., Iordan, A.E.: Development of a virtual reality simulator for an intelligent robotic system used in ankle rehabilitation. Sensors. **21**(4), 1537 (2021)
12. Shin, Y., Lee, M.: Development of sensory virtual reality interface using EMG signal-based grip strength reflection system. Appl. Sci. **14**(11), 4415 (2024)
13. Cisnal, A., Gordaliza, P., Pérez Turiel, J., Fraile, J.C.: Interaction with a hand rehabilitation exoskeleton in emg-driven bilateral therapy: influence of visual biofeedback on the users' performance. Sensors. **23**(4), 2048 (2023)
14. Liu, Q., Yang, L., Zhang, Z., Yang, H., Zhang, Y., Wu, J.: The feature, performance, and prospect of advanced electrodes for electroencephalogram. Biosensors. **13**(1), 101 (2023)
15. Etana, B.B., Malengier, B., Krishnamoorthy, J., Van Langenhove, L.: Types of EMG textile electrodes: a comparative study using PCA. Eng. Proc. **35**(1), 6 (2023)
16. Searle, A., Kirkup, L.J.P.M.: A direct comparison of wet, dry and insulating bioelectric recording electrodes. Physiol. Meas. **21**(2), 271 (2000)
17. Erickson, J.C., Stepanyan, E., Hassid, E.: Comparison of dry and wet electrodes for detecting gastrointestinal activity patterns from body surface electrical recordings. Ann. Biomed. Eng. **51**(6), 1310–1321 (2023)
18. Obukhov, A., Dedov, D., Volkov, A., Teselkin, D.: Modeling of nonlinear dynamic processes of human movement in virtual reality based on digital shadows. Computation. **11**(5), 85 (2023)
19. Tăbăcaru, G., Moldovanu, S., Răducan, E., Barbu, M.: A robust machine learning model for diabetic retinopathy classification. J. Imaging. **10**(1), 8 (2023)
20. Ahmadi, H., Mesin, L.: Enhancing motor imagery electroencephalography classification with a correlation-optimized weighted stacking ensemble model. Electronics. **13**(6), 1033 (2024)

Correlation and Regression Analysis Using Real-Time Audio and Video Conferencing

Vladislav Hinkov(✉) and Georgi Krastev

Department of Computing, University of Ruse, 7017 Ruse, Bulgaria
{vhinkov,geork}@uni-ruse.bg

Abstract. Networked multimedia is a unifying link in the parallel development of multimedia and communication technologies. The integrated interaction of technological solutions in the field of multimedia and computer networks is a condition for achieving a greater final applied effect in the presentation of information. Experimental studies of modern networked multimedia under operational conditions have a vital importance in revealing bottlenecks in their operation. On this basis, recommendations can be made to improve performance, such as performance, reliability, mode of service, etc. This publication presents an example of studying the existence of interdependence between phenomena and related quantities, by performing correlation and regression analysis using audio and video conferencing in real time.

Keywords: Research · Statistical methods · Correlation analysis · Regression analysis · Audio, and video conferencing in real time

1 Introduction

Modern computer networks provide an environment in which a person spends more of his time in the performance of his work and personal commitments, shares and receives information, communicates on social networks. Long gone is the era when the main and practically only media in the networks was text. Audio information, graphics, and video already occupy a lasting place in data traffic in computer networks. The relatmultimediaf the transfer of different media, collectively called computer multimedia, is constantly growing. This puts to the serious test the existing technologies for data exchange and requires their analysis and assessment of usability, as well as the creation of modern technologies and standards that meet not only the requirements of today, but also the needs of the future.

Networked multimedia is a unifying link in the parallel development of multimedia and communication technologies. The integrated interaction of technological solutions in the field of multimedia and computer networks is a condition for achieving a greater final applied effect in the presentation of information.

Experimental studies of modern networked multimedia under operational conditions have a vital importance in revealing bottlenecks in their operation. On this basis, recommendations can be made to improve performance, such as performance, reliability, mode of service, etc.

R. Silhavy and P. Silhavy (Eds.): CoMeSySo 2024, LNNS 1489, pp. 115–122, 2025.
https://doi.org/10.1007/978-3-031-96798-6_11

2 Exposition

2.1 Theoretical Prerequisites

The essence of modern networked multimedia is such that it bears a stochastic (probabilistic) character. When studying stochastic processes, the following main task is solved [3, 4]: *Study about the existence of interdependence between phenomena and related quantities, by performing a correlation analysis.*

The correlation analysis determines the degree of dependence between the values of two variables. In our case, we investigate the dependencies for different "factor-active rooms" quantity pairs in real-time network audio and video conferencing.

The values of the factors, which are independent variables, we denote by $X = \{x_1, x_2, \ldots x_n\}$, and the corresponding values of the dependent variable with $Y = \{y_1, y_2, \ldots y_n\}$.

In two variables (between the independent variables and the dependent variable) an ordinary correlation coefficient, also known as the Pearson – Brave coefficient, is calculated. In Table 1 the used formulae are given [3]. This correlation coefficient can take values in the interval from -1 to $+1$. When $r > 0$, the presence of a correlation relationship means that as x increases, the dependent variable y also grows. When $r < 0$ as x increases the variable y decreases. An interpretation of the different values of the Pearson correlation coefficient is:

- $|r| > 0.7$ – strong correlation.
- $0.5 < |r| \leq 0.7$ – mean correlation.
- $0.3 < |r| \leq 0.5$ – weak correlation.
- $|r| < 0.3$ – no correlation.

2.2 Linear Regression Models

The linear regression model is described in the form:

$\hat{y} = ax + b$.Where

- x are the independent variables,
- $\hat{y}$ is the approximation solution we expect to correspond to the dependent variables,
- a and b are the coefficients whose values we are looking for.

In Table 1 the formulae used are given.

The significance or credibility of the point estimate of the correlation coefficient r is checked using the procedure involving the null hypothesis $H_0 : r = 0$ and a corresponding statistical criterion. The rejection of H_0 means that the resulting estimate r is significant, and a linear correlation exists between the quantities x_i and y_i. In the opposite case, the estimate r is not significant and the linear correlation between x_i and y_i not proven. The hypothesis H_0 is tested in different ways. One of them [5] uses the student's criterion:

$$t = \frac{r\sqrt{n-2}}{\sqrt{1-r^2}}, \tag{1}$$

which under a true null hypothesis has a student's distribution with degrees of freedom $k = n - 2$.

Table 1. Formulas used.

Basic equation	$\hat{y} = ax + b$
Averages	$\overline{x} = \frac{\sum_{i=1}^{n} x_i}{n}$ $\overline{y} = \frac{\sum_{i=1}^{n} y_i}{n}$
Basic dependencies	$S_{xx} = \sum_{i=1}^{n} x_i^2 - n\overline{x}^2$ $S_{yy} = \sum_{i=1}^{n} y_i^2 - n\overline{y}^2$ $S_{xy} = \sum_{i=1}^{n} x_i y_i - n\overline{xy}$
Covariance	$q = \frac{S_{xy}}{n-1}$
Correlation coefficient	$r = \frac{S_{xy}}{\sqrt{S_{xx} S_{yy}}}$
Coefficient	$\mathrm{a} = \frac{\mathrm{S_{xy}}}{\mathrm{S_{xx}}}$
Coefficient	$\mathrm{b} = \overline{\mathrm{y}} - \mathrm{a}\overline{\mathrm{x}}$

If the alternative hypothesis is H_1: $r \neq 0$, the critical region will be two-sided. The obtained experimental (calculated) value t using the formula (1) is compared with the critical value k $t_{\frac{\alpha}{2};k}$, which is taken from a table at $\gamma = 1 - \alpha$. At $|t| \geq t_{\frac{\alpha}{2};k}$, the hypothesis H_0 is rejected at significance level α and vice versa.

3 Results

The experimental investigations were carried out with the following apparatus and program means shown in Table 2. The BigBlueButton platform is installed on both servers and is used by engineering faculties at the following addresses: https://meet.uni-ruse.bg/b/ and https://meet1.uni-ruse.bg/b/. Any discussion, lesson or consultation can be recorded and then watched offline (streaming or download). During the online session, control is in the hands of the teacher, who can give the floor to the participants or mute their microphones, as well as nominate who will be the presenter. During deliberations, a "raise your hand" button allows anyone to alert the teacher or presenter that there is a question or some other problem. The presenter, in turn, can share any documents or applications, give access to his desktop or to certain WEB content. Convenience in this case is that in the session the participants do not need to have the same program installed on their computers to work with the applications shared by the presenter - they can even have different browsers or operating systems. Control of the session can be transferred to another participant in the meeting, and the rest can make their annotations (using lines and arrows of different colors) directly on the document, but without changing its content. A separate chat panel allows everyone to write and receive messages during the session.

Table 2. Apparatus and software used.

Server	meet1	meet
CPU	2x AMD EPYC 7402 24-Core Processor	4x Intel(R) Xeon(R) CPU E7-4820 @ 2.00GHz Total Cores 8
RAM	64 GB Registered ECC DDR4	128GB DDR3 ECC Registered memory
Ethernet	2 1GbE LAN via Intel® i350-AM21	Intel® 82576 Dual-Port Gigabit Ethernet Controller
Video	ASPEED AST2500 BMC graphics	Matrox G200eW 16 MB DDR2
Storage	SATA3 (6 Gbps) (2x NVMe on PCi)	6Gb/s SAS / SATA ports (1x Samsung SSD 850)
OS	Ubuntu 18.04.6 LTS	Ubuntu 18.04.6 LTS
Software	BigBlueButton Virtual Classroom	BigBlueButton Virtual Classroom

The results of the correlation analysis performed, and the regression equations built for the relationship between the individual factors are shown from Figs. 1, 2, 3 and 4. As a result, statistically significant correlation coefficients were found for some of the factors and working regression models were made, but for others they were not.

To be able to perform such studies, it is necessary to determine in advance the changing factors, the way the data from the experiment is obtained and their processing [1, 2]. An important point is also the proper planning of experimental studies to obtain adequate results with a minimum of experimental studies.

This paper is devoted to the experimental study of this type of technological integration and is an attempt to examine multimedia technologies through the prism of computer networks and computer networks through the prism of multimedia technologies.

The correlation coefficient is $r = 0.7445$, squared – r^2 is called the coefficient of determination. It shows what percentage of the scatter of the performance variable is explained by the action of the factor variable. In this case, $r^2 = 0.5543$, i.e. 55.4% of the active rooms depend on the active users.

The tabular value of is 2.26. The resulting value for $t = 3.3456$. The null hypothesis H $t_{\frac{\alpha}{2};k}$ 0 is rejected, which means that the resulting estimate r is significant, and a linear correlation exists between "active users - active rooms".

The resulting regression equation for the relationship between "active users - active rooms" is $\hat{y} = 0.0517x + 17.3589$.

In this case, $r = -0.3199$, and $r^2 = 0.1023$, i.e. 10.2% of the active rooms depend on the active video broadcasts.

The resulting value for $t = 1.013$. The null hypothesis H_0 is assumed. The estimate r is non-significant and the linear correlation between "active video broadcasts - active rooms" has not been proven.

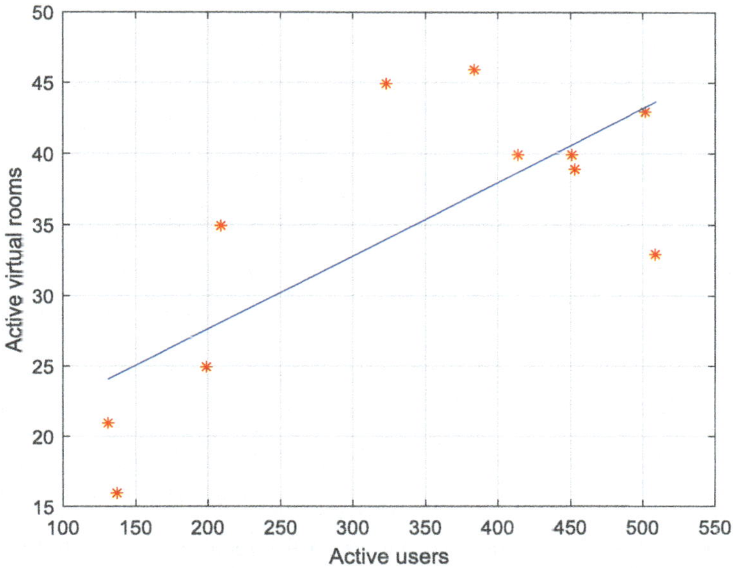

Fig. 1. Scatter diagram for "active users - active rooms" - q = 1.0837e+03, under strong correlation relationship r = 0.7445.

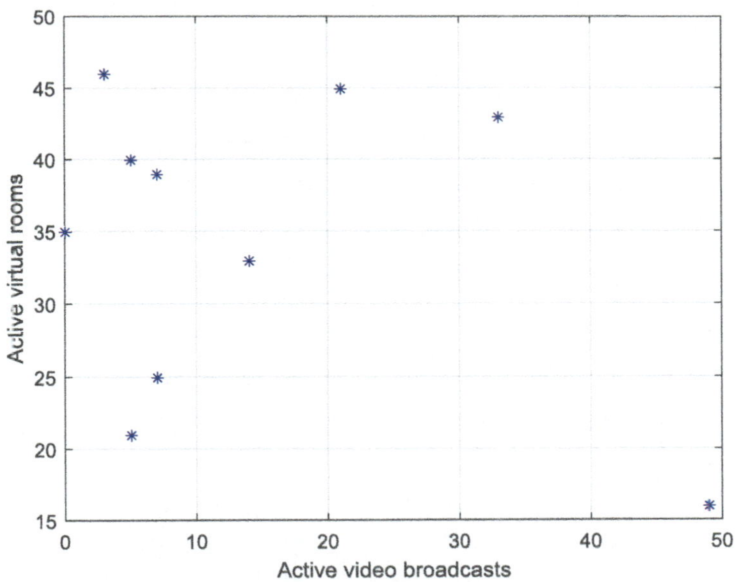

Fig. 2. Scatter diagram for "active video broadcasts - active rocms"- q = −48.6909, at weak correlation relationship r = −0.3199.

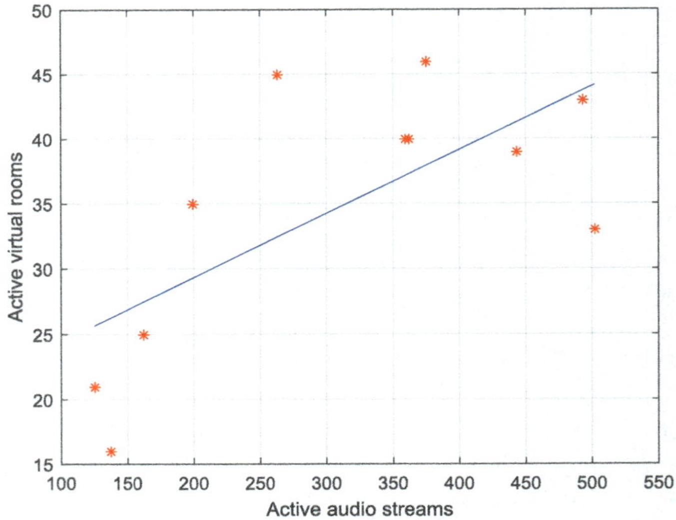

Fig. 3. Scatter diagram for "active audio streams - active rooms" - q = 972.0818, at average correlation relationship r = 0.6869.

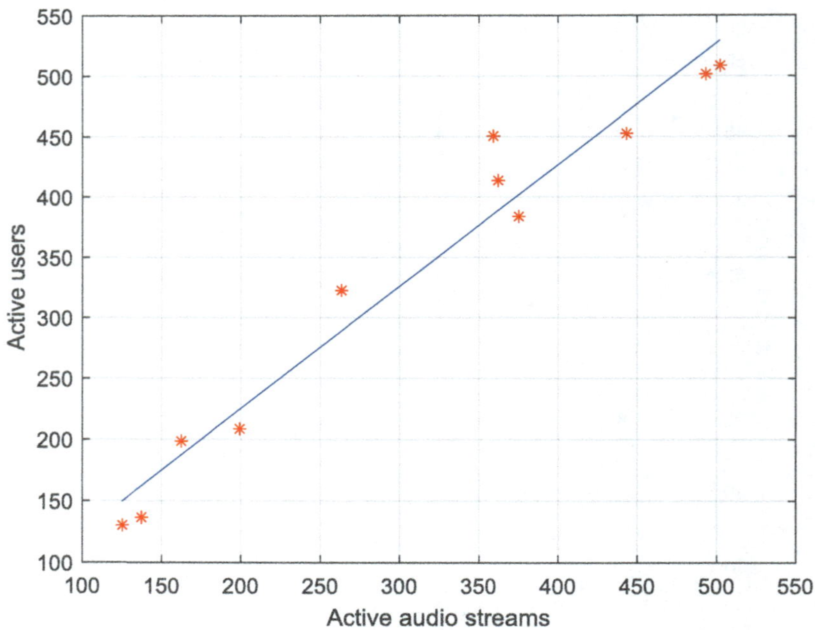

Fig. 4. Scatter diagram for "active audio streams - active users" - q = 1.9929e+04, with strong correlation relationship r = 0.9787.

In the specific case, r = 0.6869, and r^2 = 0.4718, i.e. 47.2% of active rooms depend on active audio streams.

The resulting value for t = 2.83. The null hypothesis H0 is rejected, which means that the resulting estimate r is significant, and a linear correlation exists between "active audio streams - active rooms".

The resulting regression equation for the relationship between "active audio streams - active rooms" x is = 0.0491 + 19.5503.$\hat{y}$

Obviously, in this graph there is some chaoticness and great scatter. This shows that some caution is needed in interpretations of relationships.

For this connection between "active audio streams - active users" can be described as very strong, with a coefficient: r = 0.9787, which turns out to be statistically significant for significance levels $\alpha = 0.1; 0.05; 0.01$, and r^2 = 0.9578, i.e. 95.8% of active users depend on active audio streams.

The resulting value for t = 14.29. The null hypothesis H_0 is rejected, which means that the resulting estimate r is significant, and a linear correlation exists between "active audio streams - active users".

The resulting regression equation for the relationship between "active audio streams - active users" is $\hat{y}$ =1.0068x + 24.4462.

4 Conclusions

From the preliminary analyses, through scatter diagrams, it is also clearly seen that linear regression is not always appropriate to be applied in the relationships studied by us.

In summary, the following more important conclusions can be drawn:

1. Existing statistical methods are a powerful toolkit for the processing and analysis of statistical data in research.
2. Correlation analysis does not always carry within it information about an existing causal relationship between two phenomena. The relationship between them is interpreted only for the given group, and causal relations should be subject to another type of analysis.
3. On the processes of using audio and video conferencing in real time act combined in complex relationships many factors whose action must be considered in their totality. Trying to evaluate and predict future behavior only by means of classical statistical analysis based on a single factor can lead to erroneous conclusions.
4. The fact that in some factors the linear model predicts with great error, and in others it cannot even be composed, still does not mean that there is no connection between these factors. In fact, this means that there is a weak or no linear relationship between them.

Acknowledgments. This study is financed by the European Union-NextGenerationEU, through the National Recovery and Resilience Plan of the Republic of Bulgaria, project № BG-RRP-2.013-0001-C01.

References

1. Puri, A.: Multimedia Systems, Standards, and Networks. CRC Press., ISBN 0203908449, 9780203908440 (2000)
2. Hwang, J.-N.: Multimedia Networking: From Theory to Practice. Cambridge University Press., ISBN0521882044, 9780521882040 (2009)
3. Seber, G.A.F., Lee, A.J.: Linear Regression Analysis. Wiley Series in Probability and Statistics. John Wiley & Sons., ISBN 1118274423, 9781118274422 (2012)
4. Draper, N.R., Smith, H.: Applied Regression Analysis. Wiley Series in Probability and Statistics. John Wiley & Sons., ISBN 0471170828, 9780471170822 (1998)
5. Triola, M.F., Franklin, L.A.: Business Statistics. Addison-Wesley Publishing Company (2004)

Electromagnetic Compability of LiFiMAX Devices

Štefan Počarovský[(✉)], Jozef Hallon, Adam Levák, Miloš Orgoň, and Martin Gabrišák

Institute of Multimedia Information and Communication Technologies, Slovak University of Technology in Bratislava, Ilkovicova 3, Bratislava, Slovak Republic
milos.orgon@stuba.sk

Abstract. Light-Fidelity (LiFi) is a technology that enables networked two-way optical wireless communication (OWC) and is considered as one of the possible solutions for high-speed communication indoors and later outdoors. It enables high data rates, high connection density, ultra-reliable low latency communication (URLLC) and high security. This paper focuses on the methodology to assess the EMC characteristics of devices using LiFi technology.

Keywords: LiFi · Light Fidelity · OWC · optic wireless communication · EMC charakteristic

1 EMC Wireless Data Communication Devices

Electromagnetic Compatibility (EMC) testing verifies the ability of an electronic device to operate simultaneously in an environment where other electronic devices are present. Equipment manufacturers should design equipment so that, when used under normal operating conditions, it does not interfere with the functionality of other technical equipment and unacceptably degrade telecommunications services in its environment [1]. The measured electromagnetic (EM) emissions of the equipment under test are related to the limits defined in international technical standards, which indicate the maximum permissible EM emission value. Special measurement procedures standardized in the relevant technical standards [2] are used to detect the level of interference of electrical equipment. An important requirement of the measurement procedure for EM emission measurement is that the equipment is measured in the mode when its EM emission is at its highest [2].

In the case of data communication devices, this is when their modem is in transmit mode. This is due to the fact that broadcast mode places high energy demands on the power system, which is generally associated with higher levels of EM emissions. Sometimes these are low power devices (communicating with their environment for a limited period of time and with a large repetition rate) powered by batteries, where the power supply is designed not to radiate during normal operation. When transmitting, the power supply currents increase significantly, resulting in increased EM emissions. Otherwise, these are systems that are designed for virtually continuous communication. The advantage is that such communication technology can be flexibly customized - personalized to the requirements of specific users. Devices communicate with each other

R. Silhavy and P. Silhavy (Eds.): CoMeSySo 2024, LNNS 1489, pp. 123–135, 2025.
https://doi.org/10.1007/978-3-031-96798-6_12

using well-known WiFi network technologies, long-used data communication using 2G GSM mobile networks, or modern IoT networks such as ZigBee, LoRa, Sigfox, Bluetooth Low Energy, Z-Wave and many others. All these systems are wireless and differ from each other in their range, the amount of energy consumed by a given device during communication, the frequency and volume of data transmitted, and so on [3].

To illustrate the generation of electromagnetic emissions, we have chosen data communication using 2G GSM mobile networks, because during operation the current consumed by the GSM modem can vary by several orders of magnitude and during the transmission interval the peak current can reach up to 3 A. The magnitude of this current is proportional to the radiated power. The power is proportional to the distance from the base station, the attenuation of the transmission path and also depends on the characteristics of the antenna connected to the modem. Figure 1 shows the time dependence of the modem current consumption (in GPRS mode) for typical operating modes of the MC75i modem [4]. The maximum pulse current shown causes pulsed variations in the modem supply voltage at the impedances in the power supply circuits with a wide frequency spectrum (Fig. 1). The supply voltage drops can reach up to several 100 mV (for a modem supply voltage of 3.2 V - 4.5 V). This places high demands on the design of modem power supply circuits, and if errors are made in the design and implementation of the power supply circuits, this has adverse consequences on the EMC characteristics of the overall device.

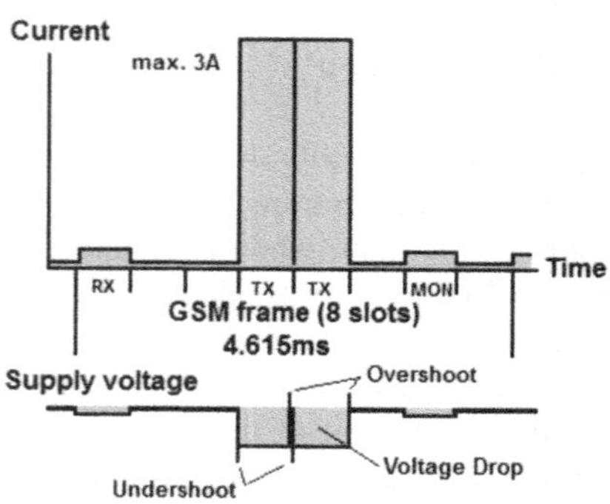

Conditions	I [mA]
TALK mode	
GSM900	460
GSM1800	275
DATA mode; GPRS class 8	
GSM900	495
GSM1800	300
DATA mode; GPRS class 10	
GSM900	680
GSM1800	500
DATA mode; GPRS class 12	
GSM900	730
GSM1800	530

Fig. 1. Time dependence of supply current, voltage and maximum supply current of the MC75i modem.

This principle, documented on the GSM modem, is universally valid, although it applies to different communication technologies to varying degrees. Due to the fact, from an energy point of view, the most commonly used switching regulators or converters produce broad-spectrum EM interference during their operation, it is only a matter of appropriate design to ensure that the current peaks do not translate into emission spectrum to a significant degree. By taking sufficient EMC measures into account, it is possible to keep both conduction and radiation emissions within a reasonable level that does not exceed the emission limits of international standards.

Reciprocally, the immunity of electrical equipment to EM disturbances is nowadays considered as part of the functional safety of systems. The standards specify minimum requirements for immunity of electrical equipment to EM disturbances that are appropriate to the specified technical environment. They specify the levels of severity of immunity tests and the functional capability criteria (i.e. the limitation of the failure rate) for assessing the impact of disturbances on the function of the equipment.

In the case of EM immunity of wireless communication devices, the situation is partially simplified. These devices communicate wirelessly and therefore do not suffer from EM disturbances whose mechanism of action is tied to long galvanic lines. It is possible to dispense with the wide range of EMC tests that test the EM immunity of equipment with cable interfaces longer than 3 m. The immunity tests can be narrowed down to the immunity test for radiated RF EM field according to EN IEC 61000-4-3 and the immunity test for electrostatic discharge according to EN 61000-4-2. In the first test, the advantage here is that the EM field is induced into the communication equipment enclosure and the cable lines simultaneously, where at the cable interfaces an interfering RF signal is generated at the respective interface impedances, which is then a potential source of a fault condition. The enclosures of such devices tend to be small in size and, in conjunction with the absence of communication lines, the EM field effect on this type of device is not a serious concern (we assume a standard power supply with satisfactory EMC characteristics for the relevant operating environment). The minimum of cable interfaces (connectors) and controls in most wireless data communication devices gives them a good ESD immunity (unless by chance some gross design error has occurred in the design of the device).

To the standard wireless communication technologies that use EM fields as a transmission medium, LiFi technology, which uses light to transmit data, has now been added. This paper is devoted to investigating how a particular representative of this technology behaves in terms of EMC.

2 Characteristics of LiFi Technology

LiFi is a new two-way wireless high-speed communication technology that uses light not only for lighting purposes but also for communication. It can be used in such a way that infrared light is used for communication in such a way that it does not restrict the illumination of the room, thus avoiding interference from visible light. LiFi technology offers a number of advantages, not least of which is the very large unregulated bandwidth available in the light spectrum, which is over 2600 times greater than the entire RF spectrum. Furthermore, there is the high energy efficiency, the ease of implementation, but also the increased security of the network implemented in this way, as light does not penetrate through opaque objects. This is also confirmed by the Institute of Electrical and Electronics Engineers (IEEE), which has released a new LiFi standard called 802.11bb [5].

For testing, we had the Oledcomm LIFIMax Kit (Fig. 2) from Oledcomm, a leading French LiFi company. This LiFiMax system is a LiFi plug and play system, so there is no need to install additional drivers or software when connecting. Up to 16 users can connect to the LiFiMax access point. The LiFiMax system is secure, with three Advanced Encryption Standard (AES) block ciphers available for use: AES-128, AES-192, or AES-256. AES ciphers are implemented in software and hardware worldwide to encrypt sensitive data [5]. The specifications of the Oledcomm LIFIMax system are as follows:

Communication Specifications LiFiMAX [5]:

- Type of light source: Infrared LED
- Download data rate: up to 100 Mbps
- Upload data rate: up to 100 Mbps
- Communication coverage: 90 degrees field of view
- Maximum number of users: 16

Electrical Specifications LiFiMAX [5]:

- Access point nominal power < 5 W
- Access point power input: PoE (IEEE 802.3af)
- Dongle nominal power: 2.5 W
- Dongle power input: USB-C

Fig. 2. Oledcomm Kit, the larger disk is the AP transmitter, the smaller disks are the LIFIMAX receiver dongles.

3 EMC Performance Testing of LiFi Devices

The tested LiFiMAX devices, which were the subject of our investigation, allow to achieve data rates up to 100 Mbps, while the data transmission is realized by infrared radiation, which does not cause interference to the surrounding electronic devices [6–9]. For this reason, in addition to the other tests performed in the aforementioned works [6–9], we proceeded to test these devices in terms of EMC.

The testing was carried out in the EMC laboratory at the Institute of Electrical Engineering, FEI STU Bratislava, which has been granted the status of Hi-Tech Centre for EMC by the Ministry of Education of the Slovak Republic and has international accreditation through SNAS.

A schematic of the testbed configuration is shown in Fig. 3. In the laboratory there is a semi-anechoic shielded chamber, the purpose of which is a significant reduction of EM interference from the external environment (by about 100 dB in a wide band of frequencies) and at the same time the chamber with its properties is close to the open area (OATS) for the propagation of EM waves over the reflective (conductive) plane.

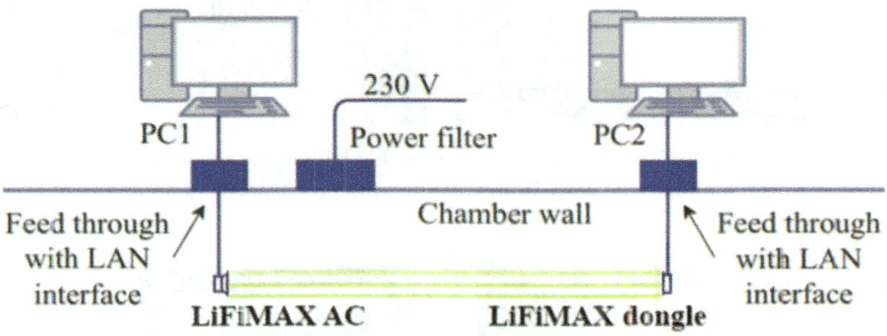

Fig. 3. LiFiMAX test bench configuration.

It should be noted that the measurement of such a communication system is problematic because the risk of the measurement being disturbed by radiation from other auxiliary electrical equipment is high. In our case, the radiation is from two computers with accessories, between which the LiFiMAX communication system mediates the data transmission. This effect has been minimized by a suitable layout of the measurement workstation together with the use of the features and capabilities of the shielded chamber. The measurement workstation was built so that only the emitter (LiFiMAX AP) and the receiver (LiFiMAX dongle) would be located in the shielded chamber, while the other equipment would be located outside the chamber. A modified measurement site was created in the chamber where the antenna and the axis of the LiFiMAX devices under test were at the foci of a hypothetical ellipse that formed an area free of obstructions that would affect the EM field propagation. The distance between the receiving antenna and the axis of the devices under test was 3 m. The antenna was 1 m above the floor and allowed measurements in both horizontal and vertical polarization. The modified measurement workstation was based on the requirements of EN 55016–2-3 (Specification of apparatus and methods for measuring radio interference and immunity - Part 2–3: Methods for measuring interference and immunity - Measurement of radiated interference) for the measurement of radiated EM emissions.

A combined TRILOG, Schwarzbeck type, VULB 9163 antenna was used for the band (30–3000 MHz). The emission spectra were measured and compared with the applicable emission limits using an R&S EMI measuring receiver, ESR 7 (100 Hz - 7 GHz), which was connected to a computer for collecting the measured data. The shielded chamber with the measurement antenna and the LiFi devices under test are shown in Fig. 4. Figure 4a presents a view of the LiFiMAX device layout in the shielded chamber, Fig. 4b presents a view of the measurement site from the measurement antenna); the computers, the PoE power system and the router were located outside the shielded chamber (Fig. 5).

Fig. 4. Location of the LiFiMAX devices in the shielded chamber: (a) view from the side of the LiFiMAX devices under test, (b) view from the side of the measurement antenna.

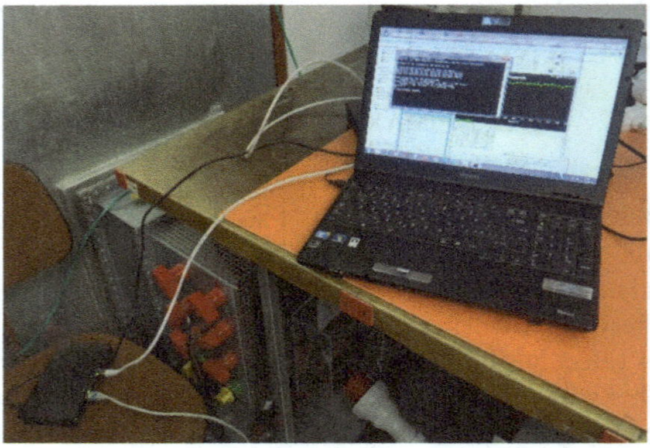

Fig. 5. Positioning the computer, PoE and router outside the shielded chamber.

The LiFiMAX transmitter was connected via a DJM Electronic Ethernet EMI filter, supporting a PoE power system, which is located on the wall of the shielded booth. The filter is designed for high-speed Ethernet/UTP communication links. There is an optical link between the input and output sides of the filter, which ensures sufficient separation of interfering signals (over 100 dB in the 10 kHz - 10 GHz band).

The LiFiMAX receiving device was powered via a shielded USB cable that was threaded into the cabin through universal opto-cable loop holes. In principle, it is a subcritical waveguide, i.e. a waveguide operating below the lowest cutoff frequency. This principle is used in EM shielding technology to create the passages needed to bring optical cables, mechanical controls inside the shielded area, for ventilation within it, etc. For a waveguide length corresponding to its diameter, a damping of about 20 dB can be achieved. In our case, the waveguide was five times longer than its diameter and thus achieved a damping of approx. 100 dB. However, for correct operation, the strict condition that no metallic element, e.g. a cable, must pass through the waveguide must be fulfilled, because it will carry the interfering signal into the shielded area.

We violated this principle because we ran a shielded metallic USB communication cable to the LiFiMAX receiving device. Therefore, EM clamps were attached on both sides of the communication cable near the wall of the shielded cabin (Fig. 6). The EM clamps EM 101 that we used consist of a binding and a separating part. However, we were only interested in its separating part, which we used as a broadband filter that significantly prevented the interfering signal from penetrating the shielded cable jacket into the cabin. In addition, the EM101 impedance separates the auxiliary device from the LiFiMAX device itself [10].

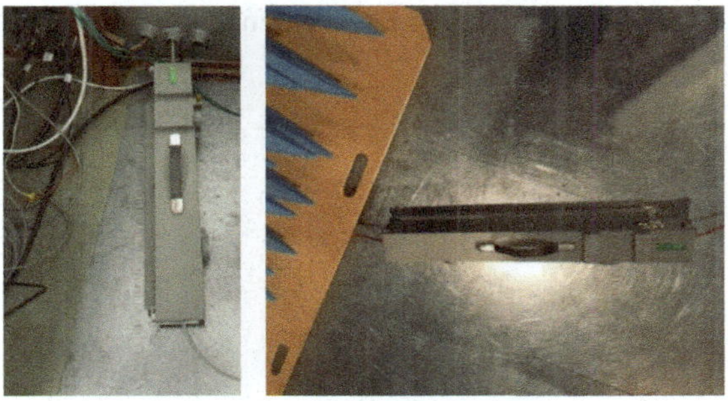

Fig. 6. Additional filtering of cables in the chamber and outside the chamber implemented by EM clamps.

These measures thus ensured that no high amplitude interfering signal penetrated into the shielded cabin. A similar arrangement of the measurement workstation has a wider application in EMC tests that are to be performed in a shielded cabin and the equipment under test needs to be connected to ancillary equipment (which has significant EM emissions, or insufficient immunity to EM fields) and is connected to the equipment under test by complicated cabling.

4 Measured Results and Their Interpretation

The objective of the emission measurements was to determine the isolated radiated emissions of the LiFiMAX device. The radiated emission measurements were performed with both vertical and horizontal polarization of the receiving antenna. The R&S EMI measurement receiver, ESR 7, has a built-in fast measurement function - FFT scan (TDEMI method), which allowed us to directly use a quasi-peak (QP) detector in the measurement, for which upper limits are defined in the standard. If the measuring receiver does not have such a function, it is necessary to use a peak detector due to the high time consumption and to measure only the peaks that are close to the limit using the quasi-peak detector. Emission measurements were performed in the following operating modes:

- measuring the background reference values of a shadow chamber in which only the LiFiMAX AP and the LiFiMAX dongle were placed and which were connected to the PC1 and PC2 computers but were switched off (in order to obtain the emission background generated by the communication cabling routed to the cabin),

- Emission measurement of LiFiMAX devices without active communication (connection of LiFiMAX AP and LiFiMAX dongle to the network without data transfer),
- measure the interference emissions of LiFiMAX devices during data transmission.

Initially, we measured the EM field penetration into the shielded cabin via the communication cabling to get a picture of the external influences that penetrate into the shielded cabin with the current configuration of the measurement workstation. In Fig. 7 it can be seen that only the signal from the VHF radio transmitters and the GSM band penetrates into the shielded cabin, but the measured values are so low that they are not meaningful from the point of view of the following measurement.

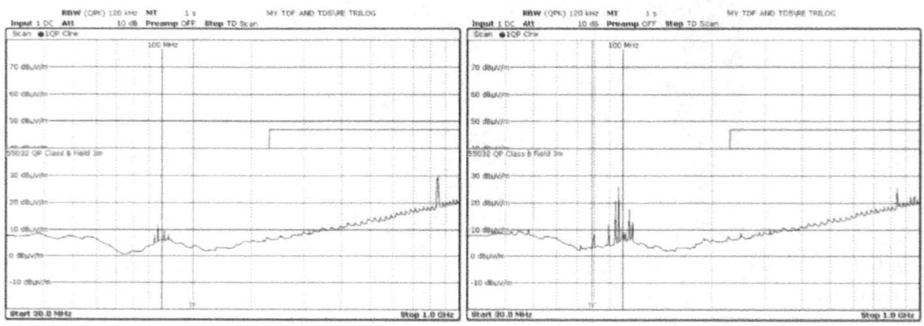

Fig. 7. EMI spectrum with disconnected PoE in horizontal (left) and vertical polarization (right).

We then operate the LiFiMAX devices, PoE and endpoint computers, but the configuration worked without active data transfer (Fig. 8). It can be seen in the measured spectra that the EM field interference values for the LifiMAX devices for both horizontal and vertical polarization of the measurement antenna did not exceed the upper limit - the red dashed line (40 and 47 dBμV/m), which defines the level of EM field interference and which must not be exceeded for devices intended for residential, commercial and light industrial environments. At low frequencies, the spectrum shows a band of broad-spectrum, high-intensity interference, which is dominant at horizontal polarization.

As higher and higher frequencies are used in information and communication systems and technologies, we examined the behavior of the tested LiFiMAX devices at higher frequencies in the band (1–3 GHz) in more detail. The threshold level was not exceeded even at frequencies above 1 GHz, as documented in Fig. 9. For measurements in the band above 1 GHz, both a peak detector and a mid-range detector with respective upper limits are prescribed [2]. What we see in the figure is the noise of the measuring apparatus with the antenna factor of the measuring antenna and the transmission of the whole measurement chain included. The contribution of the interference from the equipment under test was negligible for both polarizations.

In the EM field spectra (Fig. 10) measured during data transmission, the EM field interference values were higher and thus much closer to the upper limit. The broad-spectrum interference at low frequencies has approximately the same level, but there are

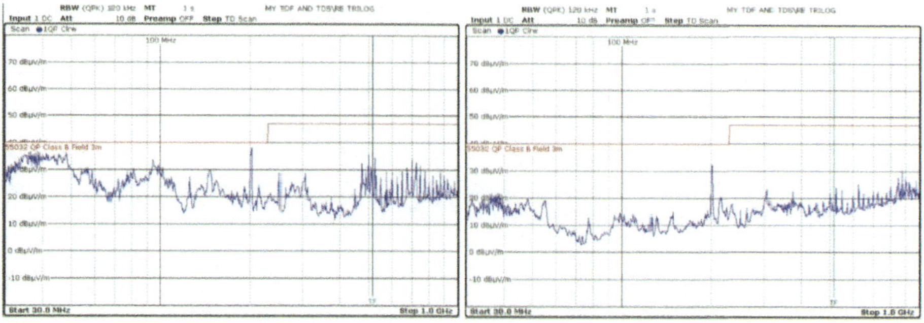

Fig. 8. EMI spectrum of LiFiMAX without operation in horizontal (left) and vertical polarization (right).

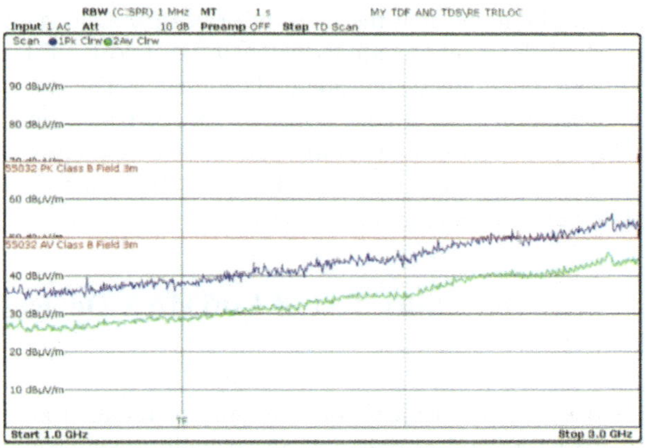

Fig. 9. EMI spectrum of the (1–3) GHz band in non-operational mode with both vertical and horizontal polarization.

an added number of peaks of the radiated EM field at higher frequencies, corresponding to the ongoing data traffic. Even with horizontal polarization, narrow peaks appeared sporadically in the (120–150) MHz band, just above the limit, as shown in Fig. 10.

Subsequently, we also measured the emissions (in the mode during data operation) in the 1 GHz - 3 GHz band and for both polarizations, but the emissions again did not significantly exceed the noise level of the measurement apparatus and therefore we do not show them (they are virtually identical to the spectrum shown in Fig. 9). The EM field peaks at the beginning of that band, where they would be expected to continue smoothly beyond 1, GHz, were not recorded because the apparatus noise for the peak detector and the RBW = 1 MHz intermediate frequency bandwidth in the upper band was higher (over 35 dBµV/m) - the measurement system was not sensitive enough to record them. The mean detector, in turn, filters the peak values of the measured field according to the time constants defined in the relevant technical standard.

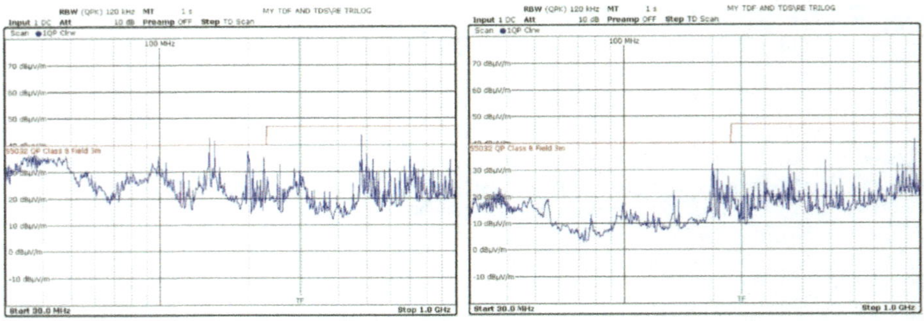

Fig. 10. EMI spectrum of LiFiMAX in operation and with horizontal (left) and vertical polarization (right).

However, it should be noted that the main priority of the emissions measurement assembly was to establish communications between the two LiFiMAX modules and isolate all support equipment outside the shielded cabin area. The recommendations for the design of the measurement set-up according to EN 55016-2-3 and EN 55032 were not strictly followed; based on this one-off measurement, it is not possible to state - due to the 2 disputed EM field peaks in the 120 MHz–150 MHz band - whether or not the equipment meets the requirements of the standard.

5 Electromagnetic Immunity Test of LiFiMAX Devices Against RF EM Field

We have also tested LiFiMAX devices for immunity to radiated high-frequency EM fields according to EN 61000-4-3. Within the chosen environment in which the electrical equipment in question is expected to operate, the standards define the maximum limit of radiation emissions and the level of immunity to which the severity of the test corresponds (in this case, to the EM field). There is a safety margin between the emission limits and the immunity test level so that equipment that is designed in accordance with the standards is compatible with each other in the relevant environment.

The test according to EN 61000-4-3 was carried out in a modified semi-anechoic shielded chamber (modification to suppress reflections from the reflective floor). The test was performed in the frequency band 80 MHz - 1 GHz with a frequency step of 1%, the test level was 3 V/m, corresponding to immunity designed for residential, commercial and light industrial environments. The workstation is shown in Fig. 11.

The test used amplitude modulation with AM parameters of 80%/ 1 kHz and the polarization of the transmitting antenna was again both vertical and horizontal. The devices were placed on a non-conductive table in the same way as for the emission measurements; both the LiFiMAX AP and the LiFiMAX dongle were placed on the table so that the light communication flux was perpendicular to the direction of EM wave propagation (Fig. 11).

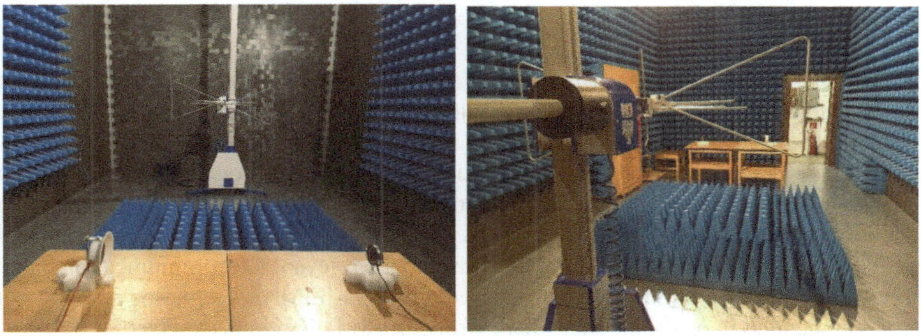

Fig. 11. Shielding chamber arrangement for electromagnet tic field immunity testing: (a) View from the LiFiMAX side, (b) View from the radiating antenna side.

During the ongoing immunity test, the transmission rate of the LiFiMAX devices on the receiving side of the communication system was recorded. A photograph of the computer display can be seen in Fig. 12, which shows a 15- to 40-s trace of the measured transmission rates when the transmit antenna is vertically polarized and an interfering radio frequency EM field is applied. The above results showed that the interference in the frequency range of 80 MHz to 1 GHz had a negative effect on the tested LiFiMAX devices - the data rate was significantly reduced (from the average value of the data rate of 93 Mbit/s to values below 50 Mbit/s), moreover, relatively often there were also several seconds of data transmission failures. Even worse was the situation with horizontal polarization, when, in addition to a significant reduction in data rates, a permanent drop in the connection occurred shortly after the measurement was started, so that the data rate was zero.

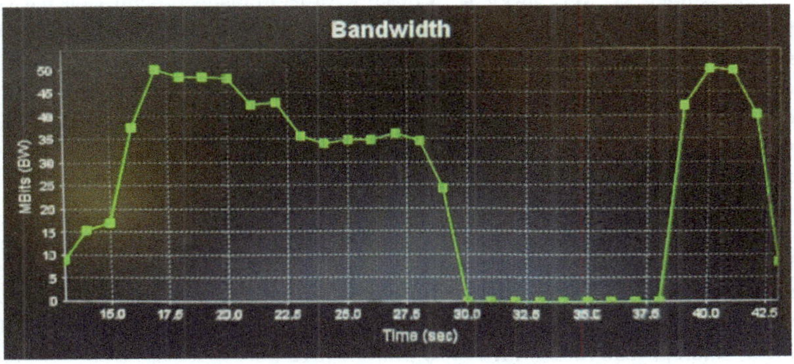

Fig. 12. Fragment of the LiFiMAX device baud rate time log during the endurance test.

Based on the results obtained when testing LiFiMAX devices, it can be concluded that the immunity to radiated high frequency EM field according to EN 61000-4-3 is very poor.

6 Conclusion

In this paper, a method for measuring EM radiation emissions and a method for testing the EM field immunity of a LiFiMAX wireless optical communication device were presented, both of which have wider applications in EMC test practice.

The measured emissions of the subject LiFiMAX device had low values. The fact that the transmission medium used for data transmission was light was a positive contributing factor. If it were a transmission using RF EM waves, the very concept of the system would predispose it to higher emissions, and not only in the relevant telecommunication band. In terms of EM field immunity, this representative of LiFiMAX technology cannot yet be recommended in environments where stronger EM radiation is expected.

Acknowledgements. This publication was based on work from the COST Action NEWFOCUS CA19111, supported by COST (European Cooperation in Science and Technology) and it was supported by the Scientific Grant Agency VEGA 1/0045/21.

References

1. Directive 2014/53/EU of the European Parliament and ot the Council of 16 April 2014 on the harmonisation of the laws of the Member States relating to the making available on the market of radio equipment. Off. J. EU. **L 153**, 62–105. Accessed 22 May 2014
2. European standard EN55032:2015 Electromagnetic compatibility of multimedia equipment – Emission requirements, CENELEC 2015
3. Drak, T., Halgoš, J., Hallon, J.: Support for assessing IoT devices in terms of electromagnetic emissions. In: Halgoš, J. (ed.) Electro-mechanical systems application in industry 2022: Zliechov, Slovakia. August 23–26, 2022. 1. vyd. Vydavateľstvo Spektrum STU, Bratislava (2022) ISBN 978-80-227-5280-0
4. Hallon, J., Bittera, M., Harťanský, R., Kováč, K.: Electromagnetic emission measurement of devices with GSM interface. In: Annals of DAAAM International Week 2016. DAAAM International, Vienna (2016) ISBN 978-3-902734-13-6
5. https://www.lifitn.com/blog/lifimaxreview
6. Róka, R., Baroňák, I., Orgoň, M., Mokráň, M., Hecl, D., Köppl, M., Letenay, J., Počarovský, Š.: Optické bezdrôtové technológie. In: Vybrané kapitoly (Optical Wireless Technologies: Selected Chapters), p. 187. Vydavateľstvo Spektrum STU (Spektrum Publishing House), Bratislava (2022) ISBN 978-80-227-5196-4
7. Köppl, M., Počarovský, Š., Orgoň, M., Hecl, D., Letenay, J.: Wireless optical data transmission via LiFiMAX technology with security using elliptical curves. In: Šilhavý, R., Šilhavý, P. (eds.) Wireless Optical Data Transmission via LiFiMAX Technology with Security Using Elliptical Curves. Software Engineering Methods in Systems and Network Systems, pp. 304–315. Springer Nature, Cham (2024) ISBN 978-3-031-54812-3
8. Hecl, D., Köppl, M., Szitkey, V., Grolmus, A., Hozlár, M., Róka, R., Baroňák, I., Počarovský, Š., Orgoň, M., Blažek, P.: Measurement of transmission characteristics of LiFiMAX. In: Šilhavý, R., Šilhavý, P. (eds.) Networks and Systems in Cybernetics, pp. 290–299. Springer, Cham (2023) ISBN 978-3-031-35316-1

9. David, H., Jakub, L., Martin, K., Andrej, G., Matúš, H., Ivan, B.: Monitoring and visualizing coverage and transmission speeds of LiFiMAX. In: Šilhavý, R. (ed.) Software Engineering Methods in Systems and Network Systems, pp. 333–345. Springer Nature, Cham (2024). ISBN 978-3-031-54812-3, ISSN 2367-3370)
10. Stroka, J.: On the proper use of the injection clamps in EN 61000-4-6 test. In: 15th International Wroclaw Symposium on EMC, EMC 2000, Wroclaw, pp. 374–378 (2000)

Data Acquisition and Preparation Algorithm for Manipulator Control in Musculoskeletal Rehabilitation System

Andrey Volkov, Kirill Patutin, Artem Obukhov(✉), and Aleksandra Nazarova

Tambov State Technical University, 392000 Tambov, Russian Federation
obuhov.art@gmail.com

Abstract. The article discusses methods of musculoskeletal rehabilitation, particularly the use of mechanotherapy with robotic manipulators. The functioning of manipulator control algorithms poses a complex technical challenge, requiring meticulous preparation and preliminary simulation modeling. The possibility of testing manipulator algorithms in musculoskeletal rehabilitation systems using motion capture data and Blender 3D simulation animation is explored. An algorithm for obtaining and processing information from virtual reality trackers for subsequent use in manipulator control systems has been developed. The algorithm addresses tasks such as tracker position determination on the human body, data validation, trajectory visualization, simulation modeling on a digital human body model, and necessary post-processing for transfer to the control system. Research results allow for improving the rehabilitation process through analysis and correction of patient movements.

Keywords: Musculoskeletal Rehabilitation · Manipulators · Control System · Motion Capture · Data Processing

1 Introduction

Human musculoskeletal disorders require comprehensive treatment involving several stages [1–3].

- Diagnosis and assessment: necessary treatment methods are determined, and an individual rehabilitation plan is developed.
- Pharmacological treatment: Depending on the diagnosis, pharmacotherapy may be prescribed to alleviate pain and inflammation, improve blood flow, and strengthen bones.
- Surgical intervention: for injuries or bone or joint deformities.
- Physiotherapy and therapeutic exercises: important for conducting physiotherapeutic procedures, massages, therapeutic exercises, muscle strengthening, and joint mobility restoration.

R. Silhavy and P. Silhavy (Eds.): CoMeSySo 2024, LNNS 1489, pp. 136–147, 2025.
https://doi.org/10.1007/978-3-031-96798-6_13

- Therapeutic exercise and rehabilitation: performing a specific set of exercises to improve joint mobility, muscle strength, and posture correction.

Modern technologies, such as electromechanical trainers, are utilized to achieve positive results in therapeutic exercise [4–6]. These trainers are part of mechanotherapy methods [7, 8], providing mechanical tissue stimulation to enhance their condition. They are individually tailored to each patient's needs, enhancing muscle strength, flexibility, and endurance. The use of mechanical devices in musculoskeletal rehabilitation is increasingly common and effective for recovery after injury or surgery.

Robotic manipulator-based trainers are prevalent, incorporating advanced technologies like artificial intelligence [9] and virtual reality [10]. These manipulator-based systems are robotic devices with a specialized interface that mechanically restricts the distal part of a person's limb. Their application reduces rehabilitation time and minimizes complications.

Various technological approaches to organizing rehabilitation processes are present in the market. One such approach involves the use of interactive trainers, which combine the advantages of robotic manipulators and virtual reality to enhance motor skills through immersion in a virtual environment, where exercises for balance, coordination, and mobility training are performed [10–14].

Control algorithms for mechanical devices must take into account natural movements, unique characteristics, and individual needs of each person to prevent incorrect activity. The trainer should ensure the absolute safety of the patient, including calibration and teaching the correct exercise technique before starting sessions.

Obtaining data for calibration and verifying its correctness can be a complex task. The trainer should correct a person's gait, not exacerbate the diagnosis or even harm the patient's health.

Let's consider the possibility of applying technologies in the rehabilitation of patients with musculoskeletal disorders of the lower extremities, specifically, methods of data acquisition and verification for controlling rehabilitation devices.

During the development of a rehabilitation system using interactive trainers, a method based on obtaining data on a person's walking movement using motion capture technology was proposed [15]. After collecting and validating the data, it needs to be processed for use in the robotic system. To ensure the safety and effectiveness of treatment, data quality and system operation must be monitored. This includes checking the accuracy and reliability of the manipulator's operation, as well as evaluating treatment outcomes.

To test the effectiveness of algorithms in robotic manipulators, various testing methods need to be applied. For example, one can start with conducting laboratory tests using simulation systems and then proceed to clinical studies on patients to evaluate treatment outcomes. It is crucial to first test algorithms in controlled laboratory conditions before moving on to real-world testing.

2 Methods

2.1 General Algorithm Description

Let's take a closer look at the algorithm for collecting and preparing data for controlling manipulators. The algorithm flowchart is presented in Fig. 1.

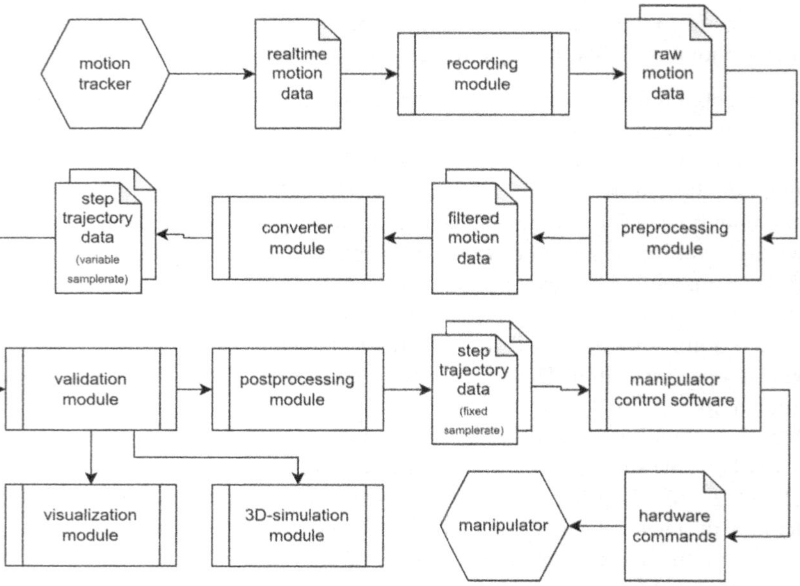

Fig. 1. Block diagram of the data collection and processing for controlling manipulators.

2.2 Data Collection and Recording

To obtain data on the movement of a person's lower extremities during walking, trackers from the HTC Vive virtual reality system [16] were used as they provide high positioning accuracy compared to other available systems [17]. Information about the position is received from the trackers in the form of absolute coordinates in a room-based reference system. Six trackers were used (three on each leg): on the waist, mid-thigh, and mid-shin. A controlled treadmill was used to monitor the person's walking speed, capable of maintaining the belt speed at a specified level. The arrangement of trackers on a person's legs is shown in Fig. 2a, and the experimental setup process is shown in Fig. 2b.

Data from all six trackers were recorded using software based on the OpenVR library [18] and saved in CSV format. For each tracker, position data in X, Y, and Z coordinates, as well as rotation angles, are saved. Data from each tracker are recorded in a separate file, and all files for one experiment are collected in a common directory.

2.3 Data Preprocessing

The next step in the data preparation process was data preprocessing. In the first step of preprocessing, the coordinate system rotation angle was corrected, which arises due to the non-ideal placement of virtual reality equipment in the room. The coordinate system was rotated so that the Z-axis points straight ahead (in the direction of the person's movement) and the Y-axis points strictly upwards. The rotation was performed through a standard matrix transformation in homogeneous coordinates. To determine

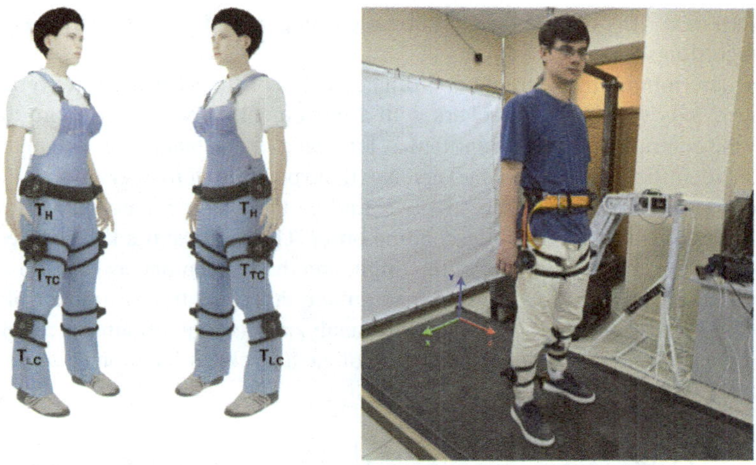

Fig. 2. Tracker arrangement scheme for recording walking data (a), experimental data recording process (b)

the necessary rotation angle, four trackers were placed at the corners of the treadmill, and their coordinates were saved in a separate file.

The next step was synchronizing all recorded data by time. Since values from each tracker were recorded independently, the measurement time for each tracker may differ slightly. To accurately plot the trajectory of movement, it is necessary to synchronize all tracker positions by time. Synchronization consists of three stages: preliminary trimming, constructing a common time scale, and interpolating data. In the preliminary trimming stage, a time interval was determined for which data was available for each tracker (considering the latest start time and earliest end time of recording). Next, the average sampling rate for all trackers was determined, and a new time scale was built for the specified interval. Then, data from each tracker is linearly interpolated based on timestamps from the new time scale.

After synchronization, the data was filtered to remove sharp value fluctuations that occur due to minor positioning errors of the trackers. Depending on the quality of the raw data, different filters may be used, such as a low-pass filter, an exponential weighted moving average (EWMA) filter, or a single burst filter that detects and eliminates sudden spikes in measurements.

2.4 Data Transformation into Trajectories for Manipulators

After preprocessing, the data was passed to the coordinate transformation module, which also consists of several stages. First, the assignment of each tracker to the right or left leg is determined. To do this, the boundaries of movement along the X-axis for each tracker were identified, and the intersection of ranges for each pair of trackers was determined. If the intersection is more than 75%, it is considered that the trackers belong to the same leg. The left leg contains trackers with X positions smaller, and the right leg contains those with larger values (see Fig. 2a). This algorithm works well on recordings containing a minimum of 8–10 s of data (several complete steps). A different method is used for shorter recordings.

After determining the extreme coordinate values of the trackers, the average point along the X-axis was found. Trackers with a movement dispersion of more than 75% less than the central point were identified as left leg trackers, and those with at least 75% more were identified as right leg trackers. Next, the position of trackers at specific points was determined for each leg. This involves finding the average Y coordinate value for each tracker and sorting them in descending order. The topmost tracker was designated as the waist point, the middle one as the thigh, and the bottom one as the shin.

This described algorithm allows for automatic determination of tracker placement on a person's body, which is convenient for analyzing a large amount of experimental data as it does not require manual indication of each tracker's location (Fig. 3).

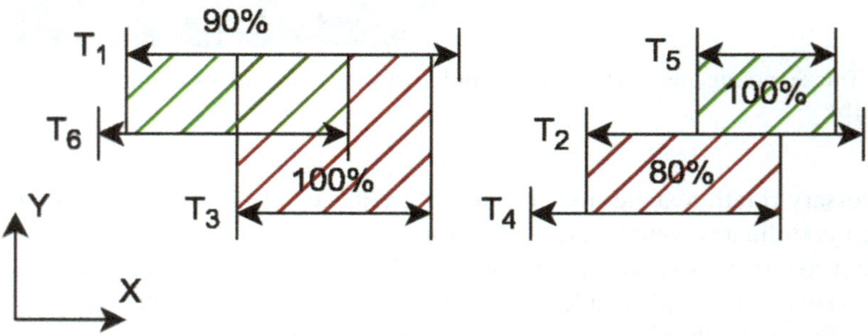

Fig. 3. Tracker segmentation scheme.

Next, trajectories of legs are transformed from linear tracker movements into angular deviations of bones. This process is considered for one leg (see Fig. 4). The tracker at the waist is denoted as T_H, the one at the middle of the thigh as T_{TC}, and the one at the middle of the shin as T_{LC}. In the first step of transformation, two missing points need to be constructed: the knee joint point and the ankle joint point, denoted as T_K and T_F respectively. The position of point T_K s determined by the formula:

$$T_K = T_H + k_H * (T_{TC} - T_H), \quad (1)$$

where k_H is the coefficient for correcting the position of the tracker on the thigh. When placing the tracker exactly in the middle of the thigh, this coefficient is equal to 2; when the tracker is shifted below the middle, it decreases, and when shifted above, it increases. Similarly, the position of the knee joint is determined:

$$T_K = T_H + k_H * (T_{TC} - T_H), \quad (2)$$

where k_L is the coefficient for correcting the position of the tracker on the shin. Next, angles of deviation of the thigh and shin are calculated using the formulas:

$$A_H = atan\left(\frac{y_{TC} - y_H}{x_{TC} - x_H}\right) + \pi/2, \quad (3)$$

$$A_H = atan\left(\frac{y_{LC} - y_K}{x_{LC} - x_K}\right) + \pi/2 - A_H, \tag{4}$$

where A_H is the thigh deviation angle, A_L is the shin deviation angle. In this case, the downward direction is taken as the zero angle value for A_H, and the direction of vector A_L angle value for $T_K T_{LC}$. Next, angular velocities of bone movement are calculated as the derivative of angle change over time:

$$v_H = \left(\frac{A_{H,i} - A_{H,i-1}}{tau_i - tau_{i-1}}\right), \tag{5}$$

$$v_L = \left(\frac{A_{L,i} - A_{L,i-1}}{tau_i - tau_{i-1}}\right), \tag{6}$$

where *tau* is the time stamp for this measurement in seconds. Similar operations are carried out for the second leg. After that, the obtained trajectory data is combined into a single array. Next, it is necessary to find the two closest points in the resulting trajectory to identify one complete step. Closeness of points is determined by the MSE metric, which is calculated cumulatively for all four angles, and then the last point with the lowest MSE value is selected from those found.

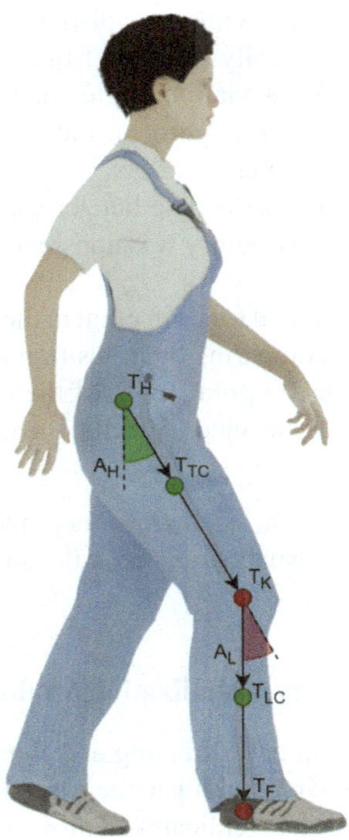

Fig. 4. Coordinate transformation.

The final operation in the coordinate transformation process is synchronizing movements between the extreme points of the trajectory. Since the command for the manipulator involves simultaneous movement of all links, it is necessary for the movement time of each link to coincide with others at this part of the trajectory (for all transitions from point i to point $i + 1$). This condition is met for all transitions except for the transition from the last point of the trajectory to the first (the step trajectory is closed), as during trimming in the previous step, movement speeds at the first and last moments may not have matched. To synchronize this last movement, it is necessary to recalculate the velocity of link movement when transitioning to the first point of the trajectory. The calculation is performed using formula (4), where the average transition time in the trajectory is substituted for *tau*. The resulting transformed trajectory is saved to a file in CSV format.

2.5 Data Validation

Despite considering all necessary conditions during trajectory creation and ensuring that the output trajectory after coordinate transformation should be correct and free of critical errors, all trajectory files undergo a validation stage before being loaded into manipulators. During this stage, several types of errors were checked for:

- The presence of points with identical values for all angles, occurring consecutively. Such points indicate transitions with zero displacement in the trajectory. These duplicated points were automatically removed from the trajectory.
- The alignment of manipulator deviation angles at the first and last points of the trajectory was verified. In case of alignment, the last point was removed from the trajectory as it results in zero displacement.
- Points with movement at a non-zero angle but zero speed were identified. If at least one such point was found, the trajectory was considered incorrect as such movement was impossible.
- The movement time was calculated for each point in the trajectory and for each manipulator segment, followed by comparing the transition time values of all segments for a single transition from point i to point $i + 1$. If the transition time differs by more than 1 ms, the movement was deemed asynchronous, and the entire trajectory was discarded as incorrect.

In addition to automated validation, each trajectory undergoes visual inspection for correctness using a trajectory visualization module and a 3D simulation modeling system.

2.6 Development of a Trajectory Visualization Module

The simplest way to animate data is by creating a skeletal animation based on it. This animation is built on the principle of creating a tree-like structure where each subsequent point is linked to the previous one and replicates all movements and rotations considering the hierarchy.

The architecture of the module for visualizing trajectories is shown in Fig. 5.

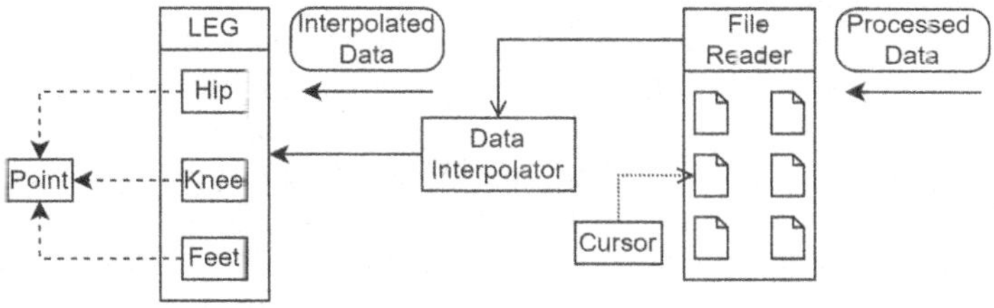

Fig. 5. Architecture of the module for visualizing recorded trajectories.

This module includes two components. The first component works with processed data provided from the previous stage after validation. It consists of two classes.

The first class, FileReader, was used to manage working with files containing data. This class gathers all data from the necessary directory, selects a specific file from this set for further processing. Since the module was developed as an independent application, data validation described earlier is repeated upon loading. If the data from the file fails the check, information about incorrect data (see Fig. 6a) and the reason for the error is displayed on the application screen. The Cursor class is responsible for navigating between files and selecting a specific file from the set, storing the current position and returning it.

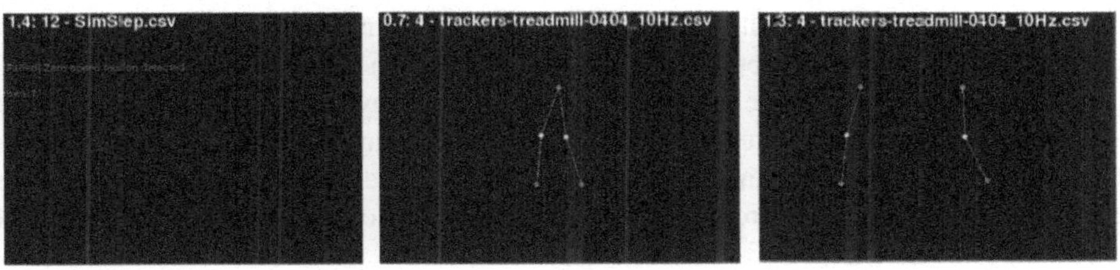

(a) (b) (c)

Fig. 6. Visualization module interface: error message (a), animation from one point (b), animation from different points (c).

The second class in this component is DataInterpolator. It regulates the transition speed from one data row to the next. This regulation follows a specific algorithm. Each record contains information about the angle position that needs to be adopted and the speed at which this angle value should be reached. Based on these values, the time it takes for a point to change its angle at the given speed is calculated. Subsequently, interpolation of values from the current angle to the new one is performed, considering the animation frame rate, which depends on the application's frame rate. This algorithm is repeated for all original data, and the resulting trajectories are sent to the visualization subsystem.

The second component of the application is the visualization system, which includes several classes, but the main ones are two classes. This component performs the rendering of skeletal animation in the user interface.

Let's consider the main classes in this subsystem. The Point class represents a part of a leg that needs to be animated. This class contains information about the point and all its possible positions, and it calculates the current position of the point on the user's screen. This calculation is necessary for skeletal animation, where all points are interconnected based on a human skeleton, implying that when one point rotates by a certain angle, the points linked to it will move to new positions. Additionally, this class handles drawing the point object in the interface based on its current coordinates.

The Leg class represents the connection between points, and it sequentially changes the position of each point, leading to leg movement animation. The animation is built according to the following algorithm: For each leg being animated, a Leg entity containing three Point points is created. Then, interpolated data is distributed among the points of each leg. Subsequently, the Leg class instructs its points to move in accordance with points from the trajectory sequentially. This process gradually reproduces the movement animation of each leg (see Fig. 3b, c). An additional function was added for drawing legs from one point (Fig. 3b) and from different points (see Fig. 3c), which can be useful in cases of identical trajectories. It should be noted that after drawing all data, playback restarts from the beginning, and to monitor this, the current time elapsed since the start of the current animation cycle is displayed in the top left corner in seconds.

2.7 Simulation Modeling

Simulating human movements can be done using programs that allow creating a model of the human body and manipulating its movements. The most common and professional programs for this purpose are the commercial program AnyBody Modeling System and AnyBody Managed Model Repositories by AnyBody Technology, and the open platform OpenSim [19, 20]. With their help, it is possible to create the most accurate model of human movement. These programs are quite complex to master and require special skills, such as knowledge of anatomy.

Therefore, for initial research, it was decided to use an open-source software product with a more simplified interface and user interaction mechanism, namely Blender 3D. This program allows loading source data (positions of points, their velocities, etc.) for animation from external sources, provides tools for animation and visualization, and includes a system for simulating the physical properties of objects. In Blender, one of the key functions is creating an armature that enables animating models. An armature in Blender can be seen as similar to a real skeleton's armature, consisting of multiple bones. These bones can be moved, and everything they are attached to or associated with will move and deform in a similar manner.

Initially, the designed simulator model needs to be animated according to the motion capture suit data. For this purpose, a set of bones is created for different parts of the model. To simplify this process, a dummy object was added to which all bones were linked to create inverse kinematics. Then, the motion of the motion capture suit's skeleton is linked to the new bones of the simulator model.

The human leg model was divided into three parts: pelvis, thigh, and shin. For the thigh and shin, as well as for the attachments of the digital simulator model, rigid body physics settings were used to simulate the impact of a real simulator on the human body. An auxiliary polygon link was used for visualizing connections between these parts. The edge vertices of these polygons were attached using hooks. The Hook modifier is used to deform a mesh, curve, or lattice using another object, achieving a soft tissue effect.

The selection of values for animating rigid body physics during the experiment was subjective and did not lead to the expected results. It was found that the obtained motion data transferred to the digital simulator model could pose a danger to humans; for example, virtual leg parts moved unnaturally around the knee joint (Fig. 7).

Fig. 7. Demonstration of visualization of the operation of a virtual simulator.

The use of interactive simulators in rehabilitation represents a promising approach to improving patients' motor skills. However, special attention must be paid to selecting data to control the system, ensuring accuracy and safety throughout the process. The conducted experiments revealed shortcomings in the current approach and identified directions for future system improvements.

2.8 Data Post-processing

The final stage before using the obtained trajectories in the musculoskeletal rehabilitation system is post-processing. At this stage, data resampling is performed by reducing the sampling frequency to a fixed value. The trajectory obtained in the previous stages has a sampling frequency tied to the polling rate of the equipment used. For a tracking system based on HTC Vive trackers using six trackers, this frequency is around 67.5 Hz. Such a high frequency is not suitable for use in the musculoskeletal rehabilitation system due to the high inertia of the mechanical part. The system simply cannot accurately perform a large number of very short movements. To ensure correct operation of the manipulators, the sampling frequency needs to be reduced to 5–10 Hz. This transformation is carried out as follows (for each segment of the manipulator):

1. In the original file, the difference between angles for each trajectory point is calculated. For the first point, this value is the difference between the first and last angles.
2. The time between trajectory points is determined based on the relationship between the calculated difference and angular velocity.
3. A column of absolute time is generated for each trajectory point.
4. A new time scale is formed from the initial to the final value of the absolute time scale, but with a new set sampling frequency.
5. The interpolation of the angle values of the trajectory is performed considering the new time scale.
6. New angular velocity values are calculated considering the new time intervals between points and the new trajectory point values. The first velocity value is calculated considering the transition from the last trajectory point to the first.
7. The calculated new trajectory points and angular velocity values are written to a file.

The resulting trajectory files are loaded into software for controlling the musculoskeletal rehabilitation system and sent to the control microcontroller.

3 Conclusion

The work considers the processes of analysis and processing of information for the musculoskeletal rehabilitation system. Due to the high safety requirements of such systems, it is necessary to ensure the correctness of the subsystems interacting with humans - robotic manipulators. To solve this task, an algorithm for obtaining and preparing data for controlling the manipulators is proposed.

Within the algorithm, all the main data processing procedures are considered. First, the process of collecting and preprocessing information from virtual reality trackers is studied. The obtained trajectories are transformed into manipulator motion trajectories, including segment angle values and angular velocities. Modules for data validation and visualization are implemented. Simulation modeling in Blender is conducted to verify the correctness of the leg movement trajectory fixed in the musculoskeletal rehabilitation system from a safety perspective. Finally, necessary software tools for post-processing information to change the sampling frequency are implemented. The research results allow improving the rehabilitation process through the analysis and correction of patient movements.

The article was carried out with the financial support of the Ministry of Science and Higher Education of the Russian Federation within the framework of the project “Development of medical VR training systems for musculoskeletal rehabilitation with biofeedback” (№124032800018-5).

References

1. Kovacevic, K.: Treatment of musculoskeletal disorders-scientific review of the literature. Int. J. Biomed. Res. Pract. **2**(1), 1–5 (2022)

2. Cento, A.S., Leigheb, M., Caretti, G., Penna, F.: Exercise and exercise mimetics for the treatment of musculoskeletal disorders. Curr. Osteoporos. Rep. **20**(5), 249–259 (2022)
3. Bhoi, D., Jain, D., Garg, R., Iyengar, K.P., Hoda, W., Vaishya, R., Jain, V.K.: Complementary and Alternative Modalities (CAM) for pain management in musculoskeletal diseases (MSDs). J. Clin. Orthop. Trauma. **18**, 171–180 (2021)
4. Cherchenko, O.V., Sheptunov, S.A.: Restoration of limb mobility using robotic mechanotherapy. In: Robots for the Rehabilitation of Disabled People Requirements in the Medium Term (2022)
5. Park, Y.L., Chen, B.R., Pérez-Arancibia, N.O., Young, D., Stirling, L., Wood, R.J., Nagpal, R.: Design and control of a bio-inspired soft wearable robotic device for ankle–foot rehabilitation. Bioinspir. Biomim. **9**(1), 016007 (2014)
6. Kuryłc, P., Cyganiuk, J., Frankovský, P., Trebuňová, M.: Wrist rehabilitation with manipulator to perform passive and active exercises. Lék. Tech. **49**(2), 52–57 (2019)
7. Khan, K.M., Scott, A.: Mechanotherapy: how physical therapists' prescription of exercise promotes tissue repair. Br. J. Sports Med. **43**(4), 247–252 (2009)
8. Van Pelt, D.W., Lawrence, M.M., Miller, B.F., Butterfield, T.A., Dupont-Versteegden, E.E.: Massage as a mechanotherapy for skeletal muscle. Exerc. Sport Sci. Rev **49**(2), 107–114 (2021)
9. Ai, Q., Liu, Z., Meng, W., Liu, Q., Xie, S.Q.: Machine learning in robot assisted upper limb rehabilitation: a focused review. In: IEEE Transactions on Cognitive and Developmental Systems (2021)
10. Chaplin, E., Karatzios, C., Benaim, C.: Clinical applications of virtual reality in musculoskeletal rehabilitation: a scoping review. Healthcare. **11**(24), 3178 (2023)
11. Brien, M., Sveistrup, H.: An intensive virtual reality program improves functional balance and mobility of adolescents with cerebral palsy. Pediatr. Phys. Ther. **23**(3), 258–266 (2011)
12. Yang, W.C., Wang, H.K., Wu, R.M., Lo, C.S., Lin, K.H.: Home-based virtual reality balance training and conventional balance training in Parkinson's disease: a randomized controlled trial. J. Formos. Med. Assoc. **115**(9), 734–743 (2016)
13. Costa, M.T.S., Vieira, L.P., de Oliveira Barbosa, E., Oliveira, L.M., Maillot, P., Vaghetti, C.A.O., Monteiro-Junior, R.S.: Virtual reality-based exercise with exergames as medicine in different contexts: a short review. Clin. Pract. Epidemiol. Ment. Health. **15**, 15 (2019)
14. Kiani, S., Rezaei, I., Abasi, S., Zakerabasali, S., Yazdani, A.: Technical aspects of virtual augmented reality-based rehabilitation systems for musculoskeletal disorders of the lower limbs: a systematic review. BMC Musculoskelet. Disord. **24**(1), 4 (2023)
15. Choo, C.Z.Y., Chow, J.Y., Komar, J.: Validation of the perception neuron system for full-body motion capture. PLoS One. **17**(1), e0262730 (2022)
16. Ikbal, M.S., Ramadoss, V., Zoppi, M.: Dynamic pose tracking performance evaluation of HTC Vive virtual reality system. IEEE Access. **9**, 3798–3815 (2020)
17. Arkhipov, A., Lopatko, A., Prokudina, N., Averin, Y., Dedov, D.: Comparison of approaches to human motion tracking in virtual reality. In: Nazarov, D., Sagintaeva, S. (eds) AIP Conference Proceedings, AIP Publishing, 2948(1), (2023)
18. Valve Corporation: OpenVR. https://github.com/ValveSoftware/openvr. Accessed 26 Jun 2024
19. Damsgaard, M., Rasmussen, J., Christensen, S.T., Surma, E., De Zee, M.: Analysis of musculoskeletal systems in the AnyBody modeling system. Simul. Model Pract. Theory. **14**(8), 1100–1111 (2006)
20. Delp, S.L., Anderson, F.C., Arnold, A.S., Loan, P., Habib, A., John, C.T., Thelen, D.G.: OpenSim: open-source software to create and analyze dynamic simulations of movement. IEEE Trans. Biomed. Eng. **54**(11), 1940–1950 (2007)

Using Markov Chain Model to Forecasting of the Agricultural Industry Development

Seyit Kerimkhulle[1(✉)], Gulnara Turtkarayeva[2], Rashid Mussaibekov[2], Nazgul Ospanova[2], Shakharzat Kuttykozhayeva[2], and Alibek Adalbek[1]

[1] Information Systems Department, L.N. Gumilyov Eurasian National University, 2, Satpayev Street, 010008 Astana, Kazakhstan
{kerimkul_sye,adalbek_a}@enu.kz

[2] Physics, Mathematics and Informatics Sciences Department, Sh. Ualikhanov Kokshetau University, 76, Abay Street, 020000 Kokshetau, Kazakhstan
{gtyrtkaraeva,rmusaibekov,nospanova,skyttykojaeva}@shokan.edu.kz

Abstract. This study examines the impact of agricultural investment on the economic growth of different sectors using a first- and fifth-order multivariate Markov chain model. Four interrelated industry series 1995–2018 were considered. The studied data were structured from the Organization for Economic Co-operation and Development (OECD) sources. Specifically, the relationship between agriculture, hunting, forestry, food products, beverages and tobacco, chemical and chemical products, wholesale and retail trade, and the rest of the economy was studied. The first- and fifth-order multivariable model of the Markov chain was proposed to predict the impact of investments allocated to agriculture on the above-mentioned industries according to the economic situation in Kazakhstan over the 1995–2018 years. The accuracy of the model was tested and analyzed. In particular, we obtain the prediction probability decreases from 0.99 to 0.58, and the accuracy of both models is 99%. Indeed, these research is a short-term probabilistic forecasting model that policymakers can use to evaluate and implement initiatives to improve the situation in a country. And the results show a significant impact of agricultural investment on sectoral economic growth, with significant variation across sectors.

Keywords: Forecasting · Markov Chain · Economic Growth · Probability · Inter-industry Relations · Gross Fixed Capital Formation

1 Introduction

Understanding the impact of agricultural investment on sectoral economic growth is vital for policymakers and stakeholders to optimize resource allocation and promote sustainable development. This study employs a multivariate Markov chain model to analyze the interaction between agricultural investment and industry output.

Accurate forecasting of economic growth in agriculture is crucial for effective policy and strategic planning. Traditional methods often overlook the complex dynamics of agricultural economies. This review investigates the potential of a multivariate Markov

R. Silhavy and P. Silhavy (Eds.): CoMeSySo 2024, LNNS 1489, pp. 148–158, 2025.
https://doi.org/10.1007/978-3-031-96798-6_14

chain model to forecast agricultural economic development. By examining research from agricultural economics, regional economics, and macroeconomics, we outline the theoretical foundations and practical applications of this approach.

In our study, we considered the relevance of these indicators to Kazakhstan. Analysing the works of previous authors on the same topic, we compared and discussed the importance of these indicators in the economic development of the country.

Analysis of Age Structure in Agriculture: Mazouch and Krejčí (2016) analyze the age structure of regional fixed capital in Czech agriculture [1]. Using a Markov chain model, they examine the age characteristics of machinery and equipment, offering insights into the modernization process. Their conclusions underscore the importance of monitoring the age of capital as a crucial indicator for forecasting future developments in the agricultural sector.

Structural Change in West German Agriculture: Hüttel and Margarian (2009) investigate structural changes in the West German agricultural sector with an empirical Markov chain model [2]. They demonstrate the significance of strategic interactions between farms in shaping regional patterns of farm size structure. Their research emphasizes the need to consider historical farm size structures to accurately predict the sector's development.

Pension Funds and Economic Growth in South Africa: Sarusi and Kapingura (2021) study the impact of pension funds on investment and economic growth in South Africa [3]. Although not directly focused on agriculture, their findings highlight the importance of efficient capital allocation for economic development. Their insights contribute to understanding the broader economic context essential for agricultural sector forecasts.

Driving Forces of China's Growth: Ding and Knight (2011) analyze the factors behind China's rapid growth, emphasizing the significance of physical and human capital formation [4]. Their results highlight the importance of investment in innovation and education for economic growth, influencing predictions of agricultural sector development in developing economies.

Comparative Evaluation of Economic Development Planning Models: Kooros and Badeaux (2007) provide a comparative analysis of economic development planning models, including the Markov chain model [5]. Their review illustrates the applicability of Markov chain models for short-term market forecasting and resource allocation planning, highlighting its potential for predicting agricultural sector development.

Trade Facilitation R&D and Productivity: Parameswaran (2009) investigates the impact of trade facilitation and R&D exposure on productivity in the Indian manufacturing industry [6]. Although not directly related to agriculture, the findings on the productivity-enhancing effects of R&D investment provide valuable insights into the factors influencing productivity and development in the agricultural sector.

Structural Change and Growth in China: Ding and Knight (2009) explore the role of structural change in China's growth, emphasizing the significance of openness, institutional reform, and sectoral shifts [7]. Their results illustrate how structural improvements have contributed to China's rapid growth, offering insights for predicting agricultural sector development in transitioning economies.

Sectoral Interrelationships and Economic Development: This article aims to understand the interrelationships and correlations between sectors such as agriculture, hunting, forestry, food products, beverages and tobacco, chemical and chemical products, wholesale and retail trade, and the overall economy. We examine the impact of investments in these sectors on the country's economic development using high-order Markov chain models to determine indices related to economic growth.

2 Methods

Forecasting is an important issue for governments and central banks to assess a country's economic development. Many methods used, ranging from classical time series models, linear regression, neural networks, probabilistic forecasting and hybrid models.

A methodological approach to structural change in regional economies: Thakur (2011) presents a methodological approach to understanding structural change in regional economies [8]. Their focus on fundamental economic structure and intersectoral linkages provides a basis for analysing regional economic dynamics and predicting the development of the agricultural sector in broader regional contexts.

Financial development and economic growth in sub-Saharan Africa: Ustarz and Fanta (2021) examine the impact of financial development on economic growth in sub-Saharan Africa from a sectoral perspective [9]. Their findings highlight the importance of financial development for the growth of the industrial sector, which has implications for predicting the development of the agrarian sector in the context of broader economic transformations.

Managing risk and resilience during a pandemic: Pamukar et al. (2023) propose an integrated interval programming and I/O knowledge model for managing risk and resilience after the COVID-19 pandemic [10]. Although not specific to agriculture, their study highlights the importance of scenario analysis for predicting the interrelationship of economic sectors and sectoral development in the face of uncertainty.

Gjika et al. (2019) used MCMC (Markov Chain Monte Carlo) simulations to model energy consumption and we use these simulations to predict different scenarios [11].

To construct a high-order multivariate Markov chain model, we consider a categorical time series d with states k and then find the intermediate and internal transition probability matrices with steps1, 2, ..., q.

The high-order (q-th order) multivariable Markov chain model is expressed as follows [12]:

$$x_{i+1}^{(j)} = \sum_{l=1}^{d} \sum_{h=1}^{q} \alpha_{jl}^{(h)} Q_h^{(jl)} x_{i-h+1}^{(l)}, \tag{1}$$

where $j = 1, 2, \ldots, d$ and $i = q - 1, q, \ldots$.

Initial probability distribution $x_0^j, x_1^j, \ldots, x_{q-1}^j$ for $j = 1, 2, \ldots, d$. Parameters α_{jl}^h satisfy the following two conditions:

$$\begin{aligned} &\text{(i)} \quad \alpha_{jl}^h \geq 0, l \leq j, l \leq d; l \leq h \leq q; \\ &\text{(ii)} \sum_{l=1}^{d} \sum_{h=1}^{q} \alpha_{jl}^h = 1, j = 1, 2, \ldots, d. \end{aligned} \tag{2}$$

The probability distribution of the time series $x_{p+1}^{(j)}$ depends on the weighted average $\alpha_{jl}^{(h)} Q_h^{(jl)} x_{p-h+1}^{(l)}$. Where $Q_h^{(jl)}$ is the h-step probability transition matrix with h steps, whose elements are the transition probabilities with h is the steps from states in sequence l at time $t = p - h + 1$ to states in sequence j-th at time $t = p + 1$. From Eq. (1) we denote [12]:

$$X_t^{(j)} = \left(\left(x_t^{(j)}\right)^T, \left(x_{t-1}^{(j)}\right)^T, \ldots, \left(x_{t-q}^{(j)}\right)^T\right)^T, \; j = 1, 2, \ldots, d.$$

Thus, we get the following matrix:

$$X_{r+1} = \begin{pmatrix} X_{t+1}^{(1)} \\ X_{t+1}^{(2)} \\ \vdots \\ X_{t+1}^{(d)} \end{pmatrix} = \begin{pmatrix} c^{(11)} & c^{(12)} & \cdots & c^{(1d)} \\ c^{(21)} & c^{(22)} & \cdots & c^{(2d)} \\ \cdots & \cdots & \cdots & \cdots \\ c^{(d1)} & c^{(d2)} & \cdots & c^{(d)} \end{pmatrix} \begin{pmatrix} X_t^{(1)} \\ X_t^{(2)} \\ \vdots \\ X_t^{(d)} \end{pmatrix} = RX_t, \qquad (3)$$

where

$$C^{(ii)} = \begin{pmatrix} \alpha_{ii}^{(1)} Q_1^{(ii)} & \alpha_{ii}^{(2)} Q_2^{(ii)} & \cdots & \alpha_{ii}^{(d-1)} Q_{d-1}^{(ii)} & \alpha_{ii}^{(d)} Q_d^{(ii)} \\ 1 & 0 & \cdots & 0 & 0 \\ 0 & 1 & \cdots & 0 & 0 \\ \vdots & \vdots & \vdots & \vdots & \vdots \\ 0 & 0 & \cdots & 1 & 0 \end{pmatrix}_{kd \times kd}.$$

If $i \neq j$, then

$$C^{(ij)} = \begin{pmatrix} \alpha_{ij}^{(1)} Q_1^{(ij)} & \alpha_{ij}^{(2)} Q_2^{(ij)} & \cdots & \alpha_{ij}^{(d-1)} Q_{d-1}^{(ij)} & \alpha_{ii}^{(d)} Q_d^{(ij)} \\ 0 & 0 & \cdots & 0 & 0 \\ 0 & 0 & \cdots & 0 & 0 \\ \vdots & \vdots & \vdots & \vdots & \vdots \\ 0 & 0 & \cdots & 0 & 0 \end{pmatrix}_{kd \times kd}.$$

The sum of the elements in each column of the matrices $Q_n^{(jk)}$ is equal to one.

For $q = 1$, model (1) is a first-order multivariate Markov chain model first proposed by Ching et al. (2002) [13]. Their model considers d categorical time series and the forecast model is constructed considering only one-step sub- and sub-transition probability matrices. Using the same symbols as above, the shape of the model is:

$$x_j^{(t+1)} = \sum_{i=1}^{d} \alpha_{ji} Q^{(ji)} x_i^{(t)}, \qquad (4)$$

For $j = 1, 2, \ldots, d$ and t =0, 1, …, where $Q^{(ji)}$ is a one-step matrix of probability of transition from i-th row to j-th row, and $x_i^{(t)}$- Probability distribution of the i-th time series at time t. Parameters α_{ji} satisfy conditions (2).

The matrix obtained from model (4) is:

$$X_{t+1} = \begin{pmatrix} X_{t+1}^{(1)} \\ X_{t+1}^{(2)} \\ \vdots \\ X_{t+1}^{(d)} \end{pmatrix} = \begin{pmatrix} \alpha_{11}Q^{11} & \alpha_{12}Q^{12} & \dots & \alpha_{1d}Q^{1d} \\ \alpha_{21}Q^{21} & \alpha_{22}Q^{22} & \dots & \alpha_{2d}Q^{2d} \\ \vdots & \vdots & \dots & \vdots \\ \alpha_{d1}Q^{11} & \alpha_{d1}Q^{11} & \cdots & \alpha_{dd}Q^{dd} \end{pmatrix} \begin{pmatrix} X_t^{(1)} \\ X_t^{(2)} \\ \vdots \\ X_t^{(d)} \end{pmatrix}.$$

For $d = 1$, model (1) is a high-order univariate Markov chain model developed by Ching et al. (2003) [14]. In their model, they considered only one time series with categorical data and 1, 2, ..., q step probability transition matrices. So, the probability distribution of the state at time $t + 1$ is $t, t - 1, \dots, t - q + 1$, where q is the probability distribution of the state at the order of the model. The form of the Markov chain model, using the above notation, is as follows:

$$x_{t+1} = \sum_{i=1}^{q} \alpha_i Q_i x_{t-i+1}, t = q - 1, q, \dots. \tag{5}$$

where Q_i is the i-th step probability transition matrix, and parameter x_{t+1} is the probability state distribution of the time series at time $t + 1$. The parameters α_i, $i = 1, 2, \dots$, q satisfy conditions (2).

3 Results

In this section, we considered the interrelationship of agriculture, food products, chemical products, wholesale and retail trade in four time series. The time series were obtained from the OECD and World Bank website [15–18], collected annually from 1995 to 2018. When working with the model, we considered data from 1995 to 2018 as a training set, and data from 1997 to 2018 as a testing set.

When working with a forecast model, the goal is not only to predict future values, but also to capture the patterns of the time series in question. Therefore, it is important to analyze time series by time and their correlation.

In this study, we have built first-order and fifth-order multivariable Markov chain models, taking into account the above economic indicators. A comparative analysis is made between these models and other classical forecasting methods. Based on accurate estimation of the models, the best ones are considered for forecasting.

A Markov chain forecasting model can work for categorical time series data. In the first step, we need to discretize our data in a reasonable and reasonable way to fit the conditions of the model. We used the first quartile and the mean to determine the breakpoints of the intervals, observing the cumulative result of the time series and their behavior. Discretization is carried out by classifying each value according to the interval in which it falls. Thus, values below the first quartile are indexed as "1", treating them as Low values; values falling between the first quartile and the mean are indexed as "2". Average values are considered and values above average are indexed as "3", considered as High values. Table 1 shows the intervals for each time series.

Table 1. Intervals considered for discretization of time series (level of confidence intervals).

	h =	LOW(1)	MEDIUM(2)	HIGH(3)
Agriculture, hunting, forestry	0	<0.1120	[0.1120; 0.4402]	0.4402
Agriculture, hunting, forestry	2	<0.2876	[0.2876; 0.7228]	0.7228
Food products, beverages and tobacco	1	<0.9663	[0.9663; 1.6239]	1.6239
Chemical and chemical products	1	<−0.4264	[−0.4264; −0.1851]	−0.1851
Chemical and chemical products	2	<0.6570	[0.6570; 2.6505]	2.6505
Wholesale and retail trade	0	<−0.1335	[−0.1335; 1.6242]	1.6242
Wholesale and retail trade	1	−0.7712	[−0.7712; −0.3522]	−0.3522
Rest of economics	0	<0.3888	[0.3888; 0.8410]	0.8410
Rest of economics	1	<0.1593	[0.1593; 0.2266]	0.2266

Table 2. Predicted reliability probability for one year of the agricultural industries.

Parameters	Y
$\hat{\beta}_{(i)}$	3.3530***
	(0.570)
Constant	2.0691***
	(0.599)
Numb of obs.	24
R-square	0.6112

Note. The dependent variable Y is the internal production of the agricultural industries. In parentheses are standard errors. *, **, *** – estimate is significant at 10%, 5%, 1% level.

We fitted a first-order multivariate Markov chain model to our data using statistical software packages. The final model is shown in Table 2.

As a result, we took domestic production in the period from 1995 to 2018. As a factor indicator, we took the investment of agriculture in the period from 1995 to 2018, and the Y-intersection was the internal production of the agricultural sector (Gross value added).

If we increase the current agricultural investment by 1 billion U.S. dollars, then the output of the agricultural sector will increase by 3353.0 million U.S. dollars, where the standard error (deviation) of the coefficient of the calculated value (3353.0) is equal to 570.2 million U.S. dollars, and the t-statistic is 5.881 is equal to That is, the confidence probability of the calculated value (3353.0) is equal to 0.001%. The calculated value (3353.0) belongs to the medium confidence interval (MEDIUM(2)). This means the deviation between the production of the agricultural sector and the consumers [19–20].

Using a fifth-order multivariate Markov chain model, we obtain following results. If we increase the investment of agriculture in the current year by 1 billion U.S. dollars, then the output of the agricultural sector will increase by 505.2 million U.S. dollars, where the standard error (deviation) of the coefficient of the calculated value (505.2) is equal to 99.9 million U.S. dollars, and the t-statistic equal to 5.881. That is, the reliability probability of the calculated value (3.353) is equal to 0.003%. The calculated value (505.2) belongs to the medium confidence interval (MEDIUM(2)). If we increase the investment of agriculture in the current year by 1 billion U.S. dollars, then the output of the agricultural sector will increase by 1295.1 million U.S. dollars, where the standard error (deviation) of the coefficient of the calculated value (1295.1) is equal to 150.9 million U.S. dollars, and the t-statistic It is equal to 8.58. That is, the reliability probability of the calculated value (1295.1) is equal to 0.0001%. The calculated value (1295.1) belongs to the medium confidence interval (MEDIUM(2)).

Now let's take a look at the food items. If we increase the agricultural investment of the current year by 1 billion U.S. dollars, then the volume of food products after one year will increase by 305.7 million U.S. dollars, where the standard error (deviation) of the coefficient of the calculated value (305.7) is equal to 55.4 million U.S. dollars, and the t-statistic is equal to 5.52. That is, the reliability probability of the calculated value (305.7) is equal to 0.0001%. The calculated value (305.7) falls within the MEDIUM(2) confidence interval.

Now let's review the chemicals and chemical products. If we increase the investment of agriculture in the current year by 1 billion U.S. dollars, then the volume of chemical products after one year will increase by 1653.8 million U.S. dollars, where the standard error (deviation) of the coefficient of the calculated value (1653.8) is equal to 457.5 million U.S. dollars, and t -statistics equal to 3.61. That is, the reliability probability of the calculated value (1653.8) is equal to 0.003%. The calculated value (1653.8) belongs to the medium confidence interval (MEDIUM(2)). If we increase the investment of agriculture in the current year by 1 billion US dollars, then the volume of chemical products after two years will increase by 745.4 million US dollars, where the standard error (deviation) of the coefficient of the calculated value (745.4) is equal to 403.4 million US dollars, and t -statistics equal to 1.85. That is, the reliability probability of the calculated value (745.4) is equal to 0.089%. The calculated value (745.4) belongs to the medium confidence interval (MEDIUM(2)).

Now let's take a look at the wholesale and retail sales. If we increase the investment of agriculture for the current year by 1 billion U.S. dollars, then the volume of wholesale and retail trade for the current year will increase by 561.7 million US dollars, where the standard error (deviation) of the coefficient of the calculated value (561.7) is equal to 96.2 million U.S. dollars, and The t-statistic is 5.84. The confidence probability of the calculated value (561.7) is equal to 0.0001%. The calculated value (561.7) falls within the MEDIUM(2) confidence interval. If we increase the investment of agriculture in the current year by 1 billion US dollars, then the volume of wholesale and retail trade after one year will increase by 614.9 million US dollars, where the standard error (deviation) of the coefficient of the calculated value (614.9) is equal to 103.8 million US dollars, and the t-statistic is equal to 5.92. That is, the reliability probability of the calculated value (614.9) is equal to 0.0001%. The calculated value (614.9) belongs to the medium confidence (MEDIUM(2)) interval [19–20].

Now let's take a look at the rest of the economy. If we increase the investment of agriculture for the current year by 1 billion U.S. dollars, then the volume of the remaining sectors of the economy for the current year will increase by 193.0 million U.S. dollars, where the standard error (deviation) of the coefficient of the calculated value (193.0) is equal to 15.4 million U.S. dollars, and t-statistics equal to 12.50. That is, the confidence probability of the calculated value (193.0) is equal to 0.0001%. The calculated value (193.0) falls within the MEDIUM(2) confidence interval.

4 Discussion

In this study, focusing on the period between 1995 and 2018, we studied the relationship between agricultural investments and production volume in various sectors of the economy. Using the first and fifth order multivariable model of Markov chain, we analyzed the effect of agricultural investments on sectoral results in different times. The results provide valuable insight into the dynamics of agricultural investments and their ripple effects across various sectors of the economy.

The results show a significant positive correlation between agricultural investment and industrial production in most sectors. Interestingly, the billion U.S. dollar increase in investment in agriculture led to a significant increase in production in the sector, demonstrating the crucial role of agriculture as a driver of economic growth and development.

First, the results obtained in the agricultural sector itself show that an increase of one billion U.S. dollars in agricultural investments this year will lead to a significant increase in agricultural production. In particular, our analysis shows that such an increase will lead to an increase in agricultural production by 505.2 million U.S. dollars in the same year. In addition, this effect will persist and even increase over time, with projected increases in agricultural output after one and two years of 1295.1 million and 3353.0 million U.S. dollars, respectively.

Second, beyond the agricultural sector, our study shows the spillover effects of increased investment in agriculture on other sectors of the economy. For example, in the food industry, a billion U.S. dollars increase in investment in agriculture leads to a significant increase in food production with an estimated increase of 305.7 million U.S. dollars after one year. Similarly, investments in agriculture have a positive effect on the production of chemicals and chemical products, increasing by approximately 1653.8 million U.S. dollars after one year and 745.4 million U.S. dollars after two years. In addition, the study determines the impact of agricultural investments on wholesale and retail trade. A billion U.S. dollars increase in investment in agriculture will lead to significant growth in wholesale and retail trade, with gains of 561.7 million and 614.9 million U.S. dollars in the current year and a year later, respectively. It shows the interrelationship of agricultural activities with trade and commerce, highlighting the role of agriculture in stimulating consumer demand and stimulating economic transactions.

Third, our analysis extends to the rest of the economy, showing the broader impact of agricultural investment beyond specific industries. The results show that an increase in investment in agriculture by one billion U.S. dollars will lead to an increase in the volume of production in the remaining sectors of the economy by 193.0 million U.S. dollars in the current year [19–20]. This highlights the role of agriculture as a catalyst for overall economic growth and diversification.

The observed relationships between agricultural investment and industry production are statistically significant, as evidenced by the high t-statistics and low probability of error. These results indicate the reliability of the estimated coefficients.

In conclusion, this study provides compelling evidence that agricultural investment has a significant impact on sectoral output in various segments of the economy. The results highlight the importance of prioritizing agricultural development as a key strategy to stimulate economic growth, increase food security and promote sustainable development.

5 Conclusion

Using empirical analysis and theoretical frameworks from various studies, policy and literature review integrating ideas from various disciplines including agricultural economics, regional economics and macroeconomics, this study concludes by gaining a comprehensive understanding of the drivers and dynamics of the agricultural economy, facilitating more accurate and informed forecasting and decision-making processes, which demonstrated the potential of multivariate Markov chain models for predicting sectoral economic development in agriculture.

In the future, the integration of advanced modeling methods, in particular the multidimensional first- and fifth-order Markov chain model, makes it possible to provide predictive data on the impact of investments directed in agriculture on the agricultural industry, on the food, beverage and tobacco industry, chemical and chemical products, wholesale and retail trade depending on the economic situation in Kazakhstan in 1995–2018.

Thus, the accuracy of the model was tested and analyzed. In particular, we find that the prediction probability decreases from 0.99 to 0.58, and the accuracy of both models is 99%. Indeed, these studies provide a short-term probabilistic forecasting model that policymakers can use to evaluate and implement initiatives to improve the situation in the country. The results show a significant impact of agricultural investment on sectoral economic growth, with significant differences between sectors.

Acknowledgments. This research was funded by the Committee of Science of the Ministry of Science and Higher Education of the Republic of Kazakhstan (Grant No. AP26196337).

References

1. Mazouch, P., Krejčí, I.: The analysis of the age structure of regional fixed capital in the agriculture. Agris online Pap. Econ. Inform. **8**(2), 89–101 (2016). https://doi.org/10.7160/aol.2016.080207
2. Huettel, S., Margarian, A.: Structural change in the West German agricultural sector. Agric. Econ. **40**, 759–772 (2009). https://doi.org/10.1111/j.1574-0862.2009.00413.x

3. Sanusi, K.A., Kapingura, F.M.: Pension funds as fuel for overall investment level and economic growth: an empirical insight from South African economy. Cogent Bus. Manag. **8**(1), 1–11 (2021). https://doi.org/10.1080/23311975.2021.1935661
4. Ding, S., Knight, J.: Why has China grown so fast? The role of physical and human capital formation. Oxf. Bull. Econ. Stat. **73**(2), 141–174 (2011). https://doi.org/10.1111/j.1468-0084.2010.00625.x
5. Kooros, S.K., Badeaux, L.M.: Economic development planning models: a comparative assessment. Int. Res. J. Financ. Econ. **7**(7), 120–139 (2007) http://www.eurojournals.com/finance.htm
6. Parameswaran, M.: International trade, R&D spillovers and productivity: evidence from Indian manufacturing industry. J. Dev. Stud. **45**(8), 1249–1265 (2009). https://doi.org/10.1080/00220380902862911
7. Ding, S., Knight, J.: Why has China grown so fast? The role of structural change. In: Role of Structural Change, Proceedings of the German Development Economics Conference, Frankfurt, vol. 7. Verein für Socialpolitik, Ausschuss für Entwicklungsländer, Göttingen (2009) https://hdl.handle.net/10419/39945
8. Thakur, S.: Fundamental economic structure and structural change in regional economies: a methodological approach Rég. Dév. **33**(4), 9–38 (2011) https://ideas.repec.org/a/tou/journl/v33y2011p9-38.html
9. Ustarz, Y., Fanta, A.B.: Financial development and economic growth in sub-Saharan Africa: a sectoral perspective. Cogent. Econ. Financ. **9**(1), 1–21 (2021) https://doi.org/10.1080/23322039.2021.1934976
10. Pamucar, D., Sarkar, B.D., Shardeo, V., Soni, T.K., Dwivedi, A : An integrated interval programming and input–output knowledge model for risk and resiliency management. Decis. Anal. J. **9**, 100317 (2023) https://www.siom.in/faculty-publication-list
11. Gjika, E., Ferrja, S., Basha, L., Kamberi, A.: Probabilistic forecasting of electricity demand using Markov chain and statistical distribution. In: International Symposium on Forecasting, Greece, Thessaloniki (2019) https://forecasters.org/events/symposium-on-forecasting/
12. Gjika, E., Basha, L., Allka, X., Ferrja, A.: Predicting the Albanian economic development using multivariate Markov chain model. In: 11th International Scientific Conference "Business and Management 2020" (2020). https://doi.org/10.3846/bm.2020.581
13. Ching, W.K., Fung, E.S., Ng, M.K.: A multivariate Markov chain model for categorical data sequences and its applications in demand predictions. IMA J. Manag. Math. **13**(3), 187–199 (2002). https://doi.org/10.1093/imaman/13.3.187
14. Ching, W.K., et al.: A higher-order markov model for the newsboy's problem. J. Oper. Res. Soc. **54**(3), 291–298 (2003). https://doi.org/10.1057/palgrave.jors.2601491
15. IOTs data: Agriculture, hunting, forestry (current U.S. dollars). In: Input-Output Tables (IOTs). Organization for Economic Co-operation and Development Stat (2021a) https://stats.oecd.org/Index.aspx?DataSetCode=IOTS_2021. Accessed 15 April 2024
16. IOTs data: Food products, beverages and tobacco (current U.S. dollars). In: Input-Output Tables (IOTs). Organization for Economic Co-operation and Development Stat (2021b) https://stats.oecd.org/Index.aspx?DataSetCode=IOTS_2021. Accessed 15 Apr 2024
17. IOTs data: Chemical and chemical products (current U.S. dollars). In: Input-Output Tables (IOTs). Organization for Economic Co-operation and Development Stat (2021c) https://stats.oecd.org/Index.aspx?DataSetCode=IOTS_2021. Accessed 15 Apr 2024
18. IOTs data: Wholesale and retail trade (current U.S. dollars). In: Input-Output Tables (IOTs). Organization for Economic Co-operation and Development Stat (2021d) https://stats.oecd.org/Index.aspx?DataSetCode=IOTS_2021. Accessed 15 Apr 2024

19. Kerimkhulle, S., Aitkozha, Z., Saliyeva, A., Kerimkulov, Z., Adalbek, A., Taberkhan, R.: Agriculture, hunting, forestry, and fishing industry of Kazakhstan economy: input-output analysis. In: Proceedings of the Computational Methods in Systems and Software, pp. 786–797. Springer International Publishing, Cham (2023). https://doi.org/10.1007/978-3-031-21435-6_68
20. Kerimkhulle, S., Saliyeva, A., Makhazhanova, U., Kerimkulov, Z., Adalbek, A., Taberkhan, R.: The input-output analysis for the wholesale and retail trade industry of the Kazakhstan statistics. In: E3S Web of Conferences, vol. 376, p. 05023. EDP Sciences (2023). https://doi.org/10.1051/e3sconf/202337605023

The Effect of Different Diacritic Settings on the Accuracy of Sentiment Analysis in Czech and Slovak Slavic Languages Upon Transfer-Learning

Ján Mojžiš(✉)

Institute of Informatics, Slovak Academy of Sciences, Bratislava, Slovakia
jan.mojzis@savba.sk

Abstract. Sentiment analysis is still actively developed scientific field of research. Vast majority of the publications (dedicated to Czech or Slovak languages) focus on the evaluation of models during training phase and perform no transfer-learning evaluations. Diacritical aspects of Slavic languages are also omitted during data preprocessing. In this paper we focus on the transfer-learning evaluation, evaluating a pre-trained sentiment analysis model in another domain. For this purpose, we selected two popular pre-trained state-of-the-art deep-learning sentiment analysis models Czert and SlovakBERT along with our traditional machine learning Maximum Entropy model. Upon evaluation we put attention on different diacritical settings of the datasets. We notice that the current two deep-learning models actually differ in the reaction to different diacritic settings. We did find robustness but also slight variations in F scores. We theorize that more attention should be paid in diacritical setting in datasets used to train models for sentiment analysis. Also, more transfer-learning evaluations of the sentiment analysis models should be performed in order to test for the robustness of the models.

Keywords: sentiment analysis · czech language · slovak language · diacritics · transfer-learning · maximum entropy

1 Introduction

As data volume continues to grow significantly, it's crucial to find efficient ways to summarize it for further processing. The literature suggests various methods for this, such as data aggregation, pattern recognition, and data mining [1–4]. This paper focuses on sentiment analysis of textual data, specifically examining how different diacritic settings affect sentiment analysis accuracy. Companies can utilize customer feedback from social media, where customers share their opinions on products or services, providing valuable feedback. This feedback helps companies understand current trends and determine which products or services are preferred by their customers.

Apart from traditional machine learning models used in sentiment analysis tasks such as Stochastic Gradient Descent (SGD) and Maximum Entropy (MaxEnt) a new models

R. Silhavy and P. Silhavy (Eds.): CoMeSySo 2024, LNNS 1489, pp. 159–168, 2025.
https://doi.org/10.1007/978-3-031-96798-6_15

based on deep-learning emerged. In this paper we consider two major modern state-of-the art deep-learning models Czert [5] for the Slovak Language and SlovakBERT [6] for the Czech language. Before we delve further into the details of sentiment analysis accuracy using different diacritical settings, we provide more information on the complexity of the Czech and Slovak Slavic languages.

Unlike many Western European languages, Slovak and Czech are among the languages with a high degree of inflection. In the Slovak language, there are four grammatical genders: animate masculine, inanimate masculine, feminine, and neuter. Each gender has four basic declension paradigms. Regarding morphology, six morphological cases are recognized. For more details, please refer to [19].

Before a meaningful sentiment analysis of texts in Czech and Slovak languages can be performed, data preprocessing is necessary. This preprocessing consists of several methods or techniques commonly recognized as natural language processing (NLP) methods. Their aim is to improve the accuracy of sentiment analysis and information retrieval by ensuring consistency in word representation.

Lemmatization is the process of reducing words to their base or root form, known as a "lemma." This helps in understanding the core meaning of the words by converting different forms of a word into a single item (feature space reduction). For example, the words "otcovi" (to the father), "otcom" (father), and "otec" (father) can all be reduced to the lemma "otec." Lemmatization always returns lexicographically correct words. Stemming, another NLP method, works similarly to lemmatization regarding feature space reduction. However, unlike lemmatization, stemming does not necessarily return a lexicographically correct word. For example, the words "otcom" and "otcovia" become "otco," which is not a lexicographically valid word in Slovak.

As we evaluate the impact of different diacritical settings on the resulting accuracy of sentiment analysis, we address several details regarding diacritics. Together, Croatia, Czechia, Slovakia, Slovenia, and Poland represent a population of over 60 million, which is 13.46% of the European Union. In all these languages, diacritics are used. Therefore, diacritics should be taken into account in NLP methods, as they further increase the word feature space.

In Czech and Slovak languages, the characters are extended by special diacritical marks. For example, ž (letter z with a caron), ô (letter o with a circumflex), ó (letter o with an acute accent) and ä (letter a with an umlaut - do not confuse with a diaeresis, which has a separate function). In general, there is not a problem with diacritics regarding NLP methods (diacritics is supported). The problem is the possible violation of consistency in a word representation due to authors of texts. The common case are texts crawled from the Web (which is the main source of many sentiment analysis datasets). In many cases the words are not written in lexicographically correct way. Aside frequent typos, we can find the same words written with and without diacritics. If not handled correctly (by preprocessing), these words can create two separate records in feature space. Thus, the aforementioned consistency violation. We address such type of preprocessing with diacritics reconstruction in Sect. 3.

In this paper we focus on the impact of diacritics in the accuracy of sentiment analysis for Czech and Slovak languages. Particularly we evaluate several settings involving the presence or absence of diacritic and the reconstruction of diacritic with the objective to evaluate the impact on sentiment analysis accuracy. For the evaluation we consider popular nowadays modern deep-learning sentiment analysis models for Czech and Slovak languages CZert [5] and SlovakBert [6], respectively and also traditional machine learning model represented by MaxEnt (implemented as Logistic in Weka[1]). Dataset is represented by one of traditional published benchmark sentiment analysis dataset made by Habernal et al. [8]. We did not consider the the RobeCzech model in our evaluations as this model is a general-purpose fill-mask model and no dedicated RobeCzech sentiment analysis model was published.

The specialty of this paper is to evaluate the models in transfer learning: pretrained models in one domain (social network) applied and evaluated in other domain (product reviews).

The remainder of this paper is organized as follows: Sect. 2 covers the related work, while Sect. 3 discusses the methodology, including data preprocessing, the dataset, and evaluation criteria. Section 4 presents and discusses the experimental results.

2 Related Works

For the scope of this paper regarding sentiment analysis we use some abbreviations. We use publicly available datasets published by Habernal et al. [8]. These datasets and their abbreviations are Czech Facebook Dataset (CFD), Czech Movie Dataset (CMD) and Czech Product Dataset (CPD).

Because special preprocessing techniques, like stemming or lemmatization, are needed when handling highly inflected synthetic languages, we have limited our review of related work to papers focused on Czech and Slovak.

It is important to note the current current state-of-the-art relevant sentiment analysis models based on BERT or RoBerta architecture Czert [5], SlovakBERT [6] or RobeCzech [7]. In these papers different applications of models are used. For the scope of this paper we reduce our focus to sentiment analysis tasks. Czert is evaluated on two Czech sentiment classification datasets (CFD and CMD) made by Habernal et al. [8]. SlovakBERT is evaluated on the Slovak version of Twitter dataset published by Mozetic [9]. RobeCzech is also evaluated on CFD by Habernal et al. [8]. Czert (B variant) achieved the macro F scores for CFD 76.55 and CMD 84.79. RobeCzech was evaluated on CFD and achieved macro F at 80.13. SlovakBERT was evaluated on Slovak Twitter dataset where it achieved macro F at 67.2 and then transferred to reviews dataset and achieved 58.3. These models belong to robust modern deep-learning neural models. They can use various sequence length (word counts) and for words that are not present in the trained vocabulary, with the use of tokenization, they can be split to sub-word tokens. Subsequently these sub-word tokens are matched in the vocabulary. Therefore, traditionally, even for Czech or Slovak languages, no stemming, lemmatization, lowercasing and other NLP techniques are required when dealing with these modern models.

Sokolová et al. [10] performed the evaluation of the stare-of-the-art model SlovakBERT together with classic machine learning models on several selected Slovak language sentiment datasets. Her primary objective, however, was not to entail the introduction

[1] https://waikato.github.io/weka-wiki/downloading_weka

of novel insights into the domain of sentiment analysis. Her aim was to contribute by presenting and disseminating a dataset (SentiSK) that underwent manual annotation, as there is a long-term shortcoming of sentiment analysis datasets in the Slovak language field. The dataset is available from the corresponding author upon "reasonable request".

Traditional machine learning methods have been used for sentiment analysis in Czech and Slovak languages. One of the first significant works in this field was by Veselovska et al. [11], focusing on text segment annotation. She not only assessed annotation quality but also evaluated the performance of sentiment analysis classifiers. Her main goal was to develop a tool to detect how news articles influenced public opinion. However, she found this challenging due to the lack of evaluative expressions in the news. Instead, she discovered that the CSFD.CZ (Czech movie reviews/comments database) and reviews from the MALL.CZ retail server (Czech Product Dataset CPD) were more expressive. She used the Naive Bayes classification approach, first lemmatizing the texts with the Morče tagger [12]. She achieved F-values of 78.1% for the CPD (Table 7 in her paper) and 75.4% for the CMD (Table 9 in her paper) [11].

Habernal et al. [8] took a more comprehensive approach by evaluating sentiment across three domains: product reviews (CPD), social media (CFD), and movie reviews (CMD). They found that the MaxEnt classifier was the most promising model for all datasets, achieving F-values of 69% for CFD (ShCl pipeline, FS3 feature set, Table 4), 75.30% for CPD (Lo pipeline, FS2 feature set, Table 6), and 78.50% for CMD (ShPe pipeline, FS2 feature set, Table 6). Despite the thoroughness and valuable results of the study, some errors were noted and corrected in our work [13].

Another notable work on sentiment analysis for Slovak language [14] was made by Krchnavy and Simko. The authors evaluated sentiment analysis on their Slovak Facebook Dataset (SFD). They compared several ML models including Naive Bayes, MaxEnt, SVM and lexicon-based approach. According to traditional metrics and class settings, they achieved F of 55.05% with MaxEnt classifier and three-class classification (MaxEnt, their Table 3). To represent the feature space, Krchnavy and Simko used a binary bag-of-words approach. Among the four evaluated models, MaxEnt was found to be the best. They employed feature space reduction through a combination of lemmatization and diacritics reconstruction, which yielded beneficial results for the accuracy of sentiment analysis. These two steps of feature space reduction were realized using online web services available at text.fiit.stuba.sk (currently out of order).

Facebook dataset was evaluated again, this time in Czech language (CFD published in [8]) along with Czech products dataset (published in [11]) by Cano and Bojar [15]. There appears to be some discrepancy in the description of the MALL.CZ dataset (we already noted details in [18]). Despite this, Cano and Bojar found that the MaxEnt model, along with the SVM, was one of the top performers. Their feature representation used a bag-of-words approach composed of n-grams (n = 1...3) with the Tf-Idf vectorizer, without stemming or lemmatization in the preprocessing stage. They achieved F-scores of 64% for the CFD dataset (NuSVM classifier, their Table 6) and 91.6% for the MALL.CZ dataset (SVM, their Table 6).

We already assessed the impact of using an emotional lexicon on sentiment analysis accuracy with the Random forest model on product reviews from Alza.sk, a leading Slovak and Czech marketplace, in our previous work [16]. By utilizing pre-structured positive and negative text segments, we aimed to enhance accuracy. We separated the reviews by language and evaluated them independently. Instead of using a bag-of-words feature representation, we applied a statistical approach. We incorporated lemmatization and tokenization as preprocessing steps on the textual data using Morphodita [17] and its models, achieving an F-score of 77.49% (Alza.sk, S2 feature set, Table 8 in [16]).

In [18] we evaluate classic machine learning model MaxEnt and its approximation SGD on Czech Facebook Dataset (CFD) with focus on features count and feature selection methods in order to improve overall F score.

In contrast to most papers related to the sentiment analysis tasks in Czech or Slovak languages, where the effort is put on training accuracy, our objective here is to evaluate two factors. First, how the different diacritic settings affect the resulting accuracy. Second, how the models perform in other domain (transfer-learning).

3 Methodology

For evaluation metric to measure sentiment analysis accuracy we use macro-averaged F score. We consider three sentiment classes positive, negative and neutral. We evaluate the following three deep-learning models: SlovakBERT (trained on Slovak Twitter [9] and Reviews dataset [6]) and Two Czert models [5] (trained on CFD and CMD). We prepare our own traditional machine learning MaxEnt model trained on CFD. Evaluation of models for transfer learning is performed CMD [8].

We evaluate also the effect of different diacritic settings. We prepare three different settings on transfer dataset: with reconstructed diacritics (denoted as DIA), original version (ORIG) and diacritics removed (NO-DIA). We prepared also three diacritic settings for our training dataset for MaxEnt model, however, we had no control over the training phase of pre-trained models SlovakBERT and Czert. Here we use only one setting – the original in which the models were trained. We evaluate Czert and SlovakBERT with transfer-learning on CPD and SFD, respectively.

Reconstruction of the diacritics was performed by Morphodita [17] with the use of word generative feature of the tool.

4 Results

In this section we present the results of transfer learning. We use pre-trained models SlovakBERT and Czert and our MaxEnt model. All models were trained in one domain and evaluation is performed in other with the use of different diacritic settings. Datasets are composed of textual records. Regarding published models Czert and SlovakBERT we performed no training thus we evaluate no diacritic settings applied to training datasets.

In Table 1 there are results of sentiment analysis with two trained Czert models; CMD and CFD transferred on CPD. Three different settings are used; restored diacritic (DIA), removed diacritic (NO-DIA) and diacritic unchanged (ORIG). Surprisingly for us, the three different settings lead to the same F in all three classes. The only notable difference here is different F measure for two models (CMD and CFD). Comparing the difference between models, the CFD model is the best for this use-case. CMD had

Table 1. Evaluations of Czert trained on CMD and CFD and evaluated on CPD. Three different diacritic preprocessing (DIA, NO-DIA and ORIG). Results are in percent. Results are for all three sentiment classes.

	CMD_{DIA}	$CMD_{NO\text{-}DIA}$	CMD_{ORIG}	CFD_{DIA}	$CFD_{NO\text{-}DIA}$	CFD_{ORIG}
F negative	19.30	19.30	19.30	32.30	32.30	32.30
F neutral	35.40	35.40	35.40	33.70	33.70	33.70
F positive	77.00	77.00	77.00	74.50	74.50	74.50
macro F	43.90	43.90	43.90	46.83	46.83	46.83

poorest results for negative class and both models had high F scores for positive class and similar scores for neutral class.

Table 2 offers the results for SlovakBERT evaluated on CPD. Similarly, in Czert case we evaluated three different diacritic settings. Although this case evaluates Slovak language trained model on Czech dataset, surprisingly the F scores between settings expose different F scores. There are slight variations on F scores between DIA and NO-DIA. We speculate that this could be due to different diacritical settings in training sets used to train SlovakBERT and Czert. Czert was probably trained with differend diacritic variations. Thus, Czert model is more robust. Compared to Czert there is only one model of SlovakBERT trained on Slovak Twitter dataset and Reviews, thus only one model evaluation is performed for SlovakBERT. Evaluation is with transfer-learning on SFD [14].

Table 2. Evaluations of SlovakBERT trained on Slovak Twitter dataset and Reviews and evaluated on CPD. Three different diacritic preprocessing (DIA, NO-DIA and ORIG).

	DIA	NO-DIA	ORIG
F negative	41.40	39.30	41.30
F neutral	11.80	10.70	11.80
F positive	82.00	81.20	81.90
macro F	45.06	43.73	45.00

In Table 3 there are results of evaluation SlovakBERT model on Czech CFD and Slovak SFD versions. Interesting that SlovakBERT expose better F scores in the neutral class for CFD. SlovakBERT should be trained in the Slovak language. However, the other two classes outperform CFD, as could be expected. This strange behavior could be due to the language impurity both in the training set of SlovakBERT and in the evaluation transfer-learning dataset. We noticed that there are at least tenths of Czech posts in the SFD dataset (all classes). But this does not answer the question that it performs better with CFD, where it should not. Regarding different diacritic settings, the F scores seem to reflect to it.

Table 3. Evaluations of SlovakBERT trained on Slovak Twitter dataset and Reviews and evaluated on CFD and SFD. Three different diacritic preprocessing (DIA, NO-DIA and ORIG).

	CFD_{DIA}	$\text{CFD}_{\text{NO-DIA}}$	CFD_{ORIG}	SFD_{DIA}	$\text{SFD}_{\text{NO-DIA}}$	SFD_{ORIG}
F negative	54.80	53.90	54.70	75.70	76.20	76.60
F neutral	49.30	47.00	49.10	12.10	11.10	11.60
F positive	68.90	68.10	68.80	73.00	73.80	74.40
macro F	57.66	56.33	57.53	53.60	53.70	54.20

In Table 4 there are results of MaxEnt trained on CFD evaluated on CPD. Together there are nine settings. Three diacritic settings for training dataset times three diacritic settings for evaluation dataset. As features we used character n-grams with setting minimum 2 and maximum 12 characters. From NLP we performed stemming (on all three diacritic settings). Empirically 2000 features were selected with the use Pearson correlations. Training CFD contained 10,000 records.

Table 4. Results of 10 times cross validation of MaxEnt on CFD. This is the training phase.

	ORIG	DIA	NO-DIA
F negative	55.00	55.20	54.70
F neutra	77.50	77.50	77.30
F positive	74.40	75.00	75.10
macro F	68.96	69.23	69.03

In Table 5 the results of evaluation MaxEnt on CPD are listed. There are variations between individual F between different diacritic settings. This is similar behavior to SlovakBERT. Surprisingly enough for us in MaxEnt results, however, is the setting DIA – NO-DIA, where model is trained on CFD with reconstructed diacritics but evaluated on CPD with diacritics removed. Here it achieved the highest F for positive class and decent for remaining two classes, resulting in the highest macro F as well.

The reason for the high scores in this case is only to be explained. We hypothesize that although the diacritic reconstruction is required in the training phase, it is more beneficial to have it removed in the final evaluation data. But more evaluations regarding the combinations of diacritic setting between training and evaluation data would be required in order to clarify this behavior.

Table 5. Results of transfer of MaxEnt trained on CFD (with 2000 features) evaluated on CPD with nine different diacritic settings (three settings on CFD and three on CPD).

CFD >	ORIG			DIA			NO-DIA		
CPD >	ORIG	DIA	NO-DIA	ORIG	DIA	NO-DIA	ORIG	DIA	NO-DIA
F negative	21.94	21.71	24.15	22.19	22.11	23.55	19.86	19.51	22.10
F neutra	35.26	35.40	33.78	35.54	35.60	35.06	34.90	34.97	34.02
F positive	64.08	64.29	60.94	63.77	63.73	66.55	56.87	56.12	65.65
macro F	40.43	40.47	39.62	40.50	40.48	41.72	37.21	36.87	40.59

5 Conclusions

In this paper we evaluate the impact of different diacritic settings on the accuracy of the sentiment analysis. We evaluate two modern, pre-trained deep-learning models represented by Czert [5] and SlovakBERT [6] and one traditional machine learning model MaxEnt. We record that Czert is more robust on the various diacritic settings compared to SlovakBERT (Tables 1, 2 and 3). We recorded surprisingly low F scores in neutral class for all diacritic variants for SlovakBERT transfer-learning model. This is in a strong contrast to its Czech CFD variant (Table 3). This behavior could be credited to language impurity at least in the evaluation transfer-learning SFD, as there are tenths of Czech posts along with majority of Slovak ones. However, it performs better with CFD. Perhaps more evaluations and attention to this aspect would answer this strange behavior.

For the case of our MaxEnt model, we discovered similar variations in F score between different diacritic settings. There seems to be also surprising behavior (Table 5) for the setting DIA – NO-DIA (train dataset diacritics reconstruction, evaluation transfer learning dataset diacritics removed). This could be due to the certain randomness in the process. More evaluations would be required to better clarify this behavior.

In future work we plan to train MaxEnt with combined diacritic settings to evaluate the possible positive impact on model's robustness. We also plan to state the optimum hyperparameters (number of features and various model's parameters) with the use of validation dataset in order to raise the macro F score in the transfer learning evaluation (Table 5 lists the current macro F scores of the transfer learning evaluation for the ME model).

To conclude our contribution, the effect of different diacritic settings on the resulting F scores should be considered when training and evaluating models for the sentiment analysis. Due to F scores variations in the SlovakBERT model, this recommendation is applicable also on modern deep-learning models built upon BERT or RoBERTa architectures.

Acknowledgments. This work was supported by the Slovak Scientific Grant Agency VEGA under the contracts 2/0135/23 "Intelligent sensor systems and data processing".

References

1. Kenyeres, M., Kenyeres, J., Budinská, I.: On performance evaluation of distributed system size estimation executed by average consensus weights. In: Recent Advances in Soft Computing and Cybernetics, pp. 15–24 (2021)
2. Kenyeres, M., Kenyeres, J.: Distributed linear summing in wireless sensor networks with implemented stopping criteria. Adv. Sci. Technol. Eng. Syst. J. **5**(2), 19–27 (2020)
3. Abiodun, O.I., Jantan, A., Omolara, A.E., Dada, K.V., Umar, A.M., Linus, O.U., Arshad, H., Kazaure, A.A., Gana, U., Kiru, M.U.: Comprehensive review of artificial neural network applications to pattern recognition. IEEE Access. **7**, 158820–158846 (2019)
4. Kenyeres, J., Kenyeres, M., Rupp, M.: Experimental node failure analysis in WSNs. In: 2011 18th International Conference on Systems, Signals and Image Processing, pp. 1–5. IEEE (2011)
5. Sido, J., Pražák, O., Přibáň, P., Pašek, J., Seják, M., Konopík, M.: Czert--Czech BERT-like Model for language representation. In: arXiv preprint arXiv:2103.13031 (2021)
6. Pikuliak, M., Grivalský, Š., Konôpka, M., Blšták, M., Tamajka, M., Bachratý, V., Šimko, M., Balážik, P., Trnka, M., Uhlárik, F.: SlovakBERT: Slovak Masked Language Model. In: arXiv preprint arXiv:2109.15254 (2021)
7. Straka, M., Náplava, J., Straková, J., Samuel, D.: RobeCzech: Czech RoBERTa, a monolingual contextualized language representation model. In: Text, Speech, and Dialogue: 24th International Conference, TSD 2021, Olomouc, Czech Republic, September 6–9, 2021, Proceedings, vol. 24, pp. 197–209. Springer International Publishing (2021)
8. Habernal, I., Ptáček, T., Steinberger, J.: Supervised sentiment analysis in Czech social media. Inf. Process. Manag. **50**(5), 693–707 (2014)
9. Mozetič, I., Grčar, M., Smailović, J.: Multilingual Twitter sentiment classification: the role of human annotators. PLoS One. **11**(5), e0155036 (2016)
10. Sokolová, Z., Harahus, M., Juhár, J., Pleva, M., Staš, J., Hládek, D.: Comparison of machine learning approaches for sentiment analysis in slovak. Electronics. **13**(4), 703 (2024)
11. Veselovská, K., Hajic, J., Sindlerová, J.: Creating annotated resources for polarity classification in Czech. In: Proceedings of KONVENS 2012 (PATHOS 2012 Workshop), Vienna, September 21, 2012
12. Votrubec, J.: Morphological tagging based on averaged perceptron. In: WDS'06 Proceedings of Contributed Papers, pp. 191–195 (2006)
13. Mojžiš, J., Krammer, P., Kvassay, M., Skovajsová, L., Hluchý, L.: Towards reliable baselines for document-level sentiment analysis in the Czech and Slovak languages. Future Internet. **14**(10), 300 (2022)
14. Krchnavy, R., Simko, M.: Sentiment analysis of social network posts in Slovak language. In: 2017 12th International Workshop on Semantic and Social Media Adaptation and Personalization (SMAP), pp. 20–25. IEEE (2017)
15. Cano, E., Bojar, O.: Sentiment analysis of Czech texts: an algorithmic survey. In: arXiv preprint arXiv:1901.02780 (2019)
16. Mojžiš, J., Kvassay, M.: Review rating prediction based on text polarity in the Czech and Slovak languages. In: Artificial Intelligence in Intelligent Systems: Proceedings of 10th Computer Science On-line Conference 2021, vol. 2, pp. 251–261. Springer International Publishing (2021)
17. Straková, J., Straka, M., Hajic, J.: Open-source tools for morphology, lemmatization, POS tagging and named entity recognition. In: Proceedings of 52nd Annual Meeting of the Association for Computational Linguistics: System Demonstrations, pp. 13–18 (2014)

18. Mojžiš, J., Kvassay, M.: In the shadow of RoBERTA: is the classical ML drawing its last breath in sentiment analysis? In: Proceedings of the Computational Methods in Systems and Software, pp. 245–252. Springer Nature, Cham (2023)
19. Wikipedia: Slovak declension. https://en.wikipedia.org/wiki/Slovak_declension. Accessed 2 Feb 2025

Toward a New Measure of Product Indistinguishability Based on Changing Relationships with Other Products

Petr Krautwurm(✉)

Prague University of Economics and Business, Winston Churchill Sq. 1938/4, 130 67 Prague, Czech Republic
petr.krautwurm@vse.cz

Abstract. This article introduces a novel method for assessing product indistinguishability, defined as how closely products approach being perfect substitutes. The method evaluates changes in product relationships when aggregated versus when maintained separately. Derived from the generalized Constant Elasticity of Substitution (genCES) function that utilizes a nesting structure, the test is applied to a dataset from a drugstore retail chain. Empirical findings demonstrate alignment with theoretical predictions, substantiating the test's validity. Additionally, the study opens up new avenues for future research aimed at enhancing the test's efficacy. These include further exploration of the theoretical underpinnings of genCES functions and the development of a more nuanced measure of product relationships that integrates the strengths of elasticity of substitution and cross-price elasticity.

Keywords: Substitutes · Complements · Cross-Price Elasticity · Elasticity of Substitution

1 Introduction

Product relationships, particularly whether they are substitutes or complements, are fundamental to economic theory, which provides many theoretical insights regarding the implications for consumer behavior. However, from an econometric standpoint, the estimation of these relationships presents numerous challenges, thereby offering an opportunity for the development of novel measures or estimation methodologies for product relationships.

This article introduces a new methodology to estimate product indistinguishability, which is defined as how closely two products approach being perfect substitutes, by examining their interactions with other products. This methodology assesses changes in the relationships among a set of products when certain items are combined into a single category, as opposed to when they remain separate. Theoretically, products nearing perfect substitutability should maintain their relationships with other products even when merged into one category.

R. Silhavy and P. Silhavy (Eds.): CoMeSySo 2024, LNNS 1489, pp. 169–182, 2025.
https://doi.org/10.1007/978-3-031-96798-6_16

This assumption aligns with the concept of perfect substitutes as an indicator of product indistinguishability, which assesses how consumers perceive different products as essentially identical.

The article specifies the prerequisites necessary to conduct this test. Currently, the test is applicable to datasets where certain groups of products maintain consistently the same prices within the group, facilitating straightforward aggregation. Subsequently, the test is employed on a dataset that meets these criteria, and the findings are juxtaposed against theoretical predictions.

2 Relationships Between Products

From a theoretical perspective, two distinct and crucial measures exist for determining the relationship between two products. One measure is based on consumer preferences, specifically the shape of the utility function, while the other stems from observable market demand. These are known as Robinson's elasticity of substitution [9,11] and the cross-price elasticity of demand [2,5], respectively.

This article conceptualizes the relationship between products based on their interactions. Thus, two products may either interact and be related, or their relationship may be indeterminate if no interaction occurs. For instance, an indeterminate relationship occurs when products are mutual economic bads, meaning the consumption of one reduces the utility derived from the other. This article focuses solely on determinable relationships, particularly examining whether products function as substitutes or complements. To this end, the notation $R \in \langle -1, 1 \rangle$ is introduced, where $R = -1$ represents perfect complements, $R = 1$ signifies perfect substitutes, and $R = 0$ indicates a Cobb-Douglas type relationship, where products are neither substitutes nor complements. This scale facilitates the comparison of relationships, enabling categorizations such as 50% substitutes.

Prior to further analysis of the elasticity of substitution and the cross-price elasticity, it is vital to note their distinct theoretical foundations. The elasticity of substitution exclusively reflects consumer preferences, independent of other influences, defining the relationship between products conditional solely on the quantities consumed. This opens a possibility to link resulting elasticity of substitution directly to the relationship R between products. Conversely, the cross-price elasticity encompasses the entire consumer decision-making process, making it less straightforward to link directly to the relationship R between products. For example, a particular value of cross-price elasticity might suggest 5% substitutes in one context and 20% substitutes in another. Thus, cross-price elasticity can only determine the sign of relationship between products.

To illustrate the behavior of these measures, it is crucial to consider the consumer optimization problem, which is modeled as follows:

$$\max_{\boldsymbol{x}} U(\boldsymbol{x}) \qquad \text{s. t.} \quad \sum_{j}^{n} P_j x_j \leq M \,. \tag{1}$$

Here, $\boldsymbol{x}$ represents a vector of quantities for n products, with x_i denoting the quantity of the i-th product, and P_i its price. The variable M represents the budget, and $U(\boldsymbol{x})$ is the utility function encapsulating consumer preferences. It is crucial to apply this optimization problem to a scenario with linear prices, as this leads to Gossen's law [14] when solved using a method of Lagrangian multipliers:

$$\frac{MU_i}{MU_j} = \frac{P_i}{P_j} \,. \tag{2}$$

Here, $MU_i = \partial U(\boldsymbol{x})/\partial x_i$ represents the marginal utility. If prices were not linear, Gossen's law, as derived from the optimization problem (1), would not be applicable.

2.1 Elasticity of Substitution

The concept of elasticity of substitution encompasses multiple variants due to the ambiguities involved in generalizing it beyond two products [6]. As noted by Blackorby and Russell [3], these elasticities are not entirely consistent with each other. Hence, this article concentrates on one of the initial and simplest forms, the Robinson's elasticity of substitution [11], defined as:

$$\sigma_{ij} = \frac{\frac{\partial(x_j/x_i)}{x_j/x_i}}{\frac{\partial(MU_i/MU_j)}{MU_i/MU_j}} = \frac{\partial \ln(x_j/x_i)}{\partial \ln(MU_i/MU_j)} \,. \tag{3}$$

Robinson's elasticity of substitution is symmetric, ensuring that $\sigma_{ij} = \sigma_{ji}$. While it can be expressed in logarithmic terms, making it suitable for regression analysis, a significant challenge arises because both marginal utility and the utility function are unobservable. To address this, it is assumed that the consumer resolves the optimization problem (1), thereby validating Gossen's law (2). This allows for the substitution of MU_i/MU_j with P_i/P_j, facilitating the estimation of the elasticity of substitution through the following regression:

$$\ln\left(\frac{x_j}{x_i}\right) = \gamma_0 + \sigma_{ij} \times \ln\left(\frac{P_i}{P_j}\right) - v \,. \tag{4}$$

This regression model is straightforward in its interpretation. Notably, the intercept may reflect the taste ratio between products, indicating the proportion in which a consumer would prefer to consume the products if their prices are the same.

In terms of the values it yields, $\sigma_{ij} \in \langle 0, 1 \rangle$ suggests the products are complements, with $\sigma_{ij} = 0$ corresponding to $R = -1$, representing perfect complements. Conversely, $\sigma_{ij} \in (1, \infty)$ implies the products are substitutes, with $\sigma_{ij} = \infty$ equating to $R = 1$, indicative of perfect substitutes.

2.2 Cross-Price Elasticity

In contrast to the elasticity of substitution, the concept of cross-price elasticity is unambiguous. It is defined based on consumer demand as follows:

$$\zeta_{ij} = \frac{\frac{\partial x_i}{x_i}}{\frac{\partial P_j}{P_j}} . \tag{5}$$

This expression generally describes own-price elasticity ζ_{ii} when $P_j = P_i$, reflecting the response of the demanded quantity to changes in its own price, and cross-price elasticity ζ_{ij} when $P_j \neq P_i$, which measures the response of the demanded quantity to changes in the price of another product. The econometric application of this formula is direct, allowing for the estimation of cross-price elasticity as follows:

$$\ln(x_i) = \alpha_0 + \zeta_{ii} \times \ln(P_i) + \sum_{j \neq i}^{n} \zeta_{ij} \times \ln(P_j) + \beta \ln(M) + u . \tag{6}$$

Regarding the interpretation of cross-price elasticity, a value of $\zeta_{ij} < 0$ indicates that the products are complements, with their relation $R < 0$. Conversely, a $\zeta_{ij} > 0$ signifies that the products are substitutes, with their relation $R > 0$. However, it is important to note that cross-price elasticity alone does not provide a precise link to the specific relationship R between products.

3 Perfect Substitutes as a Measure of Indistinguishability

As previously discussed, of the two measures, only the elasticity of substitution can be directly correlated, to a certain extent, with the product relationship R, while cross-price elasticity can only be interpreted only in terms of sign of the relationship. The relation between the elasticity of substitution and product relationship is most clearly demonstrated using the Constant Elasticity of Substitution (CES) function [1]. The CES function is characterized by its elasticity of substitution between products being constant and solely dependent on a single parameter [7]. The CES function is defined as follows:

$$U(\boldsymbol{x}) = \left(\sum_{i=1}^{n} a_i x_i^{\rho} \right)^{\frac{1}{\rho}} . \tag{7}$$

Here, the a_i values are exogenous taste parameters that determine the underlying preference for units of products. Crucially, when the elasticity of substitution formula (3) is applied to the CES function, it yields a result dependent only on the parameter ρ:

$$\sigma_{ij}(\text{CES}) = \frac{1}{1-\rho} . \tag{8}$$

In this context, the CES function is particularly useful because the parameter ρ facilitates a direct assessment of the relationship between products. If $\rho \in (0, 1\rangle$, then $\sigma_{ij} > 1$ and $R \in (0, 1\rangle$. Conversely, if $\rho \in (-\infty, 0)$, then $\sigma_{ij} \in \langle 0, 1\rangle$ and $R \in \langle -1, 0\rangle$. This relationship is highlighted when examining special cases and resulting functional forms, specifically focusing on a CES function with $n = 2$ as detailed in:

$$U(\boldsymbol{x}) = (a_1 x_1^\rho + a_2 x_2^\rho)^{\frac{1}{\rho}} = \begin{cases} a_1 x_1 + a_2 x_2 & \text{if } \rho = 1 \\ x_1^{\frac{a_1}{a_1+a_2}} \times x_2^{\frac{a_2}{a_1+a_2}} & \text{if } \rho = 0 \\ \text{MIN}\{a_1 x_1; a_2 x_2\} & \text{if } \rho = -\infty\,. \end{cases} \tag{9}$$

From the special case where $\rho = 1$ and $a_1 = a_2$, it is evident that a consumer with such a utility function is indifferent to the composition of products, focusing solely on the aggregate quantity. This occurs because the utility function is additive, resulting in products being perceived as perfectly indistinguishable if tastes are identical. If the tastes differ, the basic premise still holds; however, the less preferred product is perceived as if it were in a smaller quantity. Conversely, if $\rho \neq 1$, the product composition significantly affects consumer utility, increasingly so as ρ diverges from 1.

Consequently, the degree of substitutability also serves as a measure of product indistinguishability. The closer the products are to being perfect substitutes, the less distinguishable they are to the consumer. Thus, $R = 1$ signifies products that are perfectly indistinguishable, whereas $R = -1$ represents products that are perfectly distinguishable, reflecting the importance of product composition from the consumer's perspective.

4 Generalized CES Function

The practicality of the CES function in describing the relationship between two products has prompted researchers to explore extensions that could accommodate multiple product pairs with constant, yet differentiated, elasticities of substitution. Unfortunately, as Uzawa [13] demonstrated, it is impossible to devise a function that allows for distinct and constant elasticities of substitution across multiple pairs of products. The only feasible solution is to sort products into categories such that pairs of products within each category maintain a constant and identical elasticity of substitution, while pairs of products across categories exhibit varied elasticities. This realization led to the development of the Nested CES function [8, 12], which groups products into categories, maintaining a constant elasticity of substitution within each category but varying across them.

The Nested CES function utilizes the concept of composite goods, which reflect latent production functions [4]. To illustrate the essence of composite goods, consider a German consumer who prefers a perfectly balanced sandwich for breakfast, consisting of precisely two pieces of ham and one piece of bread. In this example, the sandwich, a breakfast bulk, acts as the composite good.

Assume, that the German consumer perceives ham and bread as perfect complements in the production of the sandwich. If the consumer has an excess of one ingredient relative to the other, the surplus becomes superfluous. Alternatively, consider a French consumer deciding between coffee with a croissant or a pain au chocolat, viewing these pastries as perfect substitutes. Here, the breakfast bulk is a different composite good, reflecting a sum of croissants and pains au chocolat.

These examples underscore that composite goods represent a specific latent production function involving original products as inputs to yield a new product. Yet, the relationship between these inputs within the composite good and other standalone products remains an open question. This issue is addressed by a specific nesting structure for three products as proposed by Lagomarsino [10], providing a logical extension of the CES function, which allows for an analysis of the relations among the components of a composite good and an intermediate product. Defining the ingredients of the composite good as x_1 and x_3, and the intermediate product as x_2, the suitable function for the analysis is the following:

$$U(\boldsymbol{x}) = \left(a_1 \left(b_1 x_1^\delta + b_2 x_3^\delta\right)^{\frac{\rho}{\delta}} + a_2 x_2^\rho\right)^{\frac{1}{\rho}} . \tag{10}$$

Here, a_i and b_i serve as taste parameters akin to those in the standard CES function. Theoretically, the parameter δ determines the relationship between x_1 and x_3, while ρ defines the relationship between the composite good $y_1 := y_1(x_1, x_3)$ and the intermediate product x_2. This function allows for an analysis of the relationships R between a component of the composite good and the intermediate product.

4.1 Relation-Preserving Property of Perfect Substitutes

In the context of the generalized CES function, three extreme cases emerge for ingredients of a composite good concerning perfect substitutes, perfect complements, and the Cobb-Douglas relationship, akin to the CES function:

$$U(\boldsymbol{x}) = \begin{cases} \left(a_1 \left(b_1 x_1 + b_2 x_3\right)^\rho + a_2 x_2^\rho\right)^{\frac{1}{\rho}} & \text{if } \delta = 1 \\ \left(a_1 \left(x_1^{\frac{b_1}{b_1+b_2}} \times x_3^{\frac{b_2}{b_1+b_2}}\right)^\rho + a_2 x_2^\rho\right)^{\frac{1}{\rho}} & \text{if } \delta = 0 \\ \left(a_1 \cdot \left(\text{MIN}\left\{b_1 x_1; b_2 x_3\right\}\right)^\rho + a_2 x_2^\rho\right)^{\frac{1}{\rho}} & \text{if } \delta = -\infty . \end{cases} \tag{11}$$

A critical aspect of this generalized function is its behavior when $\delta = 1$, indicating that x_1 and x_3 are perfect substitutes. As discussed previously, perfect substitutes can be seen as a measure for product indistinguishability or perceived sameness. The closer the products are to being perfect substitutes, the more they appear as a single product. Therefore, if x_1 and x_3 are perfect substitutes, the relationship between x_1 and x_2 should mirror that between x_3 and

x_2. Furthermore, as the products x_1 and x_3 are viewed as essentially the same by the consumer, their aggregation, weighted by tastes, should maintain the same relationship with x_2. This relation-preserving property is unique to perfect substitutes; more complementary relationships result in varying interactions between components of a composite good and an intermediate product.

Consider the earlier breakfast examples: for the French consumer, the relationship between any single component of the composite good (croissant or pain au chocolat) and coffee should resemble that between the entire composite good and coffee. This consistency does not hold for the German consumer, where the relationship between the composite ingredients and tea varies depending on their composition. This outcome stems from the inherent logic behind indistinguishability; as the proximity to perfect substitutes quantifies how indistinguishable two products are, perfect substitutes inherently describe what is essentially the same product. If one product maintains a certain relationship with another, an identical product should exhibit the same relationship.

This principle is further illustrated by the elasticity of substitution between x_1 and x_2 when $\delta = 1$ and x_1 and x_3 are perfect substitutes. The marginal rate of substitution between x_1 and x_2 is defined as follows:

$$\frac{MU_1}{MU_2} = \frac{a_1 \left(b_1 x_1 + b_2 x_3\right)^{\rho-1} b_1}{a_2 x_2^{\rho-1}}, \tag{12}$$

leading to:

$$\frac{MU_1}{MU_2} \times \frac{x_1}{x_2} = \frac{a_1}{a_2} \left(\frac{b_1 x_1 + b_2 x_3}{x_2} \right)^{\rho} \frac{b_1 x_1}{b_1 x_1 + b_2 x_3}, \tag{13}$$

and:

$$\frac{\partial \frac{MU_1}{MU_2}}{\partial \frac{x_2}{x_1}} = \frac{\partial \left(\frac{a_1 (b_1 x_1 + b_2 x_3)^{\rho-1} b_1}{a_2 x_2^{\rho-1}} \right)}{\partial \frac{x_2}{x_1}} = \frac{\frac{a_1}{a_2} \partial \left(\frac{x_2}{b_1 x_1 + b_2 x_3} \right)^{1-\rho}}{\partial \frac{x_2}{b_1 x_1}}. \tag{14}$$

As these products are perfect substitutes, it is reasonable to assume, without loss of generality, that the price ratio of the composite good ingredients will not match the ratio of their tastes $P_1/P_2 \neq b_1/b_2$, leading to the demand for only one of the products based on a higher taste-to-price ratio. In this particular case, if x_3 is zero, the elasticity of substitution derived from the first (13) and second (14) equations will mirror that of a standard CES function, confirming that:

$$\sigma_{12} = \frac{1}{1-\rho}. \tag{15}$$

Moreover, as $x_3 \to 0$, the elasticity of substitution between the components converges to that of the CES function. This coherence with the CES function behavior confirms that as products become more substitutable, the product with lesser taste-to-price ratio becomes less demanded, thus validating the result (15) with increasing accuracy as $\sigma \to 1$.

5 Test of Product Indistinguishability

Building on the theoretical underpinnings discussed earlier, this section introduces a new methodology for testing product indistinguishability, explicating the implementation process and demonstrating the test's application on a specific dataset. The method involves several steps, each crucial for understanding how product relationships are influenced by aggregation.

The test operates by initially identifying products with consistently identical prices and examining their relationships with other products. Quantities of these products are then aggregated to form a composite good. The key metric of interest is the deviation in relationships post-aggregation compared to those estimated prior, with significant deviations indicating a lesser degree of substitutability.

5.1 Same Prices

The prerequisite for applying this test is the uniformity of prices among the aggregated products. The rationale for this condition stems from both practical and theoretical considerations.

From a practical standpoint, when products are priced identically, the purchase price of one product can infer the prices of others, particularly useful when dealing with transactional data where only purchased items have recorded prices. Therefore, this assumption is crucial, especially in datasets where prices are only recorded for transactions that occur, without a universal price listing. Furthermore, if at least two products are purchased together, having consistent prices simplifies the aggregation process by providing a clear basis for combining the prices. Without it, there would be no unambiguous way of constructing a new price that would suit both products and reflect the consumer optimization problem at the same time.

Theoretically, the behavior of the Constant Elasticity of Substitution (CES) function under uniform pricing is pivotal. When prices reflect consumer tastes accurately, all degrees of substitutability $\rho < 1$ converge to a single outcome, indicating that even slightly imperfect substitutes can exhibit the purchasing patterns of perfect complements. Crucially, price uniformity implies that consumer tastes may be similar on average, assuming that firms sell products under prices that reflect tastes. As discussed in Sect. 3, this is essential for accurate product aggregation, as diverging tastes among perfect substitutes can skew the perceived quantity of indistinguishable products.

Thus, the indistinguishability test is most applicable to product pairs with consistently uniform pricing, typically found among similar products from the same brand. If price fluctuations exist but occur sporadically, the average price can be used, although this introduces potential biases in the test results.

5.2 Econometric Behavior

Exploring the test's econometric properties involves analyzing potential outcomes from observed data points and their implications for product aggregation. For illustrative purposes, consider three products: A and B (referred to as ingredients) and D (an intermediate product). The relationships among these products are quantified using standard measures such as elasticity of substitution (4) and cross-price elasticity (6). Upon aggregating A and B into a composite good, denoted G, theoretical expectations suggest that if A and B are perfect substitutes, the relationship between G and D should mirror those between A and D or B and D individually.

The price uniformity of A and B, as stipulated by the test's requirements, ensures the price ratio of A and G mirrors the one of B and G. Thus, only the variance in purchased quantities alters the relational dynamics observed. Two extreme scenarios may arise: either both products are always purchased together, or purchases alternate exclusively between the two.

Therefore, the econometric analysis must consider various purchasing patterns and their implications on product relationship estimates. From theoretical standpoint, the elasticity of substitution is always non-negative, suggesting an increasing slope in the regression model (4). However, practical discrepancies, such as the inability to apply Gossen's law when $\sigma_{ij} \geq 1$, may introduce negative slopes. For the purposes of this illustrative example, these cases may be omitted.

Thus, multiple relationships may be observed; currently, assume these are measured by the elasticity of substitution. For instance, alternating purchases of A and B allow immediate estimation of their individual relationships with D, σ_{AD} and σ_{BD}. When aggregated into G, the derived relationship σ_{GD} represents a weighted average of the individual relationships. Conversely, if A and B are consistently purchased together, the relationship σ_{GD} is theoretically expected to exceed individual relationships, because both slopes are increasing.

For cross-price elasticity, if A and B maintain a similar relationship to D, their aggregation is expected to reinforce and amplify this relationship. Conversely, if their relationships with D differ, the aggregation could lead to relationship between them.

Overall, aligning this test with theoretical expectations from earlier sections involves recognizing the nuanced role of tastes in scenarios where products are near-perfect substitutes. Consistent pricing facilitates analysis under the assumption of taste alignment; however, taste variances can introduce significant volatility in purchasing patterns, potentially leading to sparse data for products with high substitutability. Less so for products with low substitutability. This occurs because the problem is not continuous as consumers can buy only integer values of product quantities. Thus, perfect substitutes with slightly varying tastes are less likely to manifest in combined purchase data, affecting the estimability of their relational dynamics.

6 The Dataset

The methodology is evaluated using a dataset from a drugstore retail chain, encompassing the 30 most frequently purchased products. This dataset captures the quantities bought and the prices paid, linked to individual customers and specific purchasing dates throughout 2019, along with unique customer identifiers and details about personal discounts and other purchase-related characteristics.

In 2019, a total of 62,322 unique customers, each of whom purchased at least three of the 30 products on various occasions throughout the year. In total, customers purchased 643,450 products in various quantities, averaging ten products per customer over the year.

Notably, prices in this dataset are not universally fixed but are instead tied to individual consumers, reflecting personalized pricing strategies. This personalization complicates the uniformity of price data, necessitating a careful evaluation of whether products maintained consistent pricing when purchased by the same consumer.

6.1 Identified Product Groups with Consistent Pricing

Through detailed analysis, two specific product groups were identified within the dataset as having consistently identical prices among consumers who purchased them:

- **Children's Food Pouches:** This group includes four types of children's food pouches from the same manufacturer, differentiated only by flavor: Apple, Strawberry, Apricot, and Banana.
- **Hygiene Products:** This group comprises hygiene products from the same manufacturer, including two types of tissues (one with 100 tissues per package, the other divided into 10 smaller packs), kitchen sponges, and cleaning wipes.

6.2 Potential Product Groups for Analysis

Given the characteristics of these product groups, three product groups can be constructed for further analysis to test for indistinguishability:

- **Children's Food Pouches:** These can be aggregated into a single product group. Given the flavor differences, this aggregation tests whether flavor is a significant factor in consumer choice, when purchasing products for their children.
- **Tissues:** The two types of tissues differ only in packaging. This similarity might suggest a higher degree of substitutability compared to the food pouches, as packaging variations may be less influential than flavor differences in consumer decision-making.
- **Complete Hygiene Product Group:** Including all items within the hygiene product category (tissues, kitchen sponges, and cleaning wipes) forms a broader composite good. However, given the functional differences (e.g., sponges for cleaning specific surfaces versus tissues for general wiping), these products are expected to exhibit a lower level of substitutability compared to the more homogeneously used tissues.

Intuitively, flavor differences might influence consumer choice more significantly than packaging differences, suggesting that tissues might be closer to perfect substitutes than the food pouches. Moreover, if all hygiene products are considered together, this group would include items used for broader cleaning purposes, such as kitchen sponges, indicating that these products should be considered less substitutable compared to the more narrowly defined tissue group.

7 Results

This section delves into the outcomes of grouping products regarding its impact on the observed relationships between these and other products, using three distinct groups: tissues, all hygiene products, and children's food pouches, alongside a scenario with no grouping. Consumers who purchased at least two of these products together on six or more different days were included in the study. For these selected transactions, elasticity of substitution (4) and cross-price elasticity (6) regression models were estimated. The data indicated that on average, consumers purchased multiple products together on 8.5 occasions. If conditions for the use of both regression models were not met, no relationship was measured for that particular consumer and that particular product combination.

The initial, ungrouped scenario revealed 176 measurable relationships between products. Grouping tissues resulted in 184 observed relationships, and grouping all four hygiene products yielded 207 relationships. Surprisingly, grouping the children food pouches decreased the total to 159 relationships. This drop is counter-intuitive, as grouping typically adds new data points, facilitating more relationship estimations, which could not be estimated before. The decline in observed relationships for food pouches occurred because these were the most frequently bought items, allowing numerous relationships between different flavors and other products to be measured individually. When aggregated, these multiple relationships collapsed into a single category, reducing the total count. This illustrates that grouping products into composite goods can either increase or decrease the amount of observed relationships.

Given the limited data, statistically analyzing the similarity of slopes between grouped and ungrouped scenarios proved impractical due to high standard errors. Instead, this article focuses on the number of relationships that shifted between being classified as substitutes or complements due to product grouping. Specifically, changes were tracked across three metrics: elasticity of substitution, cross-price elasticity, and instances where both measures concurred on the type of relationship. The results, summarized in the Table 1, highlight the impact of grouping on these relationships. The number of retained relationships refers to those consistently classified as either complements or substitutes across both grouping scenarios. Only relationships where grouping altered or added new data were considered relevant to avoid biased results due to the renaming of products.

The data reveals that grouping food pouches significantly alters their relationships with other products, more so than when grouping tissues. This suggests

Table 1. Number of retained relations by product grouping

Product group	# Relations	EoS ret.	XPE ret.	Both ret.
Food Pouches	82	59 (72%)	55 (67%)	41 (50%)
Tissues	14	14 (100%)	12 (86%)	12 (86%)
Hygiene products	43	39 (91%)	36 (84%)	35 (81%)

that tissues are better substitutes compared to food pouches, aligning with theoretical predictions. Furthermore, the decreased number of retained relationships when grouping all hygiene products indicates a lower degree of substitutability among them compared to just tissues, supporting the initial expectations. These findings suggest that consumers differentiate less between products used for similar wiping purposes compared to how they select food pouches for their children, treating the latter as lesser substitutes.

8 Discussion

The preliminary findings affirm that the devised test for product indistinguishability yields robust outcomes, both theoretically and empirically. Grounded in microeconomic theory, the test converts latent variables into observable metrics. It leverages both the elasticity of substitution and cross-price elasticity to evaluate product relationships, each measure offering distinct merits and limitations. Notably, results thus far suggest that elasticity of substitution offers greater stability in maintaining established relationships compared to cross-price elasticity, aligning with the theoretical expectations.

Despite the promising initial outcomes, there is considerable scope for further exploration. Expanding the dataset to include a greater variety of products and potential groupings could enhance the robustness and applicability of the test. Current data facilitated the examination of only three product groups; a more extensive dataset could uncover additional groups with consistent pricing, enabling a more comprehensive comparison and validation of the test's effectiveness.

Future research could also explore methodological enhancements. One possibility is assessing the aggregation of product groups that do not share identical prices but have prices in a consistent fixed ratio. Currently, such scenarios are untestable with elasticity of substitution due to the lack of variance in the price ratio regressor. Addressing this limitation would permit the analysis of larger and more diverse product groups within the same dataset.

A critical area for improvement is the metric used to evaluate changes in product relationships. The mere count of relationships that remain classified as either substitutes or complements may not fully capture the nuances of these changes. Significant alterations in relationships that still retain their original classification might be indistinguishably counted alongside minor changes that lead to a reclassification. Developing a new metric or test that can more accurately reflect the magnitude and significance of these changes would be valuable. Possible approaches could include statistical tests for the indistinguishability of regression coefficients before and after grouping, or an entirely novel measure that combines the theoretical clarity of elasticity of substitution with the unambiguous nature of cross-price elasticity.

In conclusion, the test of product indistinguishability based on changing relationships offers significant potential for further theoretical and practical developments. The modest yet encouraging results from preliminary applications suggest a fruitful avenue for advancing the understanding of generalized Constant Elasticity of Substitution (genCES) functions and their application in analyzing product relationships. This exploration opens several promising pathways for future research, each with the potential to refine and enhance the utility of the test in economic analysis.

Acknowledgments. This work was supported by The Internal Grant Agency of Prague University of Economics and Business [VŠE IGS F4/52/2023].

References

1. Arrow, K.J., Chenery, H.B., Minhas, B.S.B., Solow, R.M.: Capital-labor substitution and economic efficiency. Rev. Econ. Stat. **43**(3), 225–250 (1961). https://doi.org/10.2307/1927286
2. Auer, J., Papies, D.: Cross-price elasticities and their determinants: a meta-analysis and new empirical generalizations. J. Acad. Mark. Sci. **48**(3), 584—605 (2019). https://doi.org/10.1007/s11747-019-00642-0
3. Blackorby, C., Russell, R.R.: Will the real elasticity of substitution please stand up? (a comparison of the Allen/Uzawa and Morishima elasticities). Am. Econ. Rev. **79**(4), 882–888 (1989)
4. Carter, M.: An expository note on the composite commodity theorem. Econ. Theory **5**(1), 175–179 (1995). https://doi.org/10.1007/BF01213649
5. Deaton, A.: Estimation of own- and cross-price elasticities from household survey data. J. Econometrics **36**(1–2), 7–30 (1987). https://doi.org/10.1016/0304-4076(87)90041-8
6. de Lima, J.E.: Alternative definitions of elasticity of substitution: review and application. Rev. Econ. Sociol. Rural. **38**(1), 119–153 (2019)
7. McFadden, D.: Constant elasticity of substitution production functions. Rev. Econ. Stud. **30**(2), 73–83 (1963). https://doi.org/10.2307/2295804
8. Keller, W.J.: A nested CES-type utility function and its demand and price-index functions. Eur. Econ. Rev. **7**(2), 175–186 (1976). https://doi.org/10.1016/0014-2921(76)90057-X
9. Knoblach, M., Stöckl, F.: What determines the elasticity of substitution between capital and labor? A literature review. J. Econ. Surveys **34**(4), 847–875 (2020). https://doi.org/10.1111/joes.12366

10. Lagomarsino, E.: Which nesting structure for the CES? A new selection approach based on input separability. Econ. Modelling **102**, 105562 (2021). https://doi.org/10.1016/j.econmod.2021.105562
11. Robinson, J.: The Economics of Imperfect Competition, 1st edn. Macmillan, London (1933)
12. Sato, K.: A two-level constant-elasticity-of-substitution production function. Rev. Econ. Stud. **34**(2), 201–218 (1967). https://doi.org/10.2307/2296809
13. Uzawa, H.: Production functions with constant elasticities of substitution. Rev. Econ. Stud. **29**(4), 291–299 (1962). https://doi.org/10.2307/2296305
14. Van Daal, J.: From utilitarianism to Hedonism: Gossen, Jevons And Walras. J. Hist. Econ. Thought **18**(2), 271–286 (1996). https://doi.org/10.1017/S1053837200003278

AI Adoption in Academia: Diverse Perspectives from Students and Teachers

Natalia Yalaeva[1(✉)], Natalia Sadykova[2], Eugenia Zherebtsova[2], Natalia Khodenkova[2], Nadezhda Chernova[1], and Natalia Katakhova[1]

[1] Russian Technological University MIREA, 78, Vernadskiy Avenue, Moscow 119454, Russia
yalaeva-nv@mail.ru

[2] Ural State Law University named after V.F. Yakovlev, Yakovlev, 21. Komsomolskaya, Yekaterinburg 620066, Russia

Abstract. The article examines the growing influence of artificial intelligence (AI) across a variety of industries, with a particular focus on higher education. AI's capabilities range from creating coherent texts and composing music to managing finances and conducting medical diagnostics, demonstrating its transformative potential. The integration of AI into education is becoming increasingly significant, leading to changes in both teaching methods and student learning. The aim of this study is to compare the use of AI technologies by students from different majors and to investigate the awareness and adoption of AI tools among university faculty members. Incorporating a quantitative approach, data were collected through online surveys from 59 English for Special Purposes (ESP) teachers and 158 students from the Russian Technological University MIREA and the Ural State Law University named after V.F. Yakovlev. The results demonstrated a higher frequency of AI use among students compared to teachers, as well as marked differences in use depending on academic discipline. The results highlight the need for targeted training and support for both students and teachers to maximise the benefits of AI in education. Integrating AI tools into the curricula, redesigning assignments and encouraging the ethical use of AI are important steps to improve the educational experience and prepare students for a technologically advanced future.

Keywords: Artificial Intelligence (AI) · Online Survey · AI Tools · AI Use · ChatGPT

1 Introduction

In recent times, there has been a rapid spread of artificial intelligence, even in the areas that were traditionally considered human's prerogative. Today we see that artificial intelligence (AI) has an enormous potential. It is capable of generating meaningful and coherent texts, writing music, managing personal finances, creating systems to optimize users' spending and savings, managing traffic, performing medical diagnostics, and much more. Innovation has a major impact on all areas of human life. "Applying the potential of artificial intelligence in various fields could open a new era of innovation: already, artificial

R. Silhavy and P. Silhavy (Eds.): CoMeSySo 2024, LNNS 1489, pp. 183–192, 2025.
https://doi.org/10.1007/978-3-031-96798-6_17

intelligence and education are increasingly seen as a possible ecosystem" [1]. Therefore, there are significant changes in the scientific and educational environment, especially in higher education. It was noted by a number of researchers, e.g. Levitan K.M [2]. and Chernova N.I [3]. Basically, such transformations come down to the proper preparation of students for successful professional activity in the realities of the new technological world. Students and lecturers "gain unprecedented control over their information space and its joint use" [4]. AI can be used to analyze student data and provide personalized recommendations for course materials and activities, as well as feedback and support to help students stay on track and make progress. "The education system must promptly and flexibly respond to modern challenges and be open to the introduction of advanced technologies that create more optimal conditions for students' learning" [5]. There have been lots of arguments around the development of artificial intelligence as having more potential to change higher education than any other technological advancement. Being given the importance of integrating AI in education, there is a need to reflect on its implications.

1.1 Literature Review

The term *artificial intelligence* is not new. It was coined in 1956 by John McCarthy. "The study [of artificial intelligence] is to proceed on the basis of the conjecture that every aspect of learning or any other feature of intelligence can in principle be so precisely described that a machine can be made to simulate it" [6]. The definition of *artificial intelligence* has dramatically changed since 1956, as there have been significant advancements in AI capabilities. "Artificial intelligence is the technological future that happens to make the lives of human beings a lot easier. It is a booming technological domain capable of altering every aspect of our social interactions. In education, AI has been seen to have already begun initiating new teaching and learning solutions that are currently under trial and undergoing restructuring in different contexts [7].

Artificial intelligence holds great promise for higher education as it has more potential to reshape teaching and learning experience. "AI has the potential to revolutionize higher education by enabling new forms of learning assessment, and personalization" [8]. "AI can also be used to analyze substantial amounts of student data to identify patterns and predict student outcomes that can inform teaching strategies and interventions" [9]. "While the use of AI in higher education is still in the preliminary stages, it has the potential to transform the way students learn and teach" [8]. However, in terms of the direct impact of AI on teaching and learning, much has been said, but little has been accomplished. Faculty members have been slower than students to adopt AI tools. "The changing student population and simultaneous technological advances should be a stimulus for evaluation and potential modification of policy and pedagogical approaches within the traditional classroom and experiential learning settings" [10]. Furthermore it is important to offer support to students and university faculty on AI technologies "by examining their perceptions and concerns, in order to reduce the gap of their expectations for promoting seamless integration and collaboration, ultimately improving the overall learning experience and harnessing the full potential of AI-driven educational tools" [10].

Thus, the purposes of the study are to compare the practice of using artificial intelligence technologies by students from different majors and to explore the awareness of artificial intelligence tools by university faculty.

2 Methods

2.1 Study Design

As digital technologies in education continue to evolve, there is a need to analyze their use at different stages of the educational process. To achieve the study's objectives, a quantitative approach was employed, involving data collection through online surveys and subsequent statistical analyses. The authors of the article worked out questionnaires to identify the AI toolkits and the practice of their application by the students of humanities and engineering specialties in educational activities and by university teachers in pedagogical activities. Data collection was realized by means of an online survey. In each questionnaire, respondents were asked to give examples of AI technologies in use, indicate the frequency of use of these software products, identify the range of tasks for which AI technologies are used, express their opinion on the advantages and limitations of integrating AI into educational and professional activities, and assess the ethical aspects of using AI systems.

2.2 Participants

The study sample comprised 59 English for Specific Purposes (ESP) teachers with experience ranging from 1 year to 35 years (10% - teachers of secondary special education, 82% - of higher education, 8% - of other directions) and 158 undergraduate and specialist students of 1–3 courses (76 law students and 82 engineering students) from the Russian Technological University MIREA, Moscow, and the Ural State Law University, Yekaterinburg.

2.3 Data Collection

Adaptive mode in Moodle is a feature that allows instructors to create personalized learning paths for their students based on their individual needs and performance. The Adaptive Mode feature in Moodle uses rule-based mechanisms that automatically adapt course content, assignments and grades according to each student's learning needs.

2.4 Questionnaire Structure

The questionnaires were organized into five main categories:

1. Demographics
2. Usage of AI
3. Benefits and Challenges
4. Perception and Ethics
5. Training and Support

2.5 Data Analysis

Collected data were analyzed using Microsoft Excel. Descriptive statistics were employed to comprehensively summarize and interpret the dataset. This methodological approach facilitated the derivation of meaningful conclusions and the effective addressing of the research questions. The analysis of respondents' answers allowed us to identify common trends and differences in the experience of practical application and in the attitude of students and teachers to the development of information and educational innovations on AI basis.

3 Outcomes

The obtained results revealed that 9 out of 10 students regularly use neural networks (from several times a week to daily use), while among the teaching staff these figures are lower (68%). The reasons for the unequal proportion of students and faculty in terms of frequency of interaction with AI are made clear in the respondents' answers to the subsequent questions.

4 Discussions

The given examples of AI services confirm that students of different fields of study and professors use the same software products to solve meta-tasks. The most popular are services assisting in information search and generating contextually relevant texts (ChatGPT) and services for online translation of texts and documents with the function of voice input, text recognition from photos and voicing of the finished product (Deepl Translate, Google Translate, Reverso).

The answers of the survey participants regarding the performance of highly specialised tasks look different. When preparing for classes, representatives of legal specialities most often use services for automated creation and management of standard documents, including those in foreign languages, and for studying the regulatory framework and judicial practice. Only 12% of future lawyers use AI technologies to solve legal situations (cases) modelled for educational purposes. These results are due to the fact that, on the one hand, law students are gaining considerable experience in applying partially search and heuristic methods of problem solving in the classroom and are more oriented towards independent search for arguments to justify their point of view, and on the other hand, AI for lawyers cannot yet cope with large multidimensional tasks (Fig. 1).

Future engineering specialists are more actively turning to AI technologies to solve complex learning tasks (28%), and almost all engineering students use AI models for programming purposes (97%).

It is indicative that engineering students are ahead of law students in terms of using AI technologies for humanities assignments, for example, for creative writing such as essays (70% and 50% respectively). But law students are more likely than engineering students to call on AI (Gamma, Tome.app) for technical issues, such as creating high-quality presentations with a large repertoire of interactive features (87% and 75%).

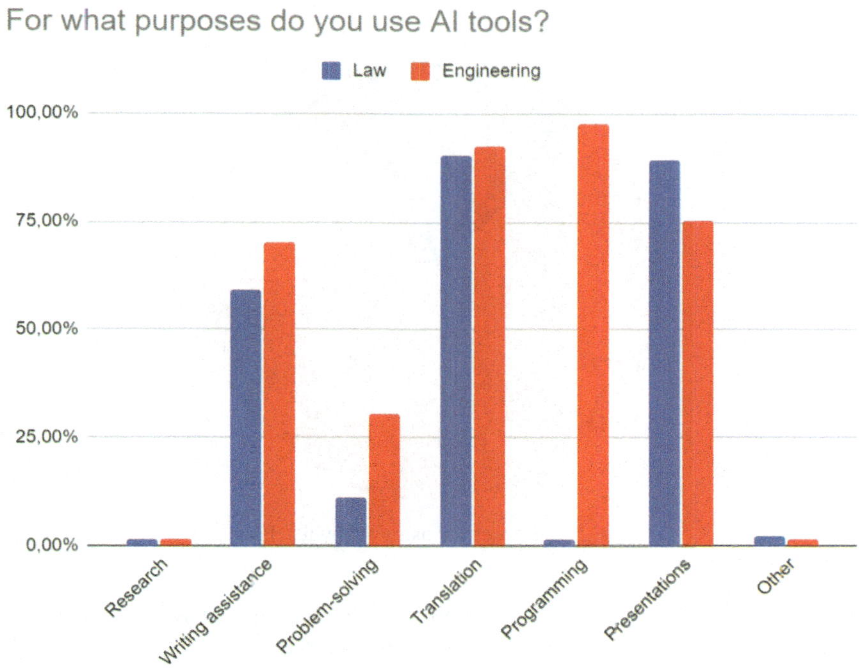

Fig. 1. Purposes for using AI by students.

Only a few 3rd year law students (1% of respondents) pointed out the potential of AI models in research activities, explaining that they significantly increase access to specialised knowledge, speed up the initial stage of scientific literature analysis, identification of key research vectors and selection of sources for citation (Fig. 2).

In pedagogical activities, the use of didactic potential of AI is aimed, first of all, at preparation for classes. This is indicated by 78% of teachers. The functionality of modern resources allows for the rapid creation of auxiliary content on the required topic. For example, the most frequently mentioned neural network for teaching English, twee.com, offers 6 sections (reading, vocabulary, writing, speaking, listening, grammar), each of which generates a wide range of different types of tasks. On the other hand, with a growing number of promising AI services designed to assess student work and provide feedback to students, only 2% of teachers use these tools. Most likely, this fact is explained by teachers' low awareness of software products appearing on the market.

The section of the questionnaire concerning aspects of academic integrity revealed different attitudes of students and teachers towards the creation of authentic and genuine works. Academic integrity reflects a set of values and principles of behaviour of participants of the educational process when performing assessed and unassessed works. 31% of lecturers categorically insist and 67% recommend paying attention to correct ways of citing and creating references to sources in the context of recognising the ideas of other authors. Students are less concerned (17%) about copyright infringement and see

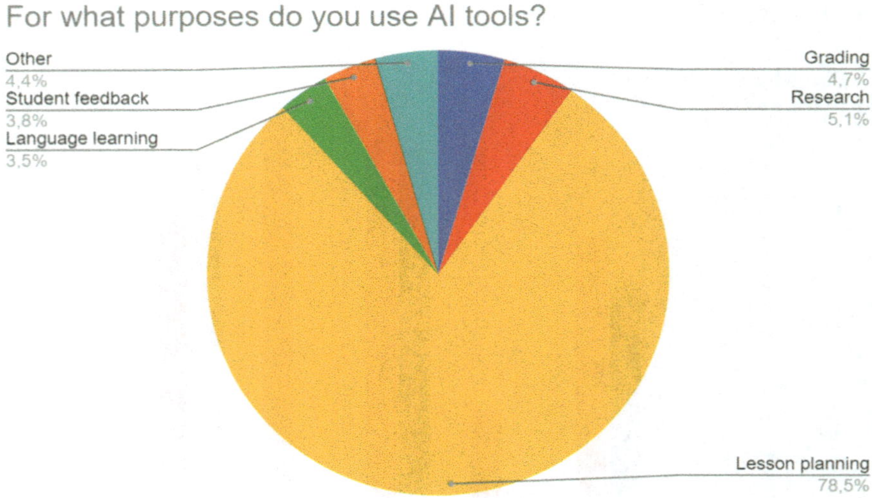

Fig. 2. Purposes for using AI by teachers.

information gathering as a means of compiling, summarising and systematising data to determine their own vectors of reasoning on a given topic.

The results presented in the graphs (see Figs. 3 and 4) also demonstrate a cardinal divergence in the opinion of students and teachers regarding the need for specialised training in modern AI technologies in order to obtain comprehensive knowledge about the developments and their potential in different industries. Of the student respondents, only 20% are willing to take advantage of pedagogical assistance. This is due to the fact that the current generations of students have spent their entire lives surrounded by the tools of the digital age and are "speakers" of the digital language of the Internet. They have spent their entire lives surrounded by tools of the digital age and they are all "native speakers" of the digital language of the Internet. "They spend most of their lives online, and the process of absorbing new digital and technological content is natural for them" [11]. At the same time, higher education teachers, especially those who were not born in the digital age, have to learn new approaches and new methods of working with artificial intelligence software by trial and error. The majority of respondents in this group (77%) agree that targeted training in AI technologies is a key strategic priority in education today.

Fig. 3. Need for support (students' answers).

Fig. 4. Need for support (teachers' answers).

5 Conclusion

The rapid spread of artificial intelligence (AI) in various fields, including education, demonstrates its significant potential to revolutionise traditional human activities. AI capabilities range from creating coherent texts and composing music to managing personal finances and conducting medical diagnostics, demonstrating its wide-ranging applications. In the education sector, the impact of AI is particularly significant, offering personalised learning experiences, optimising student support and changing teaching methodologies. As AI is implemented, higher educational establishments need to take into account new technological realities to prepare students for careers in a technologically advanced world.

The findings of the study identify common trends and differences in AI use and attitudes, highlighting the need for personalised support and training to maximise the benefits of AI in education.

To fully harness the potential of AI in education, there is a need to integrate AI tools into curricula, redesign assignments to take advantage of AI capabilities, update assessment methods, and develop students' critical thinking skills. In addition, educating teachers, promoting the ethical use of AI, and encouraging collaborative learning are critical steps towards creating an optimal AI-enabled learning environment.

By implementing these strategies, higher education institutions will be able to improve the quality of learning, better prepare students for the future, and ensure the adoption of AI in the educational process. The use of AI will not only intensify learning and improve student outcomes, but will also equip learners with the skills needed to navigate and thrive in an increasingly AI-driven world.

6 Recommendations

ChatGPT and other artificial intelligence tools are becoming more and more common in educational environments, which requires a reevaluation of traditional teaching methods. Based on our findings, we offer the following recommendations to foreign language faculty to improve the educational process for law and engineering students.

Integration of AI tools into the curriculum. Teachers should actively incorporate AI tools, such as ChatGPT, into their curricula. This may include using AI to facilitate interactive tutorials, provide instant feedback and support personalized learning paths for students. For engineering and law students, this could include AI-assisted research of various kinds or exercises to solve tech problems.

Redesigning assignments to utilize AI capabilities. Traditional assignments should be restructured to utilize the capabilities of AI tools. For example, law students could be assigned to use AI to analyze a case or draft legal documents, and engineering students could be given the task of data modeling and analysis. This will not only introduce students to AI, but also encourage them to use these tools effectively.

Revising assessment methods. Assessment strategies should be updated to reflect the skills needed in an AI-enabled educational environment.

Developing critical thinking skills. In order for students to critically evaluate and utilize the results of AI, educators should focus on developing students' critical thinking skills. This includes teaching students how to evaluate the validity of information generated by AI, understand the AI's limitations, and apply ethical considerations when using AI tools. This is particularly important for law students who need to critically evaluate legal information and engineering students who are learning to assess tech data.

Providing faculty training and support. English teachers at technical universities and law schools need ongoing professional development to effectively incorporate AI tools into their teaching practice. Educational institutions should provide training, workshops, and resources to help teachers understand how to use AI tools and incorporate them into their teaching strategies.

Promoting the ethical use of AI. Educators should emphasize the ethical use of AI in the classroom. This includes teaching students the importance of academic integrity, the potential bias of AI systems, and the responsible use of AI technologies. It is important for law students to understand the ethical implications of using AI in legal practice, and engineering students should understand the ethical aspects of developing and using AI.

Encouraging collaborative learning. AI tools can facilitate collaborative learning. Instructors need to design assignments that encourage students to work together using AI tools, fostering a collaborative learning environment that capitalizes on the strengths of each team member. This is beneficial for both law and engineering students because it prepares them for teamwork in their fields.

By adopting these recommendations, faculty at both engineering universities and law schools will be able to enhance learning, better prepare students for the future, and ensure that the integration of AI tools such as ChatGPT makes a positive contribution to the educational process.

References

1. Vezetiu, E.V., Romaeva, N.B.: Artificial intelligence as an innovative tool for implementing modern learning tools in the educational process of higher education institutions. In: Problems of Modern Pedagogical Education No. 77-2, pp. 73–77 (2022)
2. Modern Problems of Teaching Students Foreign-Language Legal Discourse: Proceedings of the All-Russian Scientific and Methodological Round Table with International Participation, Yekaterinburg, 26 February 2021. In: Levitan K.M. Yekaterinburg: Federal State Budgetary Educational Institution of Higher Education "Ural State Law University", p. 181 (2021)
3. Chernova, N.I., Ivanova, E.A., Bogush, N.B., Katakhova, N.V.: Technology for determining non-humanities university students' cognitive-and-psychological characteristics. Russ. Technol. J. **11**(3), 104–116 (2023). https://doi.org/10.32362/2500-316X-2023-11-3-104-116. EDN SMXMAL
4. Difficulties and Prospects of Digital Transformation of Education. A.Y. Uvarov, I.D. Frumin. Uvarov, I.D. Frumin; Scientific editors of the series Ya. I. Kuzminov, I.D. Frumin / A.Y. Uvarov, E. Gable, I.V. Dvoretskaya et al.. Moscow: National Research University Higher School of Economics, p. 344 (2019) doi: https://doi.org/10.17323/978-5-7598-1990-5. (RUSSIAN EDUCATION: ACHIEVEMENTS, CHALLENGES, PROSPECTS). ISBN 978-5-7598-1990-5. EDN ANYGHO.modern pedagogical education (2022). №77-2
5. Sysoyev, P.V.: Artificial intelligence in education: awareness, readiness and practice of using artificial intelligence technologies in professional activities by university faculty. Vyssh. Obraz. Ross. **32**(10), 9–33 (2023). https://doi.org/10.31992/0859-3617-2023-32-10-9-33
6. Russell, S., Norvig, P.: Artificial Intelligence: A Modern Approach, 3rd edn. Pearson, London (2010)
7. Bostrom, N.: Superintelligence: Paths, Dangers, Strategies, vol. 352. Oxford University Press, Oxford (2014) ISBN 978-0-19-967811-2
8. Shute, V.J.: AI in education: a review of research. Educ. Res. Rev. **22**, 1–22 (2018)
9. Zhang, J.J.: Big data analytics in education: a review of research. Educ. Res. Rev. **22**, 23–41 (2018)

10. Chan, C.K.Y., Lee, K.K.W.: The AI generation gap: are Gen Z students more interested in adopting generative AI such as ChatGPT in teaching and learning than their Gen X and millennial generation teachers? Smart Learn. Environ. **10**, 60 (2023). https://doi.org/10.1186/s40561-023-00269-3
11. Prensky, M.: Digital natives, digital immigrants. On Horizon. MCB University Press. **9**(5), 1–6 (2001)

Information Process Management Based on Trajectory Analysis in Multidimensional Metric Spaces

Ju. B. Shcheneva(✉), A. N. Pylkin, S. A. Bubnov, A. A. Bubnov, O. A. Bodrov, and E. S. Shchenev

Ryazan State Radio Engineering University named after V. F. Utkin, 59/1, Gagarina street, Ryazan 390005, Russia

Abstract. The article deals with the information process management features based on the trajectories analysis in multidimensional metric spaces. Particular indicators that characterize the dynamics of information process management and determine the metric space dimension are considered. The approaches choice for finding a generalized metric in n–dimensional space is justified. Based on the analyzed indicators results, the approach for constructing trajectories in multidimensional metric spaces, which is necessary to improve the information process management efficiency, is explained. #COMESYSO1120.

Keywords: Information Process · Multidimensional Metric Spaces · Dimension of the Metric Space · Particular Indicators · Generalized Indicator

1 Introduction

It is necessary to process and analyze large amounts of information which often leads to huge time costs for effective information process management. Therefore, the most promising area is the development of solutions aimed at creating mathematical, software and hardware tools that are designed to obtain fast and high-quality data processing and analysis results. Multidimensional data structures can be processed using data mining software tools [1].

One of the possible approaches to manage the information process is to build based on the trajectories analysis in multidimensional metric spaces model. To obtain management technology is necessary to provide development of classifying methods and ordering the main indicators groups for which data analysis is performed.

It is worth to note that applying decision-making methods and modeling tools to the same group of data in different metric spaces often leads to different results. This indicates that the tools used should be based on a variety of modeling methods in multidimensional metric spaces, corresponding algorithms, and data mining tools.

R. Silhavy and P. Silhavy (Eds.): CoMeSySo 2024, LNNS 1489, pp. 193–207, 2025.
https://doi.org/10.1007/978-3-031-96798-6_18

2 Methods

Initially, the information process can be defined by a certain number of indicators (partial indicators) that maximally characterize process control dynamics and determine metric space dimensions in which the model is built. By setting specific values for particular indicators and normalizing the obtained values, you can get the value of the generalized indicator, which can be obtained as the distance between points in an n–dimensional space.

While finding a generalized indicator the choice of approach based on the task requirements is determined. In the first step the distance from the studied object to the "ideal" one (with the maximum characteristics) is calculated for each case, in the second step the distance from the studied object to the "zero" one (with the minimum characteristics) is calculated.

We study a variety of metrics to correctly calculate the value of a generalized metric. The metrics are: Euclidean distance, square of Euclidean distance, Chebyshev metric, Minkowski distance, Hausdorff distance, etc. A comparative analysis of the generalized indicators obtained values and trajectories in the considered metrics allows to choose the most acceptable option for obtaining a high-quality result.

It is necessary to reduce the distance between points in the multidimensional metric space to increase the efficiency of process control for the first step and, conversely, to increase the distance when implementing the second step of the approach.

The generalized indicator in the considered metrics acts as a metric space measure. On the basis of the measure trajectories are constructed and subsequent comparative analysis of current trajectory is occurred. This trajectory characterizes a certain information process with selected during the study optimal trajectories.

The process is followed by the indicators number changing at this very moment. The multi-criteria evaluation carries out with these indicators. The volume and complexity of such criteria organizing often do not allow to effective use traditional analysis technologies, which leads to artificial intelligence (AI) algorithms Data Mining (DM) tools using.

Evaluation of the effectiveness of information process management can be carried out using multi-criteria evaluation.

As an example of building a model, integral data from educational process, medicine, manufacturing, military industry, economics, banking, marketing, telecommunications, web analytics can be considered. The rapid increase of the information amount serves as a convincing argument in favor of implementing developments. This information can be processed and analyzed in various applied fields and be aimed at creating mathematical, software and hardware tools. In its turn these tools designed for quick and efficient data processing and analysis problems solving.

Let us set the number of investigated events equal to n, where $n > 0$. This value specifies the number of private indicators that will be used later in the study. It is worth noting that number determining of private indicators affects to the dimension choice of the space in which multidimensional complex data is processed. While value n increasing, the factors that affect the information process can be presented in more details. When indicators are convoluted, the dimension of the space decreases. The increase of n value can significantly reduce the weight of an indicator, and the reduction of the n can lead to the important information loss related to the information process management effectiveness.

In accordance with the above considerations, the process under consideration is characterized by a certain number of partial indicators $w_1^*, w_2^*, \ldots, w_n^*$. Each of the private indicators is limited to minimum w_{imin}^* and maximum w_{imax}^* values that can be set based on the logic of reasoning.

It is necessary to normalize the particular indicators values of the studied object to correctly use a particular indicator in a generalized one. The most preferred method is one that converts the particular metric numeric value w_i of the studied object in segment [0; 1]:

$$w_i = \frac{w_i^* - w_{imin}^*}{w_{imax}^{*imin},}, \tag{1}$$

where w_i^* – the corresponding indicator current value; w_{imax}^*, w_{imin}^* – maximum and minimum values of the consideration indicator.

After performing obtained values normalization, it is necessary to get the generalized indicator value, which can be defined as the distance between points in *n*–dimensional space. Indicators $w_1, w_2, \ldots, w_n$ can be represented as a certain point in *n*–dimensional space. To calculate the value of the generalized indicator, necessary to select a metric that will allow getting the most optimal principle for modeling trajectories in the following way: *n*–dimensional space. Generalized metric calculating in multidimensional metric spaces can lead to different results on the same data set.

The generalized metric calculation in Euclidean metric space should start by a partial indicators set considering that define a point in the *n*–dimensional Euclidean space Rn, i.e. $W = (w_1, w_2, \ldots, w_n)$.

If this point is correlated with the studied object indicators, which has a position in the list under consideration j, then $w(j)= w_1^{(j)}, w_2^{(j)}, \ldots, w_n^{(j)}$ should be denoted, where j = 1..m, m – studied objects amount.

To calculate the generalized metric value, it is necessary to find the distance between a point w(j) and a dot w(k) (appropriate j–value and k–value objects). This is a simple Euclidean distance, which is determined by the formula:

$$d_{jk} = \sqrt{\sum\nolimits_{i=1}^{n} \left(w_{ji} - w_{ki}\right)^2}, \tag{2}$$

where w_{ji}– private *i*-th indicator of the *j*–th object, w_{ki} – private *i*-th indicator *k*–th object.

After generalized indicator calculating in the Euclidean metric for the studied objects (for considered example, the dimension of space is chosen – 12), trajectories relative to the “zero” and “ideal” objects are obtained (Fig. 1).

These dependencies make possible an obtaining of additional information on managing the information process based on the obtained trajectories analysis. Therefore, the closer distance from the considered object to the “ideal” trajectory, the more effective considered object indicators, i.e. the generalized indicator value in the analyzed area does not increase relative to the “ideal” object. And, the further distance from the trajectory of the “zero” object to the considered studied object trajectory, the more effective

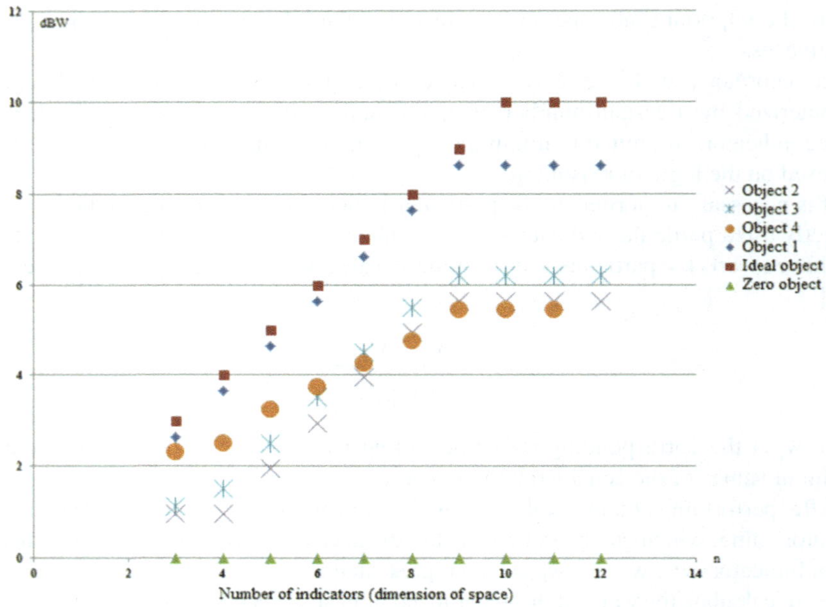

a)

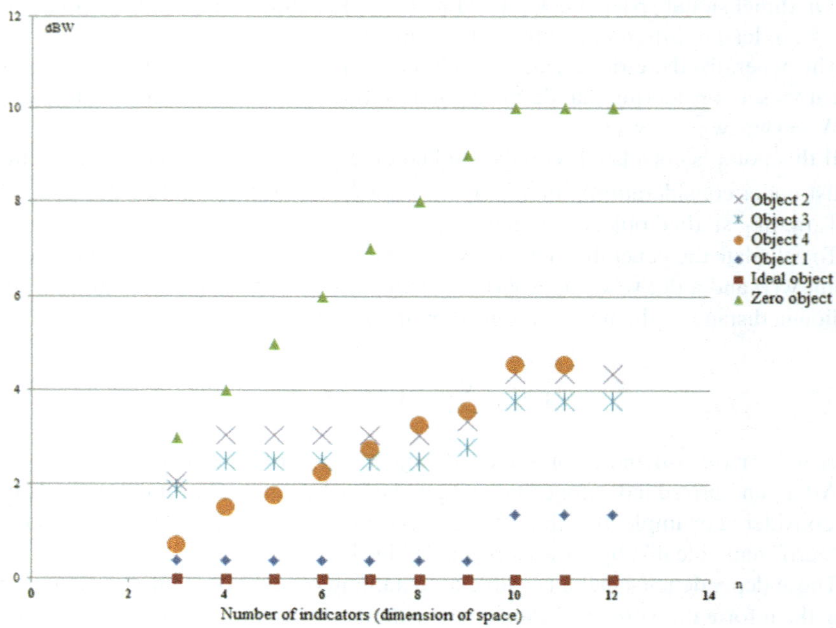

b)

Fig. 1. Trajectories relative to the "zero" (a) and "ideal" (b) objects in the Euclidean metric indicators, i.e. the generalized indicator value in the analyzed area increases relative to the "zero" object.

While analyzing trajectories in the weighted Euclidean distance metric, necessary to take into account that the variables w_{ji}, i = 1, 2,…,n ranked among themselves according to the importance degree, i.e. they are assigned weights c_i. Particular indicator weight determines the importance of taking this feature into account when classifying process is going on. While calculating the similarity measure, the weight of the metric is taken into account. The indicator weight is usually estimated by expert assessments.

The weighted Euclidean distance will be determined by the formula:

$$d_{jk} = \sqrt{\sum_{i=1}^{n} \left(w_{ji} - w_{ki}\right)^2 \bullet c_i}, \tag{3}$$

where c_i– i–th indicator weight; $c_i > 0$, $\sum_{i=1}^{n} c_i = 1$.

The generalized metric indicator calculation in the weighted Euclidean distance metric for the studied objects (for the example under consideration, the space dimension equals to 12 is also chosen) allowed to obtain trajectories relative to the "zero" and "ideal" objects (Fig. 2).

The obtained dependences show that in order to obtain additional information on information process management based on the obtained trajectories analysis. It is also necessary to take into account weights that allow to determine deviations from the "ideal" and "zero" trajectories with high accuracy thereby evaluating the information process management effectiveness. However, there may be situations where the weights of c_1, c_2,…, c_n is difficult to determine. In this case it is recommended to use the simple Euclidean distance metric. If all factors weights are the same, and one of them needs to weight increase, then this is determined by a survey of experts or policy documents.

If large weights are assigned to objects that are most distant from each other, is recommended to use the Euclidean distance squared measure

$$d_{jk} = \sum_{i=1}^{n} \left(w_{ji} - w_{ki}\right)^2, \tag{4}$$

It is considered that the measure use is particularly important for calculating the relative values of indicators [2].

After generalized indicator calculating using the Euclidean distance squared measure for the studied objects (space dimension equals to 12), trajectories relative to the "zero" and "ideal" objects are obtained (Fig. 3).

In cases where is necessary to distinguish two generalized indicators by some particular indicator, (i.e. the metric distance) should be used Chebyshev distance. The Chebyshev distance is defined as the maximum distance between the corresponding partial exponents:

$$d_{jk} = \max_i \left|w_{ji} - w_{ki}\right|, \tag{5}$$

The generalized indicator calculation in the Chebyshev metric (space dimension equals to 12) allows to obtain trajectories relative to the "zero" and "ideal" objects (Fig. 4).

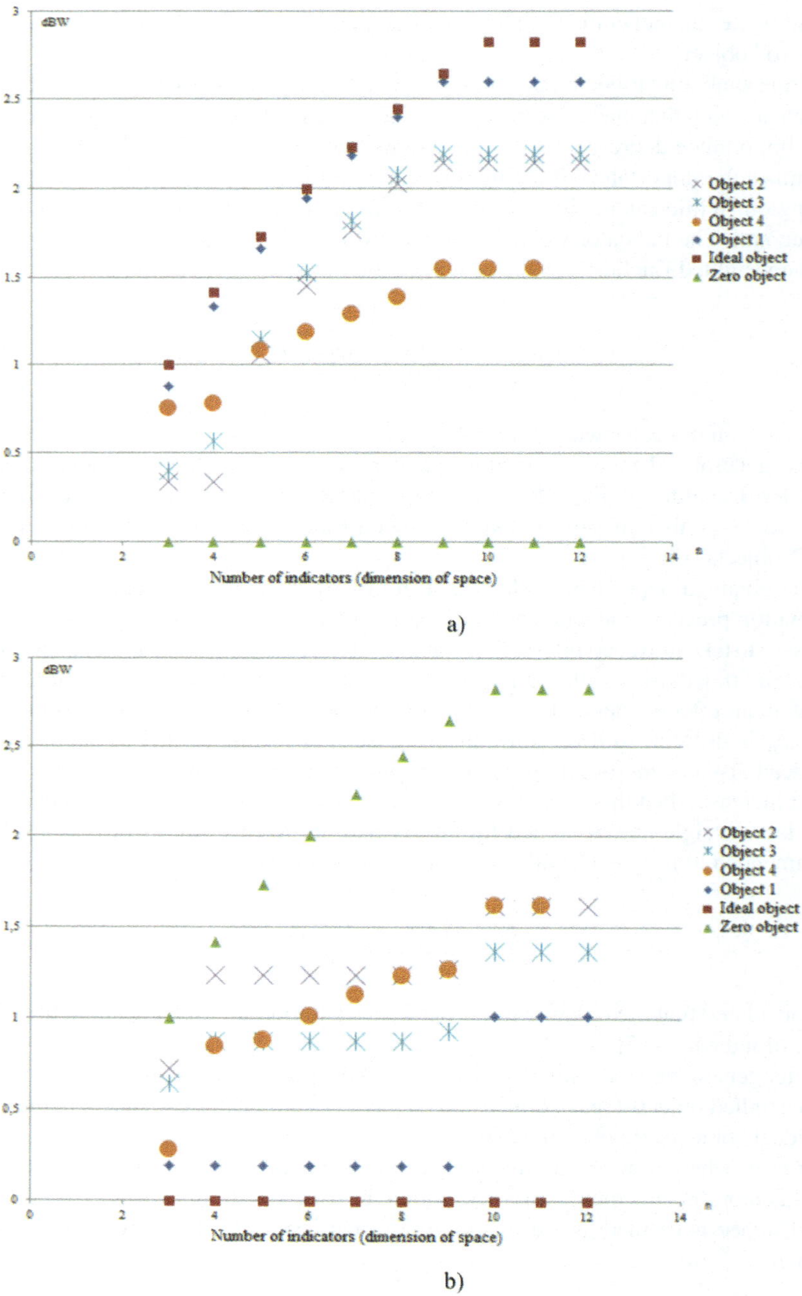

Fig. 2. Trajectories relative to "zero" (a) and "ideal" (b) objects in the weighted Euclidean distance metric

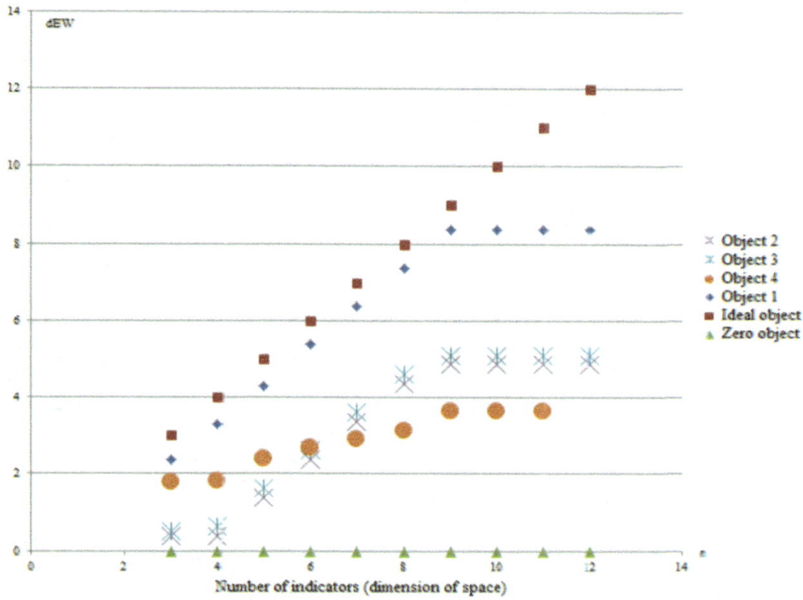

a)

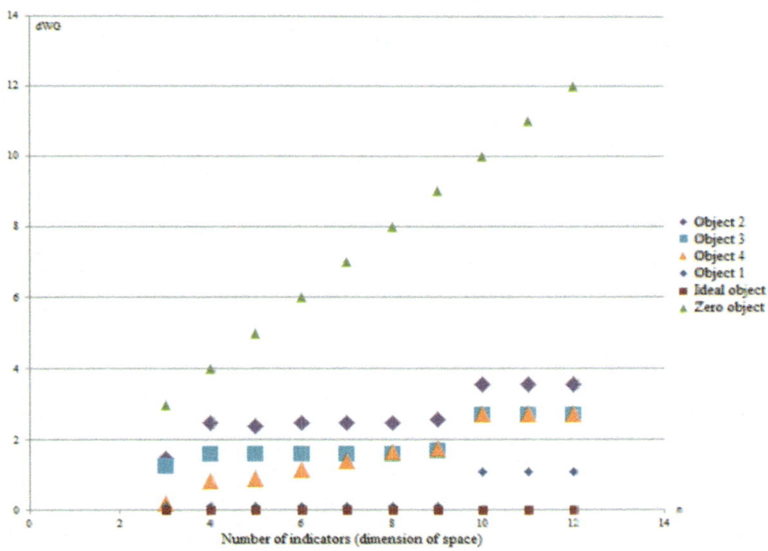

b)

Fig. 3. Trajectories relative to "zero" (a) and "ideal" (b) objects in the Euclidean distance squarec metric

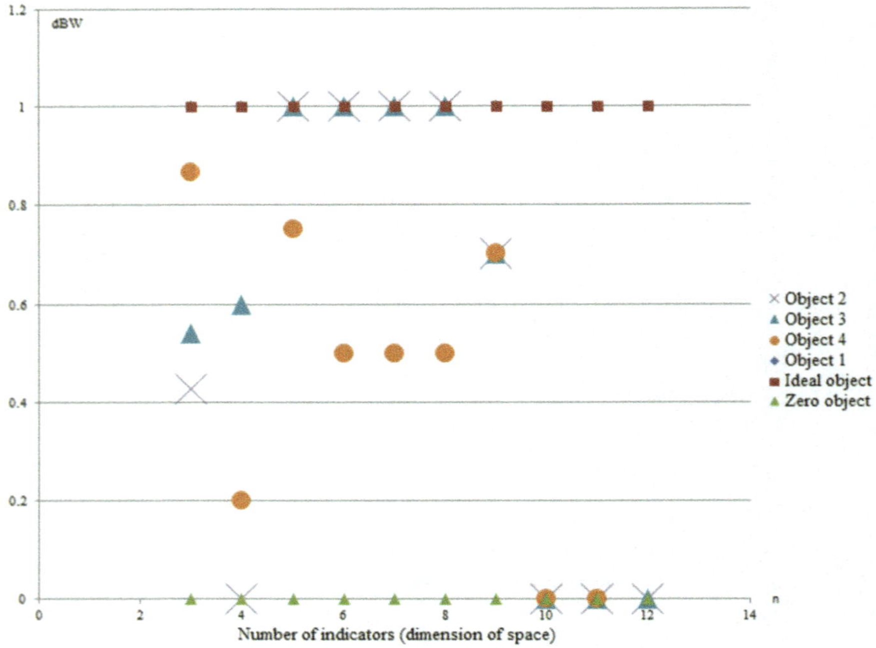

a)

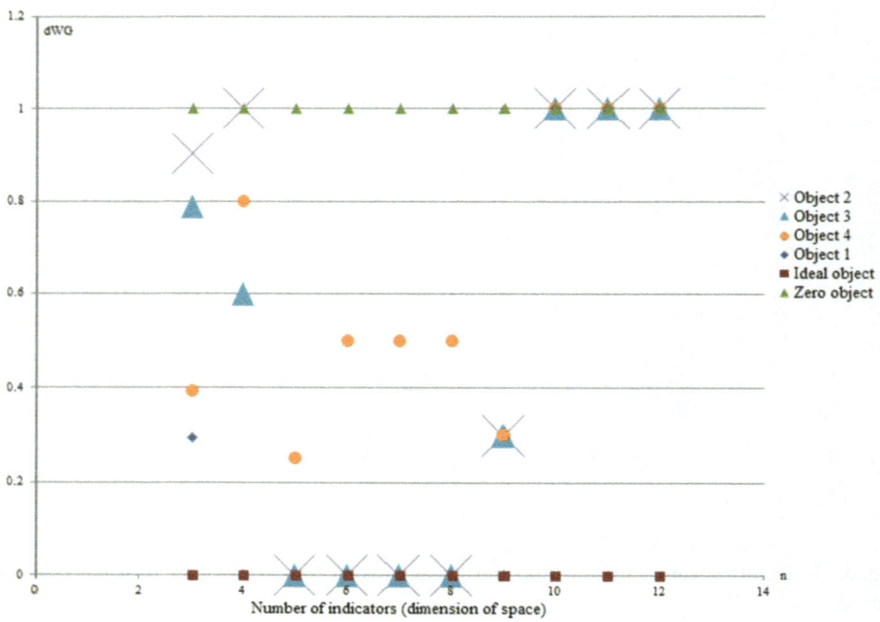

b)

Fig. 4. Trajectories relative to "zero" (a) and "ideal" (b) objects in the Chebyshev metric

After analyzing the obtained trajectories shown in Fig. 4, the main disadvantage of model constructing can be concluded: the Chebyshev distance is a rough measure of the difference due to ignoring a significant amount of available information.

A various approaches generalization to determining distance in n-dimensional space is the Minkowski distance:

$$d_{jk} = \left(\sum_{i=1}^{n} \left| w_{ji} - w_{ki} \right|^{p} \right)^{\frac{1}{p}}, \tag{6}$$

For $p = 1$, the Hamming distance (distance of city blocks, Manhattan distance) is obtained:

$$d_{jk} = \sum_{i=1}^{n} \left| w_{ji} - w_{ki} \right|, \tag{7}$$

Generalized indicator values calculating in the Hamming metric helps to obtain trajectories relative to the "zero" and "ideal" objects in the specified metric (Fig. 5).

For $p = 2$, the Minkowski distance formula corresponds to a simple Euclidean distance (look at formulas (2) and (6)). The trajectories relative to the "zero" and "ideal" objects at $p = 2$ are shown in Fig. 1.

Data analysis shows that in most cases the Hamming distance leads to the same results as in the case of using a simple Euclidean one.

At the next stage of the study objects classification in the Hausdorff metric is considered. Using the Hausdorff distance, the distance between subsets of the metric space can be finded, i.e. the largest possible distance from each point of one set to the nearest point to it of another set can be evaluated [3].

Consider the (X, d) – complete metric space. Let us select from this space the new one – $H(X)$, that is consisting of compact X subsets. For the resulting space $H(X)$, we define the metric:

- distance between $x \in X$ and $T \in H(X)$: $d(x, T) = \min\{d(x, y) : y \in T\}$;
- distance between two compact sets A and T: $d(A, T) = \max\{d(x, T) : x \in A\}$.

The Hausdorff distance from set A to set B is a function defined by formula:

$$d(\mathrm{A}, \mathrm{B}) = \max_{a \in A} \left\{ \min_{b \in B} d(a, b) \right\}. \tag{8}$$

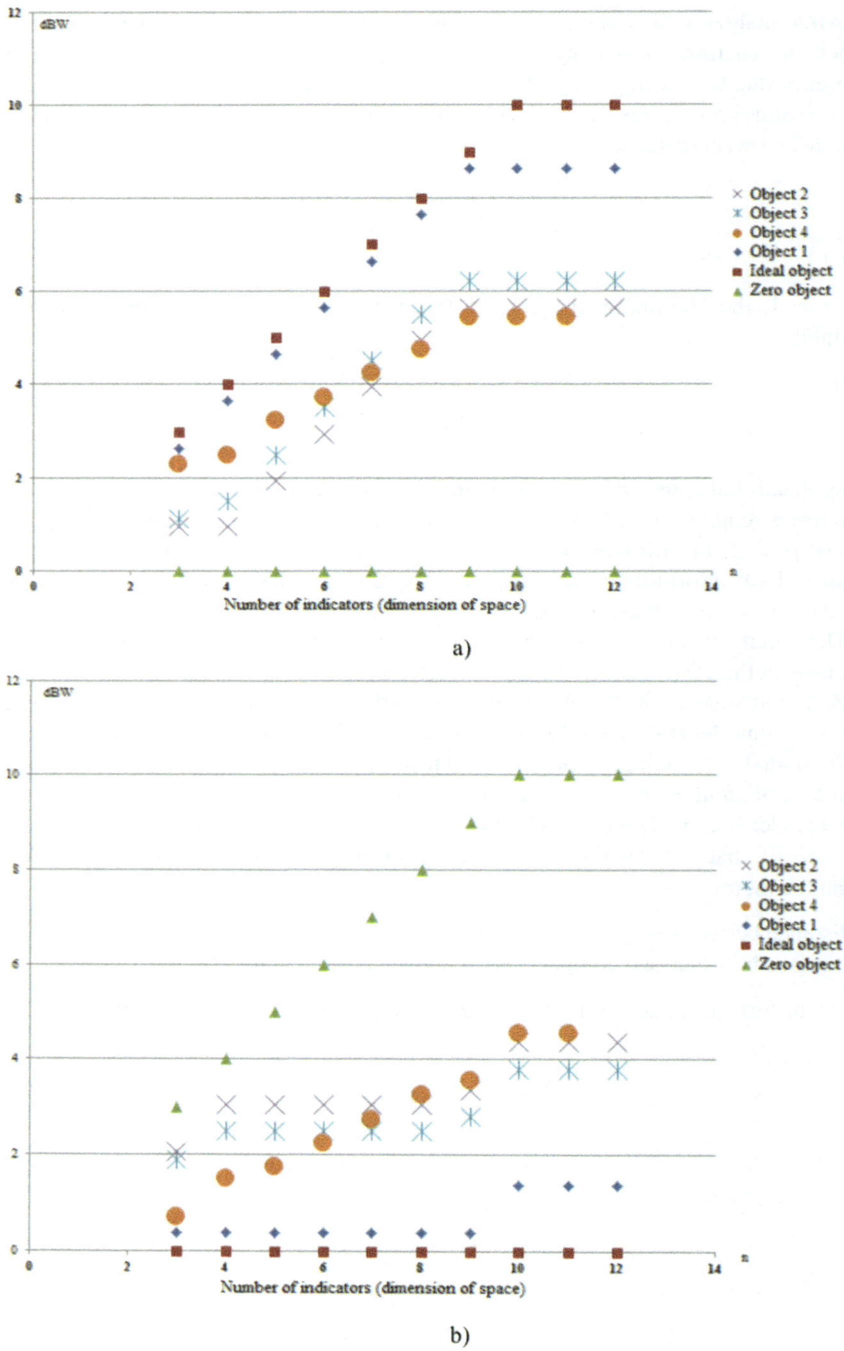

Fig. 5. Trajectories relative to "zero"(a) and "ideal"(b) objects in the Hamming metric

3 Results

Finding the Hausdorff distance algorithm can be represented by the following sequence of steps:

Step 1 Get a sample of data, analyze the obtained values, arrange them according to the studied features, and present them as the studied sets A and B.

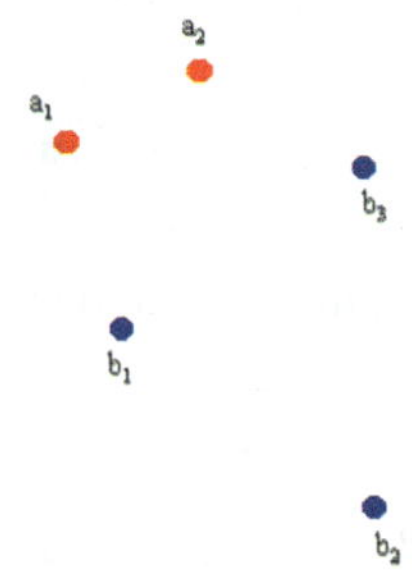

Step 2. Determine the distance between the first element of set A and all elements of set B.

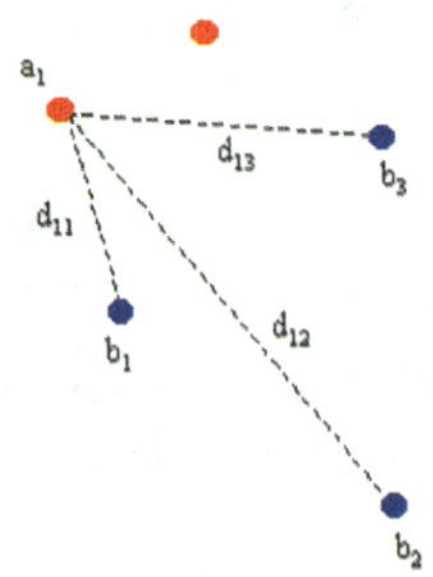

Step 3. Determine the minimum distance among the received values.

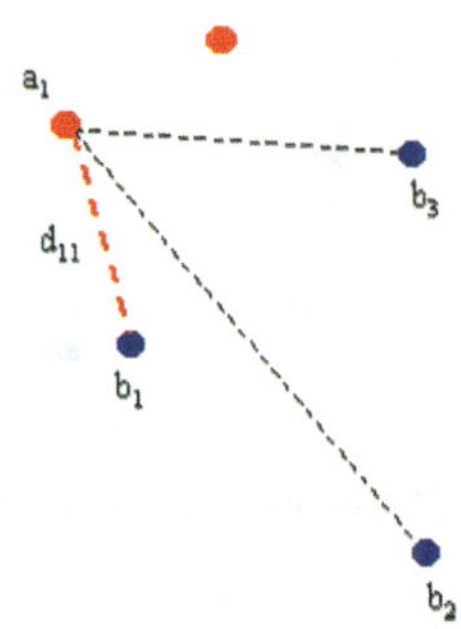

Step 4. Calculate the distance between the next element of set A and all elements of set B.

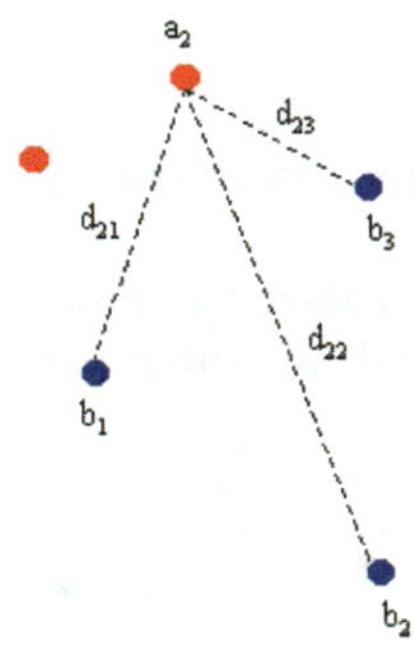

Step 5. Determine the minimum distance among the received values.

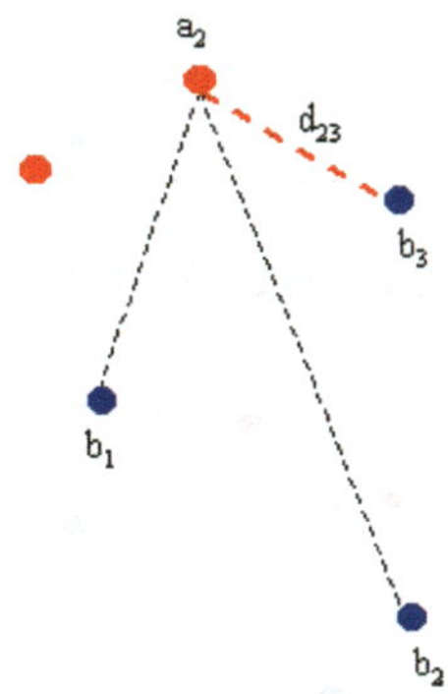

Step 6. Repeat steps 2–3 for all elements of sets A and B.
Step 7. Select the largest distance among the minimum distances obtained.

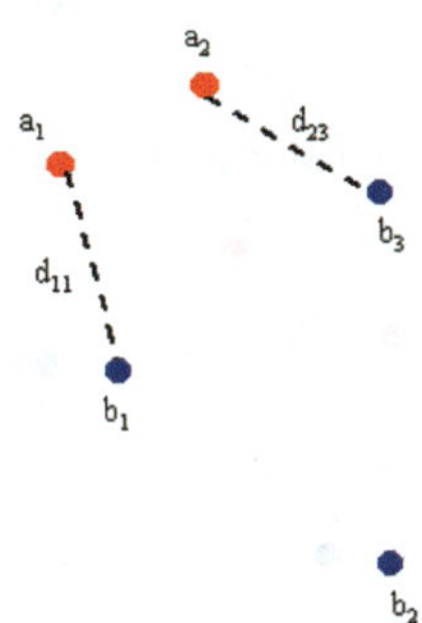

Step 8. Get the Hausdorff distance and apply the resulting value to plot the trajectories.

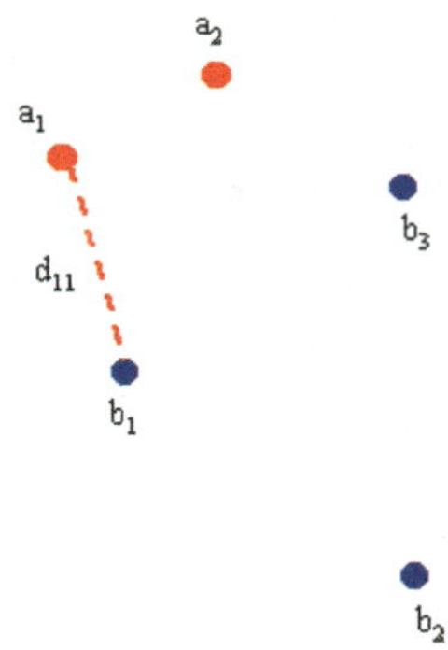

It is worth to note that the Hausdorff distance is directed or asymmetric, i.e. $d(A, B)$ in most cases is not equal to $d(B, A)$ [4].

When performing the clustering procedure, the distance function, based on the metric use, plays an important role [5].

After calculating the Hausdorff distance, using the algorithm described above for the studied objects, we obtained trajectories (Fig. 6). At this stage, the dimension of the space is 12.

4 Discussions

After partial indicators groups analyzing the studied objects and generalized indicators calculations performing that were used to obtain trajectories in multidimensional metric spaces, we conclude that due to the universality, as well as to reduce computational costs, it is advisable to use a simple Euclidean distance to obtain the generalized indicator value. To select the distance function for cluster analysis the Hausdorff metric is preferred.

To test this technique, the most promising trends in the information technologies development in the field of decision support can be chosen. These trends are directly related to the software tools development for data mining (DM) and are focused on processing multidimensional complex data structures.

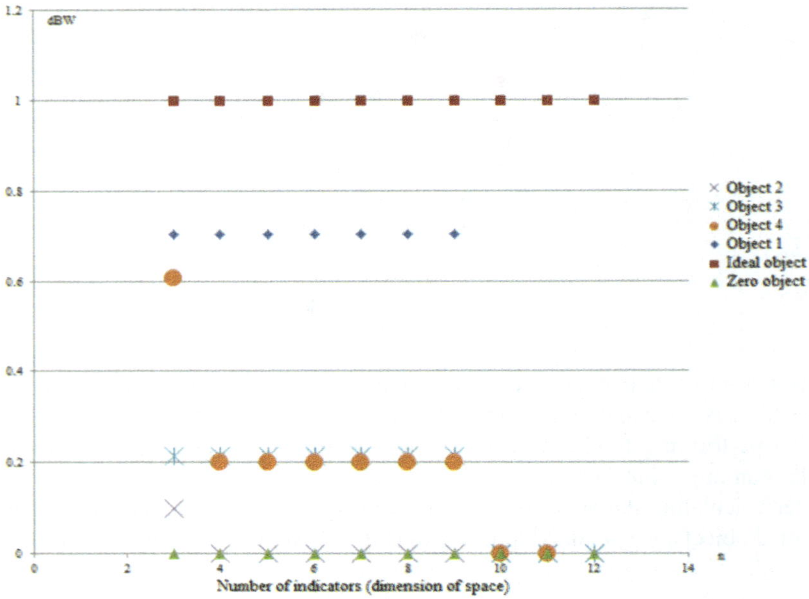

b)

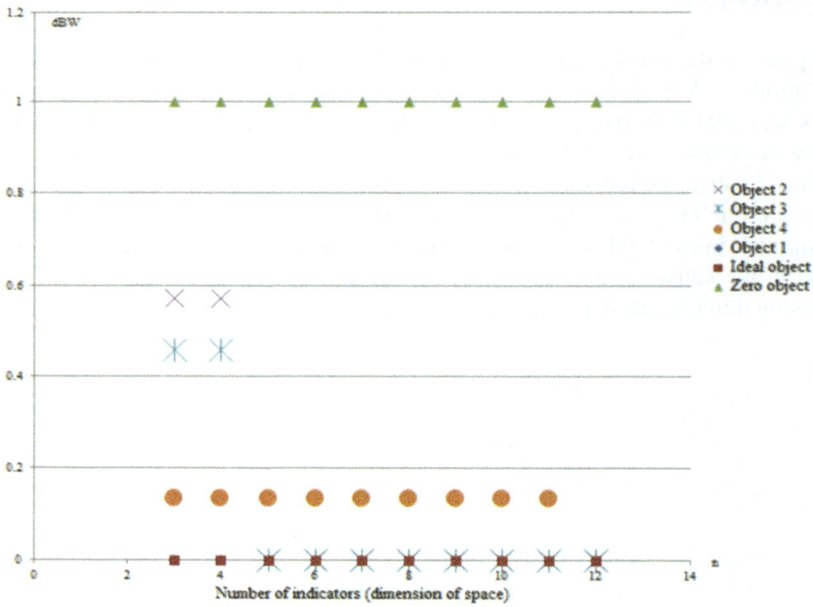

b)

Fig. 6. Trajectories relative to "zero" (a) and "ideal" (b) objects in the Hausdorff metric

5 Conclusion

Summing up the above, we can conclude the following: one of the possible approaches for managing the information process can be considered as a model construction. This construction is obtained on the trajectories analysis basis in multidimensional metric spaces. The values of partial and generalized indicators characterizes the dynamics of information process management and determines the dimension of the metric space. In the indicators number increasing process a trajectory change will be observed. Based on this process analysis a technology for managing the information process using cluster analysis methods can be developed. The uniformity of multicriteria indicators can be determined by correlation and cluster analysis methods. The article uses approaches based on a combination of these methods.

References

1. Shcheneva, Y.B.: Educational competencies development control in the training IT specialists process based on the multidimensional space trajectories analysis. In: Proceedings of the All-Russian NPC with International Participation "Information Exchange in Interdisciplinary Research II. The View of Novice Scientists", pp. 87–91, Ryazan, Russia (2023)
2. Shcheneva, Y.B., Pylkin, A.N., Shchenev, E.S.: Effectiveness indicators formation for the organizational process management of training IT specialists in higher education. In: Proceedings of International Conference "Actual Problems of Natural, Mathematical, Technical Sciences and their Teaching", pp. 207–212, Lipetsk, Russia (2022)
3. Shcheneva, Y.B., Pylkin, A.N., Shchenev, E.S., Bodrov, O.A.: Model of mastering educational competencies using data mining tools. Bull. Ryazan State Radio Eng. Univ. **84**, 119–132 (2023)
4. Shcheneva, Y.B., Pylkin, A.N.: Improving the efficiency for the organizational process management of training IT specialists at the university. In: Proceedings of Interuniversity Conference "Mathematical and Software Support of Computer Systems", pp. 10–13. Ryazan State Radio Engineering University, Ryazan, Russia (2022)
5. Demidova, L.A., Sokolova, Y.S.: Two-level classifier development for complexly organized multidimensional data of large volumes. Bull. Ryazan State Radio Eng. Univ. **2**(56), 71–82 (2016)
6. Demidova, L.A., Tishkin, R.V., Yudakov, A.A.: Clustering algorithms ensemble development based on similarity matrices of cluster labels and spectral factorization algorithm. Bull. Ryazan State Radio Eng. Univ. **4-1**(46), 9–17 (2013)

A Comprehensive Evaluation of the Efficacy and Limitations of Nutrition Software Using Machine Learning

Ricardo Arias Velásquez(✉), Bryan Díaz Dreyfus, and Eduardo Garces Rosendo

Universidad Tecnológica del Perú, Natalio Sanchez Street N° 125, Santa Beatriz, Lima, Peru
ricardoariasvelasquez@hotmail.com, {rariasv,U18306724,egarces}@utp.edu.pe
https://www.utp.edu.pe/

Abstract. Currently, non-communicable chronic diseases such as diabetes, cardiovascular diseases, and cancer are leading causes of death and disability worldwide, accounting for 70% of global deaths. Technological solutions offering recommendation and/or suggestion systems for diets or nutritional plans have played a significant role in reducing the prevalence of these diseases due to their increased usage in recent years. This study presents a systematic literature review (SLR) using the PICO and PRISMA search methodologies, focusing on the usage and precision of algorithms and classification methods in software for the automatic creation of diets and nutritional plans, and how these compare to conventional diet creation techniques. A total of 32 documents from the Scopus and IEEE databases were analyzed. The results indicate considerable progress in the field of image recognition and in the precision of classification methods in recent years, highlighting significant advancements in global studies aimed at improving population health.

Keywords: Classification · nutrition · machine learning

1 Introduction

According to the World Health Organization, issues of malnutrition due to overweight and congenital diseases are prevalent in Europe and Asia. This concern led to the inclusion of this topic as a primary point in the 2015–2020 agenda. Consequently, European regulation No. 1169/2011, specifically Article 30, was reviewed to identify the dishes and packages required for the nutrition program. A software solution is proposed to calculate nutrients and be internationally validated, incorporating crucial information for special geographic areas such as the Balkans and integrating with the EuroFIR dietary platform. However, the

Thanks to Universidad Tecnológica del Perú.

R. Silhavy and P. Silhavy (Eds.): CoMeSySo 2024, LNNS 1489, pp. 208–237, 2025.
https://doi.org/10.1007/978-3-031-96798-6_19

web platform has limitations that hinder personalized tracking due to the number of surveys required for its execution. This platform was created using the nutritional requirements of the European Union [1].

In 2022, researchers from Oxford identified two databases for nutrition assessment in foods and beverages in Australia: AUSNUT 2011–2013 and the Mintel 2014–2019 database. The developed test found that accuracy varies depending on the country of evaluation. For instance, the evaluation of the Chilean NPM nutritional classification scheme showed a 33% food labeling error, increasing the likelihood of mislabeling regarding its health potential [2].

Researchers in China demonstrated a food classification system based on image recognition with an accuracy rate of 85%, describing the attributes of the captured food [3]. This classification method resulted from combining the capabilities of two trained models to perform the task, demonstrating the model's ability to accurately identify the expected outcomes. The application of various classification methods has been comparatively studied from a machine learning perspective based on different types of processing networks [4]. However, in-depth studies for the automatic creation of nutritional plans are lacking. Due to the current variety of processing network types and their breadth, studies on systems to help patients choose the best foods for consumption are primarily oriented towards controlling glucose levels or managing general diets [5].

Food labeling stands out as a key aspect for consumers and nutritionists and should be addressed more thoroughly by researchers and nutrition professionals in both developed and developing countries. Reviews on food classification have shown advancements in applying methods and algorithms but lack a focus on creating nutritional plans to reduce body fat.

The research is structured as follows: Sect. 2 details the methodology for the systematic review using the "Preferred Reporting Item for Systematic Review and Meta-Analysis for Abstract" (PRISMA-A) [6], including selection processes, inclusion and exclusion criteria, evaluation lists, and bias considerations. Section 3 presents the results of the systematic review. Section 4 discusses the contribution of the systematic review and the limitations in implementation. Finally, the last section includes the conclusions and future work of the present research.

2 Methodology

2.1 Research Questions

In this review, we employed the PICOC methodology (Population, Intervention, Comparison, Outcome, and Context) to investigate the limitations of classification mechanisms used in automated creation of nutritional plans for weight loss. However, in alignment with recommendations from Ref. [7], which advocate expanding this methodology into six sections, we propose a modification of these aspects to streamline the evidence-based process. The following outlines the focal points of this investigation as follows:

- Population: Diet software, nutrition software.
- Intervention: Classification methods, algorithms, prediction, automation.
- Comparison: Traditional or conventional techniques, manual tasks.
- Outcome: Accuracy, efficiency, plans.
- Context: Menu recommendations, food suggestions.

Detailing research questions is crucial for systematic reviews, as mentioned in [7], as they establish scope and guide the work during the review. This study aims to explore technologies related to handling data for software that automates diet or nutritional plan creation. It investigates the classification methods or algorithms used in these technologies and examines their precision levels, metrics, and tools compared to conventional diet creation techniques. Lastly, it aims to identify the contexts in which such software has been utilized.

With these focus points defined, qualitative evaluation criteria are proposed for preliminary data selection:

- Question 1: Are there software tools that generate diets automatically?
- Question 2: Classification Methods in Automatic Diet Creation Software:
- Question 2.1: Which classification methods have been applied in these software tools?
- Question 2.2: Are there instruments or metrics to evaluate the accuracy of the classification methods used?
- Question 2.3: What numerical values have been obtained regarding the accuracy of these classification methods?
- Question 3: Comparison with Traditional Techniques:
- Question 4: How effective have these software tools been compared to traditional or conventional techniques for diet creation?
- Question 5: Precision and Limitations: What precision levels have the classification methods demonstrated in these software tools, and what limitations have been identified?
- Question 5.1: Application Contexts: In what contexts have classification methods in diet creation software or menu recommendations been utilized?

For evaluation purposes, the following search string was used: (Diet software) AND (Classification methods) AND (Traditional techniques) AND (Precision) AND ((Automatic) OR (Creation) OR (Nutrition) OR (Manual) OR (Plan)), with information consulted and downloaded on May 5, 2024, from the "ScienceDirect," "IEEE," and "Scopus" databases, detailed in Table 1.

To establish a baseline for the research, emphasis is placed on the methods, indicators, and limitations of the reviewed documents, as shown in Table 2 and Table 3.

In addition to reviewing the terminology used in related work concerning the key concepts defined at the outset of this research, an exploration of alternative terms and synonyms based on Thesauri was conducted. Table 3 presents a comprehensive summary of the identified terminologies. In this regard, the search queries comprised terms identified in Tables 1 and 3, combined using

Table 1. Terms Derived from Relevant Research

Year	Keywords
2021 [4]	Precision nutrition, Personalized nutrition, Machine learning, Deep learning, Artificial intelligence, Systematic literature review.
2022 [11]	Cardio-renal-metabolic health, Ketogenic diet, Metabolomics, Mitochondrial function, Omics, Overweight and obesity, Precision health.
2023 [8]	Precision nutrition, Data analytics, Vibrational spectroscopy, Sensors.
2023 [10]	Artificial intelligence, Clinical nutrition, Deep learning.
2023 [12]	Nutritional metabolomics, NMR, Dietary biomarkers, Diet-related disorders, Nutrition and human health.
2023 [14]	Computational tool for menu planning, preferences, nutrient intake assessment, TBCA NIA-DB, Brazilian food composition tables, Brazilian recipes and foods.
2023 [15]	Entity linking refers to linking entities in texts to unique identifiers in knowledge bases, natural language processing (NLP), machine learning.
2024 [9]	Computer vision, Food recognition, Convolutional neural network, Transformer-based methods, Food nutrient datasets, Nutrient estimation.
2024 [13]	Prediction of type 2 diabetes using machine learning in conjunction with cloud-based health analytics is a growing research area. Studies on diabetes in the Pima Indians have applied classifiers such as decision trees, support vector machines, and artificial neural networks.

Boolean operators (OR), as detailed below. For the Scopus search, the equation used was: (TITLE-ABS-KEY ("software" OR "application" OR "systems" AND ("dietary" OR "dietetic" OR "nutrition")) AND TITLE-ABS-KEY ("classification" OR "methods" OR "algorithms" OR "prediction" OR "machine learning" OR "artificial intelligence" OR "recognition" OR "classifier") AND TITLE-ABS-KEY ("traditional" OR "conventional") AND TITLE-ABS-KEY ("precision" OR "accuracy" OR "efficiency") AND TITLE-ABS-KEY ("diet" OR "plans" OR "menu" OR "meal" OR "fiid" OR "recommendation" OR "suggestion")).

Para IEEE Xplore fue la siguiente cadena: ("All Metadata": "software" OR "All Metadata": "application" OR "All Metadata": "systems") AND ("All Metadata": "dietary" OR "All Metadata": "dietetic" OR "All Metadata": "nutrition") AND ("All Metadata": "classification" OR "All Metadata": "methods" OR "All Metadata": "algorithms" OR "All Metadata": "prediction" OR "All Metadata": "machine learning" OR "All Metadata": "artificial intelligence" OR "All Metadata": "recognition" OR "All Metadata": "classifier") AND ("All

Table 2. Evolution of Trends

Year	Method	Indicators	Limitations
2021 [4]	Evidence-Based Software Engineering (EBSE)	Regression 23%, Recommendation 3%, Clustering 8%, Classification 15%	Search terms with the word "Nutri" limited results; Machine Learning was not considered.
2022 [11]	Controlled Clinical Trial	Weight, Body Mass Index (BMI), Enrollment and Retention Rate	Impact on health cannot be determined. Sample biased towards specific geographic location.
2023 [8]	Data Analytics, Principal Component Analysis (PCA)	Coefficient of Determination (R2) 0.65, Cross-validated Standard Error of Calibration (SECV) 0.25	Studies based on small datasets; Bias in slope and residual predictive deviation (RPD) not considered in model calibration.
2023 [10]		Qualitative Evaluation based on AI usage and potential	Lack of adjustable parameters or criteria for AI algorithms; Biases exist in health domains.
2023 [14]	Finite State Machine (FSM), Questionnaire	Clinical Agreement Rate 89.2%	Training set excludes cookies, pastries, dairy; Biochemical or genetic data not considered for prediction. Study only considers dietary base from one country.
2023 [15]	Few-shot Learning	Accuracy: E.L. hybrid model 89.40%, SVM 86.99%	Model fails to recognize similar terminology during classification.
2024 [9]	Search for Related Terminology	Image Database, Food Type or Category, Recognition Accuracy	Image quality affects food recognition accuracy.
2024 [13]	Dataset Training, Questionnaire	Accuracy: DTC: 71.42%, SVM 84.41%, ANN 72.72%, DP-UCE 87.41%	Lacks comparison across various classification models.

Metadata": "traditional" OR "All Metadata": "conventional") AND ("All Metadata": "precision" OR "All Metadata": "accuracy" OR "All Metadata": "efficiency") AND ("All Metadata": "diet" OR "All Metadata": "plans" OR "All Metadata": "menu" OR "All Metadata": "meal" OR "All Metadata": "food" OR "All Metadata": "recommendation" OR "All Metadata": "suggestion").

Table 3. Derived Terms and Synonyms

Keywords	Alternatives and Synonyms
Software	Computer programs, applications, systems, autonomous.
Diet	Dietetics, dietary therapy, personalized menu, plan suggestion.
Classification Method	Automatic classification, algorithm.
Traditional Technique	Conventional technique.
Consistency, Reproducibility, or Repeatability	Precision, Efficiency.
Nutrition	Food, nutritionist.

2.2 Study on Quality Assessment Process

To evaluate the quality process, criteria for selecting obtained records were defined, outlining inclusion and exclusion criteria that clearly and objectively address research questions and ensure the accuracy of collected information.

Inclusion/Exclusion Criteria for Selection. Regarding the research objectives, the importance of establishing eligibility criteria for document search and collection is noted [7]. Thus, inclusion criteria considered documents from the last 6 years (2018–2024).

Conversely, the following exclusion criteria were applied:

- Language: Articles in English from Q2 and Q1 journals.
- Document Type: Focus on scientific articles and systematic reviews.
- Method: Articles without information on classification methods or algorithms were excluded.
- During the screening process, documents on topics such as Agriculture, Livestock, and Manufacturing were preemptively excluded, as detailed in Table 4.

Preliminary Selection Process. To ensure comprehensive and transparent review, the search process was conducted in 2 phases: Identification and Selection.

The Identification phase involved selecting the following document search databases in Table 4:

- Scopus.
- IEEE Xplore.

Final Selection Process. The following points were considered:

- Review and removal of duplicate records between databases and those lacking a DOI.
- Review of Titles and Abstracts to exclude records that do not address research questions.
- Retrieval of full-text documents.
- Eligibility review based on inclusion/exclusion criteria.
- Documentation of results.

2.3 PRISMA-A [6]

In the Scopus database, 390 records were identified, and 48 records were identified in IEEE, totaling 438 records. The results were processed using screening techniques based on the PRISMA-A methodology [6], resulting in a final sample of 32 documents. The detailed application of document selection and screening under the PRISMA method is described in Fig. 1.

2.4 Checklist and Quality Assessment

In a systematic review, quality assessment is a crucial process where the condition of the obtained results is measured [7]. During this phase, we assess the validity and relevance of the studies conducted by the authors. The evaluation criteria were defined based on the alignment with the research questions. Scores for the evaluation are assigned as follows: 1 for Yes, 0 for No, and 0.5 for Partial. Additionally, quantitative assessment scores can range from 0 to 15, where 0 indicates low quality and 15 indicates high quality.

2.5 Quantitative Checking

For quantitative checking, the following detailed evaluation criteria were formulated as shown in Table 5.

2.6 Data Extraction Strategy

Following the quality assessment, the obtained information was meticulously reviewed in accordance with Table 6.

3 Results

This research offers a comprehensive overview of the current state-of-the-art and recent advancements, detailing the methodologies employed by the authors. It elaborates on the data sources used for training classification models, types of research conducted, performance indicators, and study limitations.

Table 4. Inclusion and Exclusion Criteria in Documents

PICOC	Search String	Inclusion Criteria	Exclusion Criteria
Population	"SOFTWARE" OR "APPLICATION" OR "SYSTEMS" AND ("DIETARY" OR "DIETETIC" OR "NUTRITION")	Publications from 2018 to 2024, English language	No conference papers, no book chapters, no editorials
Intervention	"CLASSIFICATION" OR "METHODS" OR "ALGORITHMS" OR "PREDICTION" OR "MACHINE LEARNING" OR "ARTIFICIAL INTELLIGENCE" OR "RECOGNITION" OR "CLASSIFIER"	Publications from 2018 to 2024, English language	No conference papers, no book chapters, no editorials, no Internet of Things, no calibration models, no parametric models
Comparison	"TRADITIONAL" OR "CONVENTIONAL"	Publications from 2018 to 2024, English language	No conference papers, no book chapters, no editorials, no production methods, no robotics, no alternative medicine, no OCR
Outcome	"PRECISION" OR "ACCURACY" OR "EFFICIENCY"	Publications from 2018 to 2024, English language	No conference papers, no book chapters, no editorials, no prototypes
Context	"DIET" OR "PLANS" OR "MENU" OR "MEAL" OR "FOOD" OR "RECOMMENDATION" OR "SUGGESTION"	Publications from 2018 to 2024, English language	No conference papers, no book chapters, no editorials, no agriculture, no manufacturing, no food safety, no spectroscopy, no sheep farming, no animal feed

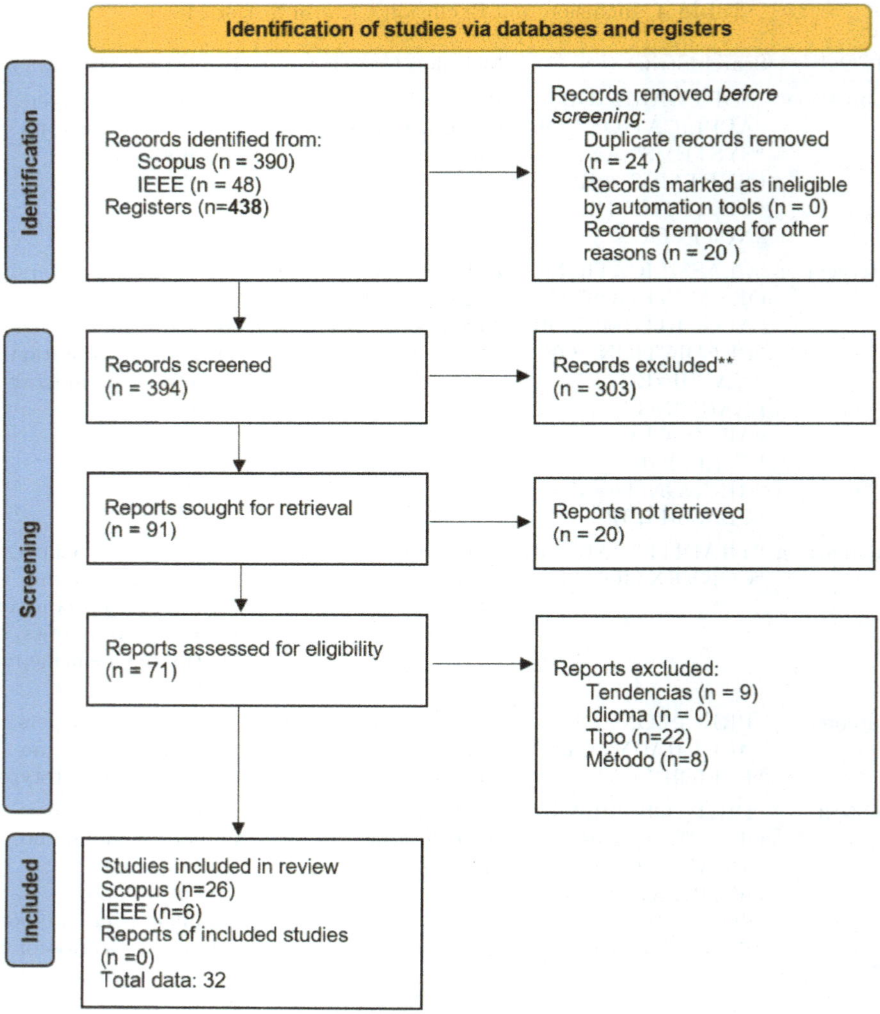

Fig. 1. PRISMA-A Flowchart

Table 5. Quantitative Check

No.	Question	Answer
1	Is the study objective clearly defined?	Yes/No/Partial
2	Is adequate justification provided for the importance of the topic?	Yes/No/Partial
3	Is the study methodology clearly described?	Yes/No/Partial
4	Is the sample size adequate and justified?	Yes/No/Partial
5	Is adequate justification provided for the importance of the topic?	Yes/No/Partial
6	Are appropriate statistical methods used for data analysis?	Yes/No/Partial
7	Are study results clearly presented with sufficient detail?	Yes/No/Partial
8	Is the discussion of results conducted within the context of existing literature?	Yes/No/Partial
9	Are study limitations identified and discussed?	Yes/No/Partial
10	Does the study include practical implications or recommendations based on findings?	Yes/No/Partial
11	Are study funding sources and potential conflicts of interest clearly declared?	Yes/No/Partial
12	Has the study been published in a peer-reviewed journal relevant to the study areas?	Yes/No/Partial
13	Is the study design suitable for the stated objectives?	Yes/No/Partial
14	Are the algorithms and classification methods used described in detail?	Yes/No/Partial
15	Is information provided on the validation or evaluation of the algorithms used?	Yes/No/Partial

3.1 Years of Publication

Our search revealed a notable upward trend in publications in recent years, underscoring the growing relevance of the research topic. Particularly noteworthy is the increase in studies focusing on diet recognition and recommendation systems since 2020. Further details on this trend are illustrated in Fig. 2.

3.2 Databases or Datasets Used by the Authors

In the publications made by the authors, various datasets used for training classification models for food recognition or recommendation systems are identified.

Table 6. Data Extraction Form

Item	Value
Methodology used in the research	YES/NO/Description
Population or problem addressed	YES/NO/Description
Type of research	YES/NO/Description
Research limitations	YES/NO/Description
Contributions	YES/NO/Description
Results of metrics or KPIs	YES/NO/Description
Comparison with conventional techniques	YES/NO/Description

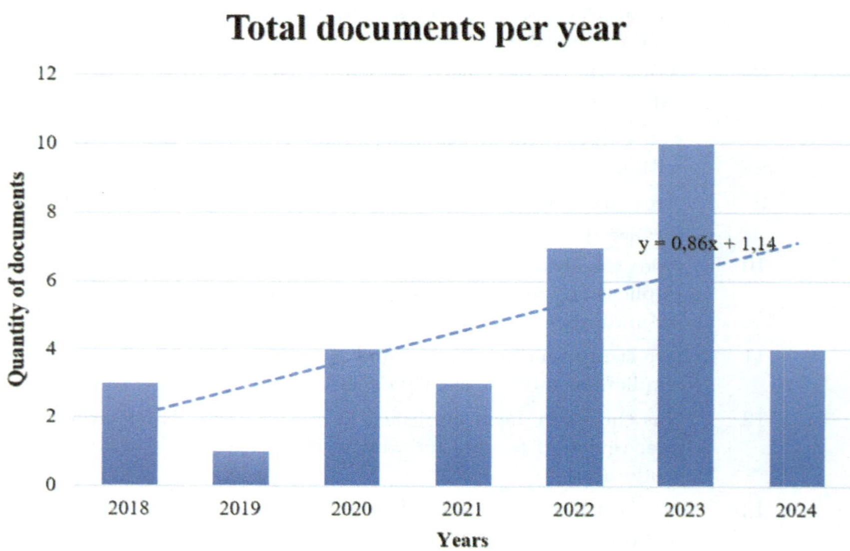

Fig. 2. Articles per year.

The details of the datasets are provided in Table 7. It was observed that there is a bias in training data segmented by countries, due to the variety of foods, ingredients, or dishes that each region possesses. It is worth mentioning that some generic datasets were observed that were not mentioned as the authors did not highlight a particular name for them.

3.3 Classification Methods or Algorithms Used by the Authors

The research published by the authors presents various types of classification methods or algorithms, with some of the most used being Yolo v5 in references [24,45], Naive Bayes [19,24,45], SVM (Support Vector Machine) [16,24,35,45]. Additionally, different metrics were found to determine the levels of precision

Table 7. Databases used by the authors

Reference	Database	Region
[24]	Halal & Haram	Muslim Population
[27]	Food Information and Price 2017 & 2020	North America
[26]	KNHANES VII	South Korea
[30]	FLIP-Canada 2010–2020 and USDA-PDPD 2022–2023	North America
[29]	FR-W (Weighed Food Record)	Brazil
[31]	Food Choices	Pennsylvania
[35]	VIREO-Food172	China and India
[22]	UNIMIB2016	Europe
[20]	Food101	Malaysia
[44]	Final Food Dataset 2022	N/A

of the classification methods. The most commonly used metrics were: Precision, Recall, F1 Score, Balanced Accuracy (BA), Accuracy, MAE, and MSE. Furthermore, the authors focused their studies on Image Recognition, Prediction Systems, and Recommendation Systems. Table 8 shows a summary of the classification methods or algorithms used in the research along with their respective metrics and study focuses in Table 8.

3.4 Publications of the Year 2018

The case of C. Liu et al. [21], introduces a food visual recognition algorithm based on deep learning with the aim of improving the accuracy and efficiency of dietary assessment. Two databases were used: UEC-256 with 256 categories and 28,357 images, and UEC-100 with 100 food categories and 8,643 images, mostly composed of Asian-origin foods.

This study, classified as Original Research, was limited to testing the system with a single brand of Smartphone, which may limit the generalization of the results. Although it is noted that the improvement time is still 5% slower than the compared models, the proposed system showed high levels of accuracy, reduction in response time, and improvement in the energy efficiency of the resources used.

The results indicate that the system achieved 77.5% accuracy in the classification of the main category (Top-1 Accuracy) and 95.2% in the classification of the top five categories (Top-5 Accuracy) in the UEC-100 test set. Algorithms such as (SURF-BoF+ColorHistogram) and DeepFoodCam(ft) were compared, highlighting the advantages of the deep learning-based approach in terms of accuracy and efficiency in dietary assessment.

Also, A. Meroni et al. [23] presents the validation of a web platform for self-recording of food and beverages. A case study was conducted with 67 participants

Table 8. Classification methods and metrics of the studies

Ref	Methods	Metrics	Focus
[24,45]	Yolo V5	Precision, Recall, F1 score	Images
In [44]	Yolo V8	mAP	Images
In [24]	Random Forest	Precision, Recall, F1 score	Images
In [19, 24,45]	Naive Bayes	Precision, Recall, F1 score	Images
In [16,24, 35,45]	SVM (Support Vector Machine)	Precision, Recall, F1 score	Images
In [24]	Multi perceptron	Precision, Recall, F1 score	
In [27]	Natural Language Processing	Coefficient of Determination (R2), MSE	Images
In [26,30]	XGBoost, LightGBM, Catboost	(BA) Balanced Accuracy, F1 score	Recommendation
In [31]	KNN (K-Nearest Neighbors), Stacked-LSTM (Long-short Term Memory)	MAE, MedAE, RMSE, MAPE	Prediction
In [33]	Decision tree	Accuracy, Loss Score, Mean Cross Entropy, Time	Recommendation
In [35]	MTA-SOM, CNN, Faster R-CNN	Precision, Accuracy, Recall, F1 Score	Images
In [20, 34,46]	SegNet, MobileNet, PspNet, Unet	Precision	Images
In [41]	FoodCam	MAE	Images
In [39,46]	EfficientDet Deep Learning, SSD Inception V2, Faster R-CNN, Inception ResNet v2	mAP, Precision, Accuracy, Recall, F1 Score	Images
In [16]	IRF (Improved Random Forest)	Accuracy, Precision, Recall, F1 Score	Recommendation
In [19]	Naive bayes with ACO, Naive bayes with PSO, Collaborative Filtering	F1, Miss-Rate, Fallout Rate, Matthews Correlation	Recommendation
In [42]	CNN, EfficientNet Lite	Accuracy	Images
In [50]	Gaussian	Correlation coefficient 98.91%	forecasting in chaotic timeseries

who completed dietary measurements using the new platform and on paper for 3 days, at the University of Sydney.

Classified as Original Research, the study highlights that the food database of the proposed platform is based on Australian foods. The ease of use of the platform for participants to record their food intake data is emphasized. Additionally, the accuracy of energy and macronutrient calculation using the platform was demonstrated.

The results showed a high Pearson product-moment correlation coefficient for energy (0.878), proteins (0.642), fats (0.682), and carbohydrates (0.792), indicating good agreement between the records made on the platform and the estimated values. A comparison was also made with a traditional estimated food diary, providing context for evaluating the accuracy and usefulness of the proposed platform compared to more conventional methods of dietary intake recording.

In the research by S. E. Kuo et al. [28], a clinical nutritional information system (CNIS) is presented. A case study was conducted with a survey directed at 12 nutritional professionals who used the system at Cheng Kung National Hospital, Taiwan.

Classified as Original Research, it is noted that the sample for the satisfaction survey was small, limited to the nutritional professionals of Cheng Kung National Hospital. Additionally, it is noted that the testing of the proposed system did not include other hospitals, which could affect the generalization of the results.

Among the findings, the standardization of nutritional assessment formats and the ease of the system to provide useful analysis information for dietary recommendation are highlighted. The results of the system quality evaluation show high scores, with an average score of 4.25 for system quality and 4.42 for user satisfaction.

Although the need for a larger sample and the inclusion of other hospitals to fully validate the effectiveness and usefulness of the CNIS is recognized, these results suggest a high level of acceptance and usefulness among the nutritional professionals who participated in the study.

3.5 Publications of the Year 2019

In another study, M. A. Subhi, S. H. Ali, and M. Abdulameer [20] implemented MobileNets, which are deep learning detection and classification methods using Google's machine learning and the TensorFlow library. A case study was conducted using 5800 images of Malaysian food collected from Google Images, with acceptable quality, used 80% for training and 20% for testing. This study is classified as Original Research.

MobileNet and Inception are highlighted as accurate and fast models for food detection and classification. A training accuracy of 97% and a test accuracy of 83% were observed for the MobileNet model, and a training accuracy of 95% and a test accuracy of 89% for the Inception V3 model.

Additionally, Deep Convolutional Neural Networks (DCNN) were compared with traditional machine learning approaches and classical image features such as SIFT, color histogram, and Gabor texture features. These findings provide a comparative perspective on the effectiveness and performance of DCNNs compared to more traditional methods of food detection and classification.

3.6 Publications of the Year 2020

Regarding the proposal by S. van Asbroeck and C. Matthys [37], the accuracy of various food and beverage image recognition APIs was evaluated. For this, 185 images were used on 7 different recognition platforms. The study was conducted using demo versions of these platforms. Among the main platforms evaluated, Calorie Mama, Bitesnap, and Foodvisor stand out for their high levels of accuracy in food and beverage recognition. The accuracy metrics for these platforms were as follows:

- Calorie Mama API: Achieved 62.9% accuracy in Top 1 and 89.6% in Top 5.
- Bitesnap: Achieved 48.9% accuracy in Top 1 and 71% in Top 5.
- Foodvisor: Achieved 46.2% accuracy in Top 1 and 71.5% in Top 5.

In addition to these platforms, others such as LogMeal, Clarifai, Google Vision API, and IBM Watson Recognition were also compared. The study highlighted the potential of the aforementioned AI platforms due to their higher levels of accuracy. This study provides a valuable comparison of various image recognition platforms for dietary assessment, highlighting the potential of AI technology in improving the accuracy and efficiency of food intake monitoring [17].

The study presented by D. Herzig et al. [18] used a smartphone with a camera combined with depth sensors to estimate the volume, macronutrients, and energy of foods. 48 test foods were used in this study, which is classified as Original Research. It is important to note that the study is based on a single smartphone model, which may bias the results towards a specific brand. Additionally, it is noted that the foods come from diets of the University Hospital of Bern in Switzerland, which may limit the generalization of the results to other populations or contexts. The system demonstrated high performance in segmentation and low processing time, highlighting its usability and effectiveness in estimating food volume. In terms of accuracy, both in absolute and relative terms, the mean absolute error (ME) of the estimated weight for all meals was 35.1 g (42.8 g; 14.0% [12.2%]), and the mean bias.

On the other hand, the study by B. Janakiraman and S. Arumugam [19] employs optimization techniques such as Ant Colony Optimization (ACO) and Particle Swarm Optimization (PSO) to provide personalized nutritional recommendations. A dataset composed of 70 records of diabetes patients obtained from the UCI Machine Learning Repository was used, classifying it as Original Research. It is noted that PSO surpasses traditional methods by 9.96% and 12.93% in terms of the F1-Measure, highlighting its effectiveness in personalized nutrition recommendation for diabetic patients. Specific results for the NP-PSO algorithm include an F1-Measure of 0.80, a Miss-Rate of 0.18, a Fallout Rate of 0.05, and a Matthews Correlation Coefficient of 0.75. The study compared three different algorithms: Collaborative Filtering (CF), Naive Bayes Probability with ACO (NP.-ACO), and Naive Bayes Probability with PSO (NP-PSO). These results provide a comparative evaluation of each algorithm's effectiveness in personalized nutrition recommendation for diabetic patients, emphasizing the superiority of the PSO-based approach in terms of accuracy and performance.

H. Bai [25] presents an algorithm for personalized diet recommendation using the Adaboost classifier along with a probabilistic matrix factorization. For this purpose, a survey on the data of a group of diabetic patients in China was employed, including their food preferences and physical data.

Classified as Original Research, the study highlights the accuracy of the proposed algorithm by considering patients' individual preferences along with their nutritional needs. However, it is noted that complete information on the training dataset used for the food database is not provided, which could limit the reproducibility and generalization of the results.

The results obtained include accuracy and distribution measurements of the proposed algorithm, although specific values are not provided in the abstract. Additionally, a comparison with recommendation algorithms based on collaborative filtering and content-based algorithms was conducted, providing a comparative context to evaluate the proposed approach's performance in relation to other dietary recommendation methods.

3.7 Publications of the Year 2021

On the other hand, the research by C. M. Blanchard et al. [36] was conducted as a randomized clinical trial using the PIQNIQ mobile application to capture dietary intake through self-report. The case study included 132 adults aged 18 to 65 years with a BMI between 18 and 30 from the Jean Mayer Human Nutrition Research Center at Tufts University, Boston, MA, USA.

Despite the short evaluation period and the design of the application and backend, which did not distinguish between manually entered data and preset options, the results were significant. Most participants were young women of white complexion, which may have influenced the results.

The study demonstrated that intuitive and technology-based methods like PIQNIQ can provide data comparable to traditional methods, suggesting a possible shift towards modern and user-friendly tools for capturing dietary data. In terms of performance metrics, self-reported total energy intake was on average 14% higher (adjusted $p = 0.03$) in the simultaneous entry group, 9% higher (adjusted $p = 0.87$) in the photo-assisted recall group, and 1% lower (adjusted $p > 0.99$) in the 24-h group. Two food capture methods using PIQNIQ were compared: simultaneous entry and photo-assisted recall. This study provides valuable evidence on the effectiveness of PIQNIQ, highlighting its potential to be a modern and user-friendly tool in capturing dietary data.

In another research by M. A. Khan, B. Smyth, and D. Coyle [38], Ensemble Topic Modeling (EnsTM) is used to support personalized recipe recommendations. A dataset was created using WebScrapping, consisting of 230,876 recipes with detailed information, including ingredients, instructions, servings, cooking time, kitchen tools, context, and nutritional information (calories, proteins, etc.). This study is classified as Original Research. The study's results highlight the good performance of the EnsTM-based approach compared to the current state-of-the-art. It was found that this approach outperforms most existing cases, suggesting its effectiveness in generating personalized recipe recommendations. The accuracy was evaluated using a participant voting measure, where a significant result was found ($F(120,97,2,47) = 14.318$, $p < 0.001$). Additionally, the research compares EnsTM-based recommenders with a conventional content-based (CB) approach, which is fundamental in many important food recommendation systems (FRS). This comparative analysis provides valuable insights into the relative effectiveness of different approaches in generating personalized recipe recommendations. In the research by M. S. N. Siemon, A. S. M. Shihavuddin, and G. Ravn-Haren [22], a new sequential transfer learning method using hierarchical clustering for food segmentation in images is proposed. The UNIMIB2016 database, containing 1027 sample images and 65 food classes in total, was used. This study is classified as Original Research.

Although the number of images used to train the model is relatively low, the research demonstrates that the proposed approach provides up to a 6% gain in accuracy compared to traditional network training. Additionally, it was found that this method works better in challenging and unseen cases.

The categorical cross-entropy prediction accuracy (CCE) reaches 88.3% on the UNIMIB2016 test set, highlighting the effectiveness of the proposed approach in accurate food segmentation in images.

Moreover, the research compares the proposed method with conventional deep learning network training approaches, providing a comparative perspective on the efficacy and performance improvement offered by the sequential transfer method based on hierarchical clustering.

3.8 Publications of the Year 2022

In [41], the proposal presents a passive, independent, multispectral stereo camera system, motion-activated and supplemented with light called FOODCAM. This system uses a three-dimensional reconstruction methodology based on stereo image pairs for estimating food portion sizes (FPSE). A case study was conducted with 5 food models (NASCO FOOD REPLICAS) and a dual-sensor (stereo) camera along with other sensors such as the PIR Sensor and an infrared projector. The results showed that lighting in the images negatively affects the final outcome, highlighting the need for a stereo camera device for three-dimensional recognition. FOODCAM offers an accurate FPSE methodology for bulk foods and irregularly shaped items, providing high-resolution RGB images for dietary assessment without interfering with user activities. In terms of performance metrics, FOODCAM achieved an average accuracy of 94.4% in volume estimation of portions, with a mean absolute error of 5.60%. This system was compared with the PICO Flexx camera, highlighting the superiority of FOODCAM in accurate food portion size estimation.

This study contributes an innovative and accurate tool for dietary assessment, using advanced stereoscopic imaging technology with structured light to enhance accuracy and facilitate monitoring of food intake. In another field, the study by J. L. Scott [40] evaluated the use of a portable camera to improve the accuracy of a 24-h dietary recall among adults. The case study included 20 adults aged 18 to 65 years.

The study highlighted several technological limitations, such as the proper placement of the camera and the potential invasion of privacy in public settings. Despite these limitations, a notable improvement in dietary assessment accuracy was observed when using portable cameras.

In terms of participant perception, 80% considered the cameras intrusive and preferred the use of voice clips as an alternative. Only 20% of participants preferred to use their smartphone cameras.

The study compared the camera-assisted recall method with the traditional 24-h recall method without camera assistance. The results suggested that while portable cameras can improve dietary assessment accuracy, there is significant resistance to their use due to privacy and intrusiveness concerns.

The research provides evidence on the feasibility and user perceptions regarding the use of portable cameras in dietary assessment, highlighting both the advantages in accuracy and the challenges related to acceptance and privacy.

In the case study by P. H. Nguyen et al. [39], the FRANI (Food Recognition Assistance and Nudging Insights) artificial intelligence (AI) mobile application was used to assess diet in adolescent girls. A case study was conducted with a sample of 36 girls aged 12 to 18 years in Vietnam, classified as Original Research.

The study identified sources of errors mainly related to the limited recipe options available in FRANI, adolescents' ability to accurately recall the foods consumed, and portion size estimation. Despite these challenges, the study contributes by demonstrating that AI-assisted dietary assessment and 24-h recalls can accurately estimate nutrient intake in adolescent girls compared to weighed records (WR).

Concordance correlation coefficients (CCC) were used to evaluate the relative validity of FRANI and 24-h recalls compared to weighed records (WR). The results showed a CCC of 0.60 for FRANI and 0.81 for WR, and 0.70 for 24-h recalls and 0.89 for WR. These findings suggest that FRANI and 24-h recalls can be useful and accurate tools for dietary assessment in adolescent girls, providing a promising alternative to traditional methods like weighed records.

In another case, Y. C. Liu et al. [48] present a multi-dish food recognition model using EfficientDet Deep Learning. A total of 4733 food images were collected and 86 items were captured with a smartphone in real environments, divided into three categories: mixed dish, single dish, and multiple dishes. This study is classified as Original Research.

Although the number of images used to train the model is relatively low, the EfficientDet model performs well in food recognition on plates, with analysis speeds of less than 5 s. The results show a Mean Average Precision (mAP) of 92%, with an accuracy of 87%, precision of 88%, recall of 97%, and F1-Score of 93%.

Additionally, other food recognition models such as SSD Inception V2 and Faster R-CNN Inception ResNet V2 were compared. These findings highlight the efficiency and precision of the EfficientDet model compared to other food recognition techniques, suggesting its potential as a valuable tool for precise and efficient dietary intake reporting.

In another case study by X. Zhang, Y. Sun, and Y. Sun [16], an intelligent approach for food quality and safety testing is utilized through a weighted Improved Random Forest (IRF) algorithm based on mutual information. A case study was conducted using a dataset to train the IRF model and Chinese vinegar for taste and quality testing. This study is classified as Original Research.

It is important to note that the analysis is based solely on one ingredient (vinegar), which may imply variations in results when testing other foods containing more elements. However, the precision of the IRF method is highlighted, with higher rating ratios than SVM, RBFNN, NB, and RF.

The study results show exceptional accuracy of the IRF model, with an accuracy of 100%, a precision of 99.83%, a recall of 96.47%, and an F1 score of 98.12%. These findings demonstrate the effectiveness of the proposed approach in food quality inspection, with significantly superior accuracy and performance compared to traditional methods.

Traditional methods such as SVM, RBFNN, NB, and RF were also compared, providing a comparative perspective on the effectiveness of the IRF-based approach relative to conventional food quality inspection techniques.

On the other hand, the study by G. Prajena, J. et al. [42] presents the use of a Convolutional Neural Network (CNN) and EfficientNet Lite to perform reliable classification of 10 types of traditional Indonesian food.

Classified as Original Research, the study relied on training the model using 200 images of different types of food. It is highlighted that the training images focused specifically on traditional Indonesian foods, providing a specific and relevant approach for the model's application.

The performance of the EfficientNet-Lite 3 model in the CNN for image recognition is particularly highlighted, with an accuracy of 90%. This result indicates a significant capability of the model to accurately classify images of traditional Indonesian foods.

Additionally, a comparison between different versions of the EfficientNet-Lite model (from 0 to 4) was conducted, providing additional information on the relative effectiveness of each version in the specific context of traditional Indonesian food classification.

In relation to the study by B. Huang et al. [47], an innovative proposal for recipe recommendation using dietary knowledge mapping and a multi-domain knowledge graph is presented. This system improves recommendations through a collaborative filtering algorithm implemented by Mahout.

The system uses a diet-based knowledge graph that includes entities such as diseases, medications, foods, and symptoms, with a total of 44,000 entities. Additionally, a Neo4j graph database is employed for knowledge storage.

Classified as Original Research, the article does not provide precision indicators for the models used in the system. However, it highlights the system's capability to offer personalized recipe recommendations and adjust ingredient proportions according to users' health profiles. This is achieved through an intelligent question-and-answer module, thus enhancing human-computer interaction capabilities. The system addresses the limitations of traditional collaborative filtering methods, such as cold start and sparse data, through the integration of a multi-domain knowledge graph. This integration allows the system to generate more precise and personalized recommendations for users, thus improving the overall user experience in recipe searching and selection.

3.9 Publications of the Year 2023

The study by G. Hu et al. [27] presents a new natural language processing (NLP) and machine learning (ML) method to analyze food label information. This case study used the "Food Label Information and Price 2017 & 2020" database from the University of Toronto and the "Health Canada's Table of Reference Amount (TRA)" for food categorization. As an original research article, it faces limitations due to the specific product information obtained from grocery retail websites and the unbalanced sample size. Nonetheless, the new NLP automation model demonstrated high effectiveness, being suitable for handling large volumes of food information. To evaluate the performance of the methods for predicting the FSANZ nutrition quality score, the coefficient of determination ($R\hat{2}$) and the mean squared error (MSE) were used. The results indicated superior performance of the new method compared to conventional techniques such as the "food label pretrained language model," the bag-of-words model, and other nutritional quality prediction methods. This underscores the ability of the new NLP and ML approach to improve the accuracy and efficiency in categorizing and predicting food nutritional quality.

In line with this, the study by G. Hu, N. Flexner et al. [30] used the NOVA food classification system. This case study employed the FLIP-Canada (2010–2020), FLIP-LAC (2018–2022), and USDA-BFPD (2023–2022) databases, which contain information on packaged foods, beverages, nutrients, ingredients, price, and other data. As original research, certain limitations were identified, including a possible reduction in method performance with new datasets that would require additional training. Additionally, limitations with language and OCR capability were found, and two of the three datasets used do not cover 100% of the food supply chain (80% for FLIP).

Fine-tuned language models showed the best performance among the different machine learning algorithms. This study highlights the power of machine learning tools to utilize available information on food package labels.

The accuracy of the NOVA system was evaluated using accuracy, balanced accuracy (BA), and F1 Score. The results were: XGBoost with an accuracy of 0.970, BA of 0.938, and F1 Score of 0.970, and the fine-tuned language models with an accuracy of 0.979, BA of 0.959, and F1 Score of 0.979. Different sources of information such as Structured Nutrient Facts, bag-of-words, and pretrained language models were compared.

In another case, the study by T. L. D. Borges et al. [29] evaluated image-assisted dietary intake through a cross-sectional observational study. The FR-

W (Weighed Food Record) data source was used to estimate records based on images (FR-PI), along with a team of 20 nutritionists specializing in food consumption assessment.

As original research, the study notes that the food-photographed record (FR-PI) captured by visually impaired persons (VIP) has only been tested on a small number of foods, which could affect analysis in a larger sample. However, the method demonstrated great potential for health professionals to understand what visually impaired persons (VIP) consume in their daily lives. The average quantity values reported by all experts were 95% for breakfast, 100% for lunch, 100% for dinner, and snacks. Additionally, images captured by visually impaired persons (VIP) were compared with traditional techniques such as diet record, diet history, 24-h recall, and food frequency. These results highlight the accuracy and usefulness of the image-based method for assessing dietary intake in visually impaired persons, providing a valuable tool for nutritionists in monitoring the daily diet of these individuals.

Regarding the proposal by A. M. Elshewey et al. [31], the Stacked-LSTM model along with K-Nearest Neighbor (KNN) were used as machine learning approaches for weight prediction in pounds based on food choices. The Food Choices database, a collection of responses from Mercyhurst University, which includes consumption preference information, was analyzed. As original research, the analysis was conducted on a database with insufficient information ($n = 126$) and highlighted a bias in the responses from Mercyhurst University students in Pennsylvania. However, the study demonstrated the effectiveness of the Stacked-LSTM model for weight prediction in pounds compared to other models. The Stacked-LSTM model's precision metrics included a mean absolute error (MAE) of 0.0087, a mean squared error (MSE) of 0.00011, a median absolute error (MedAE) of 0.006, a root mean square error (RMSE) of 0.011, and a mean absolute percentage error (MAPE) of 21.

Different methods such as Long Short-Term Memory (LSTM), Conventional Neural Network (CNN), Stacked-LSTM, CNN-LSTM, and KNN were compared. The results indicated that the Stacked-LSTM model had better performance in weight prediction based on food choice compared to the other analyzed models.

For A. Tanweer et al. [32], two versions of the 24-Hour Recall (24HR) method were used to assess dietary intake, focusing on beverage consumption among 102 participants aged 18 to 25 years from Pakistan. The traditional 24HR and digital 24HR were compared to evaluate the validity and reliability of web-based dietary recall tools.

The research revealed several important findings. The agreement between the traditional and digital methods was measured using the Kappa value, which was 0.375833 across all meals, indicating a moderate agreement. The Pearson Correlation Coefficient (PCC) was used to evaluate the correlation between the two methods for different meals. The results showed a moderate correlation for breakfast ($r = 0.465$), a low correlation for lunch ($r = 0.324$), and a moderate correlation for dinner ($r = 0.407$). Despite these promising results, the study identified limitations such as possible hesitation from participants when using unfamiliar digital interfaces, which could lead to perceptions of inconvenience or time consumption. Additionally, the age bias in the participant group (18–25 years) may affect the generalizability of the findings. The study provides valuable information on the effectiveness of digital 24HR tools in dietary assessment, specifically highlighting the reliability and validity of web-based methods compared to traditional self-report approaches, with a particular focus on beverage intake.

On another front, X. Y. Kong et al. [34] use a semantic segmentation network model based on SegNet + MobileNet to segment food images and estimate calories. The research was conducted using a sample of images based on the common yuan (CNY) with a uniform reference of objects. The dishes included both vegetarian and meat foods, with varied shapes such as granular, block, filament, flake, and combinations

The study revealed that the accuracy in food recognition depended on the inclusion of a reference object, in this case, a coin. Without this object, the accuracy level would be negatively affected. An application was developed that allows calculating the calories in foods via a smartphone, achieving a high level of accuracy with the SegNet + MobileNet model. Experimental results showed that the food recognition accuracy reached 97.82%, while calorie estimation was above 84.95%. Additionally, other segmentation network models like Unet + MobileNet and PspNet + MobileNet were compared, demonstrating that SegNet + MobileNet offered superior accuracy in the evaluated tasks. This study significantly contributes to the field of food calorie estimation, providing a practical and precise tool for users seeking to monitor their caloric intake using advanced semantic segmentation technology.

In another domain, T. Islam [43] presents an innovative approach that uses affective computing to personalize food recommendations and menu planning. Classified as Original Research, the study involved evaluating brain signals (EEG) from 18 men and 7 women, recording their reactions to images of 40 different foods. The performance and accuracy of the proposed model for food recommendation based on affecting computing are highlighted.

The model achieved an impressive 91.33% accuracy in prediction, indicating a significant ability to predict participants' food preferences based on their brain responses. This high level of accuracy suggests that the affecting computing-based approach can be an effective tool for personalizing food recommendations and menu planning more effectively and precisely.

This study highlights the potential of affective computing to improve personalization in the field of food recommendation and menu planning, which could have important implications for promoting healthier eating tailored to individual user preferences. In the study conducted by P. Kwon et al. [44], an innovative image recognition system using YOLO v8 is proposed. The study utilized the "Final Food Dataset," consisting of 3300 images of various types of foods. Although the study did not compare the performance of the YOLO v8 model with other image recognition models, it highlights the model's accuracy for food recognition.

The YOLO v8 model achieved an impressive Mean Average Precision (mAP) of 90%, indicating high accuracy in detecting foods in the dataset images used. Furthermore, it is noted that the proposed system contributes to addressing child malnutrition by providing detailed information on the nutrients of recognized foods.

Although the lack of comparison with other image recognition models and the lack of specification about the training images' origin are acknowledged, the results suggest a high potential for using this technology in nutrition awareness applications aimed at children.

In the research by M. Balpande et al. [45], an innovative approach to recommending diets and exercises based on artificial intelligence is presented.

Classified as Original Research, the study used a dataset comprising 90 types of foods, each with detailed nutritional values, such as calories, fats, proteins, iron, calcium, among others. Although it is mentioned that the food database used is small, the recommendation system offers a solution to improve physical and nutritional status through personalized diet and exercise recommendations. The performance of the recommendation system is highlighted, with accuracy varying among different machine learning algorithms used, such as Logistic Regression, SVM, Naive Bayes, CNN, and YOLOv5. Specifically, YOLOv5 achieved an accuracy of 89%, suggesting high effectiveness in recommending foods and exercises. The study also includes comparisons with other artificial intelligence-based systems, such as RFID-based systems and other gym equipment. Additionally, comparisons with different machine learning algorithms for exercise recommendation are mentioned, providing a comprehensive assessment of the proposed system's performance compared to other available technologies. Overall, this study offers a significant contribution to the field of food and exercise recommendation, using artificial intelligence to improve users' health and physical well-being.

Regarding the study by J. Li and B. Abbasi [46], an innovative approach to estimating the caloric intake of foods consumed with fingers, specifically focusing on potato chips, through the use of convolutional neural networks (CNN) is presented. Classified as Original Research, the study used a dataset consisting of 26,600 images captured during live broadcasts, where participants consumed potato chips. It is noted that the proposed system is limited to the recognition of potato chips, and difficulties in recognizing small quantities of this food are evidenced.

The novel approach of the study is highlighted, as it provides an immediate way to estimate the caloric intake of irregular foods, which are often overlooked in traditional dietary assessments. The system's accuracy is evaluated using the macro F1 score, ranging from 0.75 to 0.88 for a 3-class system and from 0.59 to 0.74 for a 4-class system. Additionally, different classification algorithms, such as InceptionResnetV2, Xception, and MobilNetV2, were compared, providing a comprehensive assessment of the proposed system's performance compared to other image recognition approaches.

In summary, this study offers a significant contribution to the field of caloric intake estimation by presenting an innovative and precise method for estimating the calories consumed from specific foods using deep learning technologies like convolutional neural networks.

4 Discussions

Based on the discussion, we present the study by S. Tarannum et al. [24], which employs the YOLO v5 classification method to identify halal foods. The research is a case study using the Halal & Haram ingredient database, labeling halal as 1 and haram as 2. The dataset was expanded using food store data available on Google. This approach is aimed at Muslim consumers, and the validation process involved feedback from three Muslim individuals.

The YOLO v5 model demonstrated remarkable performance as a binary classifier for halal and haram classification, achieving an accuracy rate of 98% when integrated with Random Forest. The accuracy of various models was as follows: Random Forest at 95%, K-Nearest Neighbors (KNN) at 97%, Naive Bayes at 76%, Support Vector Machine (SVM) at 98%, and Multilayer Perceptron at 98%. Evaluation metrics included precision, recall, and F1 score, along with a confusion matrix to determine true positives and false positives, where Random Forest recorded 196 true positives and 4 false positives. The study also involved a comparison with the opinions of three Islamic scholars regarding the classification results. This multifaceted evaluation highlights the robustness of YOLO v5 in binary classification tasks and emphasizes the high precision and recall of the model, making it a reliable tool for halal food recognition.

On the other hand, the study by I. Choi et al. [26] uses the "Latent Dirichlet Allocation (LDA)" method to extract feeding patterns from the textual information of the KNHANES VII dataset. This case study analyzes the KNHANES VII database, a survey conducted between 2016 and 2018 focusing on the dietary habits and health of the South Korean population, especially regarding metabolic diseases such as obesity and dyslipidemia. As original research, the evaluation is biased due to the comparison with the Korean population. However, a considerable improvement in predictive models for obesity and dyslipidemia is highlighted when incorporating NLP-based indices. The accuracy of the models was evaluated using balanced accuracy (BA) and F1 score. The results for obesity prediction showed that XGBoost achieved a BA of 0.528 and an F1 score of 0.496, LightGBM a BA of 0.518 and an F1 score of 0.475, and CatBoost a BA of 0.528 and an F1 score of 0.496. These findings underline the efficacy of NLP-based models in improving the accuracy of predictions related to dietary diseases.

Similarly, the proposal by C. Yaiprasert and A. N. Hidayanto [33] focuses on the use of joint Machine Learning techniques, including decision trees, logistic regression, and Naive Bayes, to personalize food recommendations in the context of the Thai restaurant and delivery market. It is original research highlighting the potential of AI to improve social well-being through personalized recommendations. The results show that decision trees achieved the best performance, with

Table 9. Trends for 2024.

Method	Limitation	Contribution	Indicators
Halal Food + Yolo V5 [45]	Study biased towards Muslim population	It was demonstrated that Yolo V5 has remarkable performance as a binary classifier for halal and haram classification.	Precision using Random Forest 95%, KNN 97%, Naive Bayes 76%, SVM 98%, and MultiPerceptron 98%.
Latent Dirichlet Allocation (LDA) [46]	Predictions related to dietary diseases in Korean population	Significant improvement in predictive models for obesity and dyslipidemia when NLP-based indices are incorporated.	Obesity prediction: XGBoost (BA 0.528)(F1 0.496), LightGBM (BA 0.518)(F1 0.475), and CatBoost (BA 0.528)(F1 0.496).
Joint Machine Learning Techniques [47]	Study based on the Thai restaurant market. Additionally, does not cover most ML techniques.	Highlights the potential of AI with Decision Trees for personalized food recommendations, emphasizing the impact of AI on social well-being.	Decision Trees: Accuracy 100%, Loss score 0.0408 ± 0.0016, Mean Cross Entropy 0.011316, Time 2.998 ms.
MTA-SOM and SGD [48]	Dissimilarity in food or biased selection of examples for training	Demonstrates improved efficiency in food recognition systems with MTA-SOM and SGD	MTA-SOM Precision 97.88%, Accuracy 99.11%, Recall 97.88%, F1-Score 96.17%, Execution Time 2.998 ms.
Arias model Yolo V5m [49]	configuration profile	Accuracy 88%	Performance of 80% with confidence interval and IoU of 60% and 40% respectively.

100% accuracy and minimal loss, with a score of 0.0408 ± 0.0016 in the probability voting metric on a test set. Additionally, the runtime for this technique was 1.55 units, highlighting its efficiency compared to other techniques.

The study acknowledges some limitations, such as bias towards the Thai restaurant market and not covering most available Machine Learning techniques. However, it emphasizes the importance of exploring the impact of AI in the food sector and the relevance of using joint models to improve recommendation personalization. Compared to other techniques, such as graphs, collaborative filtering, and content-based filtering, decision trees proved superior in terms of accuracy and efficiency. This suggests that joint Machine Learning techniques can be especially effective in personalized food recommendation environments.

The study by J. G. Kotwal [35] presents an innovative approach combining a modified time-adaptive self-organizing map (MTA-SOM) with a stochastic gradient descent (SGD) optimizer for automated food recognition. Using the VIREO-Food172 database, composed of images collected from Baidu and Google Images, this original study addresses the challenge of improving the efficiency of food recognition systems. However, a bias related to data training is recognized, which can affect the performance of the MTA-SOM model, such as imbalanced classes, food dissimilarity, and biased selection of training examples. This could influence the results when facing real data compared to test data.

The obtained results show that the MTA-SOM model achieves high accuracy and performance on the large dataset, with a precision of 97.88%, an accuracy of 99.11%, a recall of 97.88%, and an F1-Score of 96.17, with a runtime of 2.998ms. This highlights the effectiveness of combining MTA-SOM and SGD to improve the accuracy and adaptability of automated food recognition. Additionally, a comparison was made with other models, such as CNN, Faster R-CNN, SVM, WEISeR, and RLANet, providing a comparative view of the performance of different approaches in food recognition. This research significantly contributes to the advancement of assistive technologies by improving the efficiency of food recognition systems through the application of innovative techniques such as MTA-SOM and SGD (see Table 9).

5 Conclusions

This research presents a systematic literature review focusing on studies related to classification methods or algorithms used in diet recommendation systems based on the PICO and PRISMA methodology, highlighting the accuracy levels shown by the methods proposed by the authors and their contributions to nutrition and computer science. This study finds that most works focus on visual material recognition, such as images, highlighting classification models like YOLOv5 and v8 [24,44,45]. Additionally, the improvements in classification models over the years are reflected in accuracy levels of 90% or higher. However, there is an observable bias in the datasets, presenting a limitation due to the variety of food or ingredient origins, which poses a challenge for researchers in collecting data from many parts of the world. Future research could focus on compiling and unifying different databases at continental and later global levels with the aim of having a complete source of information for a solid system of dietary and food recommendations. It is also suggested to select classification models that have high levels of precision aimed at recommendation systems that can offer optimal results that contribute to improving people's health worldwide and reducing mortality rates from non-communicable diseases.

Acknowledgments. Thank to the Universidad Tecnológica del Perú.

References

1. Gurinović, M., et al.: Development, features and application of DIET ASSESS & PLAN (DAP) software in supporting public health nutrition research in Central Eastern European Countries (CEEC). Food Chem. **238**, 186–194 (2018). https://doi.org/10.1016/j.foodchem.2016.09.114
2. Dickie, S., et al.: Nutrition Classification Schemes for Informing Nutrition Policy in Australia: Nutrient-Based, Food-Based, or Dietary-Based? [En línea]. https://doi.org/10.1093/cdn/nzac112
3. Shen, Z., et al.: Machine learning based approach on food recognition and nutrition estimation. Procedia Comput. Sci., 448–453 (2020). https://doi.org/10.1016/j.procs.2020.06.113
4. Kirk, D., et al.: Precision nutrition: a systematic literature review. Comput. Biol. Med. **133**, 1 (2021). https://doi.org/10.1016/j.compbiomed.2021.104365
5. Spoladore, D., et al.: Ontology-based decision support systems for diabetes nutrition therapy: a systematic literature review. Artif. Intell. Med. **151**, 1 (2024). https://doi.org/10.1016/j.artmed.2024.102859
6. Velásquez, R.M.A.: A comprehensive evaluation and advancements in dissolved gas analysis for biodegradable oil in photovoltaic solar plants. Results Eng. **22**, 102314 (2024). https://doi.org/10.1016/j.rineng.2024.102314
7. Almeida, C., et al.: Effectiveness of nudge interventions to promote fruit and vegetables' selection, purchase, or consumption: a systematic review. Food Qual. Prefer. **116**, 1 (2024). https://doi.org/10.1016/j.foodqual.2024.105122
8. Dongdong, N., Cozzolino, D.: Possibilities on the application of vibrational spectroscopy and data analytics in precision nutrition. TrAC Trends Anal. Chem. **163**, 1 (2023). https://doi.org/10.1016/j.trac.2023.117067
9. Kaushal, S., et al.: Computer vision and deep learning-based approaches for detection of food nutrients/nutrition: new insights and advances. Trends Food Sci. Technol. **146**, 1 (2024). https://doi.org/10.1016/j.tifs.2024.104408
10. Bond, A., et al.: Artificial intelligence & clinical nutrition: what the future might have in store. Clin. Nutr. ESPEN **57**, 542–549 (2023). https://doi.org/10.1016/j.clnesp.2023.07.082
11. Du, Y., et al.: A technology assisted precision ketogenic diet intervention for cardiorenal-metabolic health in overweight or obese adults: protocol for a randomized controlled trial. Contemp. Clin. Trials **119** (2022). https://doi.org/10.1016/j.cct.2022.106845
12. Chowdhury, C.R., et al.: NMR-based metabolomics as a significant tool for human nutritional research and health applications. Food Biosci. **53**, 01 (2023). https://doi.org/10.1016/j.fbio.2023.102538
13. Prabhakar, G., et al.: User-cloud-based ensemble framework for type-2 diabetes prediction with diet plan suggestion. e-Prime Adv. Electr. Eng. Electron. Energy **7** (2024). https://doi.org/10.1016/j.prime.2024.100423
14. Coelho, K., et al.: NutriPersona: conception of a computational tool for elaboration of personalized menu from a Brazilian food composition database. J. Food Compos. Anal. **123** (2023). https://doi.org/10.1016/j.jfca.2023.105582
15. Feher, D., et al.: Few-shot entity linking of food names. Inf. Process. Manage. **60**, 5 (2023). https://doi.org/10.1016/j.ipm.2023.103463
16. Zhang, X., et al.: Application of intelligent taste analysis based on random forest algorithm in food quality inspection. Comput. Intell. Neurosci. **2022** (2022). https://doi.org/10.1155/2022/6901184

17. Höchsmann, C., et al.: Review of the validity and feasibility of image-assisted methods for dietary assessment. Int. J. Obesity **44**(12), 2358–2371 (2020). https://doi.org/10.1038/s41366-020-00693-2
18. Herzig, D., et al.: Volumetric food quantification using computer vision on a depth-sensing smartphone: preclinical study. JMIR Mhealth Uhealth **22**, 3 (2020). https://doi.org/10.2196/15294
19. Janakiraman, B., Arumugam, S.: Personalized nutrition recommendation for diabetic patients using optimization techniques. Intell. Autom. Soft Comput. **26**(2), 269–280 (2020). https://doi.org/10.31209/2019.100000150
20. Subhi, M., et al.: Deep convolutional networks for food detection and classification. J. Comput. Theor. Nanosci. **16**(5–6), 2433–2438 (2019). https://doi.org/10.1166/jctn.2019.7913
21. Liu, C., et al.: A new deep learning-based food recognition system for dietary assessment on an edge computing service infrastructure. IEEE Trans. Serv. Comput. **11**(2), 249–261 (2018). https://doi.org/10.1109/TSC.2017.2662008
22. Siemon, M., et al.: Sequential transfer learning based on hierarchical clustering for improved performance in deep learning based food segmentation. Sci. Rep. **11**, 1 (2021). https://doi.org/10.1038/s41598-020-79677-1
23. Meroni, A., et al.: Boden food plate: novel interactive web-based method for the assessment of dietary intake. J. Vis. Exp. **2018**, 139 (2018). https://doi.org/10.3791/57923
24. Tarannum, S., et al.: HALALCheck: a multi-faceted approach for intelligent Halal packaged food recognition and analysis. IEEE Access **12**, 28462–28474 (2024). https://doi.org/10.1109/ACCESS.2024.3367983
25. Bai, H., et al.: Recommendation algorithm based on probabilistic matrix factorization with Adaboost. Comput. Mater. Continua **65**(2), 1591–1603 (2020). https://doi.org/10.32604/cmc.2020.09981
26. Choi, I., et al.: An explainable prediction for dietary-related diseases via language models. Nutrients **16**, 5 (2024). https://doi.org/10.3390/nu16050686
27. Hu, G., et al.: Natural language processing and machine learning approaches for food categorization and nutrition quality prediction compared with traditional methods. Am. J. Clin. Nutr. **117**(3), 553–563 (2023). https://doi.org/10.1016/j.ajcnut.2022.11.022
28. Kuo, S., et al.: A clinical nutritional information system with personalized nutrition assessment. Comput. Methods Programs Biomed. **155**, 209–216 (2018). https://doi.org/10.1016/j.cmpb.2017.10.029
29. Borges, T., et al.: Accuracy of dietary intake assessments using food records based on photographic images captured by visually impaired people. PLoS ONE **18**, 2 (2023). https://doi.org/10.1371/journal.pone.0280725
30. Hu, G., et al.: Accelerating the classification of NOVA food processing levels using a fine-tuned language model: a multi-country study. Nutrients **15**, 19 (2023). https://doi.org/10.3390/nu15194167
31. Elshewey, A., et al.: Weight prediction using the hybrid stacked-LSTM food selection model. Comput. Syst. Sci. Eng. **46**(1), 765–781 (2023). https://doi.org/10.32604/csse.2023.034324
32. Tanweer, A., et al.: Comparing the web-based and traditional self-reported 24-hour dietary recall data in the PakNutriStudy. Comput. Methods Programs Biomed. **240** (2023). https://doi.org/10.1016/j.cmpb.2023.107682
33. Yaiprasert, C., Hidayanto, A.: AI-powered in the digital age: ensemble innovation personalizes the food recommendations. J. Open Innov. Technol. Mark. Complex. **10**, 2 (2024). https://doi.org/10.1016/j.joitmc.2024.100261

34. Kong, X., et al.: Food calorie estimation system based on semantic segmentation network. Sensors Mater. **35**(6), 2013–2033 (2023). https://doi.org/10.18494/SAM4061
35. Kotwal, J., et al.: A modified time adaptive self-organizing map with stochastic gradient descent optimizer for automated food recognition system. J. Stored Prod. Res. **107** (2024). https://doi.org/10.1016/j.jspr.2024.102314.
36. Blanchard, C., et al.: Evaluation of PIQNIQ, a novel mobile application for capturing dietary intake. J. Nutr. **151**(5), 1347–1356 (2021). https://doi.org/10.1093/jn/nxab012
37. Asbroeck, S., Matthys, C.: Use of different food image recognition platforms in dietary assessment: comparison study. JMIR Form. Res. **4**, 12 (2020). https://doi.org/10.2196/15602
38. Khan, M.A., Smyth, B., Coyle, D.: Addressing the complexity of personalized, context-aware and health-aware food recommendations: an ensemble topic modelling based approach. J. Intell. Inf. Syst. **57**(2), 229–269 (2021). https://doi.org/10.1007/s10844-021-00639-8
39. Nguyen, P., et al.: Relative validity of a mobile AI-technology-assisted dietary assessment in adolescent females in Vietnam. Am. J. Clin. Nutr. **116**(4), 992–1001 (2022). https://doi.org/10.1093/ajcn/nqac216
40. Scott, J., et al.: Feasibility of wearable camera use to improve the accuracy of dietary assessment among adults. J. Nutr. Sci. **11**, e85 (2022). https://doi.org/10.1017/jns.2022.81
41. Raju, V.B., Sazonov, E.: FOODCAM: a novel structured light-stereo imaging system for food portion size estimation. Sensors **22**, 9 (2022). https://doi.org/10.3390/s22093300
42. Prajena, G., et al.: Indonesian traditional food image recognition using convolutional neural network. In: Proceedings - 4th International Conference on Informatics, Multimedia, Cyber and Information System, ICIMCIS 2022, pp. 142–147. Institute of Electrical and Electronics Engineers Inc. (2022). https://doi.org/10.1109/ICIMCIS56303.2022.10017684
43. Islam, T., et al.: Human-behavior-based personalized meal recommendation and menu planning social system. IEEE Trans. Comput. Soc. Syst. **10**(4), 2099–2110 (2023). https://doi.org/10.1109/TCSS.2022.3213506
44. Kwon, P., et al.: Android - based smart nutrition application for children nutritional awareness. In: Proceeding - IEEE 9th Information Technology International Seminar, ITIS 2023. Institute of Electrical and Electronics Engineers Inc. (2023). https://doi.org/10.1109/ITIS59651.2023.10420296
45. Balpande, M., et al.: AI based gym trainer and diet recommendation system. In: 2023 IEEE 4th Annual Flagship India Council International Subsections Conference: Computational Intelligence and Learning Systems, INDISCON 2023. Institute of Electrical and Electronics Engineers Inc. (2023). https://doi.org/10.1109/INDISCON58499.2023.10270066
46. Abbasi, J.: Estimation of the caloric intake of food consumption using convolutional neural network. In: Proceedings - 2023 IEEE International Symposium on Multimedia, ISM 2023, pp. 177–181. Institute of Electrical and Electronics Engineers Inc. (2023). https://doi.org/10.1109/ISM59092.2023.00032
47. Huang, B., et al.: A novel recipes recommendation system based on knowledge-graph. In: 2022 7th International Conference on Intelligent Computing and Signal Processing, ICSP 2022, pp. 1408–1412. Institute of Electrical and Electronics Engineers Inc. (2022). https://doi.org/10.1109/ICSP54964.2022.9778762

48. Liu, Y., et al.: Deep-learning-assisted multi-dish food recognition application for dietary intake reporting. Electronics (Switzerland) **11**(10) (2022). https://doi.org/10.3390/electronics11101626
49. Velásquez, R.A., Lara, J.V.M., Velásquez, R.A.: Yolo algorithms to improve the detection of personal protective equipment thought industrial computer vision. In: 2023 IEEE XXX International Conference on Electronics, Electrical Engineering and Computing (INTERCON), Lima, Peru, pp. 1–6 (2023). https://doi.org/10.1109/INTERCON59652.2023.10326066
50. Velásquez, R.M.A., Lara, J.V.M.: Forecast and evaluation of COVID-19 spreading in USA with reduced-space Gaussian process regression. Chaos Soliton Fract. **136**, 109924 (2020). https://doi.org/10.1016/j.chaos.2020.109924

The Impact of Artificial Intelligence on the Enterprise Architecture

Kamelia Shoilekova(✉) and Boyana Ivanova

Angel Kanchev University of Ruse, 8 Studentska str., 7000 Ruse, Bulgaria
{kshoylekova,bivanova}@uni-ruse.bg

Abstract. The times we live in are changing dynamically, and with it, people's view of business is changing too. Over the years, companies have faced the challenge of surviving in a dynamically changing environment. This has impacted the development of managers who are a prerequisite for business people and IT professionals to start communicating with each other and overcoming numerous obstacles for better organizational development. This in turn has led to changes in the elements of enterprise architecture. In the era of artificial intelligence, the architecture of the organization will not remain untouched, and it (AI) will interfere with its most important elements.

Keywords: Business architecture · Software architecture · Enterprise architecture · Artificial intelligence · IT · diagrams · designing

1 Enterprise Architecture

Over the years, there have been many interpretations of enterprise architecture. According to [9], enterprise architecture can be seen as a "skeleton". This "skeleton" should show the main elements in the architecture of the enterprise and how they relate to each other so that problems in the enterprise can be easily defined and analyzed. According to [2], architecture defines a system's "building blocks" by specifying their properties and the way they are interconnected and designed to optimize business processes. Based on the two definitions, it can be concluded that the organization's architecture describes in detail the individual elements that make it up, defines the connections between them, and provides the ability to perform a thorough analysis and make important decisions using appropriate tools.

According to [3], enterprise architecture bridges the gap between business and IT, as its main task is to outline the relationships between business and IT in a way that describes the most important interactions between them, describes each business process in detail and, if necessary, decomposes it into sub-processes to optimize time, resources and manpower. In the era of artificial intelligence and the dynamic world in which we live, they also affect the architecture of the organization. Changes to the skeleton of the organization and the connection of individual building blocks can be coordinated not only with managers and IT specialists but also with various tools that use large language models. All this shows that organizations are broadening their perspective, which is a

R. Silhavy and P. Silhavy (Eds.): CoMeSySo 2024, LNNS 1489, pp. 238–243, 2025.
https://doi.org/10.1007/978-3-031-96798-6_20

prerequisite for their structures to become increasingly complex, the data to be processed ever more extensive and the analyses to be performed ever more sophisticated [10]. This is where artificial intelligence comes into play, because complex analyses must be presented in a form that is understandable to the average person, and artificial intelligence offers the possibility of presenting results with such accuracy and speed of response that they cannot be compared with human analyses. To integrate artificial intelligence into an organization's architecture, it is not enough to just choose the right tool; all the players (people, processes, data, and technology) need to be aligned.

2 Artificial Intelligence in the Main Elements Building the Business Architecture

In the architecture of the organization, four main elements are embedded:

- architecture of technology;
- architecture of information;
- architecture of applications;
- business architecture

If we look at the evolution of enterprise architecture, we first talked only about the architecture of technologies, whose main task was to create technological standards that would have an impact on reducing the costs of the organization. After that, the information architecture and the application architecture were gradually included. One of the main advantages of information architecture is that it introduces databases into the organization, which puts an end to the duplication of information, reducing the tasks of some of those involved in the process. The application architecture has an impact on the detailed process tracking in an organization. It aims to synchronize the processes so that they are aligned with the activities of the organization itself as well as with all external participants in them, preventing the duplication of activities. These three architectures are defined in [5] with the term system architecture [6, 11].

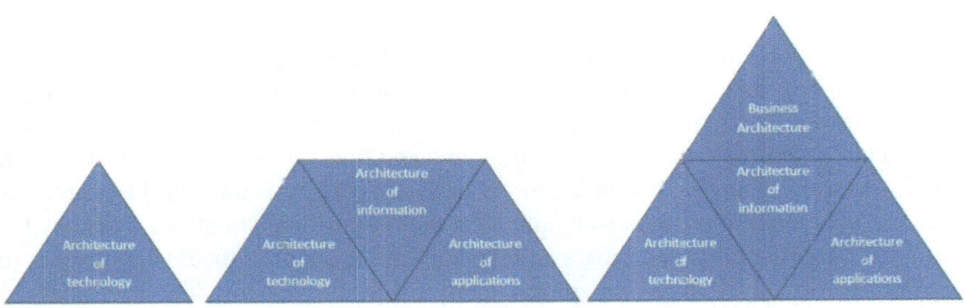

Fig. 1. The evolution of enterprise architecture.

Then comes the business architecture, the task is to present a complete picture of the organization aimed at maintaining the balance and rational implementation of goals, strategies, processes, and services embedded in the overall activity of the organization (Fig. 1).

At this stage, it cannot be said that artificial intelligence will be added to the organization's architecture as a new type of architecture. Still, it can certainly be said that artificial intelligence already plays an important role in each of the architectures under consideration. It can be expected to improve even further with the help of artificial intelligence capabilities:

- the design;
- data should be presented in a form that is easily understood by all stakeholders, improving the quality of the data and its structuring to enable fast and effective decision-making (Fig. 2).

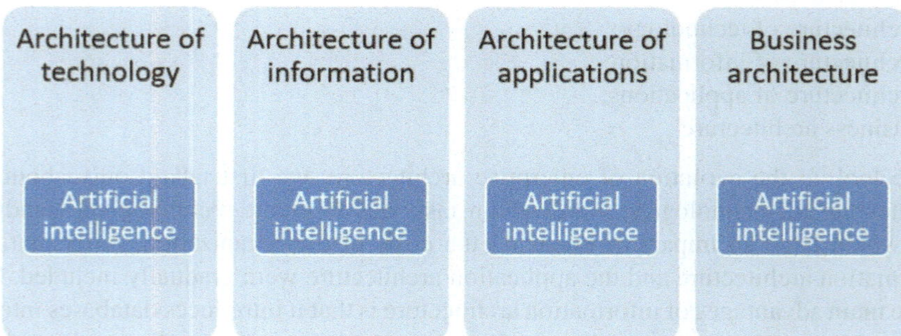

Fig. 2. The place of artificial intelligence in the enterprise architecture.

3 The Role of Artificial Intelligence in Designing the Enterprise Architecture

As architects build the overall picture of the organization and try to put the skeleton together in a way that works properly, they may make mistakes or fail to identify the most optimal option when creating the appropriate diagram [1]. With the help of artificial intelligence capabilities, the quality of diagrams and thus the overall architecture of the company can be changed, because (an AI assistant) AI assistant can help architects create projects much more precisely and alert them to critical situations that they have eliminated. Another advantage that can be taken into account at the design level is that machine learning algorithms can automatically fill in the missing data after analyzing the existing models, without having to do it manually. On the other hand, AI can provide new models and ideas that influence the company's future goals and lead to potential improvements in architectural models.

One of the most difficult tasks that are possible is mapping. AI can uncover violations that are predefined in part of the rules but have been overlooked by the architects. The key point is that the architect's approval is always required at this stage before proceeding with mapping to the implementation [4].

Another possibility that AI can offer at the design level is that architects can propose several solutions to a given problem and analyze together with all stakeholders which solution is the most appropriate. In this case, the decision depends entirely on the scenarios played out. With the help of various tools that use artificial intelligence, impact analyses are carried out, scenarios are compared and improvements are proposed based on a variety of data, which may result in a new scenario being proposed that the architect had not thought of.

4 The Role of Artificial Intelligence in Presenting Information to All Interested Parties

One of the biggest challenges in developing an organization's architecture is making the "skeleton" understandable to all interested parties, i.e. presenting models and/or diagrams to people with little technical experience. At this stage, artificial intelligence can intervene again, because one of its possibilities is to try to translate the information so that it is understandable to all those involved in the process, using an appropriate tool that translates diagrams and models into easily understandable forms.

In addition to creating models and diagrams that are aligned with the new goals, architects must also maintain an up-to-date view of the current state of the organization, including eliminating critical issues and updating changes to the organization's processes, services, goals, and objectives. All of this shows that the time required to track the entire picture of the organization is an important factor, and the picture is becoming more detailed and extensive. Again, with the help of various tools that use artificial intelligence, analyses of different scenarios can be carried out to focus the architect's attention on the need for change and optimize his time.

Over time, it is often necessary to review and optimize the organization's requirements or activities, change architects, etc. All of this requires the development of models and diagrams using different tools, which leads to different data formats, which can lead to delays in deadlines. This point can be crucial, especially when changes need to be made from the process model to the implementation of the process itself [7, 8]. To avoid confusion, it is good always to use the same file formats, but if there are differences, it is good to use a suitable tool to present the information in a way that can be easily integrated into the implementation of the process itself.

Therefore, enterprise architects need to identify potential trends, easily extract information, and perform analysis based on the vast amounts of data available to them for the respective organization to be competitive. It follows that intervention and the use of appropriate AI tools is possible:

- to direct the attention of the architects to various critical points, which will prove to be a key factor for the success of the organization;
- to propose a scenario that is not obvious to the human, aiming to show dependencies that have gone unnoticed by the architect and would lead to cost and time savings;

Up to this point, only the positive aspects of artificial intelligence have been highlighted, emphasizing that the architecture of the organization cannot be built and maintained by it (artificial intelligence) alone, but that many specialists are also needed to complete the realization of the specific goal and/or task after examining the decisions it proposes. Data security can be defined as the only drawback at this stage, as there are still no observations on whether the solutions proposed by a particular tool are available in several organizations:

- are not available in several organizations;
- whether the data provided for the different scenarios are also used in other organizations.

Despite this serious drawback, if the organization's architecture is designed and implemented according to the prescribed principles and standards, the likelihood that the information will be useful to other organizations is very low.

5 Conclusion

The introduction of artificial intelligence in enterprise architecture is proving to be a key factor in the organization's flexibility and adaptability to the rapidly changing environment. Architects can use different tools to optimize processes, time, and resources using multiple scenarios and various analyses. Based on everything that has been written so far, it can be said that thanks to artificial intelligence, changes in the organization's architecture will take place with fewer resources.

Acknowledgements. This research is supported by the Bulgarian Ministry of Education and Science under the National Program "Young scientists and Postdoctoral Students - 2".

References

1. Varzaneh, A.S., Varzaneh, M.S.: Total experience lifecycle model: based on smart university architecture. In: 2024 20th CSI International Symposium on Artificial Intelligence and Signal Processing (AISP), Babol, Iran, Islamic Republic of, pp. 1–9 (2024). https://doi.org/10.1109/AISP61396.2024.10475275
2. Bredemeyer, D.: What it Takes to be Great in the Role of Enterprise Architect. http://www.bredemeyer.com
3. Bailey, I.: A simple guide to enterprise architecture. In: Model Futures (2006)
4. Strickrodt, D.: The future of enterprise architecture and AI integration. Embrace the synergy. https://bizzdesign.com/blog/the-future-of-enterprise-architecture-and-ai-integration/
5. Kalianov, G.N.: Enterprise architecture and the instruments of its modelling (in Russian). In: Automation in Industry, vol. 7, (2004)
6. Khairina, D., Purwanto, P., Nugraheni, D.: Systematic literature review on evaluation models and methods in enterprise architecture research. Bull. Electr. Eng. Inform. **13**, 2851–2864 (2024). https://doi.org/10.11591/eei.v13i4.6943
7. Krastev, G., Voinohovska, V.: Smart mobile application for public transport schedules – data organization and program realization. In: International Congress on Human-Computer Interaction. Optimization and Robotic Applications (HORA) 2020, pp. 1–4. IEEE (2020). https://doi.org/10.1109/HORA49412.2020.9152884

8. Krastev, G., Voinohovska, V.: Smart mobile application for public transport schedules - logical model. TEM J. **9**(2), 541–545 (2020). https://doi.org/10.18421/TEM92-15
9. Lillehagen, F., Karlsen, D.: Enterprise Architectures – Survey of Practices and Initiatives. http://interopesa05.unige.ch/INTEROP/Proceedings/Industrial/IND1_Lillehagen.pdf
10. Petrova, M., Atanasova, D.: Searching algorithms and a comparison methodology, used in studying artificial intelligence. In: 11th International Conference of Education, Research and Innovation (Web of Science), pp. 1767–1775 (2018)
11. Shoilekova, K., Ivanova, B., Andreeva, M.: The meaning of business and software architecture in the enterprise architecture. In: Silhavy, R., Silhavy, P. (eds.) Data Analytics in System Engineering. CoMeSySo 2023. Lecture Notes in Networks and Systems, vol. 935. Springer, Cham (2024). https://doi.org/10.1007/978-3-031-54820-8_28

Comparison of Machine Learning Algorithms for the Identification of Pain Caused by Lumbar Disc Herniation

Christian Ovalle(✉) and Jorge Huamani Correa

Universidad Tecnológica del Perú, Lima, Peru
{dovalle,c19349,c22273}@utp.edu.pe

Abstract. The adult population has experienced low back pain at some time in their lives, especially in women between 40 and 80 years of age. Given this, the present study aims to compare Machine Learning algorithms for the identification of pain caused by lumbar disc herniation, for which several models were evaluated using key metrics such as Sensitivity (Recall), Classification Accuracy (CA), F1 Score, Area Under the Curve (AUC) and Matthews Correlation Coefficient (MCC). The results indicate that the Neuronal Network model stood out for its sensitivity with a recall of 1.000, which means an excellent capacity to identify positive cases of low back pain, as well as an AUC of 0.500 and MCC of 0.000; Gradient Boosting showed a high sensitivity Recall of 0.999, competitive F1 Score of 0.799, AUC of 0.497 and MCC of 0.002; Random Forest and k-Nearest Neighbors (kNN) obtained moderate results with metrics such as F1 Score around 0. 721–0.722 and Recalls of about 0.786–0.789, AUCs close to 0.5 and low MCCs (0.002 and 0.001 respectively); AdaBoost and the Logistic Regression model showed inferior performance compared to the previous ones, with AUC values around 0.502–0.570 and MCCs close to zero, indicating a limited ability to identify cases of low back pain correctly. In conclusion, we recognize the need to optimize and further explore the algorithms to improve the accuracy in identifying low back pain caused by disc herniation.

Keywords: Algorithms · Machine learning · Herniated disc · Lower back pain

1 Introduction

It is estimated that at least two-thirds of the adult population have experienced low back pain at some time in their lives [1], with a higher incidence in women between 40 and 80 years of age [2]. For this reason, low back pain is the second ailment that, due to its symptoms, is presented with the most excellent recurrence in medical consultations, only being surpassed by respiratory problems [3]. Likewise, the study of the global burden of diseases carried out in 2021 indicates that low back pain is the leading cause of disability in most countries [4], with a total of 619 million cases in 2020 [5]. It is also estimated that an average of 10% to 20% of sufferers develop a chronic condition [6], and between 13.4% and 18.7% are hospitalized [7]. Similarly, low back pain is the leading cause of

R. Silhavy and P. Silhavy (Eds.): CoMeSySo 2024, LNNS 1489, pp. 244–257, 2025.
https://doi.org/10.1007/978-3-031-96798-6_21

absenteeism from work globally [8], generating an economic loss of approximately $1.7 billion per year and costs to the health system of $2.2 billion globally. In this sense, there is a growing need to improve the management of low back pain to reduce its socioeconomic impact [9]. Among the most frequent lumbar conditions is a herniated disc, with a prevalence of 2 to 3% in the population [10]. This occurs when the gelatinous material in the center of an intervertebral disc tears and presses against a nerve [11], causing pain, tingling, and other discomfort that, in many cases, limits physical activity and affects the sufferer's quality of life [10]. Commonly, 85% of people with lumbar disc herniation can be treated without the need for surgery with non-invasive methods such as physiotherapy and the use of anti-inflammatory drugs [12]. In this regard, it has been shown that 63% of patients who have received appropriate physical therapy have experienced a favorable regression of the condition [13]. However, because the choice of treatment for lumbar disc herniation presents difficulties such as subjective interpretation of the diagnosis, which in many cases results in overtreatment of patients [14], it is of vital importance to implement innovative techniques that facilitate the timely treatment of lumbar disc herniation, through the accurate identification of the condition and its severity, in order to prevent complications that lead to the need for surgical intervention [15].

In this context, the emergence of machine learning (ML) has presented an innovative perspective to address medical challenges in healthcare and surgical decision-making [16]. Current medical practices incorporate ML methods that play a crucial role by extracting valuable information from data without predefined human rules [12]. Furthermore, ML helps healthcare professionals improve the quality of care and make accurate decisions based on data analysis and interpretation [17]. Due to the volume and complexity of data available in national spine registries in different countries, there is an opportunity to develop better prediction models, which is necessary to improve the quality of spine care [18]. As a reference, a study demonstrated high levels of accuracy of algorithms in disease prediction through the use of ML models to predict T-score and identify osteoporotic vertebrae based solely on conventional CT Hounsfield unit (HU) measurements, which helps spine surgeons to accurately assess the status of affected bony structures preoperatively [19]. In addition, ML techniques have achieved high accuracy in classifying conditions such as spondylolisthesis and lumbar lordosis [20].

However, although there are studies such as the one presented in the article "Medical expert and machine learning analysis of lumbar disc herniation based on magnetic resonance imaging," whose purpose is the development of machine learning models to predict specific clinical outcomes, such as the risk of lumbar disc herniation recurrence or response to treatment [15], as well as others that propose ML algorithms to predict postoperative outcomes [21] and even to assess the risk of recurrence of lumbar disc herniation and other lumbar conditions [22], present a common weakness since they base the training of machine learning algorithms on image data and patient clinical histories, in relation to this, the article "Evaluation of deep oscillation therapy for the treatment of lumbar pain syndrome using motion capture systems: A systematic review," which proposes a comparison between the results obtained by deep oscillation therapy and other conventional treatments, indicates the lack of studies that use motion capture systems to evaluate therapy in low back pain syndrome [23]. In this regard, the present research proposes to use the physiological data of the study participants captured in real-time by means of a device with sensors located in the lumbar area. This will allow better training of the ML models to be compared by providing them with accurate information that identifies patterns and trends that might go unnoticed in image-based studies or physical assessments. Other similar studies, such as the one described in "MSFF: An automated multi-scale feature fusion deep learning model for spine fracture segmentation using MRI," which has proposed an effective algorithm in spine fracture segmentation [24], present a limitation in terms of the variety of machine learning models used. In the study [25], a decision tree (DT) was used to aid in clinical decision-making, and several ML algorithms were applied to correlate lumbar disc height (LDBH) on X-ray with disc bulging or herniation. The most important factors identified were L4–5 disc height, age, and L1–2 disc height. To address this gap, our research will apply a broader set of machine learning models, including logistic regression, random forest, AdaBoost, L-Means, Gradient Boosting, and neural networks [26, 27]. By incorporating a more comprehensive range of models, we can explore different machine-learning approaches and techniques to improve accuracy and predictive ability. This will allow us to assess the relative effectiveness of different models and provide a more complete understanding of how different machine-learning approaches may influence predicted clinical outcomes [28]. On the other hand, the research described in the article "Active-Matrix Sensing Array Assisted with Machine-Learning Approach for Lumbar Degenerative Disease Diagnosis and Postoperative Assessment" demonstrated the effectiveness of a sole pressure measurement device for the diagnosis of lumbar degenerative disease through the use of the support vector machine ML model. As a field of exploration, we are exploring the possibility of testing different models to choose the most appropriate one for identifying physiological patterns that help diagnose and treat similar ailments.

In that sense, the present study proposes a comparison of Machine Learning algorithms, such as logistic regression, random forest, isolation forest, K-Means, Gradient Boosting, and neural networks [25–27], which will be evaluated based on their accuracy in identifying and classifying low back pain intensity. This research proposes a novel approach, as the training of ML models is performed from physiological data captured in real-time using a belt equipped with special sensors that allow recording the muscle contractions of the study participants. Therefore, unlike conventional studies that rely solely on imaging data and clinical history for the development of their predictive models [29], the analysis of data provided during the daily activities of study participants allows us to develop algorithms capable of identifying meaningful patterns and trends that can help improve the accuracy of diagnosis and treatment of patients with lumbar disc herniation [30]. Moreover, this novel approach not only allows us to discover how the identification of pain could complement current diagnostic and treatment techniques but also offers the possibility of exploring new data collection techniques to develop more robust predictive models for detecting and treating other ailments in medical practice.

2 Method

This study proposes an analysis of the performance of six predictive models trained with machine learning techniques to compare them based on their accuracy in identifying low back pain caused by disc herniation. For the present research, a methodology based on the CRISP-DM model is proposed, which proposes a flexible framework and is widely used as a standard for data mining and machine learning projects [31] to cover the entire process with due scientific rigor comprehensively. The present research is of a pre-experimental and applied design since it focuses on directly applying theoretical and methodological knowledge to develop innovative solutions that can be used in authentic contexts [32]. The workflow comprises, in the first place, data collection through a prototype equipped with sensors to identify low back pain. Subsequently, data preparation is performed, a stage in which an exhaustive processing of the information collected is carried out, culminating in obtaining the final dataset. The next stage is the training stage, where the training and test datasets are selected, and the models are trained using the ML techniques chosen for the study. Finally, the performance evaluation phase is carried out, where evaluation metrics will be applied to compare and analyze the performance of the machine learning techniques in identifying low back pain, allowing meaningful conclusions to be drawn regarding the effectiveness of each algorithm. Figure 1 shows the proposed workflow in greater detail for a better understanding.

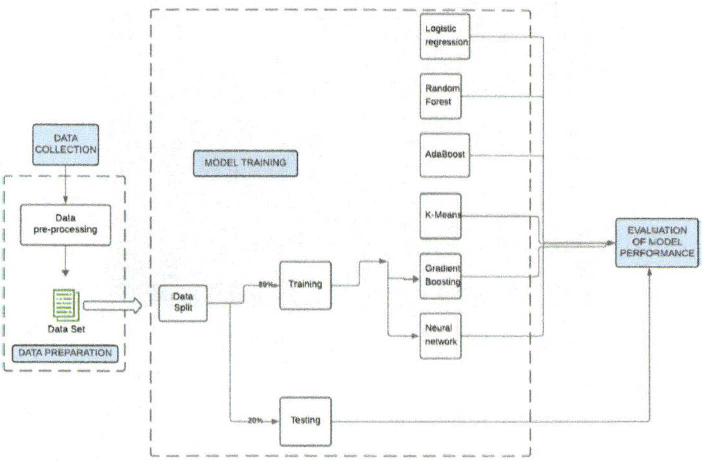

Fig. 1. Methodology workflow.

2.1 Data Collection

To experiment, we used our data obtained through a device equipped with sensors to measure muscle contractions in real-time. The study involved three people diagnosed with lumbar disc herniation of both genders and ages between 25 and 80 years, as detailed in Table 1. The participants wore the device during their daily activities for about 2–3 h daily. These sensors recorded two electromyographic signals (SECG-1 and SECG-2), representing the electrical activity of the muscles associated with the lumbar region (Fig. 2). In this regard, electromyography is the most widely used technique in rehabilitation due to the high degree of robustness of the detected signals, especially during movement, compared to other techniques [33], which is essential in this experiment. These signals were collected simultaneously within a text file for each data record, allowing a comprehensive assessment of muscle activity and its relationship with low back pain.

Table 1. Characteristics of the study participants.

Sex	Age
Female	43
Female	51
Male	33

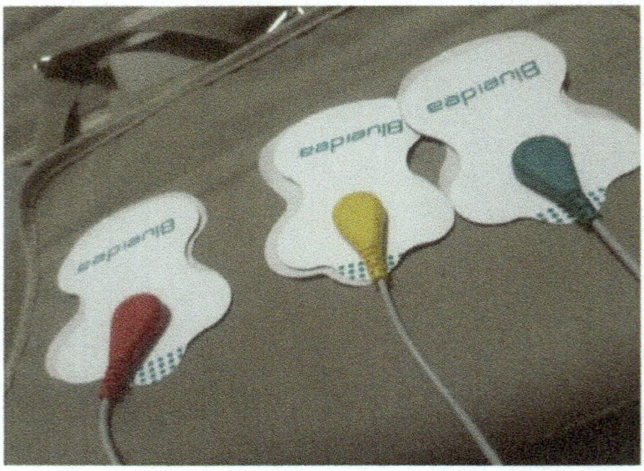

Fig. 2. Electromyographic sensors.

2.2 Data Preparation

The raw data initially collected in a text file were stored in a spreadsheet for preparation. Orange data mining software was used for this process. The data were subjected to a thorough cleaning process to eliminate incomplete, duplicate, or incorrect values that could affect the quality of the results. In addition, preprocessing techniques were applied to ensure the consistency and comparability of the variables. Subsequently, a prepared and optimized dataset was obtained, ready to be used to compare machine learning algorithms for the prediction of muscle pain, as shown in Fig. 3.

2.3 Model Training

Six machine learning techniques were used to build the predictive models, including Logistic regression, Ramdon forest, AdaBoost, K-means, Gradient boosting, and Spline neural networks. The training process was carried out using the tools provided by Orange

Date	Hour	SECG-1	SECG-2
2024-02-27 00:00:00	18:27:02	2.013298431	2.889440976
2024-02-27 00:00:00	18:27:03	0.203440314	0.73718426
2024-02-27 00:00:00	18:27:04	1.983037547	1.547114693
2024-02-27 00:00:00	18:27:06	3.289106589	2.867787153
2024-02-27 00:00:00	18:27:07	1.621162122	1.79310612
2024-02-27 00:00:00	18:27:09	2.426965316	2.945319591
2024-02-27 00:00:00	18:27:10	0.557019276	0.356106102
2024-02-27 00:00:00	18:27:11	0.184512907	0.629017259
2024-02-27 00:00:00	18:27:13	2.150015581	3.247536373
2024-02-27 00:00:00	18:27:14	2.020148601	2.73207054
2024-02-27 00:00:00	18:27:16	3.242547334	2.912768764
2024-02-27 00:00:00	18:27:17	0.045432312	0.652827394
2024-02-27 00:00:00	18:27:19	0.841039142	0.891937455
2024-02-27 00:00:00	18:27:20	2.024273534	2.933299492
2024-02-27 00:00:00	18:27:21	0.220012852	0.762449996
2024-02-27 00:00:00	18:27:23	2.423955832	2.026597323
2024-02-27 00:00:00	18:27:24	2.195568852	3.104785798
2024-02-27 00:00:00	18:27:26	2.254011833	3.045913773
2024-02-27 00:00:00	18:27:27	0.060079929	0.196485085

Fig. 3. Obtained Dataset.

software. In the first instance, the target variable was specified as "Pain," where the predicted values will be stored to identify the presence and intensity of low back pain. Subsequently, the hyperparameters of the models were adjusted according to need. Finally, the models were trained from the selected data set, establishing 80% for this purpose and the remaining 20% for the tests.

- **Logistic regression**

  Formula:

$$P(Y) = 1/1 + e\text{-}(b0 + b1X1 + b2X2 + \ldots . + bnXn)$$

Figure 4 shows the training process of the logistic regression model using Orange Data Mining software.

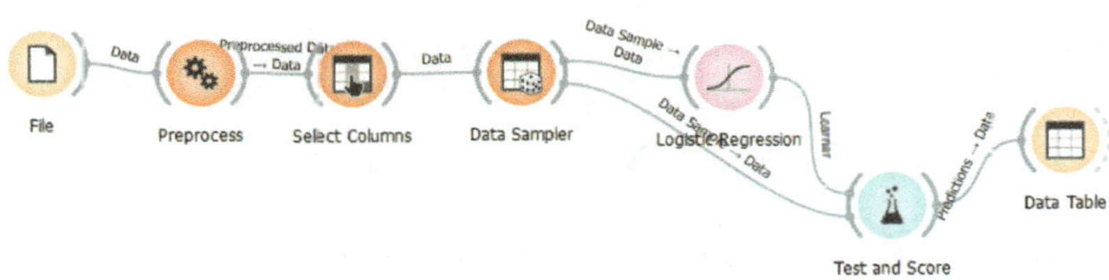

Fig. 4. Logistic regression model training.

- **Ramdon forest**

Formula:

$$y\hat{}(x) = modo(\{y\hat{}1(x), y\hat{}2(x), \ldots, y\hat{}B(x)\})$$

Figure 5 shows the training process of the Ramdon forest model using Orange Data Mining software.

Fig. 5. Random Forest model training.

- **AdaBoost**

  Formula:

$$H(x) = \text{sign}\left(t = 1 \sum \text{Tat} \cdot \text{ht}(x)\right)$$

Where, T is the total number of weak classifiers.

Figure 6 shows the training process of the AdaBoost model using Orange Data Mining software.

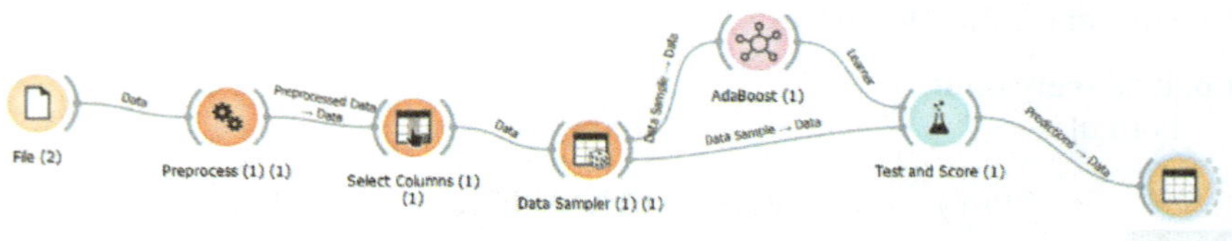

Fig. 6. K-means model training.

- **K-means**

  Formula:

$$J = N1 \sum i = 1N \|xi - mci\|2$$

Figure 7 shows the training process of the K-means model.

- **Gradient boosting**

  Formula:

$$F(x) = \sum m = 1M\, \gamma mhm(x)$$

Where:

- M is the total number of models (trees) in the ensemble.

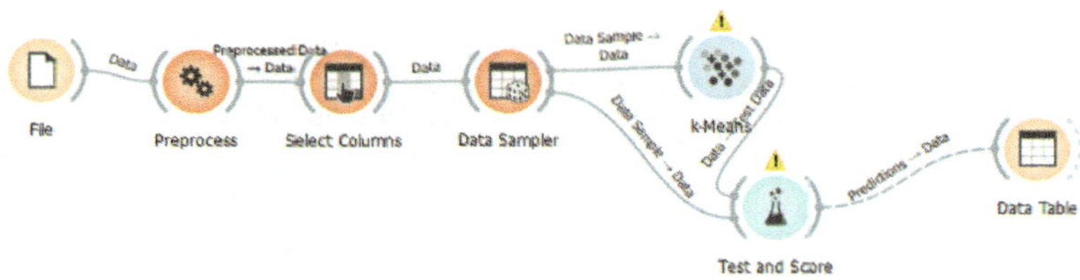

Fig. 7. K-means model training.

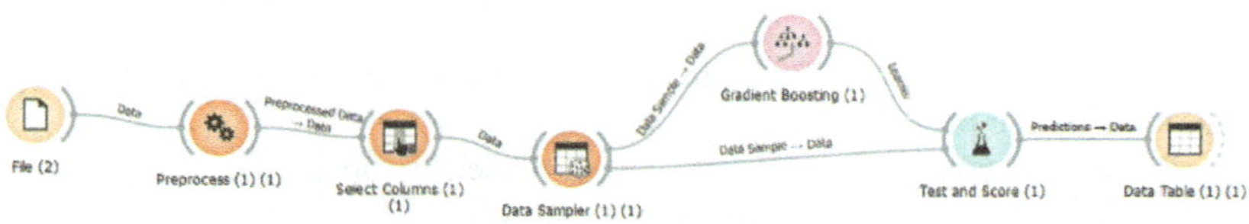

Fig. 8. K-means model training.

- Hm(x) is the m-th model (tree) in the ensemble.
- Ym is the weight of the m-th model.

Figure 8 shows the training process of the Gradient boosting model.

- **Redes neuronales**

  Formula:

$$y = f\left(\Sigma\left(w_i{}^* x_i\right)\right)$$

Where:

- y: Is the output value of the neuron.
- f: is the activation function. The activation function introduces a nonlinearity in the output of the neuron, which allows the network to learn complex relationships between the input and output data. Some common activation functions are the step function, the sigmoid function and the Re-LU function.
- Σ: Represents the sum.
- w_i: It is the weight of the connection between the input neuron i and the current neuron. The weights represent the strength of the cone-xion between neurons.
- x_i: Is the output value of input neuron i.

Figure 9 shows the training process of the Neural Network model.

2.4 Perfomance Evaluation

The trained machine learning models were evaluated based on performance metrics such as sensitivity, accuracy, F1-score, and area under the ROC curve. This process allows us to determine the best model to identify the pain caused by lumbar disc herniation.

Precision: The model's accuracy in correctly predicting the positive instances among all the optimistic predictions was evaluated. This metric, whose formula is given by:

Fig. 9. Neural networks model training.

Precision = TP / (TP + FP), is fundamental to measure the reliability of the models to correctly identify the presence of low back pain.

Recall: The capacity of the models to correctly identify positive instances is a crucial aspect of the evaluation process. This metric, Recall = TP / (TP + FN), allows us to determine the instances where low back pain is truly present, a crucial goal in our research.

F1-Score: The model's balance was evaluated in terms of its accuracy and sensitivity on a single metric. This metric, whose formula is given by F1 Score = 2(Precision x Recall/ Precision + Recall), provides an overview of the model's effectiveness. It is beneficial when a balance between avoiding false negatives and false positives is needed to identify low back pain.

Area under the ROC curve: The AUC-ROC, a comprehensive metric, was used to evaluate the model's ability to correctly distinguish between positive and negative classes at various classification thresholds. Its formula is given by:

$$\mathrm{AUC} - \mathrm{ROC} = \mathrm{i} = 1 \sum \mathrm{n} - 12(\mathrm{xi} + 1 - \mathrm{xi}) \cdot (\mathrm{yi} + \mathrm{yi} + 1)$$

The AUC is a robust metric that provides a measure of the model's performance in terms of its discrimination ability, regardless of the classification threshold. Its robustness reassures us that a higher AUC indicates a better ability of the model to discriminate between positive and negative cases, which is crucial to ensure accurate and reliable diagnosis of low back pain.

As shown in Fig. 10, the test data set was used to evaluate models in Orange data mining software.

3 Results and Discussion

3.1 Results

Based on the analysis of the resulting values after applying the selected performance metrics, as detailed in Table 2, the findings of the present research are described for each model used, where, it shows the values obtained after applying the chosen performance metrics for each trained model, while Fig. 11 shows the confusion matrix generated in the Orange software, which allows visualizing how many cases were correctly identified as positive (TP), how many cases were correctly identified as unfavorable (TN), as well as the number of false alarms (FP) and omissions (FN).

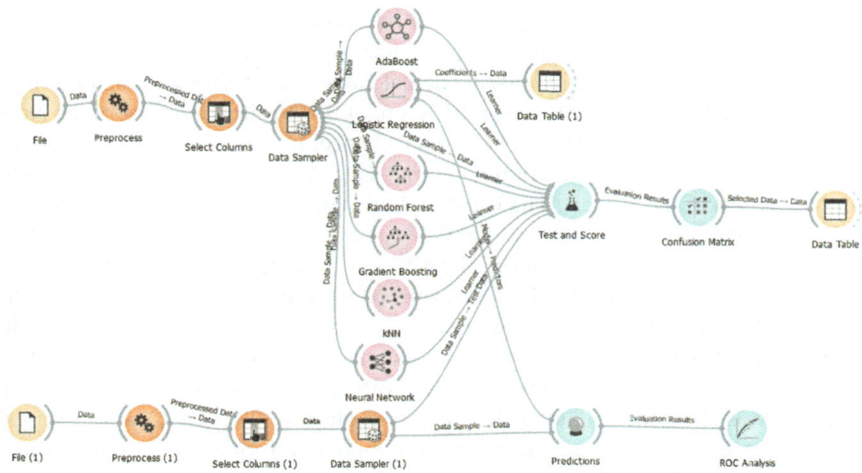

Fig. 10. Model Evaluation in Orange.

Table 2. Model Evaluation metrics.

Model	AUC	CA	F1	Prec	Recall	MCC
Random Forest	0.504	0.595	0.721	0.666	0.786	0.002
Gradient Boosting	0.497	0.665	0.799	0.666	0.999	0.002
kNN	0.502	0.596	0.722	0.666	0.789	0.001
Neural Network	0.500	0.666	0.799	0.666	1.000	0.000
AdaBoost	0.502	0.570	0.688	0.666	0.712	−0.000
Logistic Regression	0.501	0.666	0.799	0.666	1.000	0.000

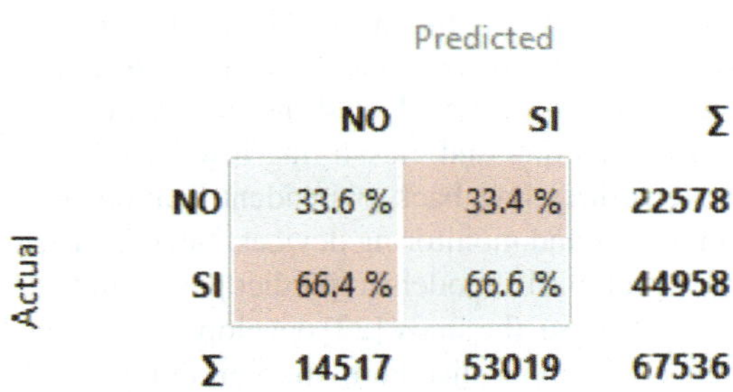

Fig. 11. Confusion matrix.

The Neural Network model showed outstanding performance in terms of sensitivity, with a perfect Recall value (1,000), indicating an excellent ability to correctly identify positive cases of low back pain. In addition, it achieved a classification accuracy (CA) of

0.666 and an F1 Score of 0.799, suggesting a good balance between accuracy and Recall. However, its AUC was 0.500, equivalent to chance, and the MCC was 0.000, indicating a null correlation between predictions and actual values. Gradient Boosting presented a high sensitivity (Recall of 0.999) and an F1 Score of 0.799, similar to that of the Neural Network, but with an AUC of 0.497, lower than chance. Its classification accuracy (CA) was 0.665, being one of the best in this metric. The MCC was also very low (0.002), similar to the rest of the models. The Random Forest model performed moderately well with an F1 Score of 0.721 and a Recall of 0.786, but its AUC of 0.504 and CA of 0.595 indicate an overall lower performance. The MCC of 0.002 reflects minimal correlation. The k-Nearest Neighbors (kNN) showed results comparable to Random Forest, with an AUC of 0.502, CA of 0.596, and F1 Score of 0.722. Its Recall was 0.789, and the MCC was near zero (0.001). AdaBoost had the worst classification accuracy (CA of 0.570) and the lowest F1 Score (0.688). Its AUC was 0.502, and the Recall of 0.712, although acceptable, was the lowest among the models evaluated. The MCC was practically zero (−0.000). Finally, the Logistic Regression model performed very similarly to the Neural Network model, with an AUC of 0.501, CA of 0.666, F1 Score of 0.799, perfect Recall (1.000) and MCC of 0.000, indicating excellent identification of true positives but zero correlation between predictions and true values. Although the Neural Network and Logistic Regression models demonstrated superior performance in terms of Recall and F1 Score, all models presented AUC values close to 0.5 and low MCC, suggesting poor performance in the ability to discriminate and correlate between predictions and true values. This underscores the need for further exploration and optimization of algorithms to improve the accurate identification of low back pain caused by disc herniation, and the potential impact of this research on healthcare.

3.2 Discussion

The results of our study make a significant contribution to the field of lumbar disc herniation diagnosis. We provide a comprehensive analysis of the effectiveness of various machine-learning techniques in classifying and predicting pain associated with this condition. Our study uniquely explores the strengths and limitations of each algorithm, shedding light on their implications for clinical practice and future research.

In addition, several studies have evaluated the effectiveness of different Machine Learning algorithms for detecting and classifying low back pain. These studies have found varied results regarding low back pain identification accuracy, using similar machine-learning techniques and monitoring devices. Such is the case of the study [21], whose purpose was to develop ML models to predict postoperative outcomes regarding lumbar disc herniation. Likewise, the study [22] developed ML models for the evaluation of the risk of recurrence of lumbar disc herniation and other lumbar conditions using algorithms whose results were favorable. The study [24] developed a practical algorithm for segmenting spine fractures, which evaluated several algorithms, highlighting the Random Forest algorithm. Finally, the study [25] applied several ML algorithms to correlate lumbar disc height (LDBH) on X-ray with disc bulging or herniation.

For a better understanding of the results of similar research, Table 3 shows the algorithms used and the results obtained through the evaluation of metrics and the AUC.

Table 3. Comparison of results with similar research.

Ref.	Algorithm	AUC	F1-SCORE	Precision	Recall
[21]	XGBoost	0.81	0.67	0.95	0.95
	SVM	0.52	0.37	0.95	0.68
	Logistic Regression	0.56	0.37	0.95	0.76
	Random Forest	0.56	0.37	0.95	0.78
[22]	Neural networks	0.88	0.85	0.82	0.89
	XGBoost	0.92	0.86	0.89	0.82
	KNN	0.8	0.83	0.76	0.91
	Random Forest	0.93	0.83	0.81	0.85
	SVC	0.68	0.69	0.68	0.71
[24]	SVM	0.79	0.94	0.93	0.93
	Random Forest	0.69	0.59	0.51	0.71
[25]	Linear Regression	0.62	0.60	0.86	0.71
	KNN	0.69	0.68	0.92	0.78
	Decision Tree	0.52	0.55	0.59	0.57
	Random Forest	0.66	0.68	0.79	0.73
	XGBoost	0.62	0.60	0.86	0.71

4 Conclusion

In this comparative study of machine learning algorithms for identifying low back pain caused by lumbar disc herniation, it was observed that the neural network and logistic regression models exhibited superior performance in sensitivity and F1 Score. These models demonstrated an outstanding ability to correctly detect positive cases of low back pain, reflecting a high level of accuracy in predicting the condition. However, the low AUC and MCC close to zero for these and other models evaluated point to a general weakness in the ability of these algorithms to effectively discriminate between positive and negative cases based on the data provided. Based on the results, it is concluded that there is an urgent need to optimize existing models or explore alternative approaches to improve the predictive ability and diagnostic accuracy in identifying low back pain caused by lumbar disc herniation. It should be emphasized that future research could focus on the inclusion of new biometric features or the application of advanced machine learning techniques that can overcome the observed limitations to provide more effective and personalized care to patients with this medical condition, thus further advancing the identification and management of low back pain caused by lumbar disc herniation.

References

1. Meucci, R.D., Fassa, A.G., Xavier Faria, N.M.: Prevalence of chronic low back pain: systematic review. Rev. Saude Publica. **49**, 1 (2015). https://doi.org/10.1590/S0034-8910.2015049005874
2. Hoy, D., et al.: A systematic review of the global prevalence of low back pain. Arthritis Rheum. **64**(6), 2028–2037 (2012). https://doi.org/10.1002/art.34347
3. Care, P.Y., et al.: Low Back Pain (2001). www.nejm.org
4. Ferreira, M.L., et al.: Global, regional, and national burden of low back pain, 1990–2020, its attributable risk factors, and projections to 2050: a systematic analysis of the global burden of disease study 2021. Lancet Rheumatol. **5**(6), e316–e329 (2023). https://doi.org/10.1016/S2665-9913(23)00098-X
5. Hubeishy, M.H., et al.: Developing a low back pain guideline implementation programme in collaboration with physiotherapists and chiropractors using the behaviour change wheel: a theory-driven design study. Implement. Sci. Commun. **5**(1), 33 (2024). https://doi.org/10.1186/s43058-024-00568-x
6. Maher, C.G.: Effective physical treatment for chronic low back pain. Orthop. Clin. North Am. **35**(1), 57–64 (2004). https://doi.org/10.1016/S0030-5898(03)00088-9
7. Fatoye, F., Gebrye, T., Mbada, C.E., Useh, U.: Clinical and economic burden of low back pain in low- and middle-income countries: a systematic review. BMJ Open. **13**(4), e064119 (2023). https://doi.org/10.1136/bmjopen-2022-064119
8. Deyo, R.A., Cherkin, D., Conrad, D., Volinn, E.: Cost, controversy, crisis: low back pain and the health of the public. Annu. Rev. Public Health. **12**(1), 141–156 (1991). https://doi.org/10.1146/annurev.pu.12.050191.001041
9. Villalba-Meneses, F., et al.: BackMov: individualized motion capture-based test to assess low back pain mobility recovery after treatment. Sensors. **24**(3), 913 (2024). https://doi.org/10.3390/s24030913
10. Costa, F., Oertel, J., Zileli, M., Restelli, F., Zygourakis, C.C., Sharif, S.: Role of surgery in primary lumbar disk herniation: WFNS spine committee recommendations. World Neurosurg. X. **22**, 100276 (2024). https://doi.org/10.1016/j.wnsx.2024.100276
11. Brinjikji, W., et al.: Systematic literature review of imaging features of spinal degeneration in asymptomatic populations. Am. J. Neuroradiol. **36**(4), 811–816 (2015). https://doi.org/10.3174/ajnr.A4173
12. Zhang, A.S., et al.: Lumbar disc herniation: diagnosis and management. Am. J. Med. **136**(7), 645–651 (2023). https://doi.org/10.1016/j.amjmed.2023.03.024
13. Wang, Y., Dai, G., Jiang, L., Liao, S.: The incidence of regression after the non-surgical treatment of symptomatic lumbar disc herniation: a systematic review and meta-analysis. BMC Musculoskelet. Disord. **21**(1), 530 (2020). https://doi.org/10.1186/s12891-020-03548-z
14. Carassiti, M., et al.: Epidural steroid injections for low back pain: a narrative review. Int. J. Environ. Res. Public Health. **19**(1), 231 (2022). https://doi.org/10.3390/ijerph19010231
15. Liu, Y., Chen, P.H.C., Krause, J., Peng, L.: How to read articles that use machine learning: users' guides to the medical literature. JAMA. **322**(18), 1806–1816 (2019). https://doi.org/10.1001/jama.2019.16489
16. Katsos, K., Johnson, S.E., Ibrahim, S., Bydon, M.: Current applications of machine learning for spinal cord tumors. Life. **13**(2), 520 (2023). https://doi.org/10.3390/life13020520
17. Berg, B., et al.: Machine learning models for predicting disability and pain following lumbar disc herniation surgery. JAMA Netw. Open. **7**(2), E2355024 (2024). https://doi.org/10.1001/jamanetworkopen.2023.55024
18. Moncada Granda, K.E., Correa Martínez, F.G.: Aplicaciones de la inteligencia artificial en cirugía. Salud ConCiencia. **2**(2), e31 (2023). https://doi.org/10.55204/scc.v2i2.e31

19. Masood, R.F., Taj, I.A., Khan, M.B., Qureshi, M.A., Hassan, T.: Deep learning based vertebral body segmentation with extraction of spinal measurements and disorder disease classification. Biomed Signal Process. Control. **71**, 103230 (2022). https://doi.org/10.1016/J.BSPC.2021.103230
20. Schönnagel, L., et al.: Predicting postoperative outcomes in lumbar spinal fusion: development of a machine learning model. Spine J. **24**(2), 239–249 (2024). https://doi.org/10.1016/j.spinee.2023.09.029
21. Ren, G., et al.: Machine learning predicts recurrent lumbar disc herniation following percutaneous endoscopic lumbar discectomy. Global. Spine J. **14**(1), 146–152 (2024). https://doi.org/10.1177/21925682221097650
22. Villalba-Meneses, F., et al.: Evaluation of deep oscillation therapy for the treatment of lumbar pain syndrome using motion capture systems: a systematic review. J. Bodyw. Mov. Ther. **38**, 180–190 (2024). https://doi.org/10.1016/j.jbmt.2024.01.010
23. Saeed, M.U., Bin, W., Sheng, J., Albarakati, H.M., Dastgir, A.: MSFF: an automated multi-scale feature fusion deep learning model for spine fracture segmentation using MRI. Biomed. Signal Process. Control. **91**, 105943 (2024). https://doi.org/10.1016/J.BSPC.2024.105943
24. Pao-Chun, L., et al.: Development of a machine learning algorithm to correlate lumbar disc height on X-rays with disc bulging or herniation. Diagnostics. **14**(2), 134 (2024). https://doi.org/10.3390/diagnostics14020134
25. Torlay, L., Perrone-Bertolotti, M., Thomas, E., Baciu, M.: Machine learning–XGBoost analysis of language networks to classify patients with epilepsy. Brain Inform. **4**(3), 159–169 (2017). https://doi.org/10.1007/s40708-017-0065-7
26. García-Bustos, S., Cárdenas-Escobar, N., Peña, I.B., Salazar, S.U., Rizzo, K.P.: Obesity in ecuadorian adults. An application of poisson regression and logistic regression. In: Proceedings of the LACCEI international Multi-conference for Engineering, Education and Technology, Latin American and Caribbean Consortium of Engineering Institutions (2019). https://doi.org/10.18687/LACCEI2019.1.1.401
27. Han, M., Liu, L., Hu, M., Liu, G., Li, P.: Medical expert and machine learning analysis of lumbar disc herniation based on magnetic resonance imaging. Comput. Methods Programs Biomed. **213**, 106498 (2022). https://doi.org/10.1016/j.cmpb.2021.106498
28. Liu, D., et al.: Active-matrix sensing array assisted with machine-learning approach for lumbar degenerative disease diagnosis and postoperative assessment. Adv. Funct. Mater. **32**(21), 2113008 (2022). https://doi.org/10.1002/adfm.202113008
29. Li, C., et al.: Sensing of joint and spinal bending or stretching via a retractable and wearable badge reel. Nat. Commun. **12**(1), 2950 (2021). https://doi.org/10.1038/s41467-021-23207-8
30. Studer, S., et al.: Towards CRISP-ML(Q): a machine learning process model with quality assurance methodology. Mach. Learn. Knowl. Extr. **3**(2), 392–413 (2021). https://doi.org/10.3390/make3020020
31. Johnston, J.M.: Distinguishing between applied research and practice. Behav. Anal. **19**(1), 35–47 (1996). https://doi.org/10.1007/BF03392737
32. Fang, C., He, B., Wang, Y., Cao, J., Gao, S.: EMG-centered multisensory based technologies for pattern recognition in rehabilitation: state of the art and challenges. Biosensors (Basel). **10**(8), 85 (2020). https://doi.org/10.3390/bios10080085
33. Kanahuati-Ceballos, M., Valdivia, L.J.: Detection of depressive comments on social media using RNN, LSTM, and random forest: comparison and optimization. Soc. Netw. Anal. Min. **14**(1), 44 (2024). https://doi.org/10.1007/s13278-024-01206-z

Physics and Forecasting of Nonstationary Time Series Based on the Moving Window Method and Neurotechnologies

M. E. Mazurov(✉), A. V. Slipchenko, and A. A. Mikryukov

Russian Economic University named after G.V. Plekhanov, Moscow, Russia
mazurov37@mail.ru

Abstract. Nonstationary time series have features that make them interesting objects of physics and technology. The nonstationarity is due to the presence of chaotic dynamics in the behavior of the time series. Chaotic dynamics is determined by the specifics of the processes in the dynamic system under study, associated with the presence of unstable components. To predict non-stationary time series, it is proposed to use the moving window method, which allows determining the structural components of the time series, attractors or characteristic figures. The attractor from the found set, which is close to the figure of the time series in the last window, is used to predict a non-stationary time series. It is established that the reliability of forecasting is directly proportional to the number of similar attractors. Software has been developed in Python to determine the attractors of a non-stationary time series. The work of the software is implemented in forecasting the financial time series of Sberbank, Brent oil price for example. The results of a computational experiment show the effectiveness of using the moving window method for forecasting non-stationary time series in semi-structured systems. #COMESYSO1120.

Keywords: Time Series · Moving Window Method · Neurotechnology · Forecasting

1 Physical Foundations of Transient Time Series

Non-stationary time series have features that make them interesting objects of physics and technology. The unsteadiness of the time series is due to the presence of chaotic dynamics in the behavior of the time series. Chaotic dynamics is due to the specifics of the processes in the studied dynamic system associated with the presence of unstable components. The presence of chaotic dynamics leads to the appearance of attractors [1]. The attractor is a compact subset of the phase space of a dynamical system, all trajectories from some neighborhood of which tend to it at a time tending to infinity. There are different formalizations of the concept of aspiration, which leads to different definitions of the attractor, specifying, respectively, potentially different sets. For all definitions, the attractor is assumed to be a closed and (completely) invariant set.

R. Silhavy and P. Silhavy (Eds.): CoMeSySo 2024, LNNS 1489, pp. 258–267, 2025.
https://doi.org/10.1007/978-3-031-96798-6_22

The main criterion for the randomness of the attractor is the exponential increase in time of small disturbances. The consequence of this is "mixing" in the system, non-periodicity in time of any of the coordinates of the system, a continuous power spectrum and an autocorrelation function decreasing in time. The dynamics on strange attractors is often chaotic: predicting the trajectory that has fallen into the attractor is difficult, since a small inaccuracy in the initial data after a while can lead to a strong discrepancy between the forecast and the real trajectory. The unpredictability of the trajectory in deterministic dynamical systems is called dynamic chaos, distinguishing it from the stochastic chaos that occurs in stochastic dynamical systems.

For most systems, it can be observed that the cloud of source points "condenses" into some marginal objects. These are the attractors. Dynamical systems that have attractors are called dissipative. The existence of an attractor leads to very important conclusions about the behavior of the system. In this case, the study of steady-state regimes is equivalent to the study of the geometric structure of the attractor. In order to be able to compare the fractal properties of various natural processes, such as river runoff, silt deposition, Hurst used a dimensionless indicator in the form of the ratio of the magnitude (R) of the accumulated deviation from the mean to the standard deviation (S) – R/S method [2.3] in the analysis of time series of observations. The dependence of the parameter (R/S) on the observation time, constructed on a double logarithmic scale, represents the process under study in the form of a fractal function. The cognitive power of the concept of fractal dimension lies in the fact that it can be used to organize the processes under study according to the properties of randomness or complexity and, thus, classify them. The general patterns of the relationship between the degree of noise of signals and their fractal properties, expressed by the Hearst indicator, are illustrated in Fig. 1, which shows the realizations of time series of observations (sample size N = 1000) having different fractal properties and, accordingly, different estimates of the Hearst indicator.

Visually, it can be determined that stationary random signals (for example, noise with a normal distribution) have maximum noise, and the noise level of fractal signals decreases with an increase in the Hurst index. For nonstationary time series, the attractors are identified as repeating shapes or repeating sections of the time series. Due to the noise level, which can often be significant, the shape of repeating shapes can vary within certain limits, which can make it difficult to identify them. It is interesting to note that in the study of the time series in the stock market, attempts have been made to classify emerging figures, and the process of identifying them is known as technical analysis [4].

The purpose of the work is to substantiate the presence of attractors in nonstationary time series (NVR) in the form of shapes repeating in shape for most practically significant non-stationary time series. The existence of attractors has been investigated using several examples. The purpose of the work was not to develop criteria for the reliability of forecasts, due to the lack of certain patterns of development of HBR residues, although it is qualitatively clear that the reliability of the forecast for attractor figures of a certain type increases with increasing length of the NVR. This follows from the Takens theorem characterizing the behavior of nonstationary chaotic time series [1]. The Takens theorem makes it possible to reconstruct the attractor of a time series and provides conditions

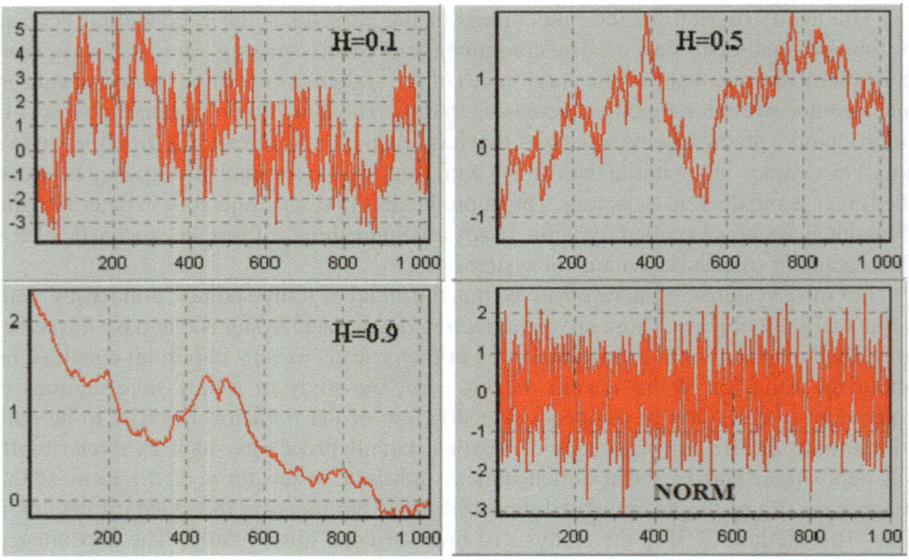

Fig. 1. Typical implementations of fractal time series of observations with H = 0.1; H = 0.5; H = 0.9 and white noise with normal distribution.

under which a smooth attractor can be reconstructed based on observations made using a sufficiently long time series.

It was found that with a sufficient number of time series attractor shapes, it becomes possible to use patterns characteristic of time processes occurring in real dynamic systems to predict [2–6]. Time series (BP) characterizing the dynamics of processes in weakly structured systems are usually non-stationary [5–9]. They are not noise types: white, pink and others [1]. Non-stationary time series are not deterministic. But they are not purely chaotic either. They can be called quasi-deterministic chaotic series. They may contain attractors or some self-similar structures, that is, they may have a quasi-fractal character. The existence of attractors and fractal properties of PVR have been noted in the literature [1]. Time series analysis is a set of mathematical analysis methods designed to identify the structure of time series and to predict them. Examples of non-stationary time series of various types are shown in Fig. 2.

The characteristic features of TS are: the presence of a trend, cyclicity, oscillatory nature in time, unsteadiness. Nonstationary time series have many practically important applications. A large number of methods for forecasting stationary time series are known [1–3]. There is no single classification of forecasting methods, it is possible to identify the main methods for forecasting financial time series: 1) Expert forecasting methods; 2) Logical modeling methods; 3) Economic and mathematical methods; 4) Statistical methods; 5) Experimental methods; 6) Factual methods; 7) Neural network forecasting methods; 8) Methods based on the use of a moving window.

The methods listed in paragraphs 1–6 are effective for predicting stationary TS, and are not suitable for predicting nonstationary time series. The methods used in the analysis of artificial neural networks [1–3], as well as methods based on the use of the moving

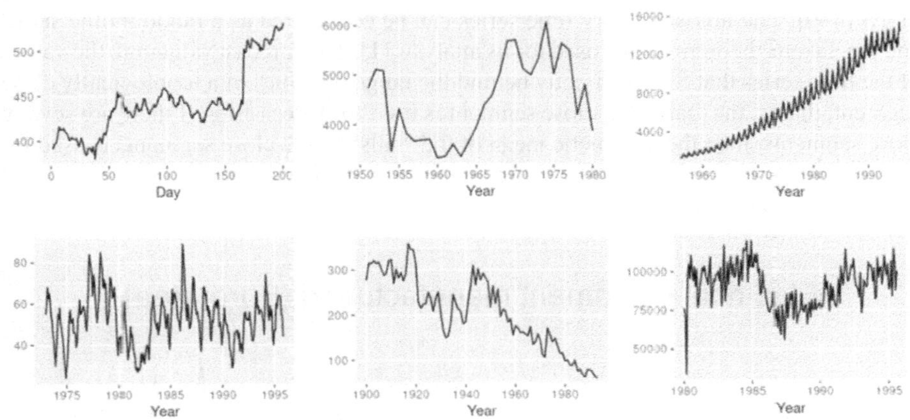

Fig. 2. Examples of nonstationary time series with trends of various types.

window method, are most suitable for predicting nonstationary time series. The methods based on the use of the moving window method will be described in more detail later.

2 Forecasting of Non-stationary Time Series Using the Moving Window Method

Nonstationary time series forecasting takes place in four stages: 1) introduction of TS values, preprocessing; 2) selection of the figure in the last window; 3) selection of the figure in the previous windows; 4) calculation of the correlation of the figure in the last window and the preceding ones. A block diagram of an algorithm for predicting non-stationary time series using the moving window method is shown in Fig. 3.

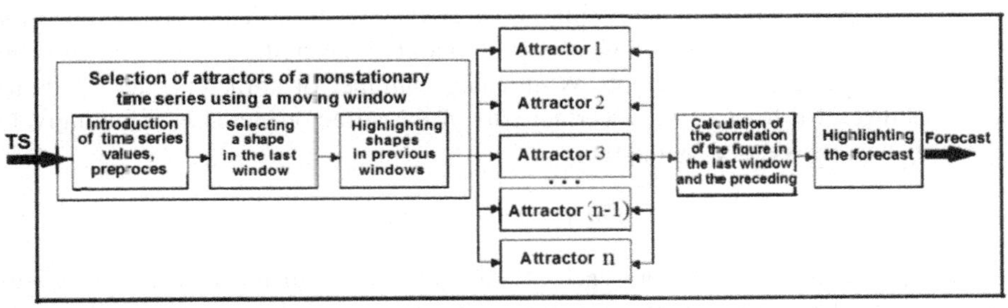

Fig. 3. A block diagram of an algorithm for predicting nonstationary time series using the moving window method.

The following technique is used for the formation of features in the algorithmic prediction of TS. Specifically repeating sections of the non-stationary time series will be called attractors or shapes. These repeating areas are areas of self-similarity or attractors of TS [4–9]. The non-stationary time series can be considered as a random time series and the Takens theorem can be used for its analysis [1]. The forecast values are the values of the time series that follow directly behind the range of values of a topologically close segment, that is, the "tail" of a close segment is used for forecasting. If there are several close segments, then the arithmetic mean of the "tails" of the close segments is used for forecasting. An example of using a sliding window to predict the production of electrical equipment is shown in Fig. 4.

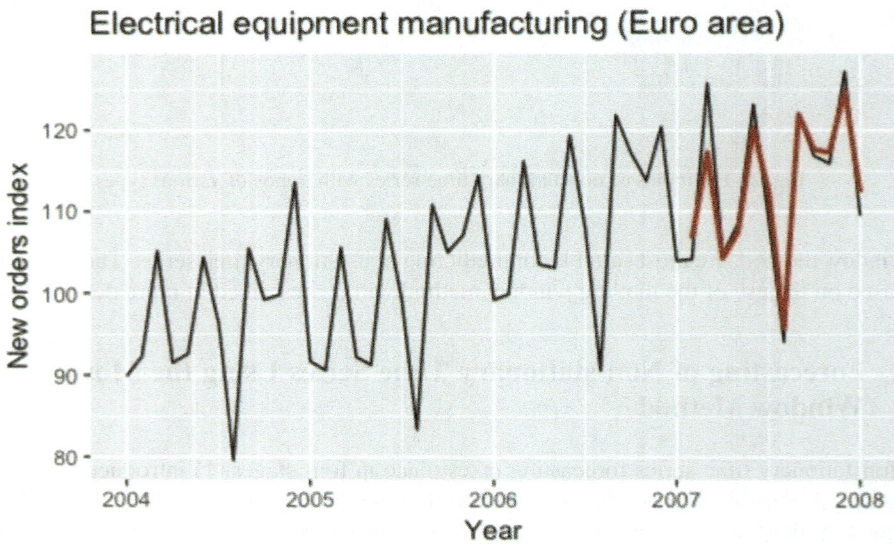

Fig. 4. An example of using a sliding window to predict the production of electrical equipment.

From the general nonstationary time series graph, it is clear that there is a repetition of nonstationary time series in a window with width T when the window shifts to the left to the beginning of the graph. With a large number of attractors, the reliability or probability of a forecast for a certain part of the attractor increases. Qualitatively, reliability can be assessed by the ratio

$$Dn + 1 = Dn + \alpha n,$$

where D_n is reliability in the presence of n attractors; αn - increase in reliability with an increase in the number of attractors by one; α - proportionality coefficient. To assess the reliability of the forecast, it is impossible to use patterns of previous nonstationary time series behavior, since changes in nonstationary time series usually do not depend on behavior in previous time periods and may be the most unexpected.

3 Computational Algorithm

3.1 Preprocessing of Time Series. Detection of Abnormal Deviations

During the preliminary analysis of the data, abnormal deviations are detected, the presence of a trend is checked, time series are smoothed and dynamics development indicators are calculated.

3.2 Smoothing the Time Series

A common technique for identifying a development trend is smoothing the time series. The essence of various smoothing techniques is to replace the actual levels of the time series with calculated levels that are subject to fluctuations to a lesser extent. This contributes to a clearer manifestation of the development trend. Sometimes smoothing is used as a preliminary step before using other methods to highlight trends.

Moving averages allow you to smooth out both random and periodic fluctuations, identify the current trend in the development of the process, and therefore are an important tool when filtering the components of a time series. The observations that are taken to calculate the average value are called the active smoothing area. For an odd value of the length of the active area g, the moving average is determined by the formula:

$$\overline{y_i} = \frac{1}{(2p+1)} \sum\nolimits_{j=i-p}^{i+p} y_j = \frac{y_{i-p} + y_{i-p+1} + \cdots + y_{i+p-1} + y_{i+p}}{2p+1},$$

where y_i is the actual value of the i-th level; $\overline{y_i}$ is the value of the moving average at time i; $2p + 1$ is the length of the smoothing interval. When using a moving average with the length of the active area $g = 2p + 1$, the first and last p levels of the series cannot be smoothed, their values are lost. Obviously, the loss of the values of the initial and last points is a significant disadvantage, because for the researcher, the latest "fresh" data has the greatest information value.

3.3 Allocation of Attractors of a Nonstationary Time Series

For an experimental study of time series (TS), it is necessary to initially find the TS trend $V(t)$ and then the TS oscillator, equal to

$$\Delta U(t_k) = U(t_k) - V(t_k),$$

where $U(t_k)$ is the given TS, $V(t_k)$ is the trend of the TS, $t_k = k\Delta t$; $(k = 1, \ldots, n)$; Δt is the time step of TS. The TS trend can be calculated using any method, for example, the smoothing method. Next, you should select a section $(t_0, t_0 + T)$ $(t_0, t_0 + T)$ of the TS and realize its movement along the time axis. The TS section containing m points in the interval $(t_C, t_0 + T)$ is selected at the end of the financial time series. Next, we divide this TS into sections $(n\Delta, n\Delta t + T)$. We produce arrays of time series points $U_k(n\cdot \Delta t, n \bullet \Delta t + T)$. Thus, algorithmic images of time series segments are created $U(n \cdot \Delta t, n \cdot \Delta t + m \bullet \Delta t)$.

To compare the proximity of the last TS segment with the algorithmic image of time series segments $U\ (n \cdot \Delta t, n \cdot \Delta t + m \bullet \Delta t)$, we use two methods.

The first method consists in a direct comparison, for which the standard deviation of the two specified segments is calculated. The smaller the deviation, the closer both segments are to each other.

The second method is to calculate the autocorrelation function of the two specified segments. If there is no correlation, then the function is zero, with full autocorrelation, the function is one.

According to the second method, the program uses an autocorrelation function to compare time series sections. The autocorrelation function is calculated using the formula

$$\sigma_j = \frac{\sum_{i=1}^{m}(U(n-l+i) \bullet U(i+j))}{\sqrt{\sum_{i=1}^{m}\left(U^2(n-l+i)\right)} \bullet \sqrt{\sum_{i=1}^{m}\left(U^2(i+j)\right)}},$$

where $i = 1, \ldots, m; j = 1, \ldots, n - m$.

A program has been developed that uses the first and second methods of comparing the contents in a moving window with the contents of the last window. In the program, the TS contained in the last interval is compared with the TS contained in intervals separated from the last by a specified number of points.

3.4 Description of the Operation of the Computational Algorithm in the Program

The algorithm provides the following procedures. 1. Finding the attractor shapes for a moving window longer in duration than the original one. 2. Finding attractor shapes for a moving window shorter in duration than the original one. 3. The found shapes are compared with the shapes found in the original moving window. For comparison, a correlation procedure is used, taking into account the equalization of the duration of the moving windows used. Let's give a description of how the computational algorithm works. Let's list the sequential execution of operations by the program: 1) We enter the values of a non-stationary time series; 2) We remove the sharply distinguished observations; 3) We smooth out the time series; 4) We highlight the trend and the variable component of the time series; 5) We calculate the TS change or determine the schedule of changes in the last window; 6) We move the window along the time series, determine the changes in TS or determine the schedule of changes in each of the windows, create a database of attractor shapes in each window; 7) Calculate the correlation function between the figure in the last window and the figures in the windows obtained when the window moves to the beginning of the row; 8) We find an algorithmic image of the TS segment among all created ones, close to the TS segment containing m points in the interval $(t_0, t_0 + T)$ at the end of TS; 9) Using this well-known algorithmic image, we find the TS points following this algorithmic image. These points will be considered predictive for points in the interval $(t_0, t_0 + T)$ at the end of the time series.

3.5 Computational Results of the Study of Attractors of Non-stationary Time Series Financial Time Series of Sberbank

Rosstat data were used to calculate the time series [18]. According to the results of the extended Dickey-Fuller test, both financial time series are non-stationary, therefore they can be the basis for further experimental steps. The time series of Sberbank for 60 days is characterized by a growing trend. The results are visualized in Fig. 5.

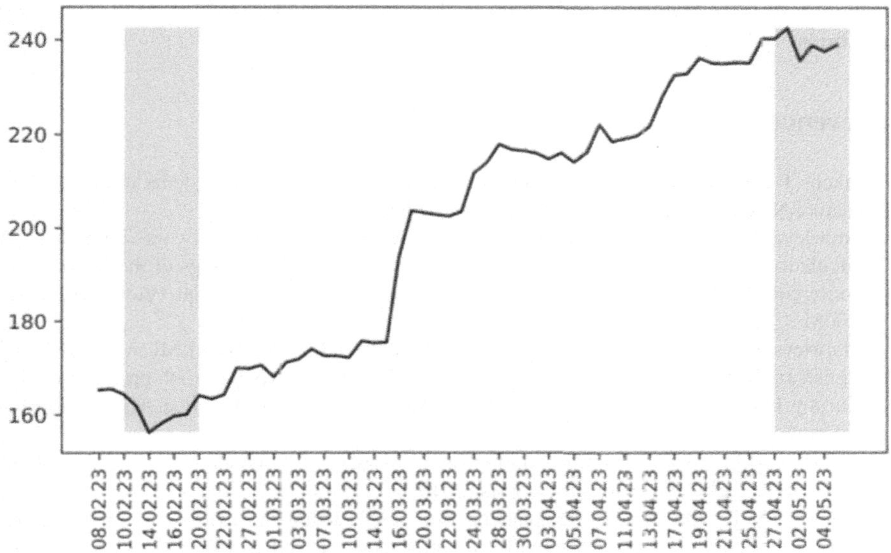

Fig. 5. The growth rate of stock quotations of Sberbank, the size of the 3-day sliding window.

A list of figures typical for the time series of shares of Sberbank is shown in Fig. 6.

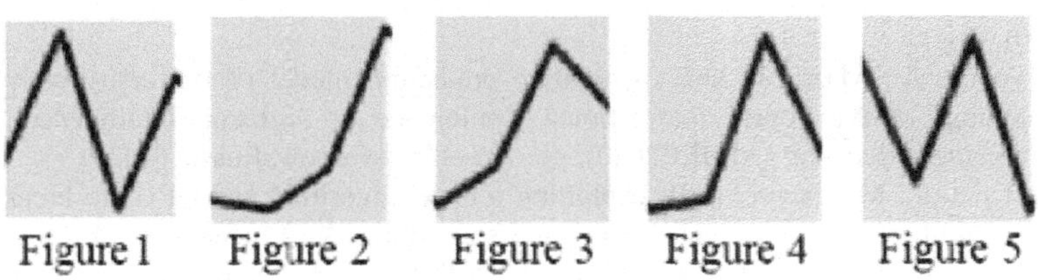

Fig. 6. List of figures typical for the time series of shares of Sberbank.

It can be seen that Figs. 2, 3 and 4 are components of each other, while Figs. 1 and 5 are a mirror image of each other, meeting more than 5 times in the time series.

The linear dimensions of the attractor shapes in real problems may exceed or be less than the selected length of the moving window. Special algorithms have been developed to find such shapes.

4 Conclusion

Nonstationary time series have features that make them interesting objects of physics. The nonstationarity is due to the presence of chaotic dynamics in the behavior of the time series. Chaotic dynamics is determined by the specifics of the processes in the dynamic system under study, associated with the presence of unstable components. The presence of chaotic dynamics leads to the appearance of attractors. A method for forecasting non-stationary time series based on the use of a moving window has been developed. Software has been developed in Python to determine the attractors of a non-stationary time series.

References

1. Takens, F.: Detection of strange attractors in turbulence. In: Dynamic Systems and Turbulence. Lecture Notes on Mathematics, vol. 898, pp. 366–381 (1981)
2. Dmitrieva, L.A., Chepilko, S.S., Kuperin, Y.A.: Method of neural networks committees in calculation of time series maximal lyapunov exponents. In: Proceedings of the International Conference “DAYS on DIFFRACTION 2008”, pp. 34–41. SpBU, Saint-Petersburg, Russia (2008)
3. Dmitrieva, L.A., Kuperin, Y.A., Mekler, A.A., Sedlinsky, I.N.: Study of EMD Modes for EEG Signals by Methods of Nonlinear Dynamics. IPRT Working Paper 225-14, pp. 1–15 (2014)
4. Murphy, D.J.: Technical analysis of futures markets. In: Theory and Practice. Alpina Publisher, Moscow, Russia (2015)
5. Mazurov, M.E.: Identification of Mathematical Models of Nonlinear Dynamical Systems: Monograph. LENAND, Moscow, Russia (2019)
6. Mazurov, M.E.: On forecasting financial time series using the method of self-organized criticality “Economics, statistics and Informatics”. Bull. UMO. **3**, 153–157 (2014)
7. Mikryukov, A., Mazurov, M.E.: Solving the classification problem based on the committee of electoral neural networks. In: Proceedings of the XXI International Scientific and Technical Conference “Neuroinformatics-2019” Part 2, pp. 186–194. MIPT, Moscow, Russia (2019)
8. Mikryukov, A.A., Mazurov, M.E.: The task of improving the university ranking based on statistical analysis methods. In: Proceedings of the Fourth International Conference of Artificial Intelligence, Medical Engineering, Education (AIMEE2020), pp. 65–75, Moscow, Russia (2020)
9. Mikryukov, A., Mazurov, M.E.: Cognitive prediction model of university activities. In: Proceedings of the Fourth International Conference of Artificial Intelligence, Medical Engineering, Education (AIMEE2020), pp. 133–145, Moscow, Russia (2020)
10. Zhu, Y., et al.: Matrix profile II: exploiting a novel algorithm and GPUs to break the one hundred million barrier for time series motifs and joins. In: Proceedings of the IEEE 16th International Conference on Data Mining (ICDM). IEEE, Barcelona, Spain (2016)
11. Box, J., Jenkins, G.: Time Series Analysis Forecast and Management, Issue 1. Mir, Moscow, Russia (2006)
12. Kendal, M.: Time Series. M.: Finance and Statistics, Moscow, Russia (1981)
13. Socks, I.: Econometrics. Introduction to Regression Analysis of Time Series. NFPC, Moscow, Russia (2002)
14. Afanasyeva, T.: Modeling of Fuzzy Trends in Time Series. UlSTU, Ulyanovsk, Russia (2013)
15. Yarushkina, N.G.: Intelligent Time Series Analysis. Textbook. Forum, Moscow, Russia (2018)

16. Tvardovsky, I.: Secrets of Stock Trading. Stock Trading on Stock Exchanges. Alpina Publisher, Moscow, Russia (2006)
17. Bestens, D.E., Berg, V.M., Wood, D.: Neural Networks and Financial Markets: Decision-Making in Trading Operations. TVP, Moscow, Russia (1997)
18. Rosstat Federal State Statistics Service: https://rosstat.gov.ru/. Accessed 7 Jul 2024

Impact of the Mine Tailings Pond Water Filling on the Hydrogeomechanical State of the Enclosing Dam

Nadezhda Kalashnik(✉)

Mining Institute Kola Science Centre of the Russian Academy of Sciences, 24, Fersman Street, Apatity 184210, Russian Federation
n.kalashnik@ksc.ru

Abstract. The object of the study is a storage facility for liquid mineral waste obtained from the processing of ore rocks. The water in the storage must be at current design levels, complying with the requirements of alluvial beach dams of sufficient length. At the same time, due to natural (rainstorms, floods) and technical (discharges of industrial waters) reasons, the degree of water saturation of the tailings pond can reach, and sometimes exceed, the maximum design value. The main goal of the study was to assess the influence of the degree of water filling of the liquid mineral waste storage facility on the hydro geo mechanical state and filtration and deformation stability of the enclosing dam. The research method consisted of constructing a 3D model of a fragment of a storage facility at one of the mining enterprises in the Kola region, and hydro geo mechanical modelling of the emerging filtration and deformation processes at different water levels in the tailings storage tank. As a result of the study, calculated indicators were obtained characterizing the resulting hydro geo mechanical state of the enclosing dam depending on the degree of water filling of the tailings storage tank. An approach is proposed to the formation of engineering solutions to ensure filtration stability and mechanical strength of dams enclosing a storage facility for liquid mineral waste, the water level in the reservoir is highly dependent on natural and man-made influencing factors.

Keywords: Tailings Dump · Mining Enterprise · Hydrogeomechanical State

1 Introduction

The storage facility for liquid mineral waste obtained during the mining and processing of ores is a container with dams enclosing it. The safety of the storage facility is largely determined not only by the state of alluvial waste deposits, but also by the degree to which the storage tank is filled with water [1–3]. The water in the tailing pond should be in the current design values (settling pond), with the formation of alluvial beaches near the enclosing dams. At the same time, due to natural (downpours, floods) and technical (discharges of industrial water) reasons, the filling of the tailing dump tank with water can reach, and sometimes exceed, the maximum design value [4–6]. In such cases, there

R. Silhavy and P. Silhavy (Eds.): CoMeSySo 2024, LNNS 1489, pp. 268–273, 2025.
https://doi.org/10.1007/978-3-031-96798-6_23

is a reduction in the lengths of alluvial beaches, sometimes up to their disappearance. Cases were noted when water reached the bottom and the upper slope of the embankment enclosing dams [5, 7].

In this regard, it is necessary to note a number of modern studies of the state of tailings dumps of mining enterprises, filtration stability and mechanical strength of the dams enclosing them [2, 6, 7]. The prevailing task is to assess the degree of water filtration through its body at different positions of the depression surface of groundwater in the tailing dump [8, 9]. The research results confirm that the depression surface formed in alluvial deposits and in the soils of embankment dams is directly related to the position of the edge and the volume of water, or the degree of filling the tailing pond with water.

Due to natural (prolonged rains, downpours, heavy snowmelt, floods, intensive run off from adjacent mountain slopes) and technical (planned and emergency discharges of industrial water) reasons, the tailing pond capacity can quickly fill up to the maximum design value, and in some cases, exceed it. This can lead to both the destruction of the enclosing dam and the overflow of water over its crest. Therefore, the main goal of the study was to assess the influence of the degree of water filling of the liquid mineral waste storage facility on the hydro geo mechanical state and filtration and deformation stability of the enclosing dam.

2 Methods and Approaches

The research methods were based on creating a 3D model of a storage facility for liquid mineral waste and modeling the emerging filtration processes in various cases of filling the storage tank with water [10, 11].

Considering that a mine's liquid mineral waste storage facility can be tens of kilometers in size, the creation of the 3D model included a waste and water tank and containment dam [1, 10]. To do this, the most important section of the storage facility was selected and the process of filling the container with waste and water was simulated (Fig. 1).

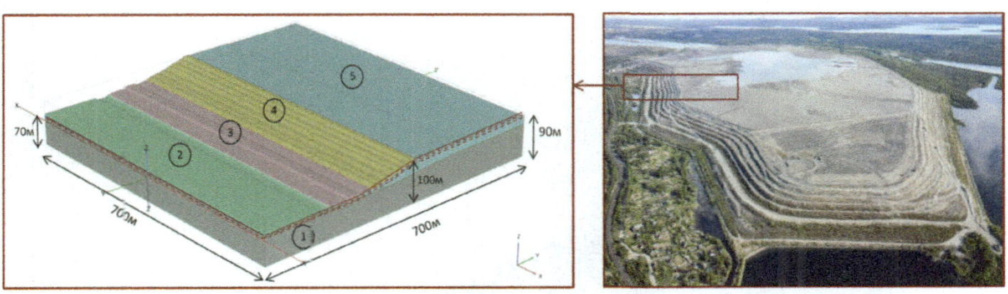

Fig. 1. Methodical Approach to Construction of 3D Model of the Most Typical Fragment of Tailing Pond.

A distinctive feature of the developed model from known ones is the possibility of implementing joint filtration and deformation calculations, which take into account hydrostatic (degree of water saturation of soils, head difference at the upper and lower slopes of the dam), hydrodynamic (formation of watercourses, flow velocity and pressure) and geomechanical (deformations and displacements of soils under gravity, as well as due to hydrostatic and dynamic water pressure) conditions and loads [11].

To calibrate the developed 3D model, the data of reconnaissance and visual survey, geological data, results of geotechnical, hydrological and geophysical surveys performed at the tailings dump of the concentration plant of one of the mining enterprises of the Kola region were used [1]. Parameters of physical and mechanical characteristics of soils used in the model and for modeling are given in [11].

The models MC (Mohr Coulomb) for rocks, HSS (Hardening Soil Small) for alluvial soils, HS (Hardening Soil) for other soils were used [12].

Computer hydro geo mechanical 3D modeling was performed for 3 rated design cases for the lower slope of the enclosing dam according to [13]:

- the first design case - a normal retaining level (H_1), between the edge and the dam - an alluvial beach, in the body of the dam - established filtration;
- the second design case -there is a high retaining level (H_2), there is practically no alluvial beach between the edge and the dam, in the body of the dam there is increased water filtration, with a possible exit at the lower slope;
- the third design (special)case - forced retaining water level (H_3), a cut on the upper slope of the dam, in the body of the dam - high water filtration with intensive output at the lower slope.

For all design cases, the groundwater level is assumed to be the same H_n.

3 Results

Calculated values were obtained that characterize the hydro geo mechanical state of the studied tailings dump for the considered combinations of loads and operating conditions. The dependence of the water filtration rate on the position of the water surface

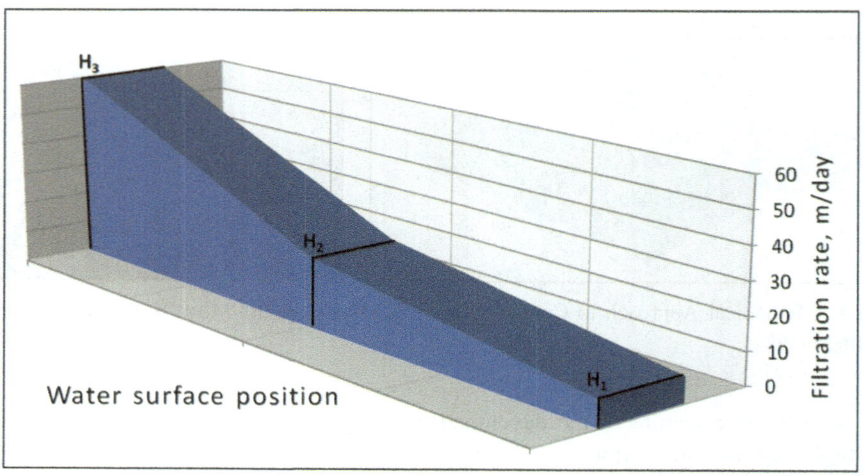

Fig. 2. Dependence of water filtration rate through the dam body on the water surface position in the tailings storage tank for design cases: H_1, H_2 и H_3.

in the tailings storage tank for design cases is shown in Fig. 2. The elevations of the water surface position for each design case are taken in accordance with the geometric dimensions shown in Fig. 1.

As follows from Fig. 2, an increase in the water surface in the tailings pond from the actual state (the first design case H_1) to the design limit (the second design case H_2) will lead to an increase in filtration rates by almost 2 times (from 7 to 13 m/day). The maximum position of the water surface in the tailing pond tank (the third design case H_3) will lead to a sharp increase in filtration rates (almost 8 times) compared to the actual one, which will lead to a significant deterioration in the hydro geo mechanical state of the dam soils.

According to [13], the reliability factor (K_{saf}) should be determined for various combinations of loads, operating conditions, reliability according to the criticality of the

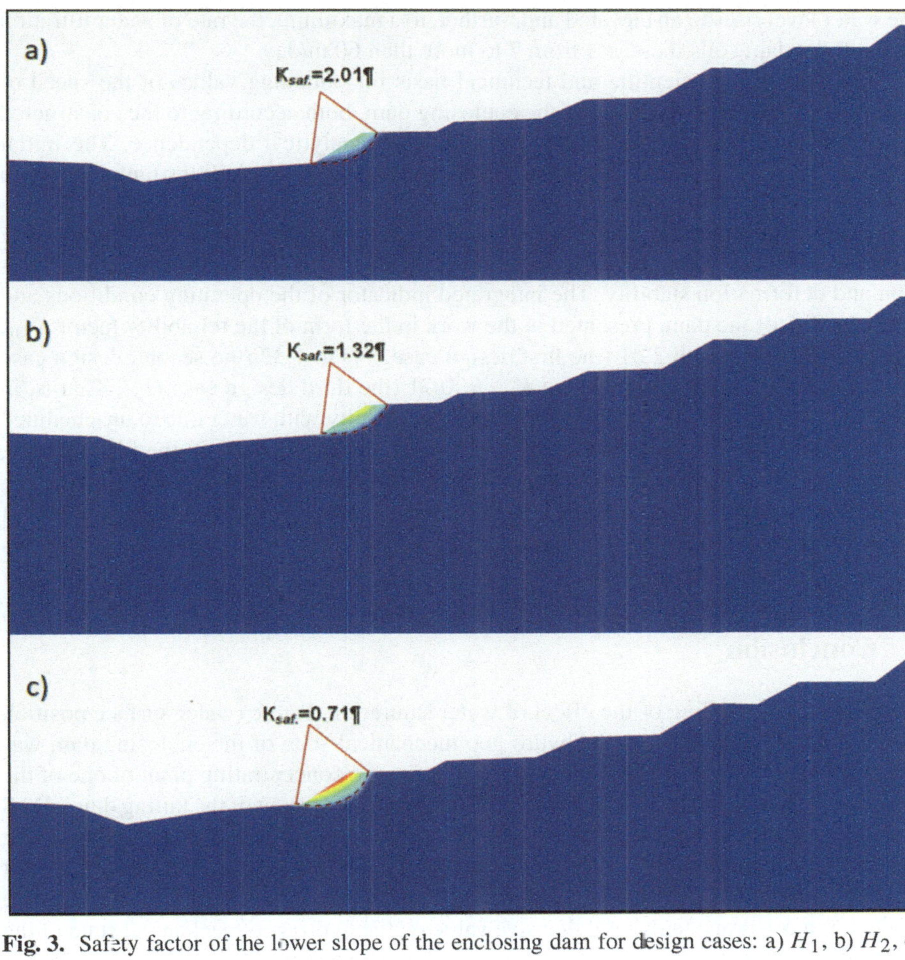

Fig. 3. Safety factor of the lower slope of the enclosing dam for design cases: a) H_1, b) H_2, c) H_3.

structure. This work defines the values of the reliability factor of the lower slope of the enclosing dam for simulated design cases as an integrated indicator of geo mechanical and hydraulic loads and operating conditions of the tailing dump (Fig. 3).

As follows from Fig. 3, an increase in the water surface (water saturation) in the tailings storage tank from the actual (first calculated case) to the increased (second calculated case) leads to a decrease in K_{saf} from 2.01 to 1.32. The maximum rise of the water surface (the third (special) design case) will lead to an extremely low value of $K_{saf} = 0.71$.

4 Discussion

The studies have established that the position of the water surface in the tailings storage tank has a decisive value on the hydro geo mechanical state of the enclosing dam. When the water level rises to an elevated and, further, to a maximum, the rate of water filtration through the dam soils increases from 7 to more than 60 m/day.

This provides a scientific and technical basis for obtaining values of the speed of filtered water through the body of the enclosing dam, both according to the constructed graph and through calculations from the obtained analytical dependence. The initial position for this will be the position of the water surface in the tailings tank relative to the height of the crest of the enclosing dam.

It should be noted that the rate of water filtration through the body of the enclosing dam largely determines the state and movement of its soils, and, thereby, their filtration and deformation stability. The integrated indicator of the operating conditions and functionality of the dam, presented in the work in the form of the reliability factor K_{saf}, sharply decreases from 2.01 (the first design case H_1) to 1.32 (the second design case H_2), and to an extremely low value $K_{saf} = 0.71$ (the third design case H_3). That is, in the case of maximum filling of the tailing dump capacity with water due to unscheduled discharges of industrial water, rainfall and floods, a sharp increase in the filtration rate through the body of the enclosing dam will occur, and, as a result, a change in the state, movement and decompression of the soils forming the dam. This gives reason to predict the formation of a hydrogeomechanical state, in which the enclosing dam may lose its filtration and deformation stability.

5 Conclusion

Computer 3D modeling of the effect of water saturation degree (water surface position in the tailing dump tank) on the hydro geo mechanical state of the enclosing dam was performed using the example of the tailings dump of a concentrating plant of one of the mining enterprises of the Kola region. A 3D model of a fragment of the tailing dump HTS was created, simulating three main cases of the position of the water surface in the tailing dump tank: the actual state, increased (without alluvial beach) and maximum filling of the tank with water. The studies have established that the position of the water surface in the tailings storage tank has a decisive value on the hydro geo mechanical state of the enclosing dam. When the water level rises to an elevated and, further, to a maximum, the rate of water filtration through the dam soils increases from 7 to more than 60 m/day, that is, practically 3 and 8 times, respectively. The reliability factor K_{saf} of the lower slope, reflecting the consideration of the existing loads and the operating conditions of the enclosing dam, sharply decreases from 2.01 (the first design case H_1) to 1.32 (the second design case H_2), and to an extremely low value $K_{saf} = 0.71$ (the third design case H_3). That is, unscheduled discharges of industrial water, rainfall, as well as floods with abundant snowmelt, which can lead to maximum filling of the tailing pond with water, will lead to a sharp deterioration in the hydro geo mechanical state of the enclosing dam. As a result, the state will change, negative movements and decompression of the soils composing the dam will occur, and the enclosing dam may lose its filtration and deformation stability.

References

1. Kalashnik, A.I.: Integrated research and monitoring of mine tailings on the Kola Peninsula. Gorn. Zh. **9**, 101–106 (2020)
2. Tao, M., Zhang, X., Wang, S., Cao, W., Jiang, Y.: Life cycle assessment on lead–zinc ore mining and beneficiation in China. J. Clean. Prod. **237**, 117833 (2019)
3. Fortuna, J., Waterhouse, J., Chapman, P., Gowan, M.: Applying practical hydrogeology to tailings storage facility design and management. Mine Water Environ. **40**, 50–62 (2021)
4. Zongjie, L., Junrui, C., Zengguang, X., Yuan, Q., Jing, C.: A comprehensive review on reasons for tailings dam failures based on case history. Adv. Civ. Eng. **2019**(9), 4159306 (2019). https://doi.org/10.1155/2019/4159306
5. Owen, J.R., Kemp, D., Lèbre, É., Svobodova, K., Pérez Murillo, G.: Catastrophic tailings dam failures and disaster risk disclosure. Int. J. Disaster Risk Reduct. **42**, 101361 (2020)
6. Song, X., Pettersen, J.B., Pedersen, K.B., Røberg, S.: Comparative life cycle assessment of tailings management and energy scenarios for a copper ore mine: a case study in Northern Norway. J. Clean. Prod. **164**, 892–904 (2017)
7. Tost, M., Hitch, M., Chandurkar, V., Moser, P., Feiel, S.: The state of environmental sustainability considerations in mining. J. Clean. Prod. **182**, 969–977 (2018)
8. Muniruzzaman, M., Karlsson, T., Ahmadi, N., Role, M.: Multiphase and multicomponent simulation of acid mine drainage in unsaturated mine waste: modelling approach, benchmarks and application examples. Appl. Geochem. **120**, 104677 (2020)
9. Kalashnik, N.A.: Influence of water filtration rate on the functionality of the mining tailings dam. J. Phys. Conf. Ser. **2388**, 012149 (2022)
10. Muller, S., Lassin, A., Lai, F., Thiery, D., Guignot, S.: Modelling releases from tailings in life cycle assessments of the mining sector: from generic models to reactive transport modeling. Miner. Eng. **180**, 107481 (2020)
11. Kalashnik, N.A.: Scenario modelling of groundwater of filtration on a enclosing tailings dam of a mining enterprise. E3S Web Conf. **376**, 03020 (2023)
12. Obrzud, R.F., Truty, A.: The Hardening Soil Model—A Practical Guidebook (2018), http://www.zsoil.com/zsoil_manual_2018/Rep-HS-model.pdf. Accessed 8 Aug 2024
13. Soil Material Dams: CP 39.13330.2012. In: Electronic Fund of Legal and Regulatory Documents. Minregion, Moscow, Russia (2012) https://docs.cntd.ru/document/1200095521. Accessed 8 Aug 2024

Systematic Review for the Automatic Image Classification of Skin Cancer Based on Machine Learning Techniques

Ricardo Arias Velásquez(✉), Denis Alonso Tineo Soto, Jeyson Alejandro Machaca Gastello, and Eduardo Garces Rosendo

Universidad Tecnológica del Perú, Natalio Sanchez street N° 125, Santa Beatriz, Lima, Peru
ricardoariasvelasquez@hotmail.com,
{rariasv,U20210018,U20208289,egarces}@utp.edu.pe
https://www.utp.edu.pe/

Abstract. The early detection and accurate classification of skin cancer, particularly melanoma, is crucial for effective treatment and patient survival. This systematic review evaluates the current state of machine learning (ML) techniques applied to the classification of skin cancer images. The review covers various ML approaches, including convolutional neural networks (CNNs), support vector machines (SVMs), and deep learning methods, emphasizing their advantages, limitations, and diagnostic accuracy. Additionally, the databases used, image preprocessing methods, and evaluation metrics are discussed. Findings highlight the necessity to improve model generalization and address challenges related to image quality and diversity. Future research areas are identified to enhance the clinical implementation of these techniques, aiming for more effective and accessible skin cancer diagnosis.

Keywords: Cancer · diagnostic · image processing · machine learning · senology

1 Introduction

Skin cancer is one of the most common and lethal types of cancer, with the number of cases increasing drastically worldwide. Early detection, however, offers a curative perspective, making rapid and accurate diagnosis of skin cancers a primary focus in current research [1]. For instance, in [2] and [3], it is shown that the main limitation for skin cancer detection is the size of the melanoma, identified at 2.5 cm. The most precise tools for detection are Support Vector Machines (SVM), Random Forest (RF), and their combination. For example, SVM + RF achieves an accuracy of 89.31%, while individually, SVM has an accuracy of 87.80% and RF 76.36%. Therefore, the use of ensemble models helps improve

R. Arias Velásquez—Thanks to Universidad Tecnológica del Perú.

R. Silhavy and P. Silhavy (Eds.): CoMeSySo 2024, LNNS 1489, pp. 274–300, 2025.
https://doi.org/10.1007/978-3-031-96798-6_24

learning. SVM is a machine learning algorithm based on statistical theory. Compared to other algorithms, SVM performs better or on par with others, solving constrained quadratic problems to differentiate between two classes, and can also address multi-class problems [4].

According to the World Health Organization (WHO), cancer is one of the leading causes of mortality globally, with over 10 million deaths annually, accounting for approximately 1 in 6 deaths worldwide. Global cancer mortality rates are projected to increase by 45% between 2008 and 2030 [5]. The fifth most common type of cancer is skin cancer, caused by the abnormal proliferation of skin cells. Despite significant advances in medical therapies and diagnostic technologies, early and precise cancer detection still faces major challenges [6]. The invasiveness, cost, and subjectivity of conventional diagnostic procedures, such as biopsies and imaging studies, can hinder timely disease identification. In this context, the use of machine learning models, like Support Vector Machines (SVM), has emerged as a promising alternative to address these challenges. These models can analyze large sets of clinical and laboratory data, identifying complex patterns that may indicate the presence of cancer. These techniques could significantly improve patients' quality of life and survival rates by enabling earlier and more accurate detection [7].

Kononenko provides a comprehensive comparison of state-of-the-art machine learning systems applied to medical diagnostic tasks, highlighting the potential of these techniques to assist in medical diagnosis [7]. Pacheco and Krohling emphasize the impact of incorporating patient clinical information in the automated detection of skin cancer, highlighting the potential to improve classification quality by integrating clinical data into the classification process. Additionally, studies such as [8] and [9] underscore the significant potential of computer-aided diagnostic systems based on imaging, particularly using deep learning algorithms like Convolutional Neural Networks (CNN), in screening and early detection of malignant melanoma. These advancements in deep learning have demonstrated remarkable performance in various image classification tasks, including medical image analysis. Furthermore, the research by Bharathi and Malleswaran shows the detection and segmentation of melanoma skin cancer in dermoscopy images using modified CNN methodologies, indicating progress in leveraging advanced neural network architectures for precise classification tasks. Additionally, Lee and Majeed present the implementation of deep learning models for classifying skin cancer, highlighting the effectiveness of convolutional networks in feature extraction from skin lesion images.

The objective of this systematic literature review (SLR) is to examine the current state of the available literature on the use of SVM in the diagnosis and detection of skin cancer. We will discuss current methods, their advantages and disadvantages, and future prospects and challenges in this field. It is anticipated that a clearer understanding of the different possibilities and ramifications of these methods will stimulate the development of more precise and effective ways to address this serious disease.

The document structured into the following sections: Methodology, describing the approach used for the systematic review; Results, presenting the key findings from the reviewed literature on the use of SVM in cancer diagnosis; Discussion, analyzing and interpreting the results; and Conclusions, summarizing the main findings and offering recommendations for future research in this field.

2 Systematic Review

The PICO framework is a valuable tool in clinical research for formulating well-focused questions that guide evidence-based practice. PICO stands for Population, Intervention, Comparison, and Outcome, and is commonly used to structure research questions and systematic reviews [12]. By following the PICO format, researchers can clearly define the patient population being studied, the intervention or exposure of interest, the comparison group, and the outcomes being measured. This structured approach helps develop research questions that are specific and answerable, leading to more effective and efficient research outcomes.

This systematic review is based on the Population, Intervention, Comparison, Outcome (PICO) methodology [13] and PRISMA [12] in the Fig. 1, structured according to the following keywords as shown in Table 1:

- Population: Cancer, diagnosis, image processing, machine learning, senology.
- Intervention: YOLO, deep learning, convolutional networks, skin, images [11].
- Comparison: Metric, accuracy, F1 score, confusion matrix.
- Outcome: Database, classifier, skin, dermis diseases, images.
- To evaluate the evidence.

The following research questions were posed to select the articles for investigation:

- Question 1.1: Is early detection achievable through image assessments?
- Question 1.2: Has the cancer stage been identified for early lesion detection in the skin?
- Question 2: Are there databases available to demonstrate the implementation of algorithms and replicate the training?
- Question 3: Are there metrics associated with accuracy, false positives, and error detection?
- Question 4: Are the results compared and discussed with cross-validation?

Using these evaluation questions, the initial search chain is structured as follows: (cancer) AND (skin) AND (image) AND (database) AND ((convolutional) OR (YOLO) OR (tensor)).

Using the previously mentioned search string, the derived terms from the keywords are presented in Table 2:

Table 1. Initial Terms for Structuring the PICO Method

PICO	Terms
Population	Cancer, skin, images
Intervention	Image classifiers, YOLO, convolutional networks, tensors [10]
Comparison	Classifier metrics
Outcome	Detection, cancer stage, images, database [13]

Table 2. Derived Terms

PICO	Terms
Aishwarya 2023 [14]	Skin cancer, YOLO, deep neural network, data augmentation, classification

2.1 Eligibility Criteria

The inclusion Criteria (IC) considers the following information (see Table 3):

- Relevance of the Topic: Studies investigating the use of machine learning techniques in the classification of skin cancer images. These studies must provide specific and relevant information about the direct application of these algorithms in detecting skin cancer.
- Type of Study: Included are empirical research articles, systematic reviews, meta-analyses, and case studies. These types of studies provide detailed data and in-depth analyses essential for a rigorous evaluation of the effectiveness of machine learning models in medical image classification.
- Language: Only studies published in English or Spanish will be considered. This ensures that articles are accessible and comprehensible to reviewers.
- Publication Date: Studies published within the last five years (from 2019 to 2024). This criterion ensures the review is based on recent technologies and methodologies, reflecting the state of the art in using machine learning for skin cancer classification.
- Accessibility: Only studies available in full text through recognized academic databases will be included. This allows for a comprehensive evaluation of the methodology, results, and conclusions of each study.

In this case, the exclusion Criteria (EC), as follows:

- Relevance of the Topic: Studies not specifically focused on the classification of skin cancer images using machine learning techniques will be excluded. These studies do not provide the specific information necessary to evaluate the focus of the review.
- Type of Study: Opinions, editorials, letters to the editor, and conference abstracts without full text will be excluded. These types of documents do not provide empirical evidence or complete methodological details, limiting their utility for a rigorous systematic review.

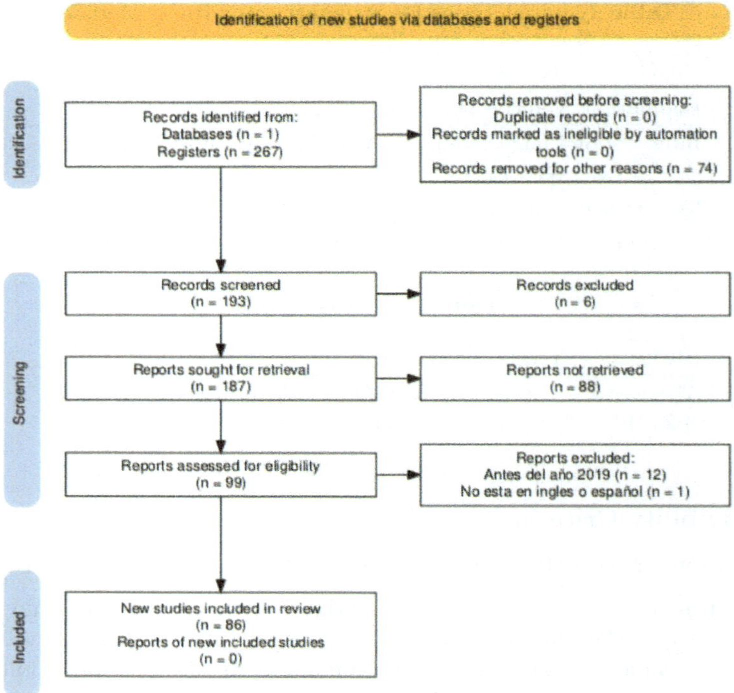

Fig. 1. PRISMA diagram

- Language: Studies published in languages other than English or Spanish will be excluded. Limiting the language facilitates the review and understanding of studies, ensuring that reviewers have the necessary resources to assess the quality and relevance of the studies.
- Publication Date: Studies published before 2019 will be excluded. Given the rapid evolution of machine learning in medical image classification, it is important to consider recent studies that reflect the most current technologies and methodologies.
- Study Quality: Studies with poor methodologies or that have not undergone peer review will be excluded. Methodological quality and peer review are crucial indicators of the reliability and validity of studies, essential for a robust systematic review.

3 Results

Several studies have explored the use of deep convolutional neural networks (DCNN) for skin cancer classification. For instance, one study proposed an interpretable deep learning system for the segmentation and classification of multiple

Table 3. Guiding PICO Question and Sub-questions

PICO Methodology	Sub-questions
P Population	Patients diagnosed with or suspected of having skin cancer. Example terms: "skin cancer", "melanoma", "dermatology", "dermatological patients"
I Intervention	Use of machine learning techniques for the automatic classification of skin cancer images. Example terms: "machine learning", "artificial intelligence", "image classification", "automated diagnosis", "deep learning"
C Comparison	Traditional or manual diagnostic methods, and other image classification algorithms. Example terms: "traditional diagnostic methods", "manual diagnosis", "clinical diagnosis", "other image classification algorithms"
O Outcome	Accuracy, sensitivity, specificity, and effectiveness of machine learning in classifying skin cancer images. Example terms: "accuracy", "sensitivity", "specificity", "diagnostic performance", "clinical effectiveness"

classes of non-melanoma skin cancers, achieving high accuracy in characterizing tissue across 12 dermatological classes [15]. A novel DCNN was designed to classify malignant versus benign melanoma with high efficiency, achieving accuracies up to 90.42% on the ISIC 2020 dataset [16]. Similarly, a hybrid deep learning approach combining multiple classifiers showed promising results, with the Xception technique achieving 90.90% accuracy [17].

Traditional machine learning (ML) techniques have also been employed for skin cancer classification. One study utilized feature extraction methods like the ABCD rule, GLCM, and HOG, followed by classification using SVM, KNN, and Naïve Bayes, achieving 97.80% accuracy with SVM [18]. To address overconfidence in deep learning models, a study applied Bayesian deep learning methods to quantify uncertainty in skin cancer classification. The proposed model achieved accuracies of 88.95% and 90.96% on two datasets, suggesting that uncertainty quantification can improve the reliability of deep learning models in medical image analysis [19].

Transfer learning has been used to enhance the performance of DL models for skin lesion classification. One study employing transfer learning with the Inception V3, VGG16, and VGG19 architectures achieved accuracies of 74%, 77%, and 76%, respectively, on the ISIC dataset [20]. A comparative study evaluated various ML algorithms, including logistic regression, k-nearest neighbors, Naïve Bayes, decision tree, random forest, and SVM, as well as CNN architectures like VGG-16, ResNet60, and InceptionV3. The study found that CNNs generally outperformed traditional ML algorithms in terms of classification accuracy [21]. Several studies have employed YOLO algorithms for detecting and classifying skin cancer with promising results. For example, one study used YOLOv3 combined with DCNNs to classify different types of skin cancer, achieving high accuracy by merging color and texture features [22]. Another study demonstrated the effectiveness of YOLO algorithms (YOLOv1, YOLOv2, and YOLOv3) in melanoma detection, achieving a mean average precision (mAP) of over 0.82 with a relatively small training set [23].

A study on the characterization of Mueller matrix elements for classifying human skin cancer using a random forest algorithm focused on color histogram feature extraction, tone moment shape extraction, and Haralick texture extraction to classify skin cancer images. The best accuracy obtained was 85.08%, suggesting the potential for further improvement with deep learning techniques [24]. The study on skin cancer classification using a random forest classifier proposed a method to classify melanoma and non-melanoma skin cancer lesions using Mueller matrix elements as predictors in an RF classifier. The model achieved an average accuracy of 93%, highlighting the predominance of circular polarization properties in the classification results [25].

Research on melanoma classification using statistical color between neighbors and mid-order pattern texture features introduced new color and texture feature descriptors for melanoma classification. The RF classifier achieved high accuracy rates of 93.27% on the ISIC 2016 dataset, demonstrating the efficacy of the proposed feature descriptors [26]. Enhancing automatic melanoma diagnosis through irregular network segmentation based on deep learning combined deep learning with conventional image processing techniques, including RF, to improve melanoma diagnosis. The approach increased recall by 11% and accuracy by 2% compared to deep learning models alone, showing the benefits of integrating RF into the diagnostic process [27].

EFFNet: a skin cancer classification model based on feature fusion and random forests, utilized feature fusion and RF for skin cancer classification, achieving an accuracy of 94.96%. The model addressed issues like imbalanced datasets and redundant information, demonstrating significant improvements over other models [28]. A study on skin cancer classification using a random forest classifier based on enhanced transfer learning and golden search optimization proposed an optimized kernel RF transfer learning model to classify 12 different types of skin cancer. The model outperformed existing methods, highlighting the efficacy of RF combined with optimization techniques [29].

A method combining CNNs with the "one-versus-all" (OVA) strategy achieved a classification accuracy of 92.09% on the HAM10000 dataset. Another model that combined Inception ResNet-v2 with a soft attention mechanism achieved an accuracy of 93.4%, an AUC of 98.40%, and an average precision of 93.71% on the HAM10000 dataset [30]. A study employing a modified VGG16 model for classifying skin cancer images into benign and malignant categories presented the following results: At epoch 10, the model achieved a maximum training accuracy of 92.55% and a minimum training loss of 0.1736 with a batch size of 32. The maximum validation accuracy was 84.56%, and the minimum validation loss was 0.3514 with a batch size of 128. At epoch 20, the maximum training accuracy was 96.98%, and the minimum training loss was 0.0801 with a batch size of 32. The maximum validation accuracy was 82.55%, and the minimum validation loss was 0.6468 with a batch size of 128. In terms of validation loss, the modified VGG16 model showed better results compared to the original VGG16 for all evaluated batch sizes. The model was trained and validated with dermoscopy images from the Kaggle dataset, consisting of 3297 skin cancer images, of which 1800 were benign and 1497 malignant. The evaluated batch sizes were 8, 16, 32, 64, and 128 [31].

Various studies have explored the use of convolutional neural networks (CNN) for skin cancer classification. For example, the AlexNet model achieved an average accuracy of around 85% in skin lesion classification tasks. Similarly, GoogLeNet (Inception v1) improved the error rate compared to previous models, excelling in melanoma classification tasks. Additionally, VGG networks, such as VGG16 and VGG19, achieved high accuracies in image classification, with VGG16 being widely used for skin lesion classification. Several variants of the ResNet network, including ResNet-34, ResNet-50, ResNet-101, and ResNet-152, showed excellent results in terms of accuracy, surpassing many other models due to their residual connections that avoid the vanishing gradient problem. DenseNet has also been frequently used due to its ability to provide stronger gradients and more diversified features, leading to better performance in image classification. EfficientNet offered better results compared to other models by uniformly scaling each dimension of the model with compound scaling coefficients, and YOLO stood out in real-time detection tasks due to its efficient architecture [32].

Various studies have examined the use of deep convolutional neural networks (CNN) for skin cancer classification. One study, for example, proposed an automated system for reproducible nevus counting by combining CNNs and three-dimensional (3D) whole-body imaging. The results showed a sensitivity of 79.04% and specificity of 91.11% for lesions ≥ 2 mm, and a sensitivity of 84% and specificity of 91% for lesions ≥ 5 mm. Additionally, a specific CNN was developed to classify malignant versus benign nevi with high efficiency. On the test dataset, this CNN achieved 90% accuracy. The comparison between automated classification and in-situ clinical evaluation by dermatologists showed moderate to substantial agreement with a Cohen's kappa of 0.56 for nevi ≥ 2 mm and 0.72 for nevi ≥ 5 mm [33].

A study employing transfer learning with the Inception V3, VGG16, and VGG19 architectures achieved accuracies of 74%, 77%, and 76%, respectively, on the ISIC dataset. Another comparative study evaluated various ML algorithms, including logistic regression, k-nearest neighbors, Naïve Bayes, decision tree, random forest, and SVM, as well as CNN architectures like VGG-16, ResNet60, and InceptionV3. The results showed that CNNs generally outperformed traditional ML algorithms in terms of classification accuracy. In another approach, YOLO algorithms have been used to detect and classify skin cancer with significant results. For example, one study used YOLOv3 combined with DCNNs to classify different types of skin cancer, achieving high accuracy by merging color and texture features. Another study demonstrated the effectiveness of YOLO algorithms (YOLOv1, YOLOv2, and YOLOv3) in melanoma detection, achieving a mean average precision (mAP) of over 0.82 with a relatively small training set [34].

Extensive research has been conducted on the use of convolutional neural networks (CNNs) for the automatic classification of skin cancer [35]. A study compared various CNN models, finding that DenseNet121 achieved an area under the curve (AUC) of 0.80 on the ISIC 2020 dataset after duplicate images were removed [36]. This study highlights the importance of a curated dataset to improve melanoma diagnosis accuracy. Furthermore, the use of EfficientNet combined with class imbalance mitigation through loss balancing has shown significant improvements in network performance [37]. The EfficientNet approach, by uniformly scaling the width and depth of the model, demonstrated better separation between melanoma and non-melanoma classes in the ISIC 2019 dataset. Additionally, the effects of label noise on CNNs for skin cancer classification were evaluated [38]. The research revealed that models trained with diagnoses from multiple dermatologists produced high-quality results on an expert-labeled test set, indicating that neural networks can learn both the characteristics identified by dermatologists and the sources of error in their decisions [39].

Significant advancements in radiogenomics and digital pathology have been emphasized, highlighting the integration of radiomic, genomic, and clinical data to enhance precision medicine. In radiogenomics, notable associations between imaging and genomic features with survival and molecular classification in breast cancer and gliomas were identified, improving the predictive capability of clinical models. Deep learning methods have also enabled the discrimination between molecular subtypes and the prediction of treatment responses in various cancer types. The study underscores the need for efficient and accessible clinical data repositories following FAIR principles to overcome the limitations of scattered and poorly indexed data. Initiatives like QIBA and IBSI promote reproducibility and multi-institutional collaboration in medical imaging research. Integrative analytics in digital pathology expands its scope by synthesizing imaging results, genomic information, and clinical data, creating unique opportunities to enhance precision medicine and evidence-based clinical practice. An evaluation of dermatologists' diagnostic accuracy in identifying melanomas and pigmented nevi using a CNN as support revealed that in the first part of the study, dermatologists achieved an average sensitivity of 59.4% (95% CI: 53.3%–65.5%), specificity of 70.6% (95% CI: 62.3%–78.9%), and accuracy of 65.0% (95% CI: 62.3%–67.6%). In the second part, integrating the CNN classifier results significantly increased sensitivity to 74.6% (95% CI: 69.9%–79.3%; P = .003), showed a positive trend in specificity (72.4%; 95% CI: 66.2%–78.6%; P = .54), and improved overall accuracy to 73.6% (95% CI: 70.9%–76.3%; P = .002). The CNN classifier alone showed a sensitivity of 84.7% (95% CI: 81.9%–87.6%), specificity of 79.1% (95% CI: 74.8%–83.4%), and accuracy of 81.9% (95% CI: 79.7%–84.2%). These results indicate that integrating the AI classifier can significantly enhance dermatologists' diagnostic accuracy. Dermatologists were more willing to change their diagnosis from benign to malignant (72%; 95% CI: 53.4%–90.9%) than from malignant to benign (28%; 95% CI: 9.1%–46.6%). Additionally, dermatologists' confidence increased when they agreed with the AI and decreased in cases of disagreement, even when they were correct.

Another study proposed an automated system based on transfer learning for skin cancer detection using a pre-trained CNN, specifically AlexNet, and data augmentation techniques. Two public datasets, DermIS and DermQuest, were used, with segmented regions of interest (ROI) images to improve classification accuracy. For the DermIS dataset, the proposed approach achieved 97.9% accuracy using ROI images and data augmentation techniques in epoch 9, compared to 97.2% accuracy using original unsegmented images. The confusion matrix generated during training showed significant improvement in classification accuracy using the ROI-based approach with data augmentation. For the DermQuest dataset, the ROI-based approach achieved 97.4% accuracy in epoch 9, while accuracy with original unsegmented images was 79.5%. The results indicate that ROI segmentation and data augmentation techniques significantly improved the classification accuracy of the proposed system in both datasets, outperforming existing methods using full images for classification.

A study evaluated a deep learning algorithm for the automatic identification of body parts in real clinical dermatological images. Using a database of 6,219 manually labeled images, the algorithm achieved an average balanced accuracy of 89%, with an average sensitivity of 79.6% and specificity of 98.4%. The body parts with the highest accuracy were the face (96.5%), hands (96.34%), and torso (95.3%), while the parts with the lowest accuracy were the female genitalia (74.8%) and the anogenital region (79.64%). The algorithm was also used to analyze the distribution of affected body parts in common diagnoses like eczema, psoriasis, and non-melanoma skin cancer (NMSC). The results showed that eczema mainly affects the head, arms, and torso; psoriasis is distributed on the torso, legs, and arms; and NMSC is mostly found on the face and torso. These distributions can help improve understanding and treatment of these dermatological conditions.

A deep learning model for irregular network segmentation in melanoma images is presented, using 487 unique annotated examples from the ISIC 2019 dermoscopic dataset. The U-Net++ architecture with an EfficientNet-b4 encoder achieved the highest intersection over union (IoU) in irregular network segmentation. Transfer learning and data augmentation techniques were used to improve model accuracy. Additionally, classic color, texture, and shape features were integrated with the deep learning results to enhance melanoma identification. This fusion approach, using the cascade generalization framework, increased melanoma recall by 11% and accuracy by 2% compared to models based solely on deep learning.

In a study on the use of machine learning in melanoma diagnosis, 2,849 high-quality dermatoscopic images of skin tumors collected between 2010 and 2014 were analyzed. Of these images, only 968 (34%) met the inclusion criteria for analysis by machine learning (ML) systems. Among the melanoma images, 335 of the 528 cases (63.4%) were excluded. The main reasons for exclusion were the absence of surrounding normal skin (40.5%) and lack of pigmentation (14.2%). The study concludes that only 36.6% of melanoma images were suitable for analysis by state-of-the-art ML systems [40]. This highlights the need to train future ML systems with larger and more representative databases of daily clinical practice. Furthermore, many of these limitations are being addressed thanks to recent advances in the scientific community [41].

In study [42], a comprehensive dermatological processing framework based on hyperspectral (HS) images was developed for the classification of pigmented skin lesions (PSL). Experimental results showed that by using the optimized SVM Linear algorithm, an average accuracy (ACC) of 78% was achieved in classifying HS test images, correctly identifying 9 out of 10 PSLs. Specifically, the images $P14_C1$ and $P102_C1$ achieved accuracies of 85% and 45%, respectively, while the image $P13_C1$ showed a low accuracy of 10% due to variability in the spectral signatures of the lesions. These preliminary results suggest the need to expand the HS database to improve the generalization of segmentation and classification models, highlighting the promising potential of the proposed framework in accurately discriminating malignant lesions in clinical settings.

Study [43] presents a hybrid model combining traditional and deep machine learning techniques for skin cancer detection. Using the DermIS dataset available on Kaggle, the proposed model achieved an accuracy of 99.4%, precision of 98.7%, recall of 98.66%, and an F1 score of 98%. Additionally, cross-validation with 1000 images from the International Skin Image Collection (ISIC) showed a detection accuracy of 98.4%. The method is based on the fusion of manually extracted features using Local Binary Patterns (LBP) and automatically extracted features using the Inception V3 technique. An Adam optimizer adjusted the learning rate, and a Long Short-Term Memory (LSTM) network was employed for binary classification of images into melanoma and benign. The results demonstrate that the proposed model significantly outperforms existing segmentation techniques and other deep learning-based methods, providing an effective and accurate tool for early skin cancer detection.

Study [44] evaluated the effectiveness of a Morphological Convolutional Neural Network (MCNN) for computer-aided diagnosis of glaucoma and melanoma. Results showed that the MCNN model achieved an Area Under the Curve (AUC) of 0.65 (95% CI: 0.53–0.74) for glaucoma classification using the ORIGA dataset, and an AUC of 0.94 (95% CI: 0.91–0.97) for melanoma classification using the ISIC dataset. Compared to popular CNN architectures (ResNet-18, ShuffleNet-V2, and MobileNet-V2), the MCNN model showed superior or comparable performance. For glaucoma classification, the best CNN performance was from MobileNet-V2 with an AUC of 0.70 (95% CI: 0.61–0.79). For melanoma classification, ResNet-18 achieved an AUC of 0.90 (95% CI: 0.86–0.96). Furthermore, the MCNN model demonstrated greater stability in classification metrics and lower variability compared to the tested CNNs. These results underline the effectiveness of combining morphological operations and convolutional layers in extracting relevant features for medical diagnosis, especially in scenarios with limited datasets.

Study [45] provides a comprehensive review of AI-based techniques for skin cancer detection and classification. The study highlights the importance of early and accurate diagnosis of skin cancer using deep learning approaches and machine learning algorithms. Techniques reviewed include Convolutional Neural Networks (CNN), Support Vector Machines (SVM), and unsupervised clustering algorithms. These methods are applied to dermatological images to identify specific features of malignant and benign lesions, improving diagnostic accuracy compared to traditional methods.

In study [46], remarkable accuracy was achieved compared to state-of-the-art methods. The proposed model's accuracy, recall, F1 score, and precision were superior by 3.78%, 12.3%, 8.97%, and 0.34%, respectively, compared to current methods. These results were based on the HAM10000 dataset, demonstrating that the MVT model is robust and effective for skin cancer classification. The model significantly improved when the number of images in all classes was increased to overcome data imbalance issues. The study identifies the lack of public datasets as a significant limitation and addresses this issue through extensive data augmentation techniques to enhance system variability and noise resistance. Additionally, the proposed model was evaluated in terms of accuracy, sensitivity, F1-score, and precision, achieving superior results compared to cutting-edge methods. Specifically, the model achieved an accuracy of 96.14%, an F1-score of 97.0%, a sensitivity of 96.50%, and a precision of 96.0%.

Study [47] proposes implementing a DCGAN model to address various challenges in classifying skin lesion images. The model achieved an overall accuracy of 93.5% after fine-tuning network parameters. Other important metrics, such as precision, recall, F1 score, and Area Under the ROC Curve (AUC-ROC), were also calculated, obtaining outstanding values. The model's precision was 0.94, recall was 0.84, F1 score was 0.87, and AUC-ROC was 0.861. The study also explored the impact of batch size and learning rate on model performance. A batch size of 128 provided better results compared to a batch size of 100. Accuracy, precision, recall, and AUC-ROC improved notably with the larger batch size. Additionally, a low learning rate helped achieve optimal model convergence. A confusion matrix was used to evaluate the generator and discriminator performance. The confusion matrix showed how the model correctly classified images of malignant and benign lesions. The generator was able to fool the discriminator by producing images that the discriminator could not easily distinguish as fake, indicating good model performance. To evaluate the quality and diversity of generated images, the Fréchet Inception Distance (FID) was used. A low FID indicates that the generated images have a probability distribution similar to real images. The FID obtained by the model was 44.41 ± 1.28, which is comparable to other advanced models like GAN and LAPGAN. The research demonstrated that a typical deep neural network-GAN model could achieve competitive performance in skin lesion classification. Experimental results were very promising, especially after applying noise reduction algorithms and image enhancement techniques. Significant improvements were observed in training after applying preprocessing and image enhancement operations, achieving a test accuracy of 93.5%.

Study [48] focuses on evaluating the performance of Convolutional Neural Networks (CNN) in skin cancer image classification compared to human dermatologists. The research was conducted to demonstrate that CNNs can consistently outperform dermatologists in classifying different types of skin cancer. The study used dermoscopic images from the ISIC archive, mainly from the HAM10000 dataset, containing images from diverse populations taken with different camera systems and publicly available and anonymous. To ensure statistical reliability of sensitivity and specificity, a sufficient number of images per skin disease were selected, limited to the most frequent differential diagnoses in the archive, including: actinic keratosis (solar keratosis), intraepithelial carcinoma (Bowen's disease), squamous cell carcinoma (akiec); basal cell carcinoma (bcc); benign keratosis (including seborrheic keratosis, solar lentigo, and lichen planus-like keratosis) (bkl); melanocytic nevi (nv); and melanoma (mel). The test dataset, consisting only of biopsy-verified images from the HAM10000 dataset, was used to evaluate the algorithm. To avoid selection biases, a random generator in Python was programmed to select 300 test images (60 for each of the five disease classes). The CNN was trained and validated using these images and its performance was compared with that of 112 dermatologists. The results demonstrated that the CNN consistently outperformed dermatologists in skin cancer classification, suggesting that CNNs can be a valuable tool in the clinical diagnosis of skin cancer, improving the accuracy and consistency in dermoscopic image classification [48].

Study [49] proposes an innovative skin cancer diagnostic system using hyperspectral imaging (HSI). The results obtained in this research are significant and demonstrate the effectiveness of this technology in the classification and detection of malignant and benign skin lesions. The study used an HSI dataset containing 76 images of skin lesions from 61 subjects, 40 of which were benign and 36 malignant. These images were acquired with an instant capture camera covering the range of 450–950 nm, distributed in 125 spectral channels. The images were collected in two hospitals in the Canary Islands, and labeling was performed by expert dermatologists and pathologists. For skin cancer image classification, three machine learning methods were employed: Support Vector Machines (SVM), Random Forests (RF), and Extreme Gradient Boosting (XGB). SVM showed a remarkable ability to perform binary classifications by finding the hyperplane that divides discrete classes in the training dataset. Multiclass classification was addressed by decomposing the problem into multiple binary classification problems. The results demonstrated that the hyperspectral imaging system effectively distinguished between different types of skin lesions. The spectral signatures of different patients were normalized to mitigate variations in lighting conditions. At the end of preprocessing, the spectral signatures contained 116 bands with values in the range of 0 to 1. Statistical analysis of the data showed a percentage distribution of skin lesions into four possible classes: Benign Epithelia (BE), Benign Melanomas (BM), Malignant Epithelia (ME), and Malignant Melanomas (MM). Synthetic RGB images and the average spectral signatures of hyperspectral pixels provided a clear representation of each type of lesion considered. In conclusion, the research highlights the efficacy of using HSI for skin cancer detection and classification, providing a non-invasive and precise tool for clinical support in dermatological diagnosis. The technology can significantly improve diagnostic accuracy and, therefore, the treatment of skin cancer patients.

The study [50] proposes integrating multiple classifiers, such as neural networks and feature-based methods, to make more accurate decisions. A neural network (NN) was developed to distinguish between melanoma and benign nevi. The NN architecture was evaluated during the training process, calculating parameters such as accuracy, specificity, sensitivity, and the Dice similarity coefficient. The NN with 72 hidden layers showed promising results on the PH2 database, achieving an accuracy of 93.33%, specificity of 94.12%, sensitivity of 92.31%, and a Dice coefficient of 92.31%.

Three methods based on Convolutional Neural Networks (CNNs)—GoogleNet, ResNet-101, and NasNet-Large—were developed and pre-trained on large datasets like ImageNet and Places365. The CNN architectures were fine-tuned using transfer learning to distinguish between different types of skin lesions. On the PH2 database, GoogleNet achieved 90% accuracy, ResNet-101 90%, and NasNet-Large 90%. On the ISIC 2019 database, the accuracies were 91.66%, 88.33%, and 88.66%, respectively. The feature-based method employed a Support Vector Machine (SVM) classifier after extracting features using Histograms of Oriented Gradients (HOG). During validation, this method achieved 93.33% accuracy on the PH2 database and 90% on the ISIC 2019 database.

A global decision-making system was proposed, based on the fusion of results from each individual method. The global system demonstrated high accuracy, achieving 95% on the PH2 database and 93% on the ISIC 2019 database. The proposed system shows that combining multiple classifiers into a single global decision system can achieve higher accuracy in classifying skin lesions than individual classifiers. The study also emphasized the importance of preprocessing images to remove hair and other noise that could interfere with lesion detection. The results are comparable or superior to those of other recent studies in the field, suggesting that integrating deep learning methods and specific features is an effective strategy for improving automatic melanoma diagnosis.

Extensive experimental studies were conducted on eleven public skin image datasets. These datasets included both dermoscopic and non-dermoscopic images, addressing the challenge of variability in image quality and format. MEL-CAP significantly outperformed several state-of-the-art models, including InceptionV3, DenseNet, VGG, MobileNet, ResNet, and EfficientNet, in terms of Matthews correlation coefficient (MCC) and area under the curve (AUC). The study concludes that combining convolutional blocks with the CapsNet architecture offers a robust and accurate solution for automatic melanoma diagnosis.

MEL-CAP not only improves diagnostic accuracy compared to traditional CNN models but also preserves the hierarchical spatial relationships crucial for reliable diagnosis. These results suggest that MEL-CAP has the potential to be a valuable tool for early melanoma detection, reducing the burden on dermatologists and improving patient outcomes [51]. Study [52] analyzes the effectiveness of two artificial intelligence (AI) approaches for classifying dermoscopic images of skin lesions as benign or malignant. A classical machine learning method based on image analysis was compared to a deep learning approach. Both models were trained using a publicly available dataset of 25,122 dermoscopic images and evaluated with an independent test set of 200 images. The results showed that the deep learning approach, based on a pre-trained CNN, outperformed the machine learning method based on a boosted tree classifier. Specifically, the deep learning model achieved 85.4% accuracy, 88% sensitivity, 75.5% specificity, and 93.6% precision, while the machine learning model achieved 73.8% accuracy, 81.1% sensitivity, 44.5% specificity, and 85.4% precision. The integration of melanoma detection algorithms with digital dermoscopic devices can facilitate faster population screening, improve patient management, and increase survival rates. This study highlights the importance of using entirely independent datasets for training and testing phases, ensuring an accurate assessment of the generalization capabilities of the developed models. The results suggest that the deep learning model not only surpassed the machine learning model in terms of accuracy and generalization but also better aligned with current melanoma detection practices, offering a potentially valuable tool for clinical practice [52].

In [21], various machine learning algorithms and deep neural networks were compared for automatic skin cancer diagnosis. The evaluated algorithms included Logistic Regression, k-Nearest Neighbors, Naive Bayes, Decision Tree, Random Forest, and Support Vector Machine, as well as CNNs like VGG-16, ResNet-50, InceptionV3, and Inception-ResNetV2. The models were trained and evaluated using the ISIC dataset, which includes dermoscopic images of benign and malignant skin lesions. The study results show that the Random Forest algorithm achieved the best accuracy (86.36%) and recall (95%) using a combination of color, texture, and shape feature descriptors. Compared to deep learning models, Random Forest proved more effective in terms of accuracy and training time, with the Inception-ResNetV2 network being the closest yet still inferior. These findings suggest that classical machine learning algorithms, when used with appropriate feature descriptors, can outperform CNNs in classifying digital images of skin lesions, especially with small datasets [21].

Study [53] analyzed various machine learning techniques for skin cancer classification using the SCSODTL-SCC (Self-Controlled Self-Organizing Dynamic Transfer Learning-Skin Cancer Classification) methodology. During the training phase (70% of the dataset), the methodology achieved an average accuracy of 98.83%, sensitivity of 95.31%, specificity of 99.27%, F-score of 95.90%, and MCC of 95.22%. In the testing phase (30% of the dataset), the average accuracy was 97.59%, sensitivity 90.70%, specificity 98.55%, F-score 91.72%, and MCC 90.52%. These results demonstrate that the SCSODTL-SCC methodology significantly improves skin cancer classification performance compared to other approaches. Additionally, the combination of feature descriptors, including Histogram, LBP, and Haralick, was the most effective for automatic detection of cancerous lesions, achieving an accuracy of 86.36% and sensitivity of 95%. Naive Bayes achieved the highest sensitivity (100%) but with much lower accuracy. The study concludes that the SCSODTL-SCC methodology is effective for skin cancer classification and that using combinations of color and texture descriptors significantly improves the accuracy and sensitivity of classification models [53].

In [54], a multi-task deep learning approach for melanoma segmentation and classification from dermoscopic images is proposed. The proposed model includes a preprocessing phase to enhance image quality and remove irrelevant details such as hair. It then uses a VGGNet-FCN-based architecture for lesion region segmentation and a super-resolution neural network to adjust image resolution before final classification. Classification is performed using a combination of pre-trained deep architectures like DenseNet, GoogleNet, and MobileNet, improving melanoma detection accuracy. The study results showed remarkable performance in both segmentation and classification of skin lesions. In the segmentation phase, performance metrics achieved 96.99% accuracy, 92.53% specificity, 97.65% precision, and 98.41% sensitivity. For classification, the results were equally impressive, with 97.73% accuracy, 99.83% specificity, 99.83% precision, and 95.67% sensitivity. These results suggest that the proposed approach is highly effective for automatic melanoma detection, significantly outperforming other existing methods and providing a valuable tool for early and accurate skin cancer diagnosis [54].

The author in [55] proposes the use of Convolutional Neural Networks (CNNs) with transfer learning to extract features from dermoscopic images, and then classify these features using a hierarchical classifier composed of Random Forest (RF), K-Nearest Neighbor (KNN), and Adaboost. The research findings indicate that the proposed transfer learning approach and hierarchical classifier deliver high performance in melanoma classification. The methodology demonstrated superior performance compared to other approaches using SVM and KNN, particularly in terms of sensitivity, specificity, and accuracy. However, limitations such as the lack of feature selection and the use of a binary classifier instead of a multiclass classifier were identified. The study concludes with proposals for future improvements, including the construction of a multiclass classifier and the implementation of metaheuristic techniques for feature selection to enhance classification accuracy and reduce classifier complexity.

In [56], the focus is on computer-assisted diagnosis and analysis of skin cancer, specifically melanoma, using dermoscopic images. The authors present statistics on skin cancer in India from 1990 to 2020, highlighting a significant increase in cases and mortality. The methods used in the study include image acquisition, preprocessing, segmentation, feature extraction, and classification. Images are obtained from public databases and processed to remove noise before segmenting the suspicious lesions. Segmentation techniques include thresholding, region growing, and deep learning. For feature extraction, color, texture, and shape of the lesions are mainly analyzed. Regarding classification, Artificial Neural Networks (ANN), CNNs, and other deep learning algorithms are employed. Results show that AI-based methods, such as CNNs, have significantly improved melanoma diagnosis accuracy, though challenges remain in accurately differentiating melanocytic from non-melanocytic lesions [56].

In [57], an approach for detecting and classifying skin cancer images, specifically melanoma, using ensemble learning based on CNNs is proposed. The main objective is to improve classification accuracy and address imbalances in melanoma classifications. Various machine learning and deep learning models, including Adaboost, Random Forest, boosted SVM, boosted GMM, and CNN ensembles, as well as transfer learning techniques, are employed. These methods are used to extract features from dermoscopic images and classify lesions into six types of melanoma: superficial spreading, nodular, lentigo maligna, acral lentiginous, subungual, and amelanotic melanoma. The study results show that ensemble models, particularly CNN-based ones, offer high accuracy in melanoma classification, with the Ensemble CNN model achieving an accuracy of 98.67%. Stratified 10-fold cross-validation is also implemented to evaluate classifier performance, using metrics such as accuracy, recall, F1-score, Jaccard index, and Kappa coefficient. Adaboost and boosted SVM models outperform other classifiers in most melanoma classes, though the Ensemble CNN model provides more consistent performance across all classes, with slight variations in amelanotic melanoma classification accuracy.

In [58], a computer-assisted classification system for quickly identifying suspicious pigmented lesions at the primary care level is proposed. This system aims to improve early melanoma identification accuracy, currently highly dependent on examiner expertise and time. A total of 133 patients with multiple skin lesions were examined by a certified dermatologist. Wide-field images of all major body parts were acquired using a consumer camera under natural lighting. Lesions were classified as "suspicious" and "non-suspicious". A total of 1759 pigmented lesions were extracted from the images. A machine learning classifier was developed and optimized to classify each lesion binary using a suspicion score. In the test set, the system achieved 100% sensitivity for dermoscopy-confirmed suspicious lesions (SPL_A) and 83.2% for non-confirmed ones (SPL_B). The sensitivity for non-suspicious lesions was 72.1%, and the overall accuracy was 75.9%.

In [59], the authors demonstrate that deep neural networks (CNNs) can surpass dermatologists in melanoma image classification. The research focuses on training a CNN model using biopsy-verified dermoscopic images of melanomas and nevi. A total of 4204 images were used for CNN training, and an additional 804 images, also biopsy-verified, were tested. A pretrained ResNet50 model was adapted with new deep learning techniques such as differential learning rates, cosine-based learning rate reduction, and stochastic gradient descent with restarts. Images were presented to dermatologists and the CNN, and the quality of each image was evaluated along with treatment recommendations. Dermatologists achieved 67.2% sensitivity and 62.2% specificity, while the CNN reached 82.3% sensitivity and 77.9% specificity. McNemar tests showed the CNN significantly outperformed dermatologists in both sensitivity and specificity with a p-value ≤ 0.001.

Study [60] proposes two innovative methods for detecting and classifying dermoscopic images of benign and malignant tumors. The KNN-PDNN method uses pretrained deep neural networks as feature extractors combined with the k-nearest neighbor (KNN) classifier. Various classifiers were tested, but KNN was selected for its high performance. The AlexGWO method optimizes the hyperparameters of the AlexNet neural network using the gray wolf optimizer (GWO). This combination aims to improve AlexNet's accuracy by tuning its parameters for optimal results. ANN achieved 81.25% accuracy, KNN 72.35%, NB 70.45%, SVM 64.92%, and DT 77.25%. CNN obtained 97.75% accuracy, with DenseNet-201 achieving the highest at 99%. Using deep neural networks as feature extractors and KNN as a classifier achieved over 99% accuracy in some models, with VGG-16 performing the best. AlexGWO optimized AlexNet, achieving 99% accuracy, significantly improving the results obtained with AlexNet as an individual classifier.

The author in [61] proposes an innovative method for skin cancer diagnosis (SCD) using a combination of artificial neural networks (ANN) and improved gray wolf optimization (IGWO). The proposed method was evaluated using two datasets: ISIC-2016 and ISIC-2017. The results show that the method can diagnose skin cancer with 97.09% accuracy on the ISIC-2016 dataset and 95.17% on the ISIC-2017 dataset, improving accuracy by at least 0.5% compared to previous methods. The combination of IGWO and ANN increased diagnostic accuracy by 2.5% compared to CNN models. The methodology outperformed other compared methods, such as ANN trained with the Levenberg–Marquardt (LM) function and models optimized with basic GWO.

The reviewed studies indicate that many ML and DL methods offer high performance in classifying skin lesions. Various techniques were used, including decision trees, support vector machines (SVM), k-nearest neighbors (KNN), artificial neural networks (ANN), among others. Convolutional neural networks (CNN) stand out as the most effective and widely used technique. It is concluded that ML and DL techniques have proven to be powerful tools for skin lesion classification. However, challenges such as variability in lesion shape, size, and location, low contrast quality between lesions and healthy skin, and visual similarity between melanoma and non-melanoma lesions remain [62].

In [63], the authors propose classifying hyperspectral skin cancer images using three popular machine learning methods: Support Vector Machine (SVM), Random Forest (RF), and eXtreme Gradient Boosting (XGBoost). Hyperspectral images acquired with a specific camera (Cubert UHD) in a wavelength range of 450 to 950 nm, distributed across 125 spectral bands, were used. SVM demonstrated to be the most suitable method for classifying hyperspectral skin cancer images, with an average accuracy above 90%. Performance metrics (accuracy, recall, specificity, false negative rate) showed that SVM outperforms RF and XGBoost in most cases. It concludes that ML algorithms, particularly SVM, are effective for classifying hyperspectral skin cancer images, especially when a small dataset is available.

In [64], the applicability of deep transfer learning for automatic classification of skin lesion images into seven possible categories is investigated. Using the HAM10000 dataset, a system was developed that accepts these images without the need for explicit feature extraction or preprocessing. Thirteen deep transfer learning models were employed: SqueezeNet, GoogLeNet, Inceptionv3, DenseNet-201, MobileNetv2, ResNet18, ResNet50, ResNet101, Xception, Inception-ResNet-v2, ShuffleNet, DarkNet-53, and EfficientNet-b0. These models were customized and retrained to classify images into the seven mentioned categories. All models achieved comparable accuracy, with ResNet101 standing out with 76.7% accuracy. The NV class with the most images achieved the highest precision (92.5%) and recall (82.5%), while the melanoma class showed 71% sensitivity but 43.1% precision. ResNet101: 76.7% accuracy, 64.3% F1 score. DenseNet201: 75.8% accuracy, 64.8% F1 score. Models slightly improved their metrics by increasing the training dataset size, with DenseNet201 reaching a 74.4% F1 score with a 90/10 data split. This proved effective for skin lesion classification, achieving high accuracy in most cases. However, limitations were observed due to dataset imbalance and the need for more images for minority classes [64].

In [65], the use of transfer learning using a set of pretrained convolutional neural networks (CNNs) for classifying skin lesions in dermoscopic images is investigated. Three models from the SeNet and EfficientNet families were selected: EfficientNetB0, EfficientNetB1, and SeResNeXt-50. These models were fine-tuned by adding pooling, normalization, dropout, and fully connected layers.

In [66], the author proposes an automated model based on deep learning for melanoma detection and classification using dermoscopic images. The model, called ADL-MDC (Automated Deep Learning-based Melanoma Detection and Classification), includes several stages: contrast enhancement, data augmentation, image segmentation, feature extraction using an optimized CapsNet (Capsule Network) with Adagrad, and melanoma classification using a sparse autoencoder (SAE) optimized by the crow search algorithm (CSO). The results showed that the ADL-MDC technique effectively classifies images with high accuracy across different training/test configurations (80:20, 70:30, and 60:40). Specific results include:

- Dataset 80:20: Average accuracy of 93.12%, recall of 93.09%, specificity of 99.01%, precision of 98.27%, and F-score of 93.10%.
- Dataset 70:30: Average accuracy of 92.37%, recall of 92.23%, specificity of 98.89%, precision of 98.06%, and F-score of 92.25%.
- Dataset 60:40: Average accuracy of 91.70%, recall of 91.67%, specificity of 98.81%, precision of 97.92%, and F-score of 91.67% [66].

In [67], the performance of two machine learning models, one based on linear regression and the other on convolutional neural networks (CNN), is compared for melanoma detection in skin images. Datasets from the International Skin Imaging Collaboration (ISIC), containing labeled images of skin lesions as melanoma (positive) and nevus (negative), were used. A total of 600 images were selected, divided into 360 for training, 120 for validation, and 120 for testing. The results for the linear regression model were:

- Linear Regression Model: Accuracy of 68%, sensitivity of 56%, specificity of 80%, false positive rate of 20%, false negative rate of 43%.
- For the CNN model: Accuracy of 70%, sensitivity of 75%, specificity of 68%, false positive rate of 30%, false negative rate of 25%.

It was concluded that the CNN model is more suitable for melanoma detection due to its higher accuracy and lower false negative rate compared to the linear regression model.

In [68], DermICNet, a dermoscopic image classification network designed for automated skin cancer diagnosis, is presented. This network includes eight convolutional layers with 3×3 filters, followed by dense layers for classification. DermICNet was evaluated using the PH2 dataset, which contains dermoscopic images of various skin lesions. The model achieved an average accuracy of 99.2% in classifying images as normal or abnormal. The best combinations of preprocessing and data augmentation improved the model's accuracy, sensitivity, and specificity: With Preprocessing and Data Augmentation: Accuracy of 99.2%, sensitivity of 99.3%, specificity of 98.9%. DermICNet demonstrated superior performance compared to other skin cancer diagnosis systems based on machine and deep learning.

In [69], an intelligent dermatological tool is proposed to automatically classify multiple subtypes of skin cancer. This tool combines multiple radiomic feature categories, including high-level features based on deep learning techniques like ResNet-50, DenseNet-201, and DarkNet-53, as well as low-level features like discrete wavelet transform (DWT) and local binary pattern (LBP). The results for low-level features showed that SVM classifiers achieved accuracies ranging from 33.5% to 70.5%, with the highest accuracy obtained using a cubic SVM classifier with DWT-A features. SVM classifiers trained with high-level features from DenseNet-201, ResNet-50, and DarkNet-53 achieved maximum accuracy of 95.6%, 95.3%, and 94.6%, respectively. Combining features from different levels significantly improved classification accuracy, achieving 99% accuracy with the cubic SVM classifier using a combination of ResNet-50, DenseNet-201, DarkNet-53, DWT-A, and LBP features.

In [70], the author examines the effectiveness of VGG16 and ResNet50 convolutional neural network (CNN) architectures in classifying images of skin lesions. The aim is to determine how these architectures, with different depths, affect the accuracy of skin cancer type classification using data from the ISIC 2019 challenge due to its data richness. Three classes (BCC, MEL, and NV) with the highest number of images were selected for the project. The images were matched in number (3323 images per class) for Dataset 1 and increased to 10911 images per class for Dataset 2 using data augmentation techniques like horizontal and vertical flipping and brightness adjustment. The results for VGG16 were:

- VGG16: Overall accuracy in Dataset 1 was 83.89%, while in Dataset 2 it was 77.38%. Accuracy slightly decreased with the increased dataset size.

- For ResNet50: Overall accuracy in Dataset 1 was 83.35%, and in Dataset 2 it was 91.23%. ResNet50 showed significant improvement with the increased dataset size.

In [71], a computer-aided diagnosis (CAD) system for skin disease detection based on a fusion of multiple features and a hybrid deep learning model is proposed. The system is designed to improve the accuracy and efficiency of diagnosis using dermoscopic images and clinical metadata. Pre-trained deep learning models such as VGG19, ResNet50, InceptionV3, DenseNet201, and Xception are used for feature extraction, which are then combined and classified using machine learning techniques. The proposed system was evaluated using various performance metrics, including precision (PRE), dice similarity coefficient (DSC), accuracy (ACC), sensitivity (SEN), specificity (SPE), and area under the curve (AUC). The results showed that the DenseNet201 model combined with logistic regression achieved an average precision of 99.94%, a sensitivity of 98.84%, a specificity of 99.97%, and an AUC of 99.94%.

Study [72] addresses the issue of data imbalance by replicating images of classes with low frequencies to balance the dataset. The analyzed transfer learning networks include VGG19, InceptionV3, InceptionResNetV2, ResNet50, Xception, and MobileNet. The results for performance without image replication were:

- VGG19: Accuracy of 67.18%.
- InceptionV3: 81.68%.
- InceptionResNetV2: 81.14%.
- ResNet50: 81.05%.
- Xception: 80.96%.
- MobileNet: 82.41%.

For performance with image replication:

- VGG19: Accuracy of 66%.
- InceptionV3: 79%.
- InceptionResNetV2: 85%.
- ResNet50: 77%.
- Xception: 90%.
- MobileNet: 88%.
- Xception was the best-performing model with an accuracy of 90.48%, the highest recall, precision, and F-measure.

In [73], an automated approach for detecting irregularities in the edges of skin lesions using dermoscopic images is proposed. This method involves extracting the skin lesion from the image, detecting the edge of the lesion, and measuring the edge irregularity. A convolutional neural network (CNN) and a Gaussian naive Bayes ensemble model are trained for automatic edge irregularity detection, providing an objective decision on whether the skin lesion edge is considered regular or irregular. The proposed approach achieved outstanding results, with accuracy, sensitivity, specificity, and F-score of 93.6%, 100%, 92.5%, and 96.1%, respectively. This demonstrates the effectiveness of the method for automatically detecting skin lesion edge irregularity, which is crucial for early melanoma diagnosis.

In [74], traditional and recent approaches in deep learning for skin cancer classification are evaluated. Traditional machine learning methods used include k-nearest neighbors (KNN), decision trees, logistic regression, artificial neural networks (ANN), and support vector machines (SVM). Additionally, deep learning-based techniques, including convolutional neural networks (CNN) with 98.1% [75], supervised, semi-supervised, and reinforcement learning, and ensemble methods are used. The analysis highlights that deep learning techniques have achieved impressive results in skin cancer classification, comparable and in some cases superior to dermatologists. However, these methods still face significant challenges, such as variability in lesion appearance and inconsistencies in image capture conditions.

4 Conclusions

The systematic review on machine learning techniques for skin cancer classification reveals significant advancements, yet also highlights important challenges. Convolutional Neural Networks (CNN) and deep learning techniques have demonstrated remarkable performance in terms of accuracy and generalization capability in classifying skin cancer images. However, variability in image quality and data diversity remain significant obstacles to effective clinical implementation. Addressing algorithmic bias and the necessity for large, well-annotated datasets to train robust models is emphasized. Additionally, integrating clinical and demographic data is crucial for enhancing diagnostic accuracy and treatment personalization. Hybrid techniques combining features from different levels have shown potential for improving diagnostic accuracy and should be further explored.

Finally, it is suggested that future research focus on developing models capable of efficiently handling image variability and integrating multiple data sources. Interdisciplinary collaboration among dermatology, informatics, and data science experts will be essential to advance the clinical application of these technologies, aiming to provide more accurate and accessible skin cancer diagnoses.

Acknowledgments. Thank to the Universidad Tecnológica del Perú.

References

1. Bhatt, H., et al.: State-of-the-art machine learning techniques for melanoma skin cancer detection and classification: a comprehensive review. Intell. Med. **3**(3), 180–190 (2023). https://doi.org/10.1016/j.imed.2022.08.004
2. Singh, V., et al.: Frontier machine learning techniques for melanoma skin cancer identification and categorization: an in-depth review. Oral Oncol. Rep. **9**, 100217 (2024). https://doi.org/10.1016/j.oor.2024.100217

3. Murugan, A., et al.: Diagnosis of skin cancer using machine learning techniques. Microprocessors Microsyst. **81**, 103727 (2021). https://doi.org/10.1016/j.micpro.2020.103727
4. Ünver, H., Ayan, E.: Skin Lesion segmentation in dermoscopic images with combination of YOLO and GrabCut algorithm. Diagnostics **9**(3), 72 (2019). https://doi.org/10.3390/diagnostics9030072
5. Camps-Valls, G., Bruzzone, L.: Kernel-based methods for hyperspectral image classification. IEEE Trans. Geosci. Remote Sens. **43**(6), 1351–1362 (2005). https://doi.org/10.1109/TGRS.2005.846154
6. Kononenko, I.: Machine learning for medical diagnosis: history, state of the art and perspective. Artif. Intell. Med. **23**(1), 89–109 (2001). https://doi.org/10.1016/s0933-3657(01)00077-x
7. Pacheco, A., Krohling, R.: The impact of patient clinical information on automated skin cancer detection. Comput. Biol. Med. **116**, 103545 (2020). https://doi.org/10.1016/j.compbiomed.2019.103545
8. Masood, A., Al-Jumaily, A.: Computer aided diagnostic support system for skin cancer: a review of techniques and algorithms. Int. J. Biomed. Imaging **1–22**, 2013 (2013). https://doi.org/10.1155/2013/323268
9. Anem, S., et al.: Leveraging convolution neural network (CNN) for skin cancer identification. Int. J. Res. Publ. Rev. **5**(4), 2150–2155 (2024). https://doi.org/10.55248/gengpi.5.0424.0955
10. Bharathi, G., Malleswaran, M.: Detection and segmentation of melanoma skin cancer in dermoscopy images using modified alexnet convolutional neural network-morphological methodology. Concurr. Comput. Pract. Exper. **34**, 25 (2022). https://doi.org/10.1002/cpe.7266
11. Lee, J., Majeed, A.: Classification of skin cancer by means of transfer learning models. Mekatronika **3**(2), 77–81 (2021). https://doi.org/10.15282/mekatronika.v3i2.7393
12. Velásquez, R.M.A.: A comprehensive evaluation and advancements in dissolved gas analysis for biodegradable oil in photovoltaic solar plants. Results Eng. **22**, 102314 (2024). https://doi.org/10.1016/j.rineng.2024.102314
13. Merlotti, A., et al.: Ester Orlandi, What is the role of postoperative re-irradiation in recurrent and second primary squamous cell cancer of head and neck? A literature review according to PICO criteria. Critical Rev. Oncol. Hematol. **111**, 20–30 (2017). https://doi.org/10.1016/j.critrevonc.2017.01.008
14. Aishwarya, N., et al.: Skin cancer diagnosis with yolo deep neural network. Procedia Comput. Sci. **220**, 651–658 (2023). https://doi.org/10.1016/j.procs.2023.03.083
15. Thomas, S., et al.: Interpretable deep learning systems for multi-class segmentation and classification of non-melanoma skin cancer. Med. Image Anal. **68**, 101915 (2020). https://doi.org/10.1016/j.media.2020.101915
16. Kaur, R., et al.: Melanoma classification using a novel deep convolutional neural network with dermoscopic images. Sensors **22**(3), 1134 (2022). https://doi.org/10.3390/s22031134
17. Bassel, A., et al.: Automatic malignant and benign skin cancer classification using a hybrid deep learning approach. Diagnostics **12**(10), 2472 (2022). https://doi.org/10.3390/diagnostics12102472
18. Karki, M.: Skin cancer detection using machine learning techniques. In: 2020 IEEE International Conference on Electronics, Computing and Communication Technologies (CONECCT), Bangalore, India, 2020, 1–5 (2020). https://doi.org/10.1109/CONECCT50063.2020.9198489

19. Abdar, M., et al.: Uncertainty quantification in skin cancer classification using three-way decision-based Bayesian deep learning. Comput. Biol. Med. **135**, 104418 (2021). https://doi.org/10.1016/j.compbiomed.2021.104418
20. Jasil, S., Ulagamuthalvi, V.: Deep learning architecture using transfer learning for classification of skin lesions. J. Ambient. Intell. Humaniz. Comput. **12**, 1–8 (2021). https://doi.org/10.1007/S12652-021-03062-7
21. Bistroń, M., Piotrowski, Z.: Comparison of machine learning algorithms used for skin cancer diagnosis. Appl. Sci. **12**(19), 9960 (2022). https://doi.org/10.3390/app12199960
22. Manoj, S., et al.: Automatic detection and categorization of skin lesions for early diagnosis of skin cancer using YOLO-v3 - DCNN architecture. Image Anal. Stereol. (2023). https://doi.org/10.5566/ias.2773
23. Nie, Y., Sommella, P., M. O'nils, Liguori, C., Lundgren, J.: Automatic detection of melanoma with yolo deep convolutional neural networks. In: E-Health and Bioengineering Conference (EHB). Iasi, Romania **2019**, 1–4 (2019). https://doi.org/10.1109/EHB47216.2019.8970033
24. Luu, N., et al.: Characterization of Mueller matrix elements for classifying human skin cancer utilizing random forest algorithm. J. Biomed. Opt. 26 (2021). https://doi.org/10.1117/1.JBO.26.7.075001
25. Khasanah, N., et al.: Skin cancer classification using random forest algorithm. J. Sci. Technol. (JST) **11**, 137–147 (2021). https://doi.org/10.30700/JST.V11I2.1122
26. Seeja, R., Suresh, A.: Melanoma classification employing inter neighbor statistical color and mean order pattern texture feature. Multimedia Tools Appl. **80**, 20045–20064 (2021). https://doi.org/10.1007/s11042-021-10685-7
27. Nambisan, A., et al.: Improving automatic melanoma diagnosis using deep learning-based segmentation of irregular networks. Cancers **15** (2023). https://doi.org/10.3390/cancers15041259
28. Ma, X., et al.: EFFNet: a skin cancer classification model based on feature fusion and random forests. PLoS ONE **18** (2023). https://doi.org/10.1371/journal.pone.0293266
29. Vidhyalakshmi, A., Kanchana, M.: Skin cancer classification using improved transfer learning model-based random forest classifier and golden search optimization. Int. J. Imaging Syst. Technol. **33**(2), 202–214 (2023). https://doi.org/10.1002/ima.22971
30. Wu, Y., et al.: Skin cancer classification with deep learning: a systematic review. Front. Oncol. **12**(893972), 1–13 (2022). https://doi.org/10.3389/fonc.2022.893972
31. Anand, V., et al.: An enhanced transfer learning based classification for diagnosis of skin cancer. Diagnostics **12**(7), 1628, 1–14 (2022). https://doi.org/10.3390/diagnostics12071628
32. Popescu, D., El-Khatib, M., El-Khatib, H., Ichim, L.: New trends in melanoma detection using neural networks: a systematic review. Sensors **22**(2), 496 (2022). https://doi.org/10.3390/s22020496
33. Betz-Stablein, B., et al.: Reproducible Naevus counts using 3D total body photography and convolutional neural networks. Dermatology **238**(1), 4–11 (2022). https://doi.org/10.1159/000517218
34. Du-Harpur, X., et al.: What is AI? Applications of artificial intelligence to dermatology. Br. J. Dermatol. **182**(1), 28–37 (2020). https://doi.org/10.1111/bjd.18880
35. Cassidy, B., et al.: Analysis of the ISIC image datasets: usage, benchmarks and recommendations. Med. Image Anal. **75**, 102305 (2022). https://doi.org/10.1016/j.media.2021.102305

36. Panayides, A.S., et al.: AI in medical imaging informatics: current challenges and future directions. IEEE J. Biomed. Health Inform. **24**(7), 1837–1850 (2020). https://doi.org/10.1109/JBHI.2020.2991043
37. Maron, R., et al.: Artificial intelligence and its effect on dermatologists' accuracy in Dermoscopic melanoma image classification: web-based survey study. J. Med. Internet Res. **22**(9), e18091 (2020). https://doi.org/10.2196/18091
38. Ashraf, R., et al.: Region-of-interest based transfer learning assisted framework for skin cancer detection. IEEE Access **8**, 147858–147870 (2020). https://doi.org/10.1109/ACCESS.2020.3014701
39. Sitaru, S., et al.: Automatic body part identification in real-world clinical dermatological images using machine learning. JDDG: Journal der Deutschen Dermatologischen Gesellschaft **21**(8), 863–869 (2023). https://doi.org/10.1111/ddg.15113
40. Nambisan, A., et al.: Improving automatic melanoma diagnosis using deep learning-based segmentation of irregular networks. Cancers **15**(1259), 1–24 (2023). https://doi.org/10.3390/cancers15041259
41. González-Cruz, C., et al.: Uso del aprendizaje automático en el diagnóstico del melanoma. Limitaciones por superar. Actas Dermosifiliogr. **111**(4), 313–316 (2020). https://doi.org/10.1016/j.ad.2019.09.002
42. Leon, R., et al.: Non-invasive skin cancer diagnosis using hyperspectral imaging for in-situ clinical support. J. Clin. Medi. **9**(6), 1662 (2020). https://doi.org/10.3390/jcm9061662
43. Mahum, R., Aladhadh, S.: Skin lesion detection using hand-crafted and DL-based features fusion and LSTM. Diagnostics **12**(2974), 1–19 (2022). https://doi.org/10.3390/diagnostics12122974
44. Canales-Fiscal, M., Tamez-Peña, J.: Hybrid morphological-convolutional neural networks for computer-aided diagnosis. Front. Artif. Intell. **6**, 1–13 (2023). https://doi.org/10.3389/frai.2023.1253183
45. Pandala, M., Periyanayagi, S.: Analysis, prediction and classification of skin cancer using artificial intelligence – a brief study and review. Scalable Comput. Pract. Exp. **24**(3), 355–370 (2023). https://doi.org/10.12694/scpe.v24i3.2241
46. Aladhadh, S., et al.: An effective skin cancer classification mechanism via medical vision transformer. Sensors **22**(11), 4008 (2022). https://doi.org/10.3390/s22114008
47. Mutepfe, F., et al.: Generative adversarial network image synthesis method for skin lesion generation and classification. J. Med. Signals Sens. **11**(4), 237–252 (2021). https://doi.org/10.4103/jmss.JMSS_53_20
48. Maron, R., et al.: Systematic outperformance of 112 dermatologists in multiclass skin cancer image classification by convolutional neural networks. Eur. J. Cancer **119**, 57–65 (2019). https://doi.org/10.1016/j.ejca.2019.06.013
49. Torti, E., et al.: Dermatologic hyperspectral imaging system for skin cancer diagnosis assistance. Sensors **24**(1399), 1–16 (2024). https://doi.org/10.3390/sensors24101399
50. El-Khatib, H., et al.: Deep learning–based methods for automatic diagnosis of skin lesions. Sensors **20**(6), 1753 (2020). https://doi.org/10.3390/s20061753
51. Pérez, E., Ventura, S.: Melanoma recognition by fusing convolutional blocks and dynamic routing between capsules. Cancers **13**(19), 4974 (2021). https://doi.org/10.3390/cancers13194974
52. Brutti, F., et al.: Artificial intelligence algorithms for benign vs. malignant dermoscopic skin lesion image classification. Bioengineering **10**(11), 1322 (2023). https://doi.org/10.3390/bioengineering10111322

53. Anupama, C., et. al.: Sand cat swarm optimization with deep transfer learning for skin cancer classification. Comput. Syst. Sci. Eng. **47**(2), 2080–2095 (2023). https://doi.org/10.32604/csse.2023.038322
54. Alenezi, F., et al.: A novel multi-task learning network based on melanoma segmentation and classification with skin lesion images. Diagnostics **13**(2), 262 (2023). https://doi.org/10.3390/diagnostics13020262
55. Bansal, P., et al.: Using transfer learning and hierarchical classifier to diagnose melanoma from dermoscopic images. Int. J. Healthcare Inf. Syst. Inform. **16**(2), 75–85 (2021). https://doi.org/10.4018/IJHISI.20210401.oa4
56. Brar, K., Shiney, O.: Computer-aided diagnosis and analysis of skin cancer from dermoscopic images in India. Curr. Med. Imaging **20**, e100423215589 (2024). https://doi.org/10.2174/1573405620666230410092618
57. AlDera, S., Othman, M.: A model for classification and diagnosis of skin disease using machine learning and image processing techniques. Int. J. Adv. Comput. Sci. Appl. **13**(5), 252–259 (2022). https://doi.org/10.14569/IJACSA.2022.0130531
58. Birkenfeld, J.S., et al.: Computer-aided classification of suspicious pigmented lesions using wide-field images. Comput. Methods Prog. Biomed. **195**, 105631 (2020). https://doi.org/10.1016/j.cmpb.2020.105631
59. Brinker, T.J., et al.: Deep neural networks are superior to dermatologists in melanoma image classification. Eur. J. Can. **119**, 11–17 (2019). https://doi.org/10.1016/j.ejca.2019.05 023
60. Magdy, A., et al.: Performance enhancement of skin cancer classification using computer vision. IEEE Access **11**, 72120–72131 (2023). https://dci.org/10.1109/ACCESS.2023.3294974
61. Lai, W., et al.: Skin cancer diagnosis (SCD) using Artificial Neural Network (ANN) and Improved Gray Wolf Optimization (IGWO). Sci. Rep. **13**, 19377 (2023). https://doi.org/10.1038/s41598-023-45039-w
62. Grignaffini, F., et al.: Machine learning approaches for skin cancer classification from Dermoscopic images: a systematic review. Algorithms **15**(11), 438 (2022). https://doi.org/10.3390/a15110438
63. Petracchi, B., et al.: Machine learning-based classification of skin cancer hyperspectral images. Procedia Comput. Sci. **225**, 2856–2865 (2023). https://doi.org/10.1016/j.procs.2023.10.278
64. Fraiwan, M., Faouri, E.: On the automatic detection and classification of skin cancer using deep transfer learning. Sensors **22**, 4963 (2022). https://doi.org/10.3390/s22134963
65. Mahbod, A., et al.: Transfer learning using a multi-scale and multi-network ensemble for skin lesion classification. Comput. Methods Programs Biomed. **193**, 105475 (2020). https://doi.org/10.1016/j.cmpb.2020.105475
66. Albraikan, A.A., et al.: Automated deep learning based melanoma detection and classification using biomedical dermoscopic images. Comput. Mat. Continua **74**(2), 2444–2458 (2023). https://doi.org/10.32604/cmc.2023.026379
67. Sujaini, H., et al.: Comparing the performance of linear regression versus deep learning on detecting melanoma skin cancer using apple core ML. Bull. Electr. Eng. Inform. **10**(6), 3110–3120 (2021). https://doi.org/10.11591/eei.v10i6.3178
68. Manikandan, S.P., et al.: DermICNet: efficient dermoscopic image classification network for automated skin cancer diagnosis. Revue d'Intelligence Artificielle **36**(5), 801–807 (2022). https://doi.org/10.18280/ria.360519
69. Attallah, O., Sharkas, M.: Intelligent dermatologist tool for classifying multiple skin cancer subtypes by incorporating manifold radiomics features categories. Con-

trast Media Mol. Imag. vol. 2021, artículo 7192016 (2021). https://doi.org/10.1155/2021/7192016
70. Bayram, B., et al.: Classification of skin lesion images with deep learning approaches. Baltic J. Modern Comput. **10**(2), 241–250 (2022). https://doi.org/10.22364/bjmc.2022.10.2.10
71. Almuayqil, S.N., et al.: Computer-aided diagnosis for early signs of skin diseases using multi types feature fusion based on a hybrid deep learning model. Electronics **11**(4009) (2022). https://doi.org/10.3390/electronics11234009
72. Jain, S., et al.: Deep learning-based transfer learning for classification of skin cancer. Sensors **21**(8142) (2021). https://doi.org/10.3390/s21238142
73. Ali, A., et al.: A machine learning approach to automatic detection of irregularity in skin lesion border using dermoscopic images. PeerJ Comput. Sci. **6**, e268 (2020). https://doi.org/10.7717/peerj-cs.268
74. Yang, G., et al.: Advancements in skin cancer classification: a review of machine learning techniques in clinical image analysis. Multimedia Tools Appl. (2024). https://doi.org/10.1007/s11042-024-19298-2
75. Lara, J.V.M., Velásquez, R.M.A.: Low-cost image analysis with convolutional neural network for herpes zoster. Biomed. Signal Process. Control **71**(Part B), 103250 (2022). https://doi.org/10.1016/j.bspc.2021.103250

Enhancing Hospital Resource Planning Through Predictive Analytics for Costs and Bed Allocations

Frederick Ayensu and Isaac Osunmakinde(✉)

Department of Computer Science, Norfolk State University, Norfolk, VA, USA
f.k.ayensu@spartans.nsu.edu, ioosunmakinde@nsu.edu

Abstract. Hospital resource planning is essential for optimizing the allocation and use of resources in healthcare settings. Accurate predictions in this area are difficult due to the complexity of healthcare systems and varying patient demands. Research has explored mathematical models, optimization techniques, and data-driven approaches to improve aspects like capacity planning, workforce scheduling, and demand forecasting. They often fall short in precision and reliability, causing inefficient resource distribution. Hence, there remains a lack of precision in forecasting specific needs such as patient cost categories and bed requirements.

This study proposes a framework of enhanced data preprocessing, integrating classifier and regressor models (including Random Forest, Decision Tree, and Neural Networks), and using techniques like SMOTE to handle class imbalance. The models are trained on publicly available data and extensively evaluated with cross-validation techniques. Comparative results with the literature show that the Random Forest Classifier achieved 93.66% best accuracy and the Random Forest Regressor showed a low Mean Absolute Error (MAE) of 0.074 with R2 of 0.9991. These findings suggest that the proposed approach enhances resource allocation accuracy and reliability, aiding healthcare administrators and policymakers.

Keywords: Hospital resource planning · machine learning model · healthcare resource allocation · cost categorization · bed utilization estimation · predictive modeling

1 Introduction

Hospital resource planning is a critical research problem focused on optimizing the utilization and availability of resources in healthcare facilities while minimizing costs and ensuring quality care [1]. It involves forecasting future demand for beds and other resources, understanding trends in patient length of stay and occupancy, and considering various factors such as patient characteristics and discharge destinations [1]. Effective resource planning is essential for efficient care delivery, minimizing wait times, improving patient outcomes, and ensuring the allocation of scarce resources aligns with demand [2]. Hospital resource planning plays a crucial role in maintaining the viability of healthcare systems, especially during times of crisis such as the COVID-19 pandemic [3].

R. Silhavy and P. Silhavy (Eds.): CoMeSySo 2024, LNNS 1489, pp. 301–327, 2025.
https://doi.org/10.1007/978-3-031-96798-6_25

Recent trends in hospital admissions, patient demographics, and healthcare utilization patterns underscore the pressing need for comprehensive hospital resource planning strategies. There has been a notable increase in emergency department visits, elective surgeries, and specialized medical procedures, leading to capacity challenges and resource constraints in many healthcare facilities [4]. In December 2022, data from U.S. Department of Health and Human Services shows that out of 2434 counties in the US, 53 exceeded their beds utilization and 219 counties had beds utilization between 85% to 100% [5] as shown in Fig. 1.

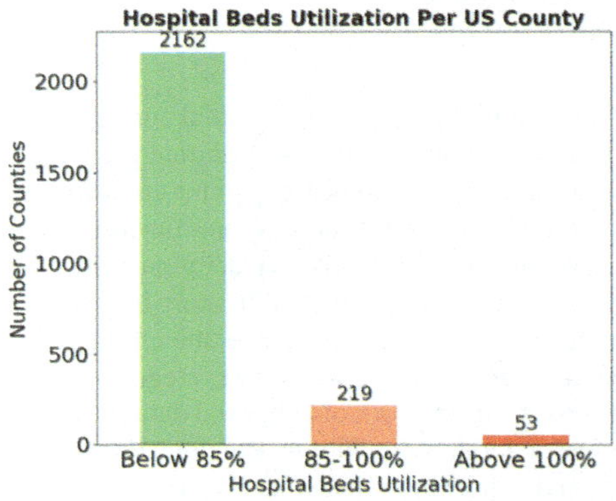

Fig. 1. County hospital beds utilization frequency.

Existing research provides valuable insights into optimal allocation of resources and hospital capacity planning, particularly in the context of critical diseases like COVID-19 [2, 3]. They showcase the application of AI, data mining, system dynamics, discrete-event simulation, and data envelopment analysis in improving healthcare resource allocation, patient flow optimization, length of stay prediction, and efficiency assessment in hospitals. These methods offer more accurate forecasting, dynamic resource allocation, and improved patient outcomes compared to traditional approaches, emphasizing the importance of leveraging advanced technologies for healthcare planning and management [7].

However, research gaps in hospital resource planning encompass limited integration of factors like patient flow patterns, resource utilization levels, and forecasted healthcare service demand [1]. While some studies focus on model development using data-driven approaches like data envelopment analysis (DEA) [6], others explore AI and data mining for resource allocation and patient care [2]. There is a need for more research to combine advanced modeling approaches, enhancing hospital capacity planning and cost estimation [1]. These gaps result in inefficiencies, mismanagement, longer wait times, and potentially adverse health outcomes, including higher morbidity and mortality rates. Addressing these critical issues requires innovative solutions leveraging AI, data mining, and advanced modeling techniques to optimize resource allocation, streamline patient flow, and enhance overall healthcare system performance.

1.1 Research Question and Contributions

Based on the above background, this research poses the question: How can integrated classification and regression models be optimized to accurately predict cost categories and bed requirements for efficient hospital resource planning? The proposed methodology addresses the research question. The contributions of this paper are as follows:

- Proposing a comprehensive framework for hospital resource planning that integrates enhanced data preprocessing, best classifier, and regressor models based on optimal hyperparameter tunning, to improve prediction accuracy for cost categories and hospital bed requirements. Thanks to the design of three integrated algorithms, each functioning as both a classifier and regressor that collectively advance the field by providing robust tools for this machine learning task.
- For model generalizability, it incorporates a stepwise SMOTE (Synthetic Minority Over-sampling Technique) to address data bias observed in class imbalance in the dataset, which enhances model performance.
- Provides comparative results on Random Forest Classifier achieving a higher accuracy of 93.66% and the Random Forest Regressor exhibits a lower Mean Absolute Error (MAE) of 0.074 with R2 of 0.9991 that shows improved performance metrics over existing literature, which contributes to the body of knowledge in hospital resource planning.

Section 2 delves into various related research papers and projects followed by a section on research gaps in the field. The methodology in Sect. 3 outlines the procedure and techniques used after which the experimental results are highlighted in Sect. 4. The conclusion and future work section summarizes the findings and suggests areas for future research.

2 Literature Review

One study employs data envelopment analysis (DEA) to gauge hospital efficiency and resource allocation effectiveness, emphasizing the method's popularity in hospital performance assessment [6]. Another study focuses on predicting hospital length of stay using machine learning algorithms, highlighting the importance of accurate forecasting for resource planning [9]. These studies, among others, provide valuable insights into various methodologies used for optimizing hospital resources. Table 1 presents a summary of the literature reviews, which detailed methodologies and suggested future work.

2.1 Research Gaps

Based on the future works suggested in Table 1, these are notable research gap areas where further investigation and development are needed to improve healthcare resource planning, allocation, and management: (1) Expand healthcare resource allocation models to cover a wider range of services beyond elective surgeries, (2) Evaluating the

Table 1. Summary of literature review.

Paper type	Problem addressed	Methodology	Future work suggested
Research paper [1]	The need to enhance resource requirement forecasting in healthcare systems	Developing a multi-absorbing state phase-type distribution	Extend the model to evaluate the impact of delayed discharge on bed availability
Research Paper [2]	Enhance resource allocation and planning for critical diseases to improve patient outcomes	AI and data mining for optimal healthcare resource allocation in critical diseases	Analyzing AI and data mining in healthcare, including real-world cases and ethics
Research paper [3]	Lacking disease-specific guidelines and hospital resource strain during COVID-19	Creating SEIR models and simulations for COVID-19 case prediction and bed analysis	Enhancing SEIR model to include reinfection, improving hospital planning for safe reopening
Research Paper [6]	Analyzing hospital efficiency and resource allocation using Data Envelopment Analysis (DEA)	Using DEA for hospital efficiency assessment and resource allocation	Improving models to incorporate hospital features for better efficiency assessment
Research paper [7]	The inefficiencies in healthcare resource planning during the Covid-19 pandemic	System modeling, distribution fitting, simulation, validation, and analysis	Simulation and ML to optimize hospital resources and forecasting demand in emergencies
Research Paper [8]	Allocating limited surgical resources in healthcare for elective surgeries	Integer programming and auction bidding for surgery allocation	Extend the healthcare resource allocation model beyond elective surgeries
Research paper [9]	Predicting the length of stay (LOS) for hospitalized patients	Five ensemble ML models, e.g. Random Forest, Gradient Boost, CatBoost	Exploring more advanced ensemble machine learning models

impact of delayed discharges on hospital bed availability, (3) Integration of additional hospital-specific features into ML models for more accurate efficiency assessment, and (4) Explore sophisticated machine learning models to improve the prediction of patient length of stay (LOS).

2.2 Selected Machine Learning Techniques

This section introduces some machine learning techniques and their associated equations. Decision trees, a supervised learning algorithm for both classification and regression, are presented with their tree structure composed of decision nodes and leaf nodes representing outcomes based on feature values [19]. Random forests, an ensemble learning method, ccmbine multiple decision trees trained on random subsets of data and features to enhance prediction accuracy and stability [20]. Neural networks, inspired by the human brain, utilize algorithms to recognize patterns and interpret sensory data, enabling machine perception, labeling, and clustering of raw input [21].

2.2.1 Information Gain

Information gain is a key metric used to decide which feature to split on at each step in building a decision tree. It measures the reduction in entropy or impurity when a dataset is split on an attribute [19]. Entropy (H) is a measure of the randomness or disorder in the data, given by:

$$H(D) = -\sum_{i=1}^{k} p_i \log_2 p_i \tag{1}$$

Where:

D is the dataset
p_i is the proportion of instances in class iii
k is the number of classes

Information Gain (IG) is calculated as:

$$IG(D, A) = H(D) - \sum_{v \in Values(A)} \frac{|D_v|}{|D|} H(D_v) \tag{2}$$

Where:

A is the attribute being considered
Values(A) are the possible values of attribute A
Dv is the subset of D for which attribute A has value v
|D| and |Dv| are the number of instances in D and Dv, respectively

2.2.2 Gini Impurity

Gini impurity is another metric to evaluate the splits in decision trees, particularly for classification tasks [22]. It measures the frequency at which any element of the dataset would be misclassified if it were randomly labeled.

$$Gini(D) = 1 - \sum_{i=1}^{k} p_i^2 \tag{3}$$

Where:

p_i is the proportion of instances in class i
k is the number of classes

2.2.3 Aggregation of Predictions

In classification tasks, the final prediction of a random forest is the mode of the predictions of all individual trees [20]:

$$\hat{y} = mode\{h_1(x), h_2(x), \ldots, h_n(x)\} \tag{4}$$

Where:

$h_i(x)$ is the prediction of the i-th decision tree for input x

For regression tasks, the final prediction is the average of the predictions of all individual trees:

$$\hat{y} = \frac{1}{n}\sum_{i=1}^{n} h_i(x) \tag{5}$$

Where:

$h_i(x)$ is the prediction of the i-th decision tree for input x
n is the number of trees

2.2.4 Multilayer Perceptron (MLP)

The Multilayer Perceptron is a class of feedforward artificial neural network (ANN). An MLP consists of at least three layers of nodes: an input layer, a hidden layer, and an output layer. The formula for calculating the output of a neuron in an MLP neural network is [23]:

$$f(x) = \sigma\left(\sum_{i=1}^{n} w_i x_i + b\right) \tag{6}$$

Where:

f(x) is the output of the neuron
σ is the activation function
w_i are the weights of the connections between the neuron's input and hidden layers
x_i are the inputs to the neuron
b is the bias term

3 Methodology

Figure 2 outlines the methodology for this research.

3.1 Data Collection and Preprocessing

The hospital and treatment center charges dataset is publicly accessible on Kaggle.com [10] and is used to build the classifier models. It contains 12 unique features and approximately 163,000 records. Additionally, the healthcare investments and length of stay dataset, also available on Kaggle.com [11], is used to build the regressor. This dataset includes 6 unique features and data from around 32 different countries. The data is sourced from CSV files and loaded into a Pandas Data Frame for initial exploration.

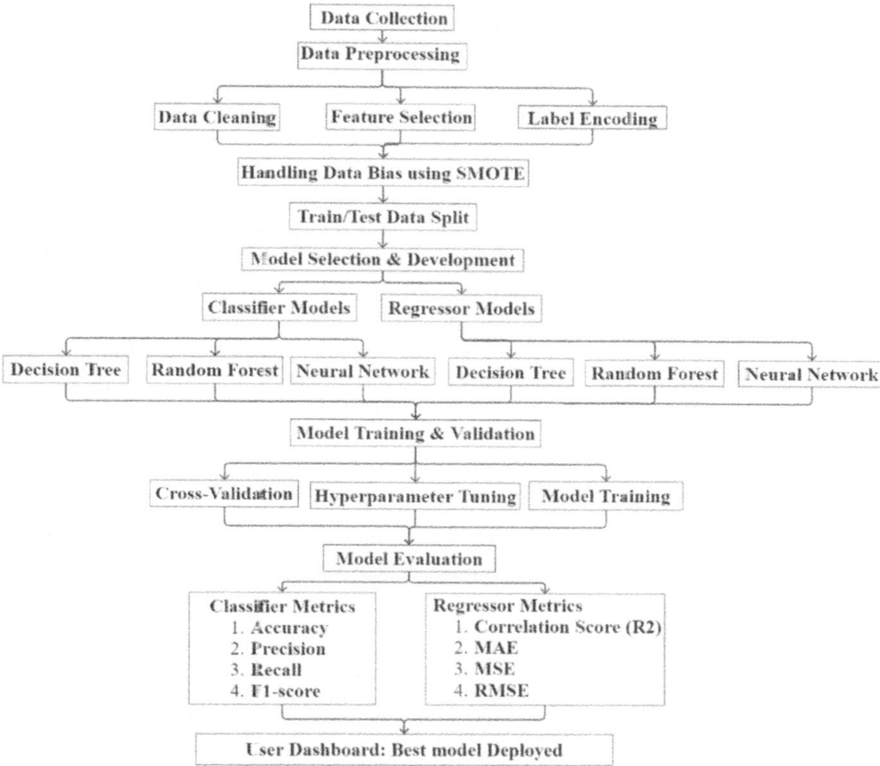

Fig. 2. Methodology flowchart.

3.1.1 Data Cleaning and Preparation

As seen in Fig. 2 data cleaning occurs during the preprocessing stage after data collection. Here, missing values are addressed by imputing them with the median for numerical attributes. This ensures that no data is lost, and the statistical properties of the dataset are maintained. Continuous variables like hospital charges are discretized into bins to facilitate classification. For instance, hospital charges are categorized into low, medium, and high based on predefined thresholds.

3.1.2 Feature Selection

For the classification task predicting medicare payment categories, relevant features including Provider_Id, Provider_State, Total_Discharges, Average_Covered_Charges, and Average_Total_Payments along with the target variable medicare_pay_cat are selected following the methodology outlined in Fig. 2. For the regression task predicting hospital bed counts, preprocessing involves selecting relevant features like Location, Time, Hospital_Stay, MRI_Units, CT_Scanners, and the target variable Hospital_Beds. Various feature selection strategies are employed [28] such as relevance-based selection where features are selected based on their relevance to the target variable. Correlation Analysis is also used where a correlation matrix is computed to examine the relationships between different features and the target variable. Features with high correlation to the target variable and low multicollinearity with other features are prioritized.

3.1.3 Label Encoding

The classification task begins with preprocessing the dataset after collection as depicted in Fig. 2. This is done by encoding categorical variables like Provider_State and medicare_pay_cat using Label Encoder from the sklearn preprocessing library. This converts categorical data into a numerical format suitable for machine learning models. For the regression task, the categorical feature Location is also encoded using LabelEncoder to transform it into a numerical format. This allows the regression models to process and learn from this feature effectively.

3.2 Handling Bias and Imbalanced Data

In addressing data bias and imbalance, various techniques can be utilized to improve the performance and accuracy of models trained on skewed datasets.

3.2.1 SMOTE

Synthetic Minority Over-sampling Technique (SMOTE) is an oversampling method that generates synthetic samples for the minority class in an imbalanced dataset by interpolating between existing samples. This helps to balance the class distribution and improve the performance of machine learning models, especially in situations where the minority class is underrepresented [12]. To address this issue, we propose a simplified approach as outlined in Algorithm 1.

Algorithm 1 Stepwise SMOTE Algorithm
Inputs

- **X**: A matrix of feature vectors, where X[i][j] represents the j-th feature of the i-th instance.
- **y**: A vector of class labels for each instance, indicating the class membership.
- **N**: The desired number of synthetic samples to generate.
- **k**: The number of nearest neighbors to use for generating synthetic samples.

Outputs

- **X_new**: The augmented matrix of feature vectors, including both the original and the synthetic minority class samples.
- **y_new:** The augmented vector of class labels, including labels for the synthetic samples.

1. Select minority class samples from the dataset [12].
2. Identify feature vectors minority for these samples.
3. For each minority sample, find its nearest neighbors.

4. Randomly select N% of neighbors to create synthetic samples.
5. Augment the dataset with synthetic samples, updating labels accordingly.

By incorporating SMOTE into the methodology, issues related to imbalanced data are addressed and the predictive capabilities of models are enhanced.

3.3 Model Selection to Advance Machine Learning

This section proposes a comprehensive approach that leverages machine learning models to accurately predict both cost categories and bed requirements, aiming to optimize hospital resource allocation. Algorithm 1 designs an integrated DT classifier and regressor.

Algorithm 1 Integrated Decision Tree (DT) Model
Inputs

- **X**: A matrix of feature vectors, where X[i][j] represents the j-th feature of the i-th instance.
- $\mathbf{y_{class}}$: A vector of class labels (cost categories) for each instance.
- $\mathbf{y_{reg}}$: A vector of numerical values (bed requirements) for each instance.

Outputs

- Trained DT classifier model for predicting cost categories.
- Trained DT regressor model for predicting bed requirements.

1. Start with the entire dataset at the root node.
2. If **classifier, do:**
   - Choose a splitting criterion such as Information Gain (IG) [13].
   - Divide the dataset into subsets based on the best split of IG.
   - Recursively apply the steps until a stopping condition is met.
   - **Prediction**: for new instance, traverse the DT model and predict class label as cost category.
3. If **regressor, do**:
   - Choose a splitting criterion such as Mean Squared Error (MSE) [14].
   - Divide the dataset into subsets based on the best split of MSE.
   - Recursively apply the steps until a stopping condition is met.
   - **Prediction**: for new instance, traverse the DT model and predict the mean value as the number of beds required.

One can see that both models use the same dataset but different target variables and criteria for splits.

Algorithm 2 designs an integrated RF classifier and regressor.

Algorithm 2 Integrated Random Forest (RF) Model
Inputs

- **X**: A matrix of feature vectors, where X[i][j] represents the j-th feature of the i-th instance.
- $\mathbf{y_{class}}$: A vector of class labels (cost categories) for each instance.
- $\mathbf{y_{reg}}$: A vector of numerical values (bed requirements) for each instance.

Outputs

- Trained RF classifier model for predicting cost categories.
- Trained RF regressor model for predicting bed requirements.

1. Start with the entire dataset.
2. If **classifier, do**:
   - Generate bootstrap samples from the original dataset [15].
   - Construct decision trees on each bootstrap sample, using subsets of features.
   - Recursively split the dataset using a chosen criterion like Information Gain.
   - **Prediction:** for a new instance, traverse each tree in the RF model and predict the class label using majority voting to determine the cost category.
3. If **regressor, do**:
   - Generate bootstrap samples from the original dataset [16].
   - Construct decision trees on each sample, using subsets of features.
   - Split the dataset based on the best Mean Squared Error (MSE) criterion, recursively
   - **Prediction:** for a new instance, traverse each tree in the RF model and predict the numerical value, then average the predictions to determine the number of beds required.

Algorithm 3 presents the design of an integrated NN classifier and regressor.

Algorithm 3 Integrated Neural Network (NN) Model Inputs

- **X**: A matrix of feature vectors, where X[i][j] represents the j-th feature of the i-th instance.
- $\mathbf{y_{class}}$: A vector of class labels (cost categories) for each instance.
- $\mathbf{y_{reg}}$: A vector of numerical values (bed requirements) for each instance.

Outputs

- Trained NN classifier model for predicting cost categories.
- Trained NN regressor model for predicting bed requirements.

1. Start with the entire dataset.
2. If **classifier, do**:
   - Design the neural network architecture, specifying layers for input, hidden and output [17].
   - Utilize activation functions like ReLU for hidden layers and softmax for the output layer.

- Compile the model by selecting a loss function like categorical cross-entropy and an optimizer, then train it on data using backpropagation with suitable epochs and batch size and evaluate with cross-validation.
- **Prediction:** For a new instance, pass the feature vector through the NN model to predict the class label.

3. If **regressor, do**:

- Design the neural network architecture with layers for input, hidden, and output [18].
- Implement activation functions like ReLU for hidden layers and linear activation for the output layer.
- Compile the model, selecting a loss function such as Mean Squared Error (MSE) and an optimizer, then train it on input data and corresponding labels using backpropagation with appropriate epochs and batch size and evaluate with cross-validation
- **Prediction:** For a new instance, pass the feature vector through the NN model to predict the numerical value.

This integrated ML approach is proposed to build and utilize both classification and regression models for effective dual hospital resource prediction tasks.

The best-performing ML models for classification and regression are integrated into a single framework capable of predicting both cost categories and bed requirements for incoming patients.

3.4 Training and Validation

3.4.1 Data Splitting

The dataset was split into training and testing sets with a 90:10 ratio. The training set was further split into training and validation sets using k-fold cross-validation (k = 5) to ensure robust model evaluation.

3.4.2 Classifier and Regressor Training

Each ML classifier in the algorithms above was trained on the training data. Cross-validation was used to tune hyperparameters and prevent overfitting.

Also each ML regressor was trained on the training data with similar cross-validation procedures to optimize performance.

3.5 Evaluation Metrics

3.5.1 Classifier Performance Metrics

The performance of classifier models is evaluated using metrics such as accuracy, precision, recall, and the F1 score [24].

i. Accuracy: the ratio of correctly predicted observations to the total observations. Given by:

$$\frac{TP + TN}{TP + TN + FP + FN} \tag{7}$$

Where:

TP is the number of true positives.
TN is the number of true negatives.
FP is the number of false positives.
FN is the number of false negatives

ii. Precision: measures the accuracy of the positive predictions. Given by:

$$\frac{TP}{TP + FP} \tag{8}$$

iii. Recall: measures the ability of the model to identify all relevant instances. Given by:

$$\frac{TP}{TP + FN} \tag{9}$$

iv. F1 score: the harmonic mean of precision and recall, providing a balance between the two metrics. Given by:

$$2 \times \frac{Precision \times Recall}{Precision + Recall} \tag{10}$$

3.5.2 Regressor Performance Metrics

Regressor models are assessed using metrics like Mean Absolute Error (MAE), Mean Squared Error (MSE), and R-squared (R^2) [25].

i. Mean Absolute Error (MAE): the average of the absolute differences between the predicted and actual values across all instances in the dataset. Given by:

$$\frac{1}{n}\sum_{i=1}^{n} | y_i - \hat{y}_i | \tag{11}$$

Where:

n is the number of instances.
y_i is the actual value.
$\hat{y}_i$ is the predicted value

ii. Mean Squared Error (MSE): the average squared difference between the estimated values and the actual value. Given by:

$$\frac{1}{n}\sum_{i=1}^{n} (y_i - \hat{y}_i)^2 \tag{12}$$

iii. R-squared (R2): the proportion of the variance in the dependent variable that is predictable from the independent variables. Given by:

$$1 - \frac{\sum_{i=1}^{n} (y_i - \hat{y}_i)^2}{\sum_{i=1}^{n} (y_i - \overline{y})^2} \tag{13}$$

Where:

$\overline{y}$ is the mean of the actual values.

These metrics collectively ensure a comprehensive evaluation of both classification and regression model performance.

4 Data Exploration

4.1 Experimental Data Analysis

The distribution of categories in the target column before label encoding reveals varying proportions among different payment ranges. Specifically, the dataset comprises 63,407 entries falling within the 5-10 k payment range, 58,444 entries within the 1–5 k range, and 41,214 entries exceeding 10 k as shown in Fig. 3. Calculating the percentages of each category relative to the total entries shows that the "5-10 k" category constitutes approximately 41.44%, while the "1–5 k" and "10 k+" categories represent approximately 38.13% and 20.43%, respectively. This distribution suggests a certain degree of imbalance, with a notable prevalence of entries in the "5-10 k" and "1–5 k" categories compared to the "10 k+" category. To address this imbalance and improve model performance, techniques like Synthetic Minority Over-sampling Technique (SMOTE) can be employed. SMOTE can help by oversampling the minority class (in this case, the "10 k+" category) and generating synthetic samples to balance the dataset, thereby enhancing the model's ability to generalize and make accurate predictions across all categories. By applying SMOTE during preprocessing, we can mitigate the impact of class imbalance and optimize the dataset for more effective machine learning model training and evaluation.

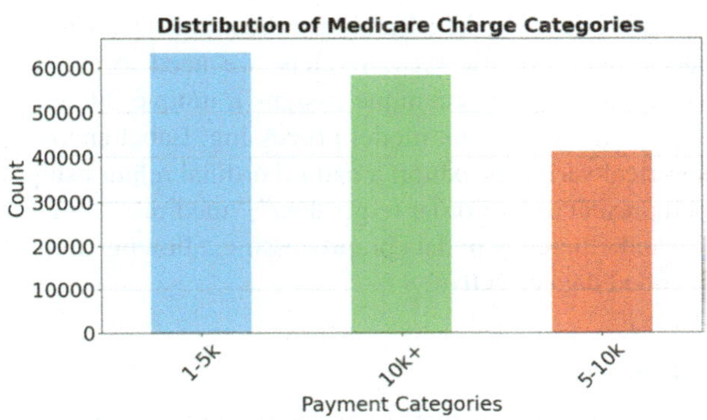

Fig. 3. Distribution of charges.

The distribution of hospital beds (in thousands) was analyzed using a histogram-based approach with 30 bins. The results revealed a diverse range of bed counts across these bins. The distribution shown in Fig. 4 highlights that most hospitals fall within lower bed count ranges, with bins such as (0.1 k - 1.9 k) and (1.9 k - 3.8 k) containing 62 and 61 hospitals, respectively. As the bed count increases, the frequency within each bin generally decreases, indicating a pattern of fewer hospitals with higher bed capacities. Notably, towards the higher end of the spectrum, bins like (31.3 k - 33.2 k) and (44.2 k - 46.0 k) registered minimal to no hospitals, emphasizing the rarity of hospitals with extremely high bed capacities. This analysis provides insights into the distribution of hospital bed capacities, crucial for understanding healthcare infrastructure and resource allocation.

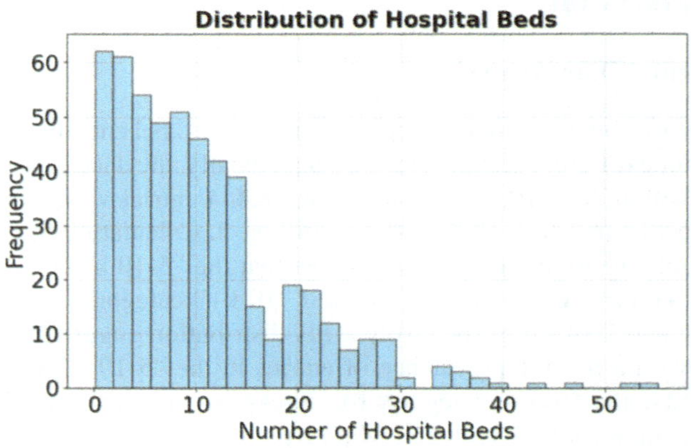

Fig. 4. Frequency distribution of hospital beds.

4.2 Getting Data Ready for Modeling

4.2.1 Encoding Categorical into Numeric Data

Some machine learning models, like Random Forest and Decision Tree are not able to handle categorical data. To use these models, we need to encode the categorical data into numerical values. This technique assigns a unique integer to each category within a feature, facilitating efficient model processing. Label encoding is particularly useful when categorical variables exhibit a natural ordinal relationship, such as in cases where categories represent a clear order (e.g., "low", "medium", "high"). This approach ensures simplicity and efficiency in data preprocessing, allowing algorithms to interpret and utilize the encoded data effectively.

4.2.2 Selecting Features

The decision to focus on specific features for modeling and analysis was driven by several considerations aimed at optimizing the predictive performance and interpretability of the machine learning model. Each chosen feature was selected based on its relevance to the target feature as seen in Figs. 5 and 6. By prioritizing these key attributes, the modeling process was streamlined to focus on the most informative aspects of the dataset, reducing noise and unnecessary complexity. This feature selection process is crucial in machine learning as it helps identify and leverage the most impactful predictors, improving model efficiency, interpretability, and overall performance. Additionally, Figs. 5 and 6 point out each selected feature and its relationship to the target variable, insights gained from the analysis is more actionable and meaningful, leading to more effective decision-making in healthcare management or related domains.

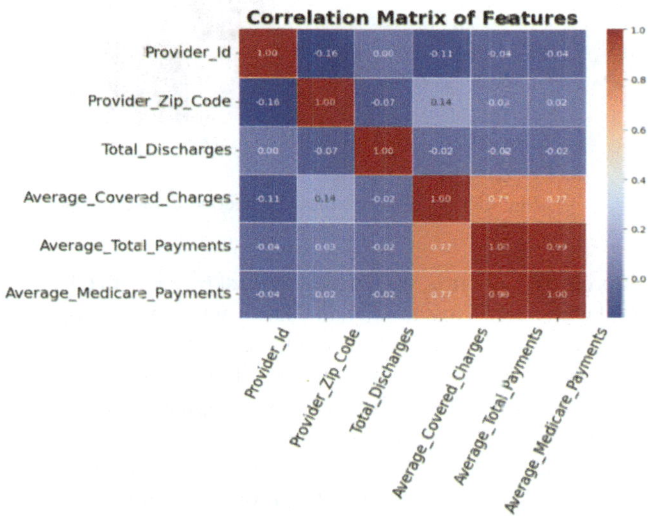

Fig. 5. Correlation Matrix of first dataset.

4.2.3 Binning

In dataset 1, the target column representing average hospital charges was discretized into three distinct bins based on predefined thresholds of $5000 and $10,000. This binning strategy was adopted to facilitate the classification and prediction of hospital charges into categorical groups, enabling machine learning models to effectively learn patterns and relationships within the data. By transforming the continuous variable into a categorical format, the prediction task becomes more interpretable and actionable for decision-making in healthcare resource allocation. This approach simplifies the complexity of the target variable, aiding in the development of robust models capable of providing valuable insights into hospital charge categorization and related analyses.

4.3 Over Sampling

The decision to oversample the first dataset using the SMOTE algorithm was motivated by the imbalanced class distribution observed in the target variable. Before applying

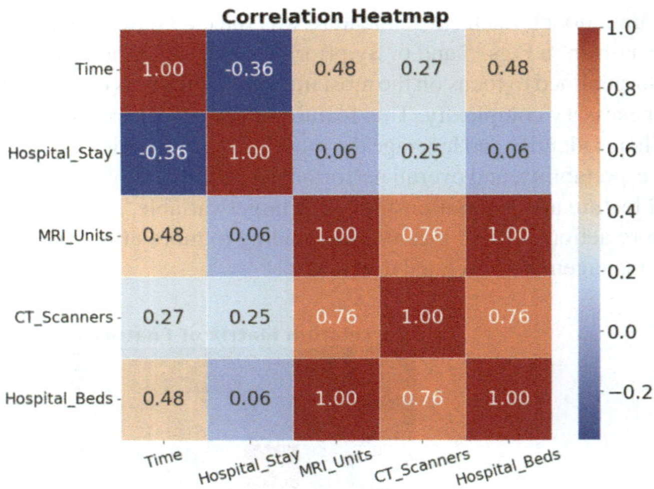

Fig. 6. Correlation Matrix of second dataset.

SMOTE, the class distribution in y_train was heavily skewed, with Class 2 having 56,984 instances, Class 0 with 52,653 instances, and Class 1 with 37,121 instances. This imbalance can lead to biased model training, where the classifier may prioritize the majority class and perform poorly on the minority classes. To address this issue, we employed the SMOTE technique. SMOTE synthesizes new samples for the minority classes by interpolating between existing minority class instances. This technique increases the representation of the minority classes to achieve a more balanced distribution. As seen in Table 2, after applying SMOTE, the class distribution in y_train_resampled was adjusted to be more uniform, thereby enhancing the classifier's ability to learn from all classes equally during training. This approach helps mitigate the bias towards the majority class and improves the overall performance and generalization of the Decision Tree model for prediction.

Table 2. Over-sampling values.

Charges category – class	Before over-sampling	After-over-sampling
<$5000–0	52,653	56,984
$5000–$10,000–2	56,984	56,984
>$10,000–1	37,121	56,984

5 Experimental Setup

In this study, two distinct datasets were employed to investigate classification and regression tasks related to hospital operations and healthcare management. The first dataset, sourced from public healthcare records, contains detailed information on hospital charges, including provider identifiers, state of operation, discharge statistics, covered charges, and total payments. The target variable of interest in this dataset is the Medicare payment category (medicare_pay_cat), which classifies reimbursement levels for medical services.

The second dataset used in this study captures essential attributes such as hospital location, time-related metrics, length of hospital stay, and counts of medical equipment like MRI units and CT scanners. This dataset is instrumental for predicting the number of hospital beds (Hospital_Beds) available at specific locations, providing valuable insights into hospital resource management and capacity planning.

For the classification experiment, the goal was to develop and evaluate machine learning classifiers capable of predicting Medicare payment categories. Three distinct models were trained on the dataset: Random Forest Classifier (RFC), Decision Tree Classifier (DTC), and Neural Network (NN). Prior to model training, the dataset underwent preprocessing steps, including encoding of categorical features using LabelEncoder, stratified splitting into training (90%) and testing (10%) sets, and application of Synthetic Minority Over-sampling Technique (SMOTE) to address class imbalance.

In the regression experiment, the objective was to build regression models for predicting hospital bed counts based on location and related attributes. Similar to the classification task, three regression models were trained: Random Forest Regressor (RFR), Decision Tree Regressor (DTR), and Neural Network (NN). Data preprocessing steps included encoding categorical location data, splitting the dataset into training (90%) and testing (10%) sets, and standardizing numerical features.

The training of each model involved initializing the respective algorithms with default parameters and optimizing the model architecture for neural network models. Model performance was evaluated using standard metrics: accuracy, confusion matrix, and classification report for classification tasks, and mean absolute error (MAE), mean squared error (MSE), root mean squared error (RMSE), and R-squared for regression tasks.

The experimental procedure included rigorous data preparation, model training, and evaluation steps to ensure robustness and reliability of the results. The selected models exhibiting superior performance based on evaluation metrics were identified for deployment in practical healthcare management scenarios, facilitating accurate predictions of Medicare payment categories and hospital bed counts.

The experimental setup was implemented using Python programming language (version 3) with essential libraries such as pandas, scikit-learn, imbalanced-learn, and matplotlib. The experiments were conducted on a computing environment equipped with 16 GB RAM and 8 CPU cores, providing sufficient computational resources for model training and evaluation.

6 Experimental Results

6.1 Regression

6.1.1 Performance Evaluations

The Random Forest model achieved the highest R-squared value (0.9991) as shown in Fig. 7, indicating excellent predictive performance and a strong fit to the data. The Decision Tree model also performed well with an R-squared value of 0.9963. However, the Neural Network model exhibited slightly lower performance with an R-squared value of 0.9739, suggesting some room for improvement in capturing the variance of the target variable.

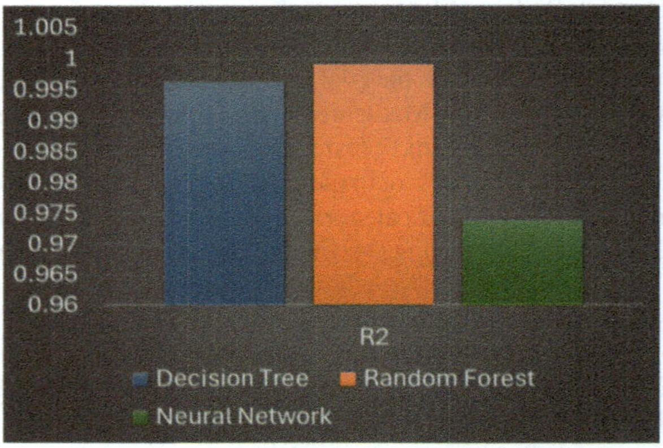

Fig. 7. Correlation score of various regressor models.

Figure 8 compares the regression error metrics across the Decision Tree, Random Forest, and Neural Network models, clear insights emerge regarding the practical performance of these models. The Mean Absolute Error (MAE) measures the average magnitude of prediction errors, with the Random Forest exhibiting the lowest MAE of 0.074 as shown in Fig. 8, indicating more accurate predictions compared to the Decision Tree (MAE = 0.15) and Neural Network (MAE = 0.9). The Mean Squared Error (MSE), reflecting the average of squared prediction errors, further underscores the superiority of the Random Forest (MSE = 0.046) over the Decision Tree (MSE = 0.29) and Neural Network (MSE = 1.47). Similarly, the Root Mean Squared Error (RMSE), representing the standard deviation of prediction errors, highlights the Random Forest's precision (RMSE = 0.215) relative to the Decision Tree (RMSE = 0.53) and Neural Network (RMSE = 1.21). Lower values across these metrics indicate more accurate and reliable predictions, with Random Forest consistently demonstrating superior performance compared to the other models.

The cross-validated R-squared scores indicate the goodness of fit of each model to the data, with higher values suggesting better performance in explaining the variance of the target variable. Table 3 shows the Decision Tree (DT) model achieved a mean R-squared

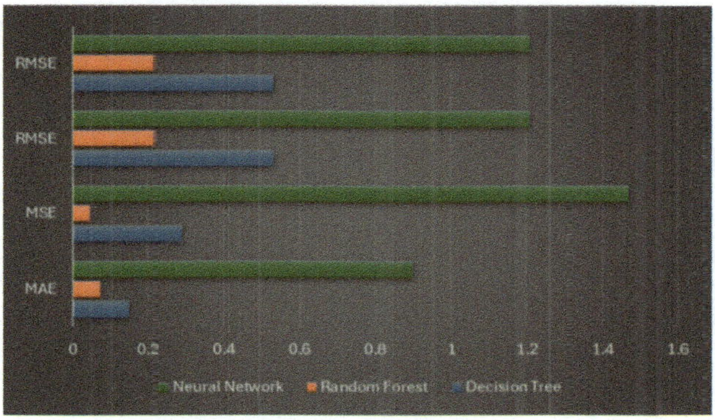

Fig. 8. Error metrics of various regressor models.

value of 0.986 with a standard deviation of 0.0205, indicating consistent performance across the five cross-validation folds. The Random Forest (RF) model demonstrated higher mean R-squared performance at 0.9964 with a smaller standard deviation of 0.0032, suggesting robust and stable predictive capability. The Neural Network (NN) model exhibited the highest mean R-squared value of 0.9975 with a standard deviation of 0.0031, implying very stable and accurate predictions across the cross-validation folds. These results suggest that while all three models perform well, the Neural Network model has a slight edge in terms of mean R-squared, closely followed by the Random Forest model. Both models also show low standard deviations, indicating more reliable performance than the Decision Tree model (Fig. 9).

Table 3. Cross-validation values for regressors.

Model	Cross-validation (R-squared)	Mean (R-squared)	Std deviation (+/−)
DT	[0.99956134 0.99282063 0.94552183 0.99245199 0.99972742]	0.986	0.0205
RF	[0.99953828 0.99827342 0.99215108 0.99933486 0.99290444]	0.9964	0.0032
NN	[0.99162096 0.99761484 0.99870988 0.99996312 0.99933013]	0.9975	0.0031

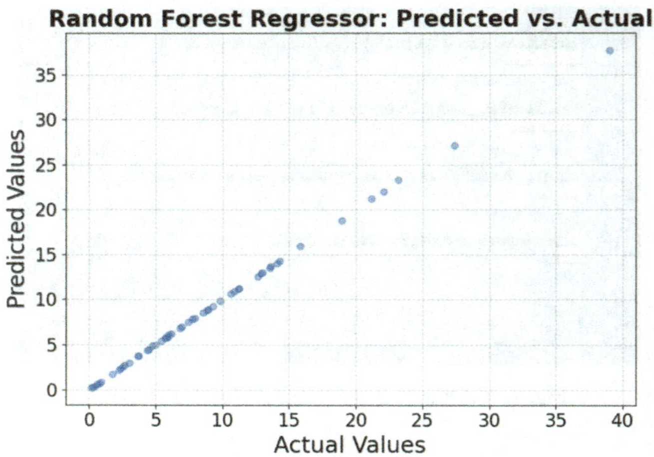

Fig. 9. Predicted vs Actual Values for RF model.

6.1.2 Test Cases on New Instances Using RF Regressor

See Figs. 10 and 11.

```
Enter values to predict Hospital Beds (Enter 'exit' to stop):
Location:  AUS
Year:  2030
Hospital Stay:  3
MRI Units:  3
CT Scanners:  10
Predicted Hospital Beds: 3.03
Location:  TUR
Year:  2026
Hospital Stay:  5
MRI Units:  5
CT Scanners:  15
Predicted Hospital Beds: 4.99
Location:  JPN
Year:  2045
Hospital Stay:  30
MRI Units:  10
CT Scanners:  50
Predicted Hospital Beds: 10.02
```

Fig. 10. Sample predictions after deployment.

6.2 Classification

6.2.1 Performance Evaluations

The performance evaluation of Decision Tree, Random Forest, and Neural Network was conducted based on their respective accuracies as shown in Fig. 12. Random Forest exhibited the highest accuracy of 93.66%, followed closely by Neural Network. The Decision Tree model achieved a slightly lower accuracy.

```
Enter values to predict Hospital Beds (Enter 'exit' to stop):
Location:  USA
Year:  2025
Hospital Stay:  3
MRI Units:  5
CT Scanners:  2
Predicted Hospital Beds: 4.99
Location:  BEL
Year:  2034
Hospital Stay:  10
MRI Units:  4
CT Scanners:  9
Predicted Hospital Beds: 3.98
Location:  EGY
Year:  2040
Hospital Stay:  15
MRI Units:  3
CT Scanners:  7
Predicted Hospital Beds: 3.01
```

Fig. 11. More sample predictions after deployment.

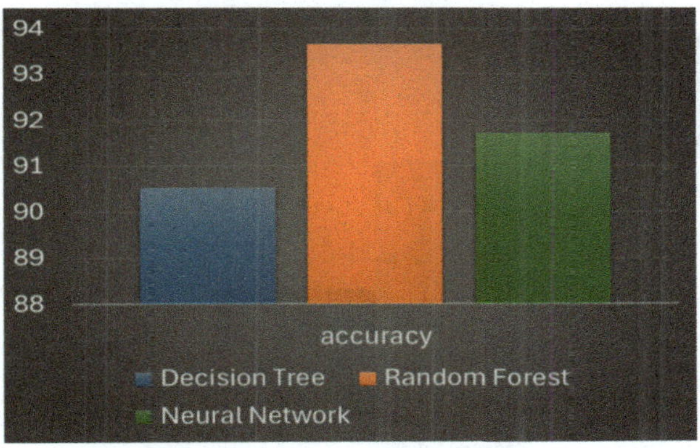

Fig. 12. Accuracies of various classifier models.

The performance of Decision Tree, Random Forest, and Neural Network was assessed based on precision, recall and F1-score metrics for prediction as seen in Fig. 13. Random Forest exhibited the highest precision across all ranges and demonstrated superior recall and F1-scores compared to the other models in most categories. Neural Network also showed competitive performance, achieving precision, recall, and F1-scores ranging from 0.9 to 0.94 across different value ranges. The Decision Tree model generally exhibited lower performance metrics compared to Random Forest and Neural Network.

In Table 4, three machine learning models are evaluated using the R-squared metric through cross-validation. The Decision Tree (DT) model achieved an average R-squared of 0.9163 with a standard deviation of 0.0009, indicating consistent performance across folds. The Random Forest (RF) model demonstrated slightly higher performance, with an average R-squared of 0.9452 and a smaller standard deviation of 0.0007. The Neural Network (NN) model yielded an average R-squared of 0.9178 with a standard deviation

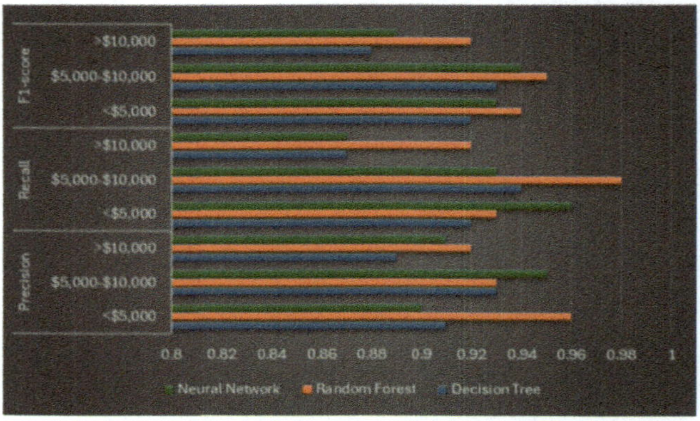

Fig. 13. Accuracy metrics of various classifier models.

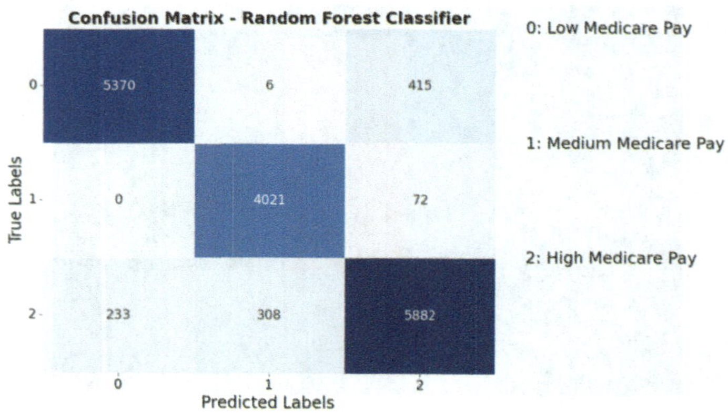

Fig. 14. Confusion Matrix for RF model.

Table 4. Cross-validation values for classifiers.

Model	Cross-validation (accuracy)	Mean (accuracy)	Std deviation (+/–)
DT	[0.91599422 0.91814741 0.91554516 0.91630743 0.91572916]	0.9163	0.0009
RF	[0.94524905 0.94603617 0.94438019 0.94577332 0.94432762]	0.9452	0.0007
NN	[0.90740253 0.9210026 0.90716584 0.92541679 0.92819538]	0.9178	0.0089

of 0.0089, showing slightly more variability in performance across folds compared to the DT and RF models. The RF model exhibited the highest mean R-squared value (Fig. 14).

6.2.2 Test Cases on New Instances Using RF Classifier

See Figs. 15 and 16.

```
Provider Id:  10020
Provider State:  VA
Total Discharges:  45
Average Covered Charges:  150
Average Total Payments:  2500
Predicted Medicare payment category: 1-5k
Provider Id:  10040
Provider State:  WA
Total Discharges:  100
Average Covered Charges:  20000
Average Total Payments:  5000
Predicted Medicare payment category: 1-5k
Provider Id:  10011
Provider State:  WA
Total Discharges:  40
Average Covered Charges:  3000
Average Total Payments:  13000
Predicted Medicare payment category: 5-10k
```

Fig. 15. Sample predictions after deployment.

```
Provider Id:  330067
Provider State:  NY
Total Discharges:  300
Average Covered Charges:  400
Average Total Payments:  8500
Predicted Medicare payment category: 5-10k
Provider Id:  50390
Provider State:  CA
Total Discharges:  10
Average Covered Charges:  200
Average Total Payments:  12000
Predicted Medicare payment category: 5-10k
Provider Id:  10078
Provider State:  AL
Total Discharges:  200
Average Covered Charges:  100
Average Total Payments:  15000
Predicted Medicare payment category: 10k+
```

Fig. 16. More sample predictions after deployment.

7 Comparative Analysis

Table 5 compares the proposed framework with the related work. The proposed framework generally outperforms the related work in terms of accuracy and regression metrics. It achieves extremely high accuracy for classification and very low error and high correlation for regression, which suggest potentially superior predictive capabilities. The

Table 5. Comparison with related work.

Articles	Problem addressed	Methodology	Data Used	Results
[9]	Prediction of length of stay category and number of days in hospitals	Logistic Regression, Random Forest, Gradient Boost, CatBoost, XGBoost	[34]	MAE for regressor: 2.783 and classifier accuracy of 83%
[26]	Predicting the length of hospital stay for patients.	Multilayer Perceptron Regressor, Decision Trees, Random Forests and Support Vector Regression.	[11]	R^2 score of 0.94 but cannot fully account for variability and complexity in data.
[27]	Forecasting the number of hospital beds.	Multiple Linear regression model	[29]	R^2 score of 0.825 but not robust to unseen data
[30]	Prediction of medical cost.	Multiple Linear Regressor, Decision-Tree, Random Forest, Gradient Boosting Regressor, Deep Neural Networks.	[31]	R^2 score of 0.93 but not robust to new and unseen data.
[32]	Patient Medical Cost Prediction and Trend Analysis	CatBoost regression	[33]	R^2 score of 0.87 but cannot handle drift in data distribution
Proposed Framework	Prediction of hospital costs and bed allocations	Comprehensive framework for hospital resource planning that integrates classifier and regressor models for Random Forest, Neural Network and Decision Tree.	[10, 11]	93.66% accuracy for classifier model and regressor had MAE of 0.074 and R^2 of 0.9991

proposed framework uses a combination of Random Forest, Neural Network, and Decision Tree models, which are also common in related work. However, the integration and possibly the hyperparameter tuning of these models might be more effective in the proposed framework, leading to better results.

Overall, the proposed framework appears to advance the state-of-the-art in hospital resource planning by achieving superior predictive accuracy and regression performance compared to existing methodologies.

8 Conclusions

Hospital resource planning has traditionally relied on manual assessments and historical data analysis. In recent times, machine learning models have been introduced to enhance predictive accuracy and efficiency. However, these models often struggle with class imbalance and do not fully utilize integrated classification and regression approaches. This research explores the prediction of hospital resource requirements using an integrated Random Forest model that combines both classification and regression tasks.

Experimental results were benchmarked against related methods and the integrated Random Forest models showed superior performance (see Tables 3, 4 and Figs. 7, 8, 12, and 13). The results obtained demonstrate that the integrated Random Forest model is more accurate for predicting both cost categories and bed requirements. Specifically, the classifier model achieved higher accuracy, while the regressor model showed a lower mean absolute error compared to existing methods. It is worth noting that the stepwise SMOTE algorithm contributed immensely to these good performances.

The insights gained from this research can help hospital administrators make more informed decisions about resource allocation, ultimately improving operational efficiency. Specifically, it will minimize the impact of delayed discharges on hospital bed availability and improve the planning on patient length of stay (LOS). Future work can expand the features used to enhance the accuracy of predictions and possibly consider hospital data from other countries to validate the model's applicability in diverse healthcare systems.

Acknowledgements. The authors gratefully acknowledge the financial support and resources made available by Norfolk State University, USA.

References

1. Garg, L., McClean, S., Barton, M.: Intelligent patient management and resource planning for complex, heterogeneous, and stochastic healthcare systems. IEEE Trans. Syst. Man Cybernet. Pt A Syst. Hum. **42**(6), 1332–1345 (2012). https://doi.org/10.1109/TSMCA.2012.2210211
2. Gautam, A.K., Tirumala Vasu, G., Mamatha, G.N.: Optimal allocation of resources and hospital capacity planning for critical diseases using AI and data mining. In: 2023 IEEE International Conference on ICT in Business Industry & Government (ICTBIG), pp. 1–6. IEEE, Indore (2023)
3. Zhang, T., et al.: Data-driven modeling and analysis for COVID-19 pandemic hospital beds planning. IEEE Trans. Autom. Sci. Eng. **20**, 1551 (2023). https://doi.org/10.1109/TASE.2022.3224171
4. Mull, H.J., et al.: Factors associated with Emergency Department visits and hospital admissions after invasive outpatient procedures in the Veterans Health Administration. JAMA Surg. **153**, 774 (2018). https://doi.org/10.1001/JAMASURG.2018.0874
5. Fillmore, N., et al.: An application to support COVID-19 occupational health and patient tracking at a Veterans Affairs Medical Center. J. Am. Med. Inform. Assoc. **27**(11), 1716 (2020). https://doi.org/10.1093/JAMIA/OCAA162

6. Wang, D., Gao, Q.: Efficiency assessment and resource allocation for hospitals by data envelopment analysis. In: 2017 3rd International Conference on Information Management (ICIM), pp. 289–293. IEEE, Chengdu (2017)
7. Abid, A., Tlili, M., Maaroufi, F., Korbaa, O.: A management analysis tool to support healthcare resource planning in public hospitals during the COVID-19 pandemic: a case study. In: 2023 International Conference on Innovations in Intelligent Systems and Applications (INISTA), pp. 1–6. IEEE, Hammamet (2023)
8. Viswanadham, N., Balaji, K.: Resource allocation for healthcare organizations. In: 2011 IEEE International Conference on Automation Science and Engineering, pp. 543–548. IEEE, Trieste (2011)
9. Ramurs, V.H., Shreyas, C., Bangalore, R.A.: Prediction of length of stay in hospitals using ensemble machine learning models. In: 2023 2nd International Conference on Futuristic Technologies (INCOFT), pp. 1–6. IEEE, Belagavi (2023)
10. Hospital Charges in America: https://www.kaggle.com/datasets/dhirajnirne/hospital-charges-in-america. Accessed 24 Apr 2024
11. Healthcare Investments and Length of Hospital Stay: https://www.kaggle.com/datasets/babyoda/healthcare-investments-and-length-of-hospital-stay. Accessed 24 Apr 2024
12. Elreedy, D., Atiya, A.F., Kamalov, F.: A theoretical distribution analysis of synthetic minority oversampling technique (SMOTE) for imbalanced learning. Mach. Learn. **113**, 4903 (2023). https://doi.org/10.1007/s10994-022-06296-4
13. Thakkar, A., Lohiya, R.: Attack classification using feature selection techniques: a comparative study. J. Ambient. Intell. Humaniz. Comput. **12**(4) (2021). https://doi.org/10.1007/S12652-020-02167-9
14. Rahul, G.A., Bansal, A., Roy, K.: Solar energy prediction using decision tree regressor. In: 2021 5th International Conference on Intelligent Computing and Control Systems (ICICCS), pp. 489–495. IEEE, Madurai (2021). https://doi.org/10.1109/ICICCS51141.2021.9432322
15. Mohandoss, D.P., Shi, Y., Suo, K.: Outlier prediction using random forest classifier. In: 2021 IEEE 11th Annual Computing and Communication Workshop and Conference (CCWC), pp. 27–33. IEEE, Las Vegas (2021)
16. Genuer, R., Poggi, J.-M., Maindonald, J.H.: Random Forests with R, p. 89 (2021). https://doi.org/10.1111/INSR.12454
17. Ghaffari, A., Savaria, Y.: CNN2GATE: toward designing a general framework for implementation of convolutional neural networks on FPGA. arXiv preprint arXiv:2004.04641 (2020)
18. Madhiarasan, M., Louzazni, M.: Analysis of artificial neural network: architecture, types, and forecasting applications. J. Electr. Comput. Eng. **2022**, 1 (2022). https://doi.org/10.1155/2022/5416722
19. Charbuty, B., Abdulazeez, A.M.: Classification based on decision tree algorithm for machine learning. JASTT. **2**, 20 (2021). https://doi.org/10.38094/JASTT20165
20. Ibomoiye, D., Sun, Y.: A survey of ensemble learning: concepts, algorithms, applications, and prospects. IEEE Access. **10**, 99129 (2022). https://doi.org/10.1109/ACCESS.2022.3207287
21. Alam, M.S., Samad, M.D., Vidyaratne, L.: Survey on deep neural networks in speech and vision systems. Neurocomputing. **417**, 302 (2020). https://doi.org/10.1016/J.NEUCOM.2020.07.053
22. Daniya, T., Geetha, M., Kumar, S.: Classification and regression trees with Gini Index. Adv. Math. Sci. J. **9**(10), 8237–8247 (2020). https://doi.org/10.37418/AMSJ.9.10.53
23. Naskath, J., Sivakamasundari, G., Siddiqua Begum, A.A.: A study on different deep learning algorithms used in deep neural nets: MLP SOM and DBN. Wirel. Pers. Commun. **128**(4), 2913–2936 (2022). https://doi.org/10.1007/s11277-022-10079-4

24. Vujovic, Ž.Đ.: Classification model evaluation metrics. Int. J. Adv. Comput. Sci. Appl. **12**(6) (2021). https://doi.org/10.14569/IJACSA.2021.0120670
25. Plevris, V., Solorzano, G., Bakas, N., Ben Seghier, M.B.: Investigation of performance metrics in regression analysis and machine learning-based prediction models. In: 8th European Congress on Computational Methods in Applied Sciences and Engineering. ECCOMAS, pp. 1–5 (2022). https://doi.org/10.23967/eccomas.2022.155
26. Gadri, D.: IA pour la santé: cas d'étude 'Prédiction de la durée de séjour hospitalier'. Univ-tebessa dz (2023)
27. Nagpal, S., Athavale, V.A., Saini, A.K., Sharma, R.: Indian health care system is ready to fight against COVID-19 a machine learning tool for forecast the number of beds. In: 2020 Sixth International Conference on Parallel, Distributed and Grid Computing (PDGC), pp. 61–65. IEEE, Waknaghat (2020)
28. Liu, Y., Mu, Y.: Daily activity feature selection in smart homes based on Pearson correlation coefficient. Neural Process. Lett. **51**, 1771 (2020). https://doi.org/10.1007/S11063-019-10185-8
29. Hospitals and beds in India (Statewise). https://www.kaggle.com/datasets/dheerajmpai/hospitals-and-beds-in-india. Accessed 4 Jul 2024
30. Eswari, G.S., Gayathri Devi, I., Rupesh Kumar, C.K., Ramesh, U.V., Lakshmanarao, A.: An efficient medical cost prediction system with machine learning and deep learning model. In: 2023 First International Conference on Advances in Electrical, Electronics and Computational Intelligence (ICAEECI), pp. 1–5. IEEE, Tiruchengode (2023)
31. Medical Cost Personal Datasets (2024). https://www.kaggle.com/datasets/mirichoi0218/insurance. Accessed 4 Jul 2024
32. Rao, A.R., Jain, R., Singh, M., Garg, R.: Machine learning models for patient medical cost prediction and trend analysis using open healthcare data. In: 2023 IEEE 3rd International Conference on Electronic Communications, Internet of Things and Big Data (ICEIB), pp. 292–296. IEEE, Taichung (2023)
33. Statewide Planning and Research Cooperative System (2024). https://www.health.ny.gov/statistics/sparcs/. Accessed 5 Oct 2022
34. Hospital Inpatient Discharges (SPARCS De-Identified). State of New York (2015). https://health.data.ny.gov/Health/Hospital-Inpatient-Discharges-SPARCS-De-Identified/82xm-y6g8. Accessed 4 Jul 2024

Artificial Neural Networks in Evaluation of Clinical Parameters of Actinic Dermatitis

P. E. Konnov[1(✉)], V. V. Eskov[2], T. V. Voronyuk[2], and I. S. Samoilenko[2]

[1] FSBEE HE "Samara State Medical University" of the Ministry of Health of Russia, 89, Chapaevskaya Street, Samara 443099, Russia
konnoff@yandex.ru

[2] BU VO KHMAO-Yugra "Surgut State University", 1, Lenin Avenue, Surgut 628400, Russia

Abstract. The study of three groups of patients with chronic actinic dermatitis showed the presence of type 1 uncertainty. In this case the groups are clinically different, but the samples are statistically the same. To solve this problem used an artificial neural network was used in two special modes: chaos of initial weights of all diagnostic parameters and multiple reconfigurations of the neural network. This results clear division of samples, but also a ranking of their features. This made it possible to find the main diagnostic features based on the ranking of average weights $<w_i>$ all diagnostic features. As a result, we found the most important features - order parameters (the problem of system synthesis was solved).

Keywords: chronic actinic dermatitis · stochastics · type 1 uncertainty

1 Introduction

Quite often in dermatology a situation arises when a clinical doctor can diagnose characteristics of disease (e.g. staging), but methods of mathematical statistics cannot reveal differences between samples of measured diagnostic parameters. Methods of modern deterministic and stochastic science may be limited in this case. Then it is necessary to create new approaches to describe biosystems. Such approaches should go beyond deterministic and stochastic science according to W. Weaver [1]. Now we present this idea in the form of theory of chaos-self-organization [2–7]. In this theory the above situation is designated as uncertainty of the first type [8–13]. Statistics do not show differences, but the clinical picture reveals them. We prove that new theory of chaos-self-organization and its methods can solve this problem, i.e. reveal the uncertainty of the first type [7–13].

To solve this problem, theory of chaos-self-organization proposes to use artificial neural networks in two special modes. We now offer a specific example based on clinical signs from the field of dermatology. We emphasize that this problem cannot be solved in principle by modern deterministic and stochastic science methods [10–15], [16].

R. Silhavy and P. Silhavy (Eds.): CoMeSySo 2024, LNNS 1489, pp. 328–334, 2025.
https://doi.org/10.1007/978-3-031-96798-6_26

2 Object and Methods

For a long time, 5 clinical signs were recorded in three groups of patients with chronic actinic dermatitis. These options were: x_1 – presence of erythema (E); x_2 – registration of papules (infiltration) (I); x_3 –presence of excoriation (E_x); x_4 – lichenification (L); x_5 - classification according to eczema area and severity index (EASI). According to the latter sign, 3 groups were examined (18 people in each group) from the standpoint of the clinic. In other words, doctors clearly separated these three groups of patients in the clinic.

After proving the presence of non-parametric statistics, descriptive statistics (calculation of medians) and pairwise comparison of samples of all these 5 parameters (for all 3 groups) were performed. With the Wilcoxon signed-rank test $p_{ij} \geq 0.05$, we considered the presence of a statistical match (such a pair of samples may have a common general population). Samples $x_i(t)$ then they were not separated (statistically).

As a result, for all pairs we used an artificial neural network in the mode of chaotic assignment of initial weights w_{i0} of all diagnostic features x_i and multiple repeated adjustments of the artificial neural networks. As a result, we received samples of finite weights w_i and found their average values $<w_i>$. This processing of weight samples w_i allowed us to rank all diagnostic features in descending order of magnitude. In fact, we find the order parameters, i.e. main diagnostic features. It is absolutely impossible to find them by the methods of modern science because of the uncertainty of the first type.

3 Results

First of all, we note that almost all samples showed the absence of parametric distributions x_i. As a result, we used nonparametric statistics. The results of these calculations are presented in Table 1 (in the form of medians and percentiles).

Next, we perform a pairwise comparison of all samples of these clinical diagnostic features. When comparing the samples of these traits in pairs, we obtained the criteria p_{ij}, which (when $p_{ij} \geq 0.05$) could show statistical agreement between the samples. Obviously, at $p_{ij} \geq 0.05$ these two samples may have a common population. In this case, we can talk about the statistical coincidence of these two compared samples.

As a result, Table 2 was obtained, in which there were only 15 pairs of comparisons for three groups (with different clinical parameters of chronic actinic dermatitis), i.e. for couples 1-2, 2-3 and 1-3. Some pairs in Table 2 showed the presence of statistical matches between the samples. For example, this first clinical sign shows us (see Table 2) the uncertainty of the first type (no differences). In this case, all three pairs of diagnostic parameters showed statistical agreement for all three pairs of difference criteria $p_{ij} \geq 0.05$.

Since such statistical coincidences indicate uncertainty of the first type, we used the artificial neural networks in two special modes. These modes result from the analysis of electroencephalograms of any person. Any electroencephalogram sample shows the Eskov-Zinchenko effect in form of inability to arbitrarily repeat this sample (statistically) [8–13].

It turned out that 15 electroencephalograms samples of one person (at rest) can coincide with each other with a frequency (probability) of no more than $p_{ij} \leq 0.3$.

Table 1. Descriptive statistics of clinical parameters of chronic actinic dermatitis in men (n = 18).

	Min	Max	5-th percentile	Median	95-th percentile
Erythema					
Group 1	1	3	1	3	3
Group 2	2	3	2	3	3
Group 3	1	3	1	3	3
Papules					
Group 1	1	3	1	2	3
Group 2	2	3	2	3	3
Group 3	1	3	1	3	3
Excoriation					
Group 1	0	3	0	1	3
Group 2	1	3	1	2	3
Group 3	2	3	2	3	3
Lichenification					
Group 1	0	3	0	1	3
Group 2	1	3	1	2	3
Group 3	2	3	2	3	3
Eczema area and severity index					
Group 1	4.5	9.9	4.5	6.3	9.9
Group 2	22.1	29.8	22.1	26.4	29.8
Group 3	5.3	39.8	5.3	36.2	39.8

Table 2. Pairwise comparison of five clinical parameters of chronic actinic dermatitis in men (nonparametric Wilcoxon signed-rank test ($p \geq 0.05$)).

	X1 (Erythema)	X2 (Papules)	X3 (Excoriation)	X4 (Lichenification)	X5 (Eczema area and severity index)
Group 1 - Group 2	**0.32**	0.023	0.015	0.008	0.000
Group 1 - Group 3	**0.48**	0.047	0.000	0.000	0.000
Group 2 - Group 3	**0.65**	**0.65**	0.011	0.005	0.003

Usually, more than 70% of pairs of neighboring electroencephalograms samples (or any other parameters of the functions of the human body) do not statistically coincide during their continuous recording. This means the absence of electroencephalograms statistical stability for any person and proves the uniqueness of the samples. Statistics does not work with unique samples, since there is no future forecast for such biosystems [13–15], 16, 17]. All deterministic and stochastic science works only with repetitive processes. The deterministic and stochastic science does not study unique systems [13–15].

As a result, the Eskov-Zinchenko effect represent the uniqueness of statistical models of the tremorograms, tappinggrams as parameters of human motor functions. As a result the effect was proved [10–15] for any parameters x_I (functions of the human body) [10–15].

Thus, statistical chaos (the uncertainty of the future due to the lack of repetition of electroencephalograms samples) and the modeling of reverberations of brain biopotentials (in the form of electroencephalograms) are the basic properties of the brain. We introduced these two special properties into the artificial neural networks work [6–11]. Until now, this has not been used in the operation of the artificial neural networks (in the mode of many repeated settings of the network).

We emphasize that in a living person, the electroencephalograms never shows zero activity (in biological brain, his $x(t) \neq$ const); otherwise it will be the brain of a dead person (with $dx/dt = 0$). Therefore, biopotentials of brain generate continuous reverberations and these continuous reverberations ($dx/dt \neq 0$) were introduced into the work of the artificial neural networks [2–7].

We asked randomly (from the interval $w_{i0} \in (0, 1)$) initial weights w_{i0} all diagnostic features x_i and forced the artificial neural networks to repeatedly tune. It was equivalent to the reverberations of brain biopotentials, $dx/dt \neq 0$, and at each stage of tuning we set new values w_{i0}.

Such modes of operation in artificial neural networks allowed to separate all samples for all three pairs of comparisons (1-2, 2-3, and 1-3) and to obtain final values of neuron network weights. The number of such iterations is $n \geq 100$ during the operation of artificial *neural* networks, i.e. the values w_i formed their own samples.

As a result, we statistically processed the samples of all w_i (for all x_i) and received the final Table 3. In this Table 3 we have presented the average weights $<w_i>$ all x_i for all comparison pairs of these three groups. In Table 3, the largest values ($<w_5> > 0.2$) takes the fifth feature, i.e. it is the order parameter for all comparison pairs (for all features x_i). Let us emphasize that in all our studies with artificial neural networks it is always possible to find the main diagnostic signs (after ranking $<w_i>$).

It is obvious that the artificial neural networks not only helped to get rid of the type 1 uncertainty, but also made it possible to find the main diagnostic feature (according to the average weights $<in_i>$). It turned out to be x_5 - eczema area and severity index, i.e. this feature is an order parameter in diagnosing the degree of chronic actinic dermatitis disease.

Numerous studies with different samples of parameters of the human body in biomedicine have allowed us to identify type 1 uncertainty. In this case, statistics may not show differences between samples, but the clinic gives clear differences between

Table 3. Mean value of weights (arb.u.) obtained by training artificial neural network solving binary classification task (50 iterations).

	Erythema	Papules	Excoriation	Lichenification	Eczema area and severity index
Group 1/Group 2	0.182	0.207	0.207	0.201	0.200
Group 1/Group 3	0.199	0.189	0.196	0.191	0.222
Group 2/Group 3	0.194	0.203	0.207	0.192	0.204

groups of patients. A paradox arises: the clinic gives differences, but the statistics do not.

In our case with chronic actinic dermatitis, the groups differ in the severity of the diseases and their duration. In the case of chronic actinic dermatitis, we identified three such groups that showed differences not for all pairs of comparisons of dynamic features x_1. This proves the type 1 uncertainty, i.e. statistics does not work here [6–15], 16, 17]. We emphasize that, in addition to chronic actinic dermatitis, we obtained similar results in cardiology, industrial ecology, rehabilitative medicine, etc. [6–15].

To eliminate this uncertainty, we used the artificial neural networks in two special modes: chaos of initial weights w_{i0} of all diagnostic parameters x_i and repeated reverberations (repeated artificial neural networks settings). As a result, we obtain samples of finite weights w_i all these x_i after numerous artificial neural networks calculations.

Such a procedure follows from the Eskov-Zinchenko effect for any electroencephalograms samples of any person. In a calm state, a person cannot generate statistically matching electroencephalograms samples. At the same time, neural networks cannot be at rest, $dx/dt \neq 0$. Brain biopotentials continuously generate mismatched electroencephalograms samples (statistical chaos) [10–15].

We include these two properties of the neural networks of the brain in the work of the artificial neural networks and get a new quality of work of the artificial neural networks. In this case, the artificial neural networks not only classify the samples which eliminate type 1 uncertainty, but also detect the order parameters, i.e. the main diagnostic sign (in the form of x_5- eczema area and severity index).

After numerous settings of the neural network (in the chaos mode of the initial weights w_{i0}) we have obtained samples of finite weights w_i (after many iterations of artificial neural networks tuning). As a result, we statistically processed these samples and ranked their average values $<w_i>$. From here we find the average values $<w_i>$ (according to their samples).

We emphasize that it is absolutely impossible to find the order parameters (main diagnostic features) in the general case (with type 1 uncertainty) within the framework of traditional science. Statistics in this case does not work. The correctness of this statement is proved not only by uncertainty of the first type, but also by the uniqueness (statistical non-repeatability) of any sample of any parameter of the human body.

As a result, we come to the need to use theory of chaos-self-organization methods in medical informatics and biocybernetics. The neural network in two new modes solves the problem of system synthesis and at the same time it simulates work of biological brain. Recall that the artificial neural networks was created on the basis of brain neural networks modeling principle, and we have now implemented this principle.

4 Conclusion

The proof of the predominance of non-parametric distributions in the chronic actinic dermatitis parameters was performed by us on the example of five diagnostic features. It is traditionally believed that parametric distributions predominate in biomedicine, but this is not the case and we demonstrate it using the examples of chronic actinic dermatitis.

Moreover, it turns out that uncertainties of the first type appear quite often. In this case, the statistics may not show differences between the samples, but the groups differ clinically. In our case, we are talking about three groups of patients with chronic actinic dermatitis according to five clinical parameters.

In this case, artificial neural networks can be used in two special modes: chaos of initial weights ($w_{i0} \in (0.1)$) of diagnostic features and multiple reconfigurations of the artificial neural network. This simulates the work of the neural network of biological brain (chaos and electroencephalograms reverberations). In other words, we brought the artificial neural networks closer to the work of real neural networks of the human brain in two main modes of artificial neural networks operation, and received a new quality of artificial neural networks. In fact, we simulated the work of real brain neural networks.

It turned out that after numerous iterations of the artificial neural networks, we can obtain samples of finite values of weights w_i of diagnostic parameters x_i. The resulting samples w_i can be processed statistically and obtain their average values $<w_i>$. As a result, we can rank these $<w_i>$ and choose the largest of them. These highest values $<w_i>$ are order parameters, i.e. the main diagnostic features and can be used in the clinic to differentiate chronic actinic dermatitis.

References

1. Weaver, W.: Science and complexity. Am. Sci. **36**(4), 536–544 (1948)
2. Gazya. G.V., Eskov, V.V., Filatov, M.A.: The state of the cardiovascular system under the action of industrial electromagnetic fields. Int. J. Biol. Biomed. Eng. **15**, 249–253 (2021). https://doi.org/10.46300/91011.2021.15.30
3. Khadartseva, K.A., Filatov, M.A., Melnikova, E.G.: The problem of homogenous sampling of cardiovascular system parameters among migrants in the Russian North. Hum. Ecol. **7**, 27–31 (2020)
4. Filatov, M.A., Ilyashenko, L.K., Kolosova, A.I., Makeeva, S.V.: Stochastic and chaotic analysis of students' attention parameters of different ecological zones. Hum. Ecol. **7**, 11–16 (2019)
5. Filatova, D.Y., Bashkatova, Y.V., Filatov, M.A., Ilyashenko, L.K.: Parameter evaluation of cardiovascular system in schoolchildren under the conditions of latitudinal displacement. Hum. Ecol. **4**, 30–35 (2018)
6. Eskov, V.M., Bazhenova, A.E., Vochmina, U.V., Filatov, M.A., Ilyashenko, L.K.: N.A. Bernstein hypothesis in the description of chaotic dynamics of involuntary movements of person. Russ. J. Biomech. **21**(1), 14–23 (2017)

7. Eskov, V.M., Filatova, O.E., Ivashenko, V.P.: Computer identification of compartmental neuron circuits. Meas. Tech. **37**(8), 967–971 (1994)
8. Eskov, V.M.: Cyclic respiratory neuron network with sub cycles. Neural Netw. World. **4**(4), 403–416 (1994)
9. Eskov, V.M., Filatova, O.E.: Problem of identity of functional states in neuronal networks. Biophysics. **48**(3), 497–505 (2003)
10. Eskov, V.M.: Hierarchical respiratory neuron networks. Model. Meas. Contr. **48**(2), 47–63 (1995)
11. Eskov, V.M., Eskov, V.V., Gavrilenko, T.V., Vochmina, Y.V.: Formalization of the effect of "repetition without repetition" discovered by N.A. Bernstein. Biophysics. **62**(1), 143–150 (2017)
12. Eskov, V.M.: Evolution of the emergent properties of three types of societies: the basic law of human development. Emerg. Complex. Self. Org. **16**(2), 107–115 (2014)
13. Gavrilenko, T.V., Es'kov, V.M., Khadartsev, A.A., Khimikova, O.I., Sokolova, A.A.: The new methods in gerontology for life expectancy prediction of the indigenous population of Yugra. Adv. Gerontol. **27**(1), 30–36 (2014)
14. Eskov, V.M., Filatova, O.E.: Respiratory rhythm generation in rats: the importance of inhibition. Neurophysiology. **25**(6), 348–353 (1995)
15. Garaeva, G.R., Eskov, V.M., Eskov, V.V., Gudkov, A.B., Filatova, O.E., Khimikova, O.I.: Chaotic dynamics of cardiointervals in three age groups of indigenous people of Ugra. Hum. Ecol. **2015**(9), 50–55 (2015)

A Two-Stage Dimension Reduction and Clustering Framework for Financial Behavior and Socio-Demographic Profiling

Alejandro Aybar-Flores[1], Rocío Maehara[1(✉)], Luis Benites[2], and Miguel Muñoz[1]

[1] Departamento de Ingeniería, Universidad del Pacífico, Lima, Peru
{a.aybarf,ma.munozc}@alum.up.edu.pe, rp.maeharaa@up.edu.pe
[2] Departamento Académico de Administración, Universidad del Pacífico, Lima, Peru
le.benitess@up.edu.pe

Abstract. Financial inclusion (FI) is a critical component of global financial advancement. Despite technological progress, about 1.5 billion people in emerging economies lack access to formal financial systems. Understanding financial decision-making behaviors and preferences is essential for enhancing comprehension of financial inclusion. This study focuses on Peru, a country with relatively low FI levels. Using data from the 2019 National Survey of Demand for Financial Services and Financial Literacy (NSDFS), we implemented a two-stage clustering methodology with dimension reduction techniques and clustering algorithms to uncover social profiles within the Peruvian population. Our findings identified three clusters based on financial behaviors and socio-demographic characteristics. The optimal configuration, utilizing Isomap reduced to 2 dimensions combined with the K-means++ algorithm, achieved a mean aggregated score of 0.832, yielding the best results among the other dimension reduction and clustering techniques considered. The clusters highlighted disparities in financial access, emphasizing the need for targeted interventions. These insights can aid policymakers and regulators in developing strategies to enhance FI in Peru, underscoring the value of clustering techniques in addressing financial inclusion challenges.

Keywords: Financial Behaviour · Socio-demographic Characteristics · Unsupervised Learning · Clustering · Dimension Reduction · Peru

1 Introduction

In recent decades, financial inclusion (FI) has emerged as a component of economic advancement and a concern for policymakers and researchers globally [1]. It broadly means enabling the unbanked to utilize the formal financial system [2], offering opportunities to access financial services [3]. Numerous studies state

R. Silhavy and P. Silhavy (Eds.): CoMeSySo 2024, LNNS 1489, pp. 335–351, 2025.
https://doi.org/10.1007/978-3-031-96798-6_27

FI plays a vital role in economic welfare and reducing poverty [1], supporting monetary and financial stability, enhancing economic policy transmission, and facilitating economic functioning [1]. Despite technology advances, approximately 1.5 billion people in emerging economies lack access to formal savings or credit [4], making FI a critical societal priority.

Many of the global population still lacks financial services for saving, credit, loans, equity, and insurance [4]. Peru, the focus of this study, is among the less developed Latin American countries regarding FI, with high levels of limited inclusion [5]. In 2017, 43% of Peruvians over 15 had an account, compared to 54% in Latin America and the Caribbean. Peru recently improved, with overall account ownership rising from 35% in 2015 to 40% in 2018. Despite progress, Peru's informal economy accounted for about 45% of total economic activity in 2018, with nearly 75% of the working-age population in informal employment over the past decade [5].

Understanding FI requires a thorough analysis of individual financial behaviors, which includes the people involved, the methods employed, and the timing of financial activities [6]. Analyzing demand-side data is essential to understand FI, focusing on financial decision-making behaviors and socio-demographic factors [6]. A growing body of literature highlights the association between demand-side factors and FI across regions worldwide, emphasizing socioeconomic and demographic characteristics and economic behaviors and preferences [7]. Therefore, a more nuanced analysis of data based on identifiable social groups with distinct behaviors and preferences could yield significant benefits. This approach would enable researchers to formulate targeted recommendations that can guide policymakers and inform strategies for financial regulators effectively [6]. In this sense, unsupervised techniques such as clustering algorithms emerge as attractive alternatives for categorizing individuals into homogeneous groups [8], leveraging their financial behaviors and socio-demographic traits, among other factors.

Nonetheless, traditional clustering methods often need to improve with high-dimensional data due to issues related to dimensionality and their design limitations for such spaces. Additionally, these methods need help with datasets containing numerical and categorical data [9]. Researchers commonly address these challenges by incorporating a dimensionality reduction step into the clustering process. This approach not only resolves the issues posed by high-dimensional data but also showcases the effectiveness of artificial intelligence in data processing and optimization [9]. Consequently, this method effectively handles complex, multifaceted data and underscores the critical role of AI, particularly in an era where data is a vital economic asset within the financial industry.

Despite the growing importance of these methodologies, no prior study has explored the feasibility of applying dimensionality reduction techniques alongside unsupervised clustering algorithms in Peru to identify distinct population segments based on their financial behaviors and socio-demographic characteristics. Therefore, the primary objective of this study is to develop and implement a two-stage approach based on dimension reduction and clustering to uncover specific social patterns or profiles within the Peruvian population: firstly, dimension reduction techniques such as Isomap, PCA (Principal Components Analysis), and t-SNE were utilized to reduce the dimensionality of the dataset, transforming high-dimensional categorical data into a lower-dimensional space, this stage is followed by the application of clustering algorithms, including K-means, K-means++ and K-means – Hierarchical clustering (K + H), to the transformed data to identify distinct clusters, the clustering performance is evaluated by estimating an aggregated score using the Silhouette score, Calinski-Harabasz index, and Davies-Bouldin index, ensuring robust and meaningful clusters. This proposed two-stage clustering framework was performed to assess its effectiveness in profiling financial behaviors and sociodemographic characteristics within the Peruvian population. The secondary objective is to unveil the most explanatory financial and sociodemographic features within each identified cluster.

The remainder of the paper is structured as follows: Section 2 reviews relevant literature. Section 3 outlines materials and methods, including the dataset and methodology. Section 4 presents the results and evaluation of the two-stage technique. Section 5 presents the concluding remarks from the findings.

2 Literature Review

Various dimensionality reduction and clustering techniques perform better on specific datasets, with no definitive evidence favoring one over another. Thus, multiple approaches are used in academic and industry settings [9].

Clustering, an unsupervised learning technique, groups similar records within clusters, uncovering data patterns for applications like data analysis, pattern recognition, and machine learning [10]. The effectiveness of clustering and classification methods depends on the problem and dataset [10]. Clustering benefits sectors like computer science, biology, medicine, and business [11]. As such, this technique is a valuable tool that complements regression-style studies and serves diagnostic and data reduction purposes [12]. In finance, it aids credit scoring, stock market analysis, portfolio selection, and trade strategy development by grouping similar financial data for better decision-making [11].

From another perspective, enhancing clustering efficiency in identifying meaningful patterns can be achieved through dimension reduction techniques, simplifying data while retaining essential characteristics [13]. Dimensionality reduction techniques (DRTs) are extensively used in the financial sector to manage complex data effectively. They simplify data while retaining essential characteristics, making it more manageable for analysis and modeling [13].

Principal Component Analysis (PCA) is commonly used but limited to nonlinear variables. Nonlinear methods like Multidimensional Scaling (MDS), Locally Linear Embedding (LLE), Isomap, Kernel PCA, Self-Organizing Maps (SOM), and Auto-Associative Neural Networks (AANN) are suitable for data compression and various applications [13].

In that sense, clustering algorithms enhanced by dimensionality reduction detect money laundering, with PCA preprocessing network features for k-means and DBSCAN algorithms, validated by random forest [14]. t-Stochastic Neighbour Embedding (t-SNE) and Uniform Manifold Approximation and Projection

(UMAP) aid in visualizing and identifying transaction anomalies and differentiating bank stability, aiding investment decisions, with UMAP being considered superior to PCA and having advantages over t-SNE [14]. These techniques also streamline features in credit scoring models, improving accuracy and efficiency [15]. In stock market analysis, they identify key components driving market movements. Integrating these techniques with machine learning algorithms enhances performance, highlighting the importance of selecting suitable dimension reduction methods for various data types and applications [15].

3 Methodology

3.1 Study Sample

The data utilized in this study originates from the National Survey of Demand for Financial Services and Financial Literacy (NSDFS), conducted in Peru during July and August 2019 [16]. The survey provided valuable perspectives on the extent of financial inclusion, financial habits, and economic adaptability among the participants, presenting a thorough picture of Peru's financial environment [16]. The study sample comprises 1,205 Peruvian citizens over 18, spanning various socioeconomic strata and residing in urban and rural areas. The variables considered for the study are shown and described in Table 1.

It is noteworthy to indicate that several steps were undertaken to prepare the features for computational analysis. The variable reflecting respondents' ownership of multiple accounts was labeled as "account access". Transfers, bill payments, deposits, purchases, and remittances defined the "account usage" variable. The "education level" variable was simplified by merging various categories into a single higher-education group. For the "occupation" variable, categories like "not employed" and "uncompensated worker" were combined, while the "other" category included military jobs and incapacitated circumstances. The "income" variable ranges were combined into a single higher-income group. Finally, imputation for the "occupation" and "income" variables was carried out exclusively using Multiple Correspondence Analysis (MCA).

3.2 Experimental Design for Two Stage Clustering

The methodology proposed is an adaptation of the method presented in [9], which suggests an explainable three-stage deep clustering process for financial customer profiling. Due to the characteristics of our data, which are entirely categorical and of lower dimensionality than those proposed, we only reference Stages 2 and 3 of their proposed methodology, adapting it to our needs as our Stage 1 and Stage 2 framework as shown in Fig. 1. This adaptation involves utilizing t-SNE, Isomap, and PCA for dimensionality reduction and applying a suitable clustering algorithm to identify meaningful clusters within the data, depicted in Fig. 1.

The framework initiates with the proposed dimensionality reduction techniques to lower the dimensionality of high-dimensional data. Additionally, clustering is performed on the original high-dimensional data to assess its performance without reduction. The parameter combinations used in the experiments are provided in Sect. 4.1. Likewise, dimensions are typically set to two or three for practical reasons. First, conventional clustering algorithms perform effectively in

Table 1. Operational definition of the financial activity and socio-demographic factors used in the study.

Category	Variable	Definition	Categories
Financial Behaviour	Access to financial credits	Can obtain loans or credit from formal financial institutions.	1 = Yes 0 = No
Financial Behaviour	Access to bank accounts	Whether individuals have a savings account, a time deposit account, or savings in cooperatives.	1 = Yes 0 = No
Financial Behaviour	Usage of bank accounts	Conducted operations involving funds over the past year.	1 = Yes 0 = No
Socio-demographic	Zone	Area of residence of the respondent	1 = Rural 2 = Urban
Socio-demographic	Gender	Gender of the respondent	1 = Male 2 = Female
Socio-demographic	Usage of electronic devices	The individual made payments for services last year using smartphones or computers.	1 = Yes 0 = No
Socio-demographic	Socioeconomic level	Socioeconomic status of the respondent	1 = SEL E 2 = SEL D 3 = SEL C 4 = SEL B 5 = SEL A
Socio-demographic	Education level	The highest level of education that the respondent has.	1 = No education / Preschool / Elementary school / Incomplete high school 2 = Complete high school / Incomplete technical education 3 = Complete technical education 4 = Incomplete higher education 5 = Higher education/Postgraduate studies
Socio-demographic	Work / Occupation	Occupation of the respondent	1 = Unemployed 2 = Employer 3 = Self-employed 4 = Technical/professional employee 5 = Laborer 6 = Domestic worker 7 = Farmer/Countryman 8 = Dedicated to household
Socio-demographic	Income	Households monthly income of the respondent	1 = < 300 2 = 300 - 600 3 = 601 - 1200 4 = 1201 - 2400 5 = 2401 - 4800 6 = > 4801
Socio-demographic	Age	Age of the respondent	1 = 18 - 19 2 = 20 - 29 3 = 30 - 39 4 = 40 - 49 5 = 50 - 59 6 = 60 - 69 7 = 70 - 79

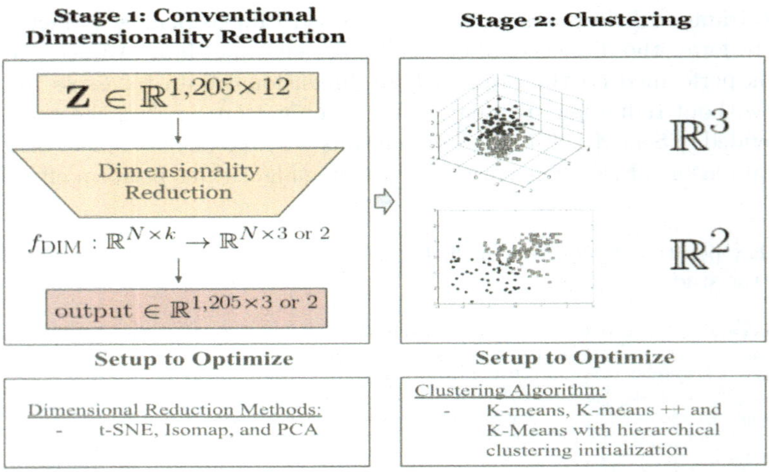

Fig. 1. Proposed Two-Stage Clustering Methodology adapted from [9].

low-dimensional spaces. Second, clusters become intuitively visualizable to non-experts—a crucial feature in the financial services industry [9]. The optimal configurations for Stages 1 and 2 are identified by maximizing the combined scaled score for each configuration set. The three clustering metrics—Silhouette score, Calinski-Harabasz score, and Davies-Bouldin score—are scaled and equally weighted. The Davies-Bouldin score is adjusted to match the directional orientation of the Silhouette and Calinski-Harabasz scores. All scores undergo MinMax scaling to ensure fair relative weighting in the aggregation. The optimization approach for the second and third stages is denoted by $C1$ and $C2$, representing the configuration sets optimized in Stages 1 and 2, respectively. The score set S is defined as $\{\text{Silhouette}, \text{Calinski-Harabasz}, 1 - \text{Davies-Bouldin}\}$.

$$\sum_{s \in S} \text{MinMax}(s \mid C_1, C_2). \tag{1}$$

3.3 Stage 1: Conventional Dimensionality Reduction

t-distributed Stochastic Neighbor Embedding

The t-SNE algorithm transforms data point affinities into probabilities, making it sensitive to local data structure and effective for identifying clusters [17]. The conditional probability $p_{j|i}$ represents the likelihood that point x_j is a neighbor of point x_i in the high-dimensional space and is calculated as follows:

$$p_{j|i} = \frac{\exp(-\|x_i - x_j\|^2 / 2\sigma_i^2)}{\sum_{k \neq i} \exp(-\|x_i - x_k\|^2 / 2\sigma_i^2)},$$

where σ_i^2 is the variance of the Gaussian centered at x_i, it then minimizes the Kullback-Leibler divergence between this distribution and a similar distribution in low-dimensional space, preserving the local structure [17].

Isometric Mapping
Isometric Mapping (Isomap) [18] is a non-linear dimensionality reduction algorithm combining PCA and multidimensional scaling (MDS) principles. Isomap computes the shortest geodesic distances and applies MDS to generate low-dimensional representations:

$$J(Y) = \min_{Y} \|\tau(D_G) - \tau(D_Y)\|_F,$$

where $\tau(D_Y)$ denotes the Euclidean distance inner product matrix, and $||_F$ is the Frobenius norm.

Principal Component Analysis
PCA transforms high-dimensional data into a new coordinate system, capturing data variation with fewer dimensions. It identifies orthogonal vectors of data points (principal components), retaining those with large variance [19].

3.4 Stage 2: Clustering

K-Means
K-means clustering partitions a dataset into K clusters, minimizing within-cluster variances [22]. It iteratively assigns data points to the nearest centroid and updates centroids until convergence:

$$J = \sum_{i=1}^{K} \sum_{\mathbf{x} \in S_i} ||\mathbf{x} - \mu_i||^2,$$

where S_i is the set of points in cluster i, μ_i is the centroid, and $\mathbf{x}$ denotes an individual observation in the dataset.

K-Means++
K-Means++ improves K-Means by selecting initial centers that are as far apart as possible [20]. It starts with a random center and selects subsequent centers based on distance from existing ones, speeding up convergence.

K-Means with Hierarchical Clustering
K-Means with hierarchical clustering enhances K-Means by using hierarchical clustering to determine better initial centers [21]. This method reduces data through partitions to create a structured dataset, leading to more effective initial centers for K-Means and faster convergence.

Performance of Metrics
Three metrics evaluate unsupervised clustering performance: Silhouette score, Calinski-Harabasz score, and Davies-Bouldin score [9,23]. The Silhouette score

assesses clustering quality by comparing cohesion within clusters to separation between clusters [24]:

$$s(i) = \frac{b(i) - a(i)}{\max(a(i), b(i))},$$

where $a(i)$ is the average distance within the cluster, and $b(i)$ is the average distance to the nearest cluster. The Calinski-Harabasz score evaluates the ratio of between-cluster to within-cluster dispersion [25]:

$$CH = \frac{\mathrm{Tr}(B_k)}{\mathrm{Tr}(W_k)} \cdot \frac{(N-k)}{(k-1)},$$

where $\mathrm{Tr}(B_k)$ and $\mathrm{Tr}(W_k)$ are the traces of the between-group and within-cluster dispersion matrices, N is the number of observations, and k is the number of clusters. The Davies-Bouldin Index measures the average similarity ratio of each cluster to its most similar cluster [26]:

$$DB = \frac{1}{k} \sum_{i=1}^{k} \max_{j \neq i} \left(\frac{d_i + d_j}{d_{ij}} \right),$$

where d_i and d_j are average distances within clusters, and d_{ij} is the distance between centroids. This study uses an aggregated metric score to evaluate clustering performance, as Eq. 1 outlined. Using the proposed dimensionality reduction techniques and clustering algorithms, the number of clusters K varies from 2 to 10. The best two-stage clustering approach is selected based on the outcomes of these experiments.

Feature Importance

Feature importance in tree-based models (such as Random Forest) is calculated by measuring the average decrease in impurity brought by each feature across all trees. The importance score $FI(i)$ for feature i is given by:

$$FI(i) = \frac{1}{T} \sum_{t=1}^{T} \sum_{j \in N_t} \Delta I(s(j)) \cdot \Bbbk_{[v(j)=i]},$$

where T is the total number of trees, $\Delta I(s(j))$ is the impurity decrease at node j, and $\Bbbk_{[v(j)=i]}$ indicates if feature i is used at node j. Feature importance was using the *one vs. rest* approach [27], in which the model is trained in a series of binary classifications, where each class is distinguished from all other classes.

4 Results and Evaluation

4.1 Experiments of the Two-Stage Clustering Process

Table 2 shows the variations in dimension reduction and clustering configurations used in the study. It details the total experiments, dimensions for reduction, t-SNE parameters (perplexity and early exaggeration), PCA components,

Table 2. Parameter Variations for the Two-Stage Clustering Framework

Clustering Technique	Reduction Techniques	Total of Experiments	Number of Dimensions	Parameter Variations	Values
K-means	t-SNE	486	2 - 3	Perplexity	10 - 20 - 30 - 40 - 50 - 60 - 70 - 80 - 90
K-means++				Early exaggeration	12 - 14 - 16 - 18 - 20 - 22 - 24 - 26 - 28
K-means + Hierarchical				Number of clusters	2 - 3 - 4 - 5 - 6 - 7 - 8 - 9 - 10
K-means	PCA	54	2 - 3	Number of components	2 - 3
K-means++					
K-means + Hierarchical				Number of clusters	2 - 3 - 4 - 5 - 6 - 7 - 8 - 9 - 10
K-means	Isomap	486	2 - 3	Number of neighbors	5 - 10 - 15
K-means++				Eigen solver	auto - arpack - dense
K-means + Hierarchical				Number of clusters	2 - 3 - 4 - 5 - 6 - 7 - 8 - 9 - 10
K-means	None	27	None		
K-means++				Number of clusters	2 - 3 - 4 - 5 - 6 - 7 - 8 - 9 - 10
K-means + Hierarchical					

Isomap parameters (neighbors and eigen solver), and cluster numbers for K-means, K-means++, and K-means with hierarchical clustering. It also specifies the experiments evaluating clustering performance without dimension reduction.

Considering the parameter combinations outlined in the previous table, 1053 clustering experiments were conducted across Stage 1 (dimensionality reduction) and Stage 2 (clustering). These included four test sets: Set 1 had 54 t-SNE configurations with varying perplexity, early exaggeration parameters, 2 or 3 dimensions, and 9 cluster counts (2 to 10), totaling 486 experiments (162 per clustering algorithm). Set 2 involved 6 PCA configurations with different component counts, two dimensions, and 9 cluster counts, resulting in 54 experiments (18 per algorithm). Set 3 included 27 Isomap configurations for neighbors and eigen solvers, two dimensions, and 9 cluster counts, leading to 486 experiments (162 per algorithm). Set 4 tested direct clustering without dimensionality reduction, with 9 cluster counts, totaling 27 experiments (9 per algorithm).

Furthermore, Fig. 2 shows visualizations of a subset of experiments using the specified dimensionality reduction and clustering techniques, illustrating the impact of different parameter combinations on cluster formation. Overall, each dimensionality reduction technique and clustering algorithm combination exhibits distinct strengths: Isomap preserves the overall data structure, PCA reveals more linear and less dispersed clusters, and t-SNE maintains local and global structures, providing valuable insights into the Peruvian population's characteristics.

From another perspective, to assess the relative performance of different parameter combinations, aggregated scores from all experiments were calculated as outlined in Sect. 3.4. The top 10 configurations are shown in Table 3.

The optimal two-stage clustering configuration involves Isomap, followed by K-means++ clustering, with the highest aggregated scaled score of 0.832. All top configurations also apply Isomap, indicating its robustness in preserving data structure. Configurations with K-means++ consistently outperform other clustering algorithms, highlighting K-means++'s effectiveness in centroid initialization. In addition, Fig. 3 shows the results of using Isomap for dimensionality

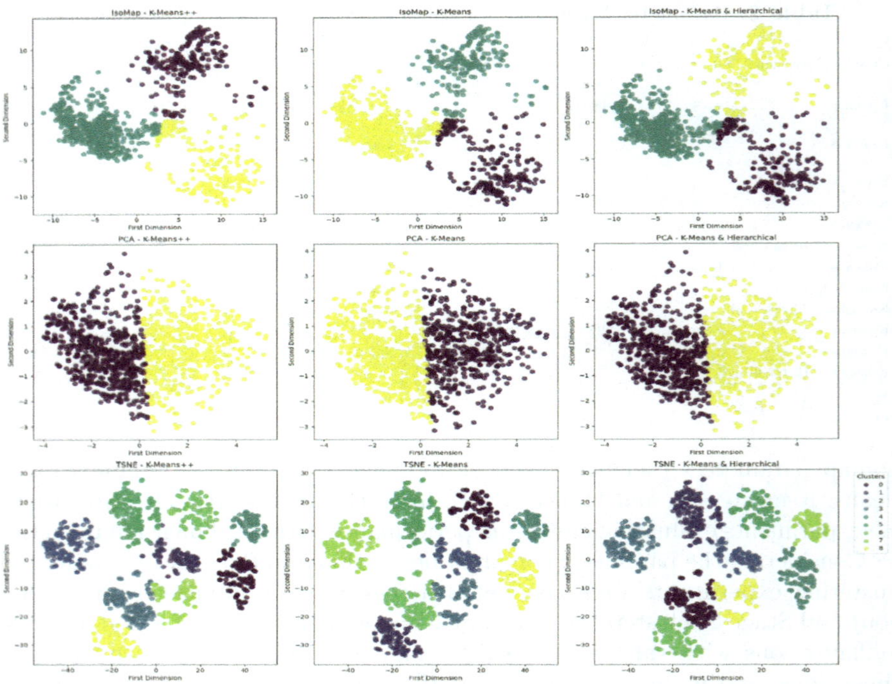

Fig. 2. Visualization of a subset of experiments of the two-stage framework.

Table 3. Top 10 aggregated scaled scores of the configurations of the two-stage clustering framework experiments

	Stage 1: Dimension Reduction					Stage 2: Clustering				
	Parameter values					Scaled scores				
Rank	Technique	Neighbors	Eigen Solver	Final Dimension	Clustering Method	Number of Clusters	Silhouette	Calinski-Harabasz	Davies-Bouldin	Agg. **Score**
1	Isomap	5	arpack	2	K-Means++	3	0.641	0.854	1.00	0.832
2	Isomap	5	auto	2	K-Means++	3	0.641	0.854	1.00	0.832
3	Isomap	5	dense	2	K-Means++	3	0.641	0.854	1.00	0.832
4	Isomap	5	arpack	2	K-Means	3	0.641	0.854	0.999	0.831
5	Isomap	5	auto	2	K-Means	3	0.641	0.854	0.999	0.831
6	Isomap	5	dense	2	K-Means	3	0.641	0.854	0.999	0.831
7	Isomap	5	arpack	2	K + H	3	0.641	0.853	0.996	0.830
8	Isomap	5	auto	2	K + H	3	0.641	0.853	0.996	0.830
9	Isomap	5	dense	2	K + H	3	0.641	0.853	0.996	0.830
10	Isomap	5	arpack	2	K + H	4	0.559	0.919	0.963	0.814

reduction and clustering into three clusters with K-means++, identified as the optimal number. Visual inspection reveals that the Isomap + K-means++ clusters are distinct and well-separated. This demonstrates that the proposed two-stage approach effectively differentiates clusters by reducing data dimensionality before applying the clustering algorithm.

Moreover, it is relevant to compare the overall performance of all combinations within the two-stage framework based on the mean aggregated scaled

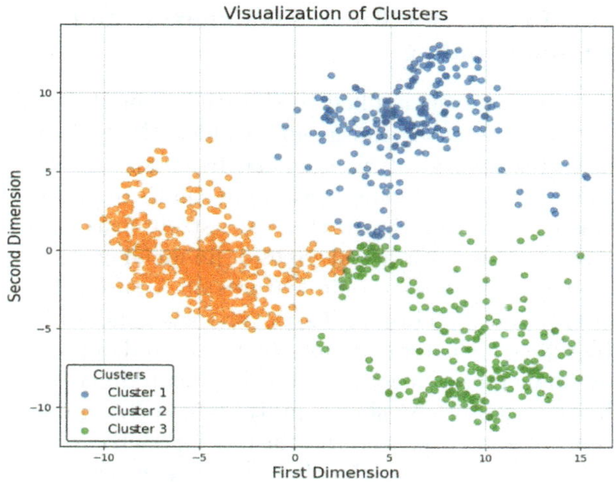

Fig. 3. Optimal distribution of clusters based on the two-stage framework.

Table 4. Mean aggregated scores for the dimension reduction and clustering techniques proposed in the study

Clustering techniques	**K-means Clustering**		**K-means++ Clustering**		**Hierarchical + K-means**		
Only Stage 2	0.1113		0.1158		0.1114		
	Dimension Input		**Dimension Input**		**Dimension Input**		
DRTs	2D	3D	2D	3D	2D	3D	**Mean (± STD)**
Isomap only	0.7188	0.5502	**0.7206**	0.5525	0.7147	0.5493	0.6257 ± 0.0885
PCA only	0.5227	0.3624	0.5238	0.3630	0.5195	0.3553	0.4255 ± 0.0901
t-SNE only	0.6855	0.4921	0.6861	0.4926	0.6772	0.4877	0.5797 ± 0.1023
Mean (±1STD)	0.6423 ± 0.1049	0.4683 ± 0.0962	0.6435 ± 0.1051	0.4694 ± 0.0969	0.6372 ± 0.1036	0.4641 ± 0.0991	

scores and the performance of those only applying clustering algorithms. Table 4 presents the mean aggregated scaled scores for all the experiments.

Based on the aggregated mean scores, configurations employing Isomap for dimensionality reduction followed by K-means++ clustering in the 2D input space emerged as the most effective, achieving a mean score of 0.7206. This demonstrates the superiority of the K-means++ algorithm in combination with the Isomap technique for enhancing cluster formation and differentiation. The results revealed that dimensionality reduction techniques significantly improved clustering outcomes compared to the baseline methods (those experiments only applied clustering algorithms). Specifically, the mean aggregated score of K-means++ clustering in the 2D input space was consistently above 0.6, outperforming both K-means and Hierarchical + K-means clustering methods. This trend was observed across various configurations, underscoring the critical role of dimensionality reduction in facilitating more precise clustering.

4.2 Clustering Results for Financial Behavior and Socio-Demographic Profiling

The two-stage cluster analysis applied to the study sample to divide the Peruvian population into different social groups yielded three clusters, as shown in 4.1. Table 5 illustrates each cluster's financial behavior and socio-demographic profiles. Additionally, mean difference tests were conducted using the Chi-Square test to assess cluster differentiation within each study feature.

All three clusters exhibit varying degrees of financial engagement, each characterized by distinct sociodemographic features on a particular level. Nonetheless, it is noteworthy that all clusters display an even gender distribution. These differences highlight each cluster's unique challenges and opportunities in accessing and utilizing financial services. Additionally, mean difference tests were conducted on each study factor. As shown in Table 5, each significance value indicates a significant difference between the clusters for each corresponding variable used in the study, suggesting that the clusters are distinct regarding all specific characteristics.

Cluster 1 (C1) exhibits a high level of financial inclusion, with most individuals having access to and actively using bank accounts. This engagement with formal financial services is supported by higher educational attainment, which enhances financial literacy. Individuals in this cluster predominantly belong to higher socioeconomic levels and possess medium to high income, allowing them to navigate financial opportunities effectively. The cluster is also marked by a significant presence of self-employed workers and technical professionals, indicating economic stability. Most individuals are urban residents and young adults, facilitating access to financial institutions and technology. High technology usage further enables interaction with digital financial services.

Conversely, Cluster 2 (C2) demonstrates significant financial exclusion, with many individuals lacking access to bank accounts and formal financial services. Educational attainment is generally low, limiting financial literacy and understanding of financial products. The socioeconomic profile shows a concentration in lower income levels and low social strata, leading to a reliance on informal financial mechanisms. Nonetheless, this cluster is also notable for having a significant number of self-employed individuals, technical professionals, and individuals engaged in domestic chores, reflecting a mix of economic progress and limited earning potential. This cluster is also characterized by individuals in early to mid-adulthood who reside in urban areas, reflecting a dynamic and potentially mobile population. Technology usage is minimal, restricting access to digital financial services and information.

On the other hand, Cluster 3 (C3) displays moderate financial inclusion, with a reasonable proportion of individuals having access to bank accounts and engaging with credit services. Educational backgrounds in this cluster are diverse, ranging from high school completion to higher education, which facilitates better economic opportunities and financial needs. Socioeconomically, individuals in this cluster are situated in middle-income levels and medium social strata. The occupations include self-employed individuals and technical/professional roles,

Table 5. Financial and sociodemographic characteristics of the identified clusters.

Variables	Categories	C1		C2		C3		Mean diff.
		n	%	n	%	n	%	
Access to bank accounts	Yes	174	72.80	214	29.40	53	22.20	<0.01 ***
	No	65	27.20	513	70.60	186	77.80	
Usage of bank accounts	Yes	237	99.20	209	28.70	90	37.70	<0.01 ***
	No	2	0.80	518	71.30	149	62.30	
Access to financial credits	Yes	110	46.00	56	7.70	75	31.40	<0.01 ***
	No	129	54.00	671	92.30	164	68.60	
Education level	No education/Preschool/Elementary school/Incomplete high school	7	2.90	146	20.10	65	27.20	<0.01 ***
	Complete high school/Incomplete technical education	57	23.80	315	43.30	87	36.40	
	Complete technical education	59	24.70	121	16.60	41	17.20	
	Incomplete higher education	47	19.70	67	9.20	14	5.90	
	Higher education/Postgraduate studies	69	28.90	78	10.70	32	13.40	
Socioeconomic Level	SES A	25	10.50	15	2.10	3	1.30	<0.01 ***
	SES B	102	42.70	141	19.40	26	10.90	
	SES C	88	36.80	286	39.30	57	23.80	
	SES D	17	7.10	217	29.80	53	22.20	
	SES E	7	2.90	68	9.40	100	41.80	
Gender	Male	138	57.70	343	47.20	122	51.00	<0.01 ***
	Female	101	42.30	384	52.80	117	49.00	
Work/Occupation	Unemployed	14	5.90	49	6.70	16	6.70	<0.01 ***
	Employer	17	7.10	25	3.40	11	4.60	
	Self-employed	95	39.70	344	47.30	108	45.20	
	Technical/professional employee	60	25.10	91	12.50	34	14.20	
	Laborer	9	3.80	50	6.90	7	2.90	
	Domestic worker	4	1.70	13	1.80	8	3.30	
	Farmer/Countryman	0	0.00	6	0.80	13	5.40	
	Household chores	8	3.30	89	12.20	21	8.80	
	Student	13	5.40	45	6.20	15	6.30	
	Unspecified occupation	19	7.90	15	2.10	6	2.50	
Income	<300	4	1.70	49	6.70	48	20.10	<0.01 ***
	300–600	23	9.60	112	15.40	44	18.40	
	601–1200	59	24.70	309	42.50	80	33.50	
	1201–2400	79	33.10	179	24.60	41	17.20	
	2401–4800	54	22.60	63	8.70	22	9.20	
	>4801	20	8.40	15	2.10	4	1.70	
Age	18–19	11	4.60	45	6.20	11	4.60	<0.01 ***
	20–29	100	41.80	201	27.60	63	26.40	
	30–39	57	23.80	178	24.50	57	23.80	
	40–49	36	15.10	144	19.80	61	25.50	
	50–59	24	10.00	93	12.80	29	12.10	
	60–69	11	4.60	60	8.30	13	5.40	
	70–79	0	0.00	6	0.80	5	2.10	
Zone	Urban	227	95.00	727	100	49	20.50	<0.01 ***
	Rural	12	5.00	0	0.00	190	79.50	
Usage of electronic devices	Yes	219	91.60	0	0.00	11	4.60	<0.01 ***
	No	20	8.40	727	100.00	228	95.40	
Cluster size		239	19.80	727	60.30	239	19.80	
Total sample		1,205						

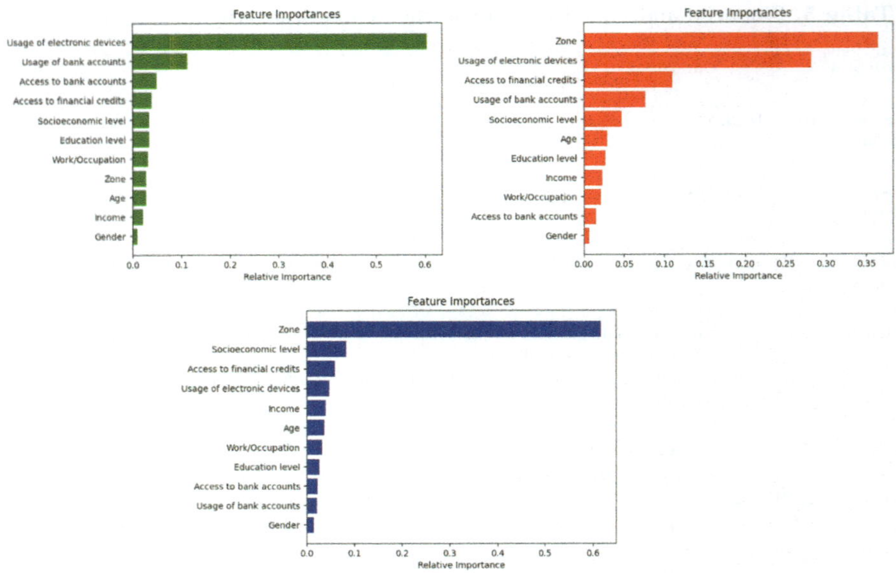

Fig. 4. Variable importance plots for Cluster 1 (**a**), Cluster 2 (**b**) and Cluster 3 (**c**).

promoting financial participation. There is also a notable engagement with electronic devices to engage in financial activities, enhancing access to information and financial services. This cluster has a significant rural population, which may affect access to financial institutions and services.

From another perspective, it is relevant to identify the most explanatory sociodemographic characteristics within each cluster generated by the proposed two-stage methodology to effectively measure how important a variable is and assess the differences within each group based on the feature importance method described in Sect. 3. The feature importance scores were calculated for each feature by developing a Random Forest model for each identified cluster. The results are presented in Fig. 4. The most important variables are at the top of each plot, and the less important ones are at the bottom.

The cluster's feature importance analysis reveals distinct characteristics and exhibits significant trends. C1 is predominantly driven by "Usage of electronic devices", with "Usage of bank accounts", "Access to bank accounts" and "Access to financial credits' following in importance; the lesser role of features such as "Socioeconomic level" and "Education level" in this cluster also suggests a focus on digital and financial behaviors. In C2, the importance is more evenly distributed. While "Zone" is notably significant, "Usage of electronic devices", "Access to financial credits" and "Usage of bank accounts" also emerge as significant factors, showcasing this cluster's emphasis on technology and financial access. On the other hand, in C3, "Zone" is remarkably the most important feature, indicating a solid geographical influence. This cluster also places moderate importance on "Socioeconomic level" and "Access to financial credits," suggesting economic factors are significant but secondary to geographic location. Conversely, several similarities and low-importance features can be observed. "Gender" consistently ranks as one of the least significant variables, indicating that it does not play a major role. Additionally, while "Socioeconomic level" and "Education level" are moderately important in some clusters, they do not dominate the feature importance rankings. This suggests that while these factors are relevant, they are not the primary drivers of cluster differentiation.

5 Conclusions

This study proposed a two-stage dimensionality reduction and clustering methodology to identify social clusters of financial behavior and socio-demographic characteristics in Peru. Significant indicators were found within each cluster. The optimal configuration involved using Isomap with 5 neighbors and an arpack eigen solver to reduce data to 2 dimensions, followed by K-means++ clustering.

We identified 3 clusters: C1 with high financial inclusion, supported by higher education, socioeconomic status, and urban residency; C2 with a significant financial exclusion, low access to financial services, lower education, minimal technology usage, and a presence of domestic workers and students; C3 with moderate financial inclusion, diverse educational backgrounds, middle-income levels, self-employed and professional roles, and a significant rural population. The study's approach can support financial inclusion policies by providing insights into economic and demographic factors influencing financial behavior and suggesting unsupervised models' potential in improving financial access and usage.

However, the study had limitations. The sample size may not fully capture Peru's financial behavior and socio-demographic characteristics. A larger, more representative sample would enhance analysis robustness. The reliance on self-reported data may introduce biases, and future research should include longitudinal data and diverse sampling methods. Testing additional combinations of dimensionality reduction and clustering methods, validating the non-linear reduction method against others, and incorporating geospatial analysis could further improve the findings and inform public policies.

References

1. Herrerias, R., Alvarez, C.: Financial Behavior and Degrees of Financial Inclusion. Available at SSRN (2023)
2. Demirgüç-Kunt, A., Singer, D.: Financial Inclusion and Inclusive Growth: A Review of Recent Empirical Evidence. Policy Research Working Paper, No. 8040, World Bank (2017)
3. Hannig, A., Jansen, S.: Financial Inclusion and Financial Stability: Current Policy Issues. ADBI Working Paper 259, Asian Development Bank Institute, Tokyo (2010)
4. Fiserv: Financial Inclusion: The Road to Global Progress - The Value of Broader Access to Financial Services, Aided by Technology. White Paper (2024)

5. Aurazo, J., Gasmi, F.: Financial inclusion transitions in Peru: The role of labor informality. TSE Working Papers, No. 2023-007, Central Reserve Bank of Peru, Lima (2023)
6. Saraswati, D., Kusumaningrum, D.: Cluster analysis of financial behaviors and preferences among the bottom of the pyramid group. Jurnal Manajemen Dan Kewirausahaan (JMK) **21**(1), 51–63 (2019)
7. Khan, F., Siddiqui, M., Imtiaz, S.: Role of financial literacy in achieving financial inclusion: a review, synthesis and research agenda. Cogent Bus. Manag. **9**(1), 2034236 (2022)
8. Bester, T., Rosman, B.: Towards financially inclusive credit products through financial time series clustering. arXiv:2402.11066 (2024)
9. Choi, I., Koh, W., Koo, B., Kim, W.C.: Network-based exploratory data analysis and explainable three-stage deep clustering for financial customer profiling. Eng. Appl. Artif. Intell. **128**, 107378 (2024)
10. Mulyaningsih, S., Heikal, J.: K-means clustering using principal component analysis (PCA) Indonesia multi-finance industry performance before and during Covid-19. APMBA (Asia Pac. Manage. Bus. Appl.) **11**(2), 131–142 (2022)
11. Tang, G., Tian, R., Wu, B.: An overview of clustering methods in the financial world. In: Proceedings of the 2022 7th International Conference on Financial Innovation and Economic Development (ICFIED 2022), pp. 524–529 (2022)
12. Kok, C., Puigvert, J. M.: Euro Area Banking Sector Integration - Using Hierarchical Cluster Analysis Techniques. Working Paper Series, No. 627, European Central Bank (2006)
13. Mohd, Z., Yusof, R., Mesbahi, E.: Dimension reduction and clustering of high dimensional data using auto associative neural networks. Int. J. Comput. Appl. **72**(11), 31–37 (2013)
14. Velykoivanenko, H., Korchynskyi, V.: Application of clustering in the dimensionality reduction algorithms for separation of financial status of commercial banks in Ukraine. Univ. J. Account. Finance **10**, 148–160 (2022)
15. Bakry, A., Alsharkawy, A., Farag, M., Raslan, K.: Combating financial crimes with unsupervised learning techniques: clustering and dimensionality reduction for anti-money laundering. Al-Azhar Bull. Sci. **35**, 10–22 (2024)
16. Superintendencia de Banca Seguros y AFP (SBS): Encuesta Nacional de Demanda de Servicios Financieros y Nivel de Cultura Financiera (2019). Accessed 18 Dec 2022
17. Maaten, L.v.d., Hinton, G.: Visualizing data using t-SNE. J. Machine Learn. Res. **9**(86), 2579–2605 (2008)
18. Tenenbaum, J.B., de Silva, V., Langford, J.C.: A global geometric framework for nonlinear dimensionality reduction. Science **290**, 2319–2323 (2000)
19. Jolliffe, I.T.: Principal Component Analysis, 2nd edn. Springer, New York (2002)
20. Liu, L., et al.: Application of K-Means++ algorithm based on t-SNE dimension reduction in transformer district clustering. In: Asia Energy and Electrical Engineering Symposium, pp. 74–78 (2020)
21. Lu, J.F., Tang, J.B., Tang, Z.M., Yang, J.Y.: Hierarchical initialization approach for K-Means clustering. Pattern Recogn. Lett. **29**(6), 787–795 (2008)
22. Xu, R., Wunsch, D.C.: Clustering. 3rd edn. Wiley (2009)
23. Baarsch, J., Celebi, M.E.: Investigation of internal validity measures for K-means clustering. In: Proceedings of the International Multiconference of Engineers and Computer Scientists, vol. 1, pp. 14–16. International Association of Engineers Hong Kong (2012)

24. Rousseeuw, P.J.: Silhouettes: a graphical aid to the interpretation and validation of cluster analysis. J. Comput. Appl. Math. **20**, 53–65 (1987)
25. Caliński, T., Harabasz, J.: A dendrite method for cluster analysis. Commun. Stat. **3**(1), 1–27 (1974)
26. Davies, D.L., Bouldin, D.W.: A cluster separation measure. IEEE Trans. Pattern Anal. Mach. Intell. **PAMI-1**(2), 224–227 (1979)
27. Bishop, C.: Pattern Recognition and Machine Learning, 1st edn. Springer, New York (2006)

The Unforeseen Blackout and Cyber Risk Management for Business Continuity: Network-Oriented Computational Analysis

Jack Caneva Silvagni[1], Oliver Norp[1], Emma Rietze[1], Linas Šileris[1], Nick Barelds[1], Charlotte Hoffmans[1,2], Jan Treur[1,2(✉)], and Peter H. M. P. Roelofsma[1]

[1] Risk Management and Cybersecurity Group, Cybersecurity Center of Expertise, The Hague University of Applied Sciences, The Hague, The Netherlands
j.treur@vu.nl, p.h.m.p.roelofsma@hhs.nl

[2] Social AI Group, Department of Computer Science, Vrije Universiteit Amsterdam, Amsterdam, The Netherlands

Abstract. This paper analyses a scenario describing a leading tech company that had to deal with an unforeseen blackout. This blackout caused a major disruption to electricity, internet services, telecommunication networks and water supply. The company's cyber risk management for business continuity plans were tested in this multifaceted challenge. For the computational analysis, an adaptive temporal-causal network model was developed, addressing the scenario. This included adaptive elements such as connection weights and speed factors. The adaptive network model shows how disruptions and responses can unfold. A series of simulation experiments were conducted. These gave an insight into cause-and-effect relationships between different factors involved. A systematic What-If analysis was also provided to assess the impact that changing one factor may have on all other states in the system. All of these provided an enhanced understanding of the scenario and the interconnectedness between the factors involved.

Keywords: Blackout · cyber risk management · business continuity · adaptive network model

1 Introduction

Business continuity planning has become a crucial process that supports companies and organizations with tackling unforeseen blackouts and disasters. The necessity of robust business continuity and disaster preparedness plans is evident from analysing historical case studies and the attributable damages. In 2003, the Northeast USA blackout affected 50 million people, causing $6 billion in damages and exposing gaps in continuity planning (Goodrich, 2005). Two years later, Hurricane Katrina in USA resulted in

R. Silhavy and P. Silhavy (Eds.): CoMeSySo 2024, LNNS 1489, pp. 352–375, 2025.
https://doi.org/10.1007/978-3-031-96798-6_28

$4.6 billion in commercial revenue losses, showcasing the critical need for disaster preparedness (Burton & Hicks, 2005). Furthermore, in 2016, the Northern Lincolnshire and Goole NHS Foundation Trust had to spend around $100 million in order to recover their services from a ransomware attack, highlighting the need for resilient IT infrastructure (Rock, 2024). Another example of being ill prepared for major disaster is the massive disruption caused by Hurricane Harvey in 2017, which cost $5.9 billion in commercial revenue damages and was even expected to reduce the GDP of Houston's economy by as much as 0.25 percent for the third quarter of that year (Burton & Hicks, 2017). In this paper, a scenario will be examined where a large tech company faces an unforeseen blackout despite the presence of robust business continuity plans. This blackout disrupts multiple essential business operations, which exposes gaps in the company's risk management strategies. A comprehensive review and enhancement of these strategies will be given to strengthen the business' continuity planning.

Through the use of network-oriented modelling (Treur, 2016, 2020), this paper aims to demonstrate a simulation on business resilience. Network-oriented modelling is a way to break down processes that are complex, based on temporal-causal networks. In this way, it is possible to break down complex processes such as cognition, emotion or sensory processing, but also other complex systems like in this paper in business resilience. This approach allows to create a temporal dimension for cyclical causal connections (Treur, 2016). Considering the cyclical nature of how corporations and organizations manage reemerging blackouts, temporal-causal network modelling approach is suitable for structuring network models that could be tested in a simulated environment. Therefore, a conceptual network model utilizing the above-mentioned approach was developed in order to analyse the temporal and causal aspects of managing unforeseen blackouts and disruptions. Additionally, Plan-Do-Check-Act (PDCA) continuous improvement strategy was applied into the scenario and later incorporated into the model to test out the reliability and relevance of this technique that is essential for effective business continuity and resilience.

This work is part of a research programme on computational analysis of safety and security in and through cyberspace. In (Abromaityte et al., 2024; Akers et al., 2024; Bouma et al., 2024; Capra et al., 2024; Hoffmans et al., 2024; Jeffery et al., 2024; Keijzer et al., 2024) various other themes within cybersecurity were addressed. Moreover, in (Roelofsma et al., 2024) it was analysed how an AI coach can support safety and security for medical teams in hospitals.

2 Background Knowledge and Scenario Description

In this simulated scenario, a blackout strikes a leading tech company through vital systems and infrastructure, disrupting electricity, telecommunication networks, internet services, and water supply. The blackout is originated by severe weather conditions which lead to the critical infrastructure failure. As consequence, the blackout results in a total collapse of the company's essential utilities. Because the company relies heavily on technology and modern equipment, significant challenges arise in order to maintain its operations, despite preparedness for isolated incidents of the same nature. This highlights the company's need for adaptability in its business continuity plans.

In order to understand the relationship between different concepts used during this simulation, a few will be outlined below:

2.1 Business Continuity and Resilience

Business continuity and business resilience have overlapping objectives. Business continuity is the ability for an organization to continue operations and processes within an acceptable timeframe after a disruption. In business continuity, organizations focus on the anticipation of disruptions reducing the risk of loss and increase the ability to continue to operate. A typical business continuity plan may contain a framework and procedures consisting of business impact analysis, risk analysis, training and testing (Fani & Subriadi, 2018).

On the other hand, business resilience refers to the ability of an organization to adapt and deliver its objectives if the environment or circumstances change, affecting the core business operations. To be resilient, an organization needs to be able to anticipate, cope and adapt (Duchek, 2019). Additionally, unforeseen situations that may be internally or externally originated, that could potentially harm the organization. These situations may include environmental disasters, financial crises, terrorism and other rare events (Hepfer & Lawrence, 2022).

Companies that need to maintain business continuity, usually adopt extensive planning and measures if critical infrastructure fails for internal or external reasons unexpectedly. These measures involve mitigating vulnerable infrastructure that prevents the company from continuing operations effectively (Duchek, 2019). As an example, a company relying on electricity may incorporate generators as backup, and different types of GPS-based telecom devices.

2.2 Plan-Do-Check-Act (PDCA)

PDCA is a common continuous improvement technique used to improve modern management processes and systems. The technique consists of four steps:

1. Plan
2. Do
3. Check
4. Act

In this scenario, PDCA is used with business continuity, Plan refers to the ability of the company to assess risks and scenarios to determine mitigation actions in case of a severe disruption. Do, refers to regulating control measures to ensure effective continuity. Check, refers to updates, tests and drills in order to maintain the functionality of the contingency plan. Lastly, Act, where improvement feedback is implemented in order to improve the business continuity plan.

3 The Network Modelling Approach: Concepts and Methodology

The temporal-causal network modelling approach (Treur, 2016, 2020) is a comprehensive modeling approach for designing models for complex processes. This approach captures the dynamics of adaptive systems through temporal and causal relationships.

It is particularly effective to analyse how factors or states influence each other over time, such as in the case of this paper, with a blackout crisis faced by a tech company. By integrating first-order and second-order adaptation, the introduced network model captures both immediate responses and long-term strategies, enhancing the company's resilience.

In network modelling, particularly within the context of temporal-causal networks, the concepts 'temporal' and 'causal' contribute to the overall understanding and functionality of the model. Temporal refers to the dimension of time within the network model. It encompasses how the states and behaviours of nodes in the network change and evolve over time. Temporal aspects in network modelling include:

1. **Timing of events:** This refers to the specific points in time at which events occur within the network and how these timings impact the network's dynamic behaviour. This includes the precise moments when nodes are activated, when certain interactions occur, and when state transitions are observed.
2. **Duration:** This involves the measurement of time that certain states or interactions persist within the network. For instance, this could refer to the duration of signal transmission, the length of time a node remains in a particular state, or the persistence of an effect following a causal event.
3. **Temporal patterns:** This includes the identification and modelling of sequences and rhythms of events and interactions. Such patterns may include cyclical processes, periodic events, and temporal dependencies, all of which influence the network's behaviour over extended periods.

In temporal-causal networks, the temporal dimension ensures that the model accurately reflects patterns of real-world dynamics where the timing of interactions and changes is crucial. For example, in this scenario, a tech company experiences a wide scale blackout that disrupts service of essential utilities. The timing of back up generators being utilised and the subsequent energy they supply to the business operations in a short time is critical for maintaining appropriate operations capacity.

The causal dimension in network modelling is defined to be the explicit representation of cause-and-effect relationships between nodes in the network. This involves understanding how one state or event influences another, leading to changes within the system. The causal aspect encompasses several features:

1. **Causal links:** These are directed connections within the network that signify the direction and nature of influence between nodes. A causal link $A \rightarrow B$ from node A to node B implies that changes in node A cause changes in node B. Nodes such as A and B are also called states; they have activation levels $A(t)$ and $B(t)$ that change over time t.
2. **Strength of Influence:** This relates to the quantification of the magnitude or weight of the causal effect, or in other words, the connectivity weight $\omega_{A,B}$. For simulating a conceptual network model. For this to work, the intensity needs to be specified how one node influences another, where this influence can be positive, negative, or neutral.
3. **Causal Mechanisms:** These are the underlying processes or rules that define how influence is exerted from one node to another, such as mathematical combination

functions $\mathbf{c}_B$ that specify the aggregation of the impacts. For the network model of this scenario, combination functions **alogistic** and **stepmod** were utilised, see Table 1.
4. **Timing**: Speed factors $\boldsymbol{\eta}_B$ specify how fast B changes upon aggregated impact on it.

In temporal-causal networks, causal relationships are explicitly modelled to capture the dynamic between different nodes. This allows for a detailed overview and understanding of how changes spread through the network and how various factors interact to produce complex behaviours.

Table 1. Combination functions used.

Concept	Representation
Advanced logistic sum $\mathbf{alogistic}_{\sigma,\tau}(V_1, \ldots, V_k)$	$\left[\frac{1}{1+e^{-\sigma(V_1+\cdots+V_k-\tau)}} - \frac{1}{1+e^{\sigma\tau}}\right](1+e^{-\sigma\tau})$
Stepmodulo $\mathbf{stepmod}_{\rho,\delta}(V_1, \ldots, V_k)$	time t, repetition interval $\boldsymbol{\rho}$, tipping point $\boldsymbol{\delta}$ 0 if $t \mod \rho < \delta$ else 1

3.1 First-Order Adaptation

Adaptations involve changes in the basic characteristics of the network, such as connection weights, combination functions, and speed factors. These adaptations are driven by direct interactions and experiences within the network.

Connection Weights: Refers to adjustments in the strength of connections between nodes based on their interactions. For instance, in a neural network, Hebbian learning principles suggest that simultaneous activation of neurons strengthens their connection. If a connection weight $\boldsymbol{\omega}_{A,B}$ is adaptive, this can be specified by adding a self-model state $\mathbf{W}_{A,B}$ to the network that represents the value of $\boldsymbol{\omega}_{A,B}$.
Speed factors: Changes in how quickly nodes respond to causal impacts, reflecting the flexibility or rigidity of the state change. If a speed factor $\boldsymbol{\eta}_B$ is adaptive, this can be specified by adding a self-model state $\mathbf{H}_B$ to the network that represents the value of $\boldsymbol{\eta}_B$.

In a temporal-causal network model, first-order adaptations allow for the representation of plasticity, which refers to how network structure evolves in response to direct experiences and interactions. In the considered scenario, where the tech company experiences wide scale blackout, the blackout has direct interactions with the states and nodes of the base level, such as electricity supply and internet connection, which over time propels the tech company to evolve their countermeasures (both in the base level and the first-order self-model level) to better manage an energy blackout.

3.2 Second-Order Adaptation

Second-order adaptations involve changes in the adaptation principles themselves, governing how first-order adaptations occur. This meta-level adaptation addresses the evolution of the rules and mechanisms that drive the changes in the network's basic characteristics.

1. **Adaptability of Learning Rates**: Adjusting how quickly learning and adaptations occur based on the network's history and context. For instance, the stronger the total disruption level is, the quicker the network will adapt to increase the regulation capacity. For example, if a first-order self-model state $\mathbf{W}_{A,B}$ has an adaptive learning speed factor $\mathbf{\eta}_{\mathbf{W}A,B}$, then this can be modeled by adding a second-order self-model state $\mathbf{H}_{\mathbf{W}A,B}$ to the network that represents the value of $\mathbf{\eta}_{\mathbf{W}A,B}$.
2. **Persistence Factors:** Modifying how persistent changes in the network structure are, which can depend on factors such as stability of the environment or the accumulation of experiences.

Second-order adaptations are crucial for modelling metaplasticity, regulating the capacity for change of the first-order adaptations in the base level. In a neural context, metaplasticity refers to how the brain adjusts its plasticity based on prior activity, ensuring that learning processes remain adaptive and efficient. In a social network, second-order adaptations might involve changes in how quickly trust develops between individuals based on previous interactions and broader social dynamics.

Any smooth dynamical system has a canonical representation as a network and any smooth (multi-order) adaptive dynamical system has a canonical representation as a (multi-order) self-modeling network, as is shown in (Treur, 2021; Hendrikse et al., 2023). Therefore, compared to adaptive dynamical systems in general, the network-oriented modeling approach used does not introduce any fundamental limitations concerning what it can model. This has also been confirmed by applications to complex adaptive processes in different disciplines, such as mental models in cognitive science, organisational learning in management science, synchronisation and bonding in social psychology, and teamwork in health science; e.g., (Treur and Van Ments, 2022; Canbaloğlu et al., 2023; Hendrikse, et al., 2025; Roelofsma et al., 2024).

4 The Introduced Second-Order Adaptive Network Model

In this section, the design of the network model is described. Figure 1 shows the network picture. In Table 2, the 18 states in the network model are explained and elaborated upon. These explanations are substantiated through desk research. In Table 3, the connections between all states are explained, along with indications of being positive or negative connections.

In the design of the network model, the PDCA cycle was incorporated into the different levels of the network model of the tech company simulation scenario, see Fig. 2. In the Plan phase, due to the possibility of an occurrence of disruptions of a widespread blackout (X1), the company identifies vital utilities (X2, X3, X4, X5) to ensure that there are sufficient mitigations against the most vital utilities included in the company's contingency plans such as continuity plans. The company activates the planned mitigations once the blackout begins, initiating the Do phase, which is illustrated by the connecting lines from the solutions X6, X7, X8, X9 towards the Total Disruption

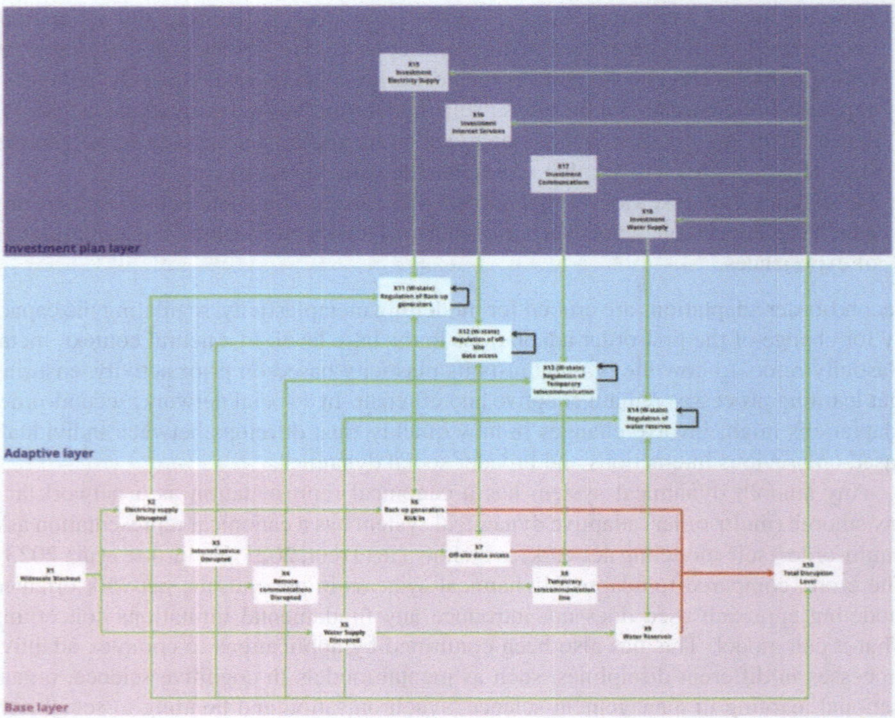

Fig. 1. Adaptive network model design.

Table 2. Base level states and their explanations.

State	Explanation
X1: wide scale blackout	Large-scale power outages have the potential to result in substantial business interruption losses (Eyer & Rose, 2018). This, combined with the fact that the tech sector is almost entirely reliant on a power supply (Petermann et al., 2011, p. 78), means that almost all critical operations will come to a halt in the case of a company-wide blackout
X2: electricity supply disrupted	Loss of electrical power to a facility brings everything to a halt. This can damage sensitive equipment, disrupt ongoing tasks, and create uncomfortable working conditions for employees. For a tech company, which relies heavily on computer-based information processing and electronic communications, a disruption of the electricity supply is especially detrimental since these aspects of the business operations are not possible without a power supply (Petermann et al., 2011, p.74)

(continued)

Table 2. (*continued*)

State	Explanation
X3: internet service disrupted	Without internet connectivity, a tech company loses its connection to the outside world. This means a disruption of the communication and information flow; companies lose connection to clients, partners, and employees, which hinders collaboration and the delivery of services (World Economic Forum, 2022)
X4: remote communications disrupted	A blackout can also disrupt phone lines, internal networks, and other means of remote communication. Disruptions in the communication network hinder collaboration, and can also lead to a loss of productivity and revenue (Chen et al., 2019)
X5: water supply disrupted	A blackout can disrupt the water supply if water pumps are reliant on power. A disruption of the water supply means a loss of access to clean running water. Even a short disruption of the water supply can cause dire consequences for companies, forcing them to slow down their business operations (Sjöstrand et al., 2021)
X6: back-up generators kick in	Generators are backup power systems that automatically start supplying electricity when the main power source fails. In emergency power planning, providing essential facilities with permanent backup power is essential (Hordeski, 2005, p. 18). Backup power systems are a critical component in enhancing the electrical system's resilience during emergencies
X7: off-site data access	Storing critical data in the cloud ensures that it remains accessible even during a power outage at the main facility. In contemporary disaster recovery planning, storing data in the cloud is vital; it equips businesses with a dependable solution that shields their essential operations and data, which improves their resilience against a possible power outage (Tatineni, 2023)
X8: temporary telecommunication line	Communication resilience is achieved through a diversification of the communication channels. This allows the system to utilise alternative resources to maintain service when the main power source fails (Mauthe et al., 2016)
X9: water reservoir	A water reservoir can provide a temporary supply of clean running water if the main supply is disrupted due to a power outage. a resilient water system is able to draw on multiple, diversified water sources like water reservoirs (Homer et al., 2021)
X10: total disruption level	Severity and scope of disruption

Table 3. First- and second-order adaptation states and their explanations.

State	Explanation
X11: regulation of back-up generators	Having a backup system in place allows critical functions to continue in the event of a major power failure (Hordeski, 2005, p.15). Regular maintenance of this system plays a critical role in ensuring the reliability of the generators during a power outage (Gilani et al., 2020). This state X11 is a self-model **W**-state representing the weight of the connection X2 → X6
X12: regulation of off-site data access	Creating an alternative system for storing and accessing critical data ensures business continuity and minimises data loss risks in case of a blackout. Regular hardware and software maintenance of this system, like security protocols and patch management procedures, will ensure the smooth operation and security of data storage infrastructure (Khan, 2016). This state X12 is a self-model **W**-state representing the weight of the connection X3 → X7
X13: regulation of temporary telecommunication line	The resilience of business operations against a power outage can be improved by maintaining the communications infrastructure (Castillo, 2014). Thus, by doing this, the company ensures that communication is still available in the case of a blackout.). This state X13 is a self-model **W**-state representing the weight of the connection X4 → X8
X14: regulation of water reservoir	Being proactive in the management, upkeep, and replacement of the water infrastructure's assets will enhance the resilience of the system (Homer et al., 2021), and will ensure the provision of high-quality back-up water sources in the case of a blackout.). This state X14 is a self-model **W**-state representing the weight of the connection X5 → X9
X15: investment back-up generators	The investment in a backup power systems is a critical component in enhancing the electrical system's resilience during emergencies (Faraji et al., 2019). This state X15 is a self-model $\mathbf{H_W}$-state representing the speed factor of **W**-state X11
X16: investment internet services	Upgrading internet can minimise downtime and ensures the company remains connected during disruptions (Tatineni, 2023). This state X16 is a self-model $\mathbf{H_W}$-state representing the speed factor of **W**-state X12

(*continued*)

Table 3. (*continued*)

State	Explanation
X17: investment temporary telecommunication line	Upgrading telecommunication lines allows for the utilisation of alternative resources to maintain service during a power outage (Mauthe et al., 2016). This state X17 is a self-model $\mathbf{H_W}$-state representing the speed factor of **W**-state X13
X18: investment water reservoir	Upgrading water infrastructure ensures a more reliable water supply and minimises the impact of power outages on water access (Homer et al., 2021). This state X18 is a self-model $\mathbf{H_W}$-state representing the speed factor of **W**-state X14

Level X10 node. The company is able to check the total effectiveness of the Plan and Do phases in its Check phase, which is conducted in Total Disruption Level X10 node. The incoming connections from the disruption states X2, X3, X4, X5 that increase the Total Disruption Level state X10 value and the subsequent incoming connections from the solution states X6, X7, X8, X9 that decrease X10 state value provide a simplistic, yet effective method to CHECK whether the PLANNED solutions are DOING the intended function of mitigating the blackout consequences. Finally, the Act phase commences in the Investment Level (second-order adaptation level), where the results from the Check phase directly influences the strength of connectivity and speed factor of the investment states X15, X16, X17, X18 that strengthen the regulation states X11, X12, X13, X14, subsequently improving the Plan and Do phases of the PDCA continuous improvement strategy.

4.1 First- and Second-Order Adaptation Level

The first-order adaptation relates to plasticity of the solutions that can be made operational at the base level. The second-order adaptation involves changes in the principles that control how the network will adapt and refer to the metaplasticity component of the network model. These adaptations regulate the capacity to adapt based on higher-level dynamics and experiences.

4.2 Regulation of Solutions

Each regulatory state (X11, X12, X13, X14) at the first-order adaptation level ensures that adaptive measures are effective and continuously improved over time. These states do not only link to base level states to monitor the situation but also to themselves to automatically persist and strengthen their regulatory impact.

- **X11: Regulation of Backup Generators**
  Ensures reliability through regular testing and maintenance. This state positively influences X6 (Backup generators kick in) by enhancing the preparedness and operational

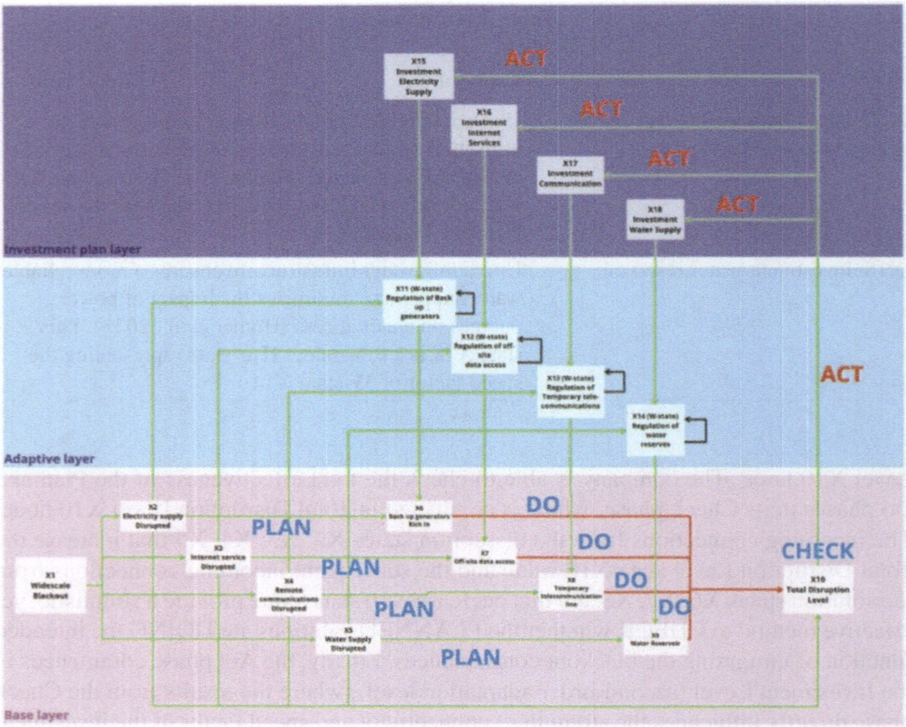

Fig. 2. How the network model relates to stages of PDCA.

efficiency of the backup generators, mitigating the impact of electricity disruptions (X2).

- **X12: Regulation of Off-Site Data Access**
  Ensures continuous data availability during disruptions by implementing secure and reliable access protocols for off-site data access X7, thus minimising the operational impact of internet disruptions (X3).
- **X13: Regulation of Temporary Telecommunication**
  Ensures the establishment and maintenance of alternative communication methods such as satellite phones and backup internet connections. This state impacts X8 (Temporary Telecommunication Lines) by ensuring communication channels remain open during disruptions, addressing remote communication issues (X4).
- **X14: Regulation of Water Reserves**
  Ensures water availability through reservoirs and alternative sources. This state positively impacts X9 (Water Reservoir) by guaranteeing a steady water supply, thereby mitigating the effects of water supply disruptions (X5).

Each of these regulatory states is influenced by the company's investments in resilience and regulation measures at the second-order adaptation level:

- **X15: Investment in Electricity Supply**

- **X16: Investment in Internet Services**
- **X17: Investment in Communications**
- **X18: Investments in Water Supply**

These investments strengthen the regulatory states at the first-order adaptation level, ensuring that adaptive measures are robust and effective in mitigating disruptions.

As discussed in Sect. 3, self-modeling for networks (also called network reification) involves extending the base network by adding specific self-model states that represent characteristics of the network's structure, such as connectivity, aggregation, and timing. This process allows the network to have explicit representations of its structural characteristics, which can then be adapted. In the considered scenario, network self-modeling is crucial for showcasing the dynamic and multilevel nature of the company's response to the blackout. By incorporating self-model states X11 to X14 (first-order) and X15 to X18 (second-order), the model allows for plasticity of the organisation but also higher-order adaptations that adjust the adaptation principles themselves, subsequently enhancing the network's overall resilience.

In the Appendix section all connections are explained and the full specification of the network model can be found by role matrices.

5 Simulation Results

Several simulations were conducted in order to analyse the tech company's response to the blackout. These simulations used varied initial values and settings for the other network characteristics to observe different possible outcomes.

In Simulation 1, see Fig. 3, the first step was assessing the X10 disruption level. In simulation 1 the X10 disruption level during the first blackout is at a max value of 1 on the vertical axis with a duration of 50 time units on the time axis. When assessing the simulation graph it shows that the X10 disruption level is at max value 1 during the duration of the entire blackout indicated by the X1 blackout. This implies that the company was not prepared for the initial first blackout. Another factor that is present during the first blackout is that the solutions X6, X7, X8, and X9 are low on the vertical axis, indicating that the company has yet to invest in the solutions.

During the second blackout, the X10 disruption level on the vertical axis still reaches 1 but the time duration of X10 drastically decreases. This indicates that the solutions which the company has introduced are effectively working, but the initial disruption of the blackout still affects the company. In the second blackout the implemented solutions (X6, X7,X8,X9) are working more effectively. The vertical axis value of the solutions at the start of the second blackout is now 0.6 compared with the starting value of 0 during the first blackout. This shows an increase in preparedness from the company, since the first initial blackout, thereby decreasing the effect of the disruption level. This is a recurring trend within further future blackouts. It shows that the ever-increasing solutions (X6, X7, X8, X9) consistently decrease the effects of the total disruption X10 level during blackouts X1. The one consistency that does reoccur, is the start of the blackout X1 will always disrupt to spike since it takes a short while for the solutions systems to respond to the blackout.

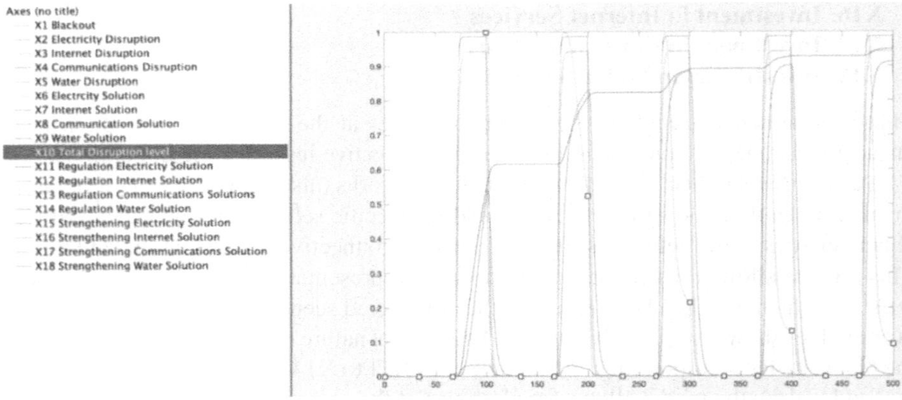

Fig. 3. Simulation 1.

Several other simulations and explanations can be found in the Appendix section found as Linked Data at URL.

https://www.researchgate.net/publication/382562944.

6 What-If Analysis

A What-If analysis was performed with varying input and model characteristics, like excitability thresholds and connection weights. By doing this, the effects of these changes on the network's outcomes were analysed. The following If factors were chosen.

If factor 1: Connection weight from X10 to X15, X16, X17, X18 (they represents the speed of X11, X12, X13, X14) (consider short-term and long-term impact)

- If factor 1 option 1 Low: 0.0-0.05
- If factor 1 option 2 High: 0.05-0.1

The investment nodes have an adaptive principle, in which the connectivity weight of the total disruption level affects the speed factor of the investment nodes. So the If factor 1 option 1 is when the connectivity weight is low (0.0-0.05), meaning that the speed of investments will also be low. In the option 2, the connection weight is high, hence the speed of investments will also increase. Essentially, in option 1, the tech company is very slow at investing into regulation processes, and in option 2, the tech company is very fast at investing.

If factor 2: Base connection weight of each Disruption X2, X3, X4, X5 onto each **W**-states X11, X12, X13, X14 of Regulation changes.

- If factor 2 option 1 Low: $0.0 - 1.0$
- If factor 2 Option 2 High: 1.0-2.0

The regulation nodes have an adaptive principle, in which the connectivity weight and speed factor of incoming connections from each Disruption strengthen the regulation process. So the If-factor 2 option 1 is when the connection weight is low (0.0 – 1.0), then the regulation will also be low. For option 2, the connection weight is high, meaning that regulation process will also be higher than for option 1. Essentially, in option 1 the tech company's regulation processes are weak at adapting to the disruptions, and in option 2, the tech company's regulation processes are strong at adapting to the disruptions.

If factor 3: Connectivity Combination Function Parameter threshold values of each W-state X11, X12, X13, X14 for Regulation changes.

- If factor 3 option 1 Low: 0.0–0.4
- If factor 3 option 2 High: 0.4–0.8

The regulation nodes have a threshold function, which refers to reaching a certain value which will force the node to adapt. Essentially, the threshold value tells at which point the state will tip over and adapt incoming connections. So the If-factor 3 option 1 is when the threshold parameter (Omega) is low (0.0-0.4), meaning that regulation requires less input to be activated. The If-factor 3 option 2 is when the threshold is high (0.4-0.8), meaning there is a higher requirement of input values in order to activate regulation processes. Essentially, in option 1, the tech company's regulation processes have lower threshold for activation, hence they are more responsive to incoming disruptions and investments, and in option 2, the tech company's regulation processes have higher threshold for activation, hence they are less responsive to incoming disruptions and investments.

For all combinations of If factor options in total $2^3 = 8$ simulations have been performed. For graphs of all of them, see the Appendix at https://www.researchgate.net/publication/382562944

As a next step, to make a risk assessment, probabilities for each If factor option have been estimated as shown in Table 4. Based on these values, the probabilities of the combinations of If factor options were determined by the product rule p(A and B) = p(A)*p(B), thereby assuming independence, see the first 4 columns in Table 5.

Table 4. Overview of If factor options and their labels, value intervals and probabilities.

If factors	If factor options	If factor option labels	If factor option value intervals	If factor option probabilities
If factor 1	If factor 1 option 1	low	0.0–0.05	0.75
	If factor 1 option 2	high	0.05–0.1	0.25
If factor 2	If factor 2 option 1	low	0.0–1.0	0.75
	If factor 2 option 2	high	1.0–2.0	0.25
If factor 3	If factor 3 option 1	low	0.0–0.4	0.25
	If factor 3 option 2	high	0.4–0.8	0.75

After the probabilities are calculated, a risk assessment can be developed based on the effects the probabilities will have on the different If factors. In regard to determining the impact grade for each What Factor, some logical reasoning is required. For What Factor 1,

Table 5. Risk assessment.

If factor 1 options	If factor 2 options	If factor 3 options	If factor option combination probabilities	What factor 1 probability	What factor 2 probability
If factor 1 option 1: Low (0.75)	If factor 2 option 1: low (0.75)	If factor 3 option 1: low (0.25)	0.141		X 0.141
		If factor 3 option 2: high (0.75)	0.422	X 0.422	
	If factor 2 option 2: high (0.25)	If factor 3 option 1: low (0.25)	0.047		X 0.047
		If factor 3 option 2: high (0.75)	0.141	X 0.141	
If factor 1 option 2: High (0.25)	If factor 2 option 1: low (0.75)	If factor 3 option 1: low (0.25)	0.047		X 0.047
		If factor 3 option 2: high (0.75)	0.141	X 0.141	
	If factor 2 option 2: high (0.25)	If factor 3 option 1: low (0.25)	0.016		X 0.016
		If factor 3 option 2: high (0.75)	0.047	X 0.047	
Sum of probabilities				0.75	0.25
Impact Indicator				9	7
Risk Assessment				6.75	1.75

in which the Total Disruption Level X10 simulation value was observed, if the disruption level does not decrease at a great enough rate, the tech company's business operations capacity would be greatly diminished, as well as its security capacity, indicating a critical effect, hence the impact grade of 9 was given. The same could be argued for What Factor 2, in which the Solutions responsiveness X6, X7, X8, X9 was observed. If the solutions outputted energy does not counteract the lost capacity that the tech company experiences during the blackout, it can also cause additional operational and security issues, hence the high impact grade of 7.

7 Discussion

This paper analysed a scenario describing a leading tech company that was severely impacted by an unforeseen blackout. This blackout caused a major disruption to electricity, internet services, telecommunication networks and water supply. The limits of the company's cyber risk management for business continuity plans were tested in this multifaceted challenge. A temporal-causal network model was developed, based on the scenario which represented the connections within the scenario. The states showed how disruptions and responses could unfold.

A series of simulations were created using the software environment in MatLab,. These graphs gave an insight into cause-and-effect relationships between states. A What-If analysis was also provided to assess the impact that changing one state could have on all other states. Both of these provided an enhanced understanding of the scenario and the interconnectedness between states.

The modeling approach (Treur, 2020), which included network self-modeling (or reification), provided a framework for modeling adaptive networks. This approach demonstrated adaptability and effectiveness across multiple levels of adaptation. Network reification was particularly crucial in this scenario, as it involved extending the base network by adding explicit self-model states that represent characteristics such as connectivity, aggregation, and timing. These self-model states enabled the network to adapt its structure dynamically, which improved its resilience and response to disruptions.

Appendix

In Table A1 an overview can be found of the connections and their explanation.

Table A1. Connections and their explanation.

Connection	Explanation
X1 → X2, X3, X4, X5 (positive)	A wide scale blackout disrupts electricity, internet, telecommunication and water supply
X2 → X6, X10, X11 (positive)	A disruption of the electricity supply causes back-up generators to kick in, and heightens the total disruption level. It also causes the need for regulation of the back-up generators
X3 → X7, X10, X12 (positive)	A disruption of the internet service causes the use of off-site data access, and heightens the total disruption level. It also causes the need for regulation of off-site data access

(*continued*)

Table A1. (*continued*)

Connection	Explanation
X4 → X8, X10, X13 (positive)	A disruption of remote communications causes the use of temporary telecommunication lines, and heightens the total disruption level. It also causes the need for regulation of temporary telecommunication
X5 → X9, X10, X14 (positive)	A disruption of the water supply causes the use of a water reservoir, and heightens the total disruption level. It also causes a need for regulation of water reserves
X6 → X10 (negative), X11 (positive)	The kicking in of backup generators lowers the total disruption level, and causes the need for regulation of the back-up generators
X7 → X10 (negative), X12 (positive)	The use of off-site data access lowers the total disruption level, and causes the need for regulation of off-site data access
X8 → X10 (negative), X13 (positive)	The use of temporary telecommunication lines lowers the total disruption level, and causes the need for regulation of temporary telecommunication
X9 → X10 (negative), X14 (positive)	The use of a water reservoir lowers the total disruption level, and causes the need for regulation of water reserves
X10 → X15, X16, X17, X18 (positive)	The severity of the total disruption level impacts the available investments for electricity supply, internet services, communication, and water supply
X11 → X6 (positive), self-loop to X11	The **W**-state 'regulation of back-up generators' has adaptive principles which facilitate its regulation process, and it adapts based on the connections incoming from X6 and X15. It also connects to itself, to ensure that the regulation process persists. This connection indicates that X11 is a **W**-state for a connection to X6
X12 → X7 (positive), self-loop to X12	The **W**-state 'regulation of off-site data access' has adaptive principles which facilitate its regulation process, and it adapts based on the connections incoming from X7 and X16. It also connects to itself, to ensure that the regulation process persists. This connection indicates that X12 is a **W**-state for a connection to X7

(*continued*)

Table A1. (*continued*)

Connection	Explanation
X13 → X8 (positive), self-loop to X13	The **W**-state 'regulation of temporary telecommunication' has adaptive principles which facilitate its regulation process, and it adapts based on the connections incoming from X8 and X17. It also connects to itself, to ensure that the regulation process persists. This connection indicates that X13 is a **W**-state for a connection to X8
X14 → X9 (positive), self-loop to X14	The **W**-state 'regulation of water reserves' has adaptive principles which facilitate its regulation process, and it adapts based on the connections incoming from X9 and X18. It also connects to itself, to ensure that the regulation process persists. This connection indicates that X14 is a **W**-state for a connection to X9
X15 → X11 (positive)	Investments into electricity supply facilitate the strengthening of back-up generator regulations. This indicates that X15 is an $\mathbf{H_W}$-state for **W**-state X11
X16 → X12 (positive)	Investments into internet services facilitate the strengthening of off-site data access regulations. This indicates that X16 is an $\mathbf{H_W}$-state for **W**-state X12
X17 → X13 (positive)	Investments into communications facilitate the strengthening of temporary telecommunication regulations. This indicates that X17 is an $\mathbf{H_W}$-state for **W**-state X13
X18 → X14 (positive)	Investments into the water supply facilitate the strengthening of water reserves regulations. This indicates that X18 is an $\mathbf{H_W}$-state for **W**-state X14

Role Matrices

See Tables A2, A3, A4, A5 and A6 for all role matrices.

Table A2. Base connectivity **mb**.

mb	Base connectivity								
X1	WBs	X1							
X2	EDs	X1							
X3	IDs	X1							

(*continued*)

Table A2. *(continued)*

mb	Base connectivity								
X4	CDs	X1							
X5	WDs	X1							
X6	ESs	X2							
X7	ISs	X3							
X8	CSs	X4							
X9	WSs	X5							
X10	DLs	X2	X3	X4	X5	X6	X7	X8	X9
X11	$\mathbf{W}_{ED,ES}$	X2	X11	X15					
X12	$\mathbf{W}_{ID,IS}$	X3	X12	X16					
X13	$\mathbf{W}_{CD,CS}$	X4	X13	X17					
X14	$\mathbf{W}_{WD,WS}$	X5	X14	X18					
X15	$\mathbf{HW}_{ED,ES}$	X10							
X16	$\mathbf{HW}_{ID,IS}$	X10							
X17	$\mathbf{HW}_{CD,CS}$	X10							
X18	$\mathbf{HW}_{WD,WS}$	X10							

Table A3. Connection weights **mcw**.

State nr.	State name	mcw						
X1	WBs	5						
X2	EDs	1						
X3	IDs	1						
X4	CDs	1						
X5	WDs	1						
X6	ESs	X11						
X7	ISs	X12						
X8	CSs	X13						
X9	WSs	X14						
X10	DLs	1	1	1	1	−1	−1	−1	−1
X11	$\mathbf{W}_{ED,ES}$	1	1	1				
X12	$\mathbf{W}_{ID,IS}$	1	1	1				

(continued)

Table A3. (*continued*)

State nr.	State name	mcw							
X13	$\mathbf{W}_{CD,CS}$	1	1	1					
X14	$\mathbf{W}_{WD,WS}$	1	1	1					
X15	$\mathbf{HW}_{ED,ES}$	0.05							
X16	$\mathbf{HW}_{ID,IS}$	0.05							
X17	$\mathbf{HW}_{CD,CS}$	0.05							
X18	$\mathbf{HW}_{WD,WS}$	0.05							

Table A4. Speed factors and initial values **ms** and **iv**.

State nr.	State name	Speed factor	Initial value
X1	WBs	1	0.001
X2	EDs	0.7	0
X3	IDs	0.7	0
X4	CDs	0.7	0
X5	WDs	0.7	0
X6	ESs	0.3	0
X7	ISs	0.3	0
X8	CSs	0.3	0
X9	WSs	0.3	0
X10	DLs	1	0
X11	$\mathbf{W}_{ED,ES}$	X15	0
X12	$\mathbf{W}_{ID,IS}$	X16	0
X13	$\mathbf{W}_{CD,CS}$	X17	0
X14	$\mathbf{W}_{WD,WS}$	X18	0
X15	$\mathbf{HW}_{ED,ES}$	0.3	0
X16	$\mathbf{HW}_{ID,IS}$	0.3	0
X17	$\mathbf{HW}_{CD,CS}$	0.3	0
X18	$\mathbf{HW}_{WD,WS}$	0.3	0

Table A5. Combination function weights **mcfw**.

State nr.	State name	**alogistic**	**stepmod**
X1	WBs	1	0.5
X2	EDs	1	
X3	IDs	1	
X4	CDs	1	
X5	WDs	1	
X6	ESs	1	
X7	ISs	1	
X8	CSs	1	
X9	WSs	1	
X10	DLs	1	
X11	$\mathbf{W}_{ED,ES}$	1	
X12	$\mathbf{W}_{ID,IS}$	1	
X13	$\mathbf{W}_{CD,CS}$	1	
X14	$\mathbf{W}_{WD,WS}$	1	
X15	$\mathbf{HW}_{ED,ES}$	1	
X16	$\mathbf{HW}_{ID,IS}$	1	
X17	$\mathbf{HW}_{CD,CS}$	1	
X18	$\mathbf{HW}_{WD,WS}$	1	

Table A6. Combination function parameters **mcfp**.

State nr.	State name	1 Steepness $\boldsymbol{\sigma}$	2 Threshold $\boldsymbol{\tau}$	1 Repetition $\boldsymbol{\rho}$	2 Tipping point $\boldsymbol{\tau}$
X1	WBs			100	70
X2	EDs	5	0.4		
X3	IDs	5	0.4		
X4	CDs	5	0.4		
X5	WDs	5	0.4		
X6	ESs	5	0.4		
X7	ISs	5	0.4		
X8	CSs	5	0.4		
X9	WSs	5	0.4		

(*continued*)

Table A6. (*continued*)

State nr.	State name	1 Steepness **σ**	2 Threshold **τ**	1 Repetition **ρ**	2 Tipping point **τ**
X10	DLs	5	0.4		
X11	$\mathbf{W}_{ED,ES}$	5	0.4		
X12	$\mathbf{W}_{ID,IS}$	5	0.4		
X13	$\mathbf{W}_{CD,CS}$	5	0.4		
X14	$\mathbf{W}_{WD,WS}$	5	0.4		
X15	$\mathbf{HW}_{ED,ES}$	5	0.4		
X16	$\mathbf{HW}_{ID,IS}$	5	0.4		
X17	$\mathbf{HW}_{CD,CS}$	5	0.4		
X18	$\mathbf{HW}_{WD,WS}$	5	0.4		

References

Abromaityte, M., Vos, T., Lucassen, B., Kruger, Q., Dubero, E., Van den Hout, N.J., Bouma, D., Treur, J., Roelofsma, P.H.M.P.: Computational analysis of user experience of password-based authentication systems. In: Proceedings of the 16th International Conference on Intelligent Human Computer Interaction, IHCI'24. Lecture Notes in Computer Science, vol. 15557, pp. 341–352. Springer Nature, Cham (2024). https://doi.org/10.1007/978-3-031-88705-5_29

Akers, D., Blomquist, T., Niedziela, J., Šebok, S., Van den Hout, N.J., Bouma, D., Treur, J., Roelofsma, P.H.M.P.: Organizational response to apt attacks: computational analysis by behavioral network modelling. In: Proceedings of the 16th International Conference on Intelligent Human Computer Interaction, IHCI'24. Lecture Notes in Computer Science, vol. 15557, pp. 411–425. Springer Nature, Cham (2024)

Bouma, D., Hoffmans, C., Van den Hout, N.J., Zwarts-Wojtowicz, N., Treur, J., Roelofsma, P.H.M.P.: Cyber security in hospitals: a network-oriented model for behavioural learning of employees during phishing simulations. In: Proceedings of the 17th International Conference on Computational Intelligence in Security for Information Systems, CISIS'24. Lecture Notes in Networks and Systems, vol. 957, pp. 98–111. Springer Nature, Cham (2024)

Burton, L.M., Hicks, M.J.: Hurricane Katrina: Preliminary Estimates of Commercial and Public Sector Damages. Marshall University Center for Business and Economic Research. https://www.marshall.edu/cber/files/2021/04/2005_09_XX_Katrina-Estimates.pdf (2005)

Burton, L.M., Hicks, M.J.: Hurricane Harvey: Preliminary Estimates of Commercial and Public Sector Damages on the Houston Metropolitan Area. CBER Data Center. Ball State University Center for Business and Economic Research. https://projects.cberdata.org/reports/HurricaneHarvey2017.pdf (2017)

Capra, A.M.B., Dekker, E., Hoffmann, L., Hoffmans, C.C.A., Mueller, T., Van den Hout, N.J., Treur, J., Roelofsma, P.H.M.P.: Computational analysis of human factors in spear-phishing attacks: an adaptive network model. In: Proceedings of the 16th International Conference on Intelligent Human Computer Interaction, IHCI'24. Lecture Notes in Computer Science, vol. 15557, pp. 353–366. Springer Nature, Cham (2024)

Hoffmans, C.C.A., Van der Does, L., Van der Kolk, M., Van Rooyen, T., Barelds, N., Treur, J., Roelofsma, P.H.M.P.: Analysing the rise of biohacking in a tech-driven city: assessing risks through adaptive network modeling. In: Proceedings of the 16th International Conference on Intelligent Human Computer Interaction, IHCI'24. Lecture Notes in Computer Science. Springer Nature, Cham (2024)

Castillo, A.: Risk analysis and management in power outage and restoration: a literature survey. Electr. Pow. Syst. Res. **107**, 9–15 (2014)

Canbaloğlu, G., Treur, J., Wiewiora, A.: Computational Modeling of Multilevel Organisational Learning and Its Control Using Self-Modeling Network Models. Springer Nature, Cham (2023). https://doi.org/10.1007/978-3-031-28735-0

Dahles, H., Prabawa Susilowati, T.: Business resilience in times of growth and crisis. Ann. Tour. Res. **51**, 34–50 (2013)

Duchek, S.: Organizational resilience: a capability-based conceptualization. Bus. Res. **13**, 215–246 (2019). https://doi.org/10.1007/s40685-019-0085-7

Eyer, J., Rose, A.: Mitigation and resilience tradeoffs for electricity outages. Econ. Disast. Clim. Change. **3**, 61–77 (2018). https://doi.org/10.1007/s41885-018-0034-5

Fani, S.V., Subriadi, A.P.: Business continuity plan: examining of multi-usable framework. Procedia Comput. Sci. **161**, 275–282 (2018)

Faraji, J., Babaei, M., Bayati, N., Hejazi, M.A.: A comparative study between traditional backup generator systems and renewable energy based microgrids for power resilience enhancement of a local clinic. Electronics. **8**(12), 1485 (2019). https://doi.org/10.3390/electronics8121485

Galaitsi, S.E., Pinigina, E., Keisler, J.M., Pescaroli, G., Keenan, J.M., Linkow, I.: Business continuity management, operational resilience, and organizational resilience: commonalities, distinctions, and synthesis. Int. J. Disast. Risk Sci. **14**, 713–721 (2023). https://doi.org/10.1007/s13753-023-00494-x

Gilani, M.A., Kazemi, A., Ghasemi, M.: Distribution system resilience enhancement by microgrid formation considering distributed energy resources. Energy. **191**, 116442 (2020)

Goodrich, J.N.: The Big American blackout of 2003. J. Travel Tour. Mark. **18**(2), 31–37 (2005). https://doi.org/10.1300/j073v18n02_03

Hendrikse, S.C.F., Treur, J., Koole, S.: Modeling emerging interpersonal synchrony and its related adaptive short-term affiliation and long-term bonding: a second-order multi-adaptive neural agent model. Int. J. Neural Syst. **33**(7), 2350038 (2023)

Hendrikse, S.C.F., Treur, J., Koole, S.L. (eds.): New Analysis and Modeling Directions in Social Interaction Science: Emergent Multimodal Interpersonal Synchrony, Affiliation, and Bonding. Springer Nature, Cham. To appear (2025)

Chen, H.Y., Das, A., Ivanov, D.: Building resilience and managing post-disruption supply chain recovery: Lessons from the information and communication technology industry. Int. J. Inf. Manag. **49**, 330–342 (2019). https://doi.org/10.1016/j.ijinfomgt.2019.06.002

Hepfer, M., Lawrence, T.B.: The heterogeneity of organizational resilience: exploring functional, operational and strategic resilience. Org. Theory. **3**(1), 1–29 (2022). https://doi.org/10.1177/26317877221074701

Homer, J.S., Rice, J.S., Mongird, K., Oikonomou, K.: Coordination and Planning for Water and Power System Resilience. U.S. Department of Energy. https://www.pnnl.gov/main/publications/external/technical_reports/PNNL-31880.pdf. Accessed 10 Jun 2024 (2021)

Hordeski, M.F.: Emergency and Backup Power Sources: Preparing for Blackouts and Brownouts. The Fairmont Press, Inc (2005)

Isniah, S., Purba, H.H., Debora, F.: Plan do check action (PDCA) method: literature review and research issues. J. Sist. Dan Manaj. Indus. **4**(1), 72–81 (2020). https://doi.org/10.30656/jsmi.v4i1.2186

Jeffery, E., Malsagov, U., Tijssen, N., Zeisig, A., Hoffmans, C., Treur, J., Roelofsma, P.H.M.P.: Computational analysis of the effectiveness of the intelligence cycle for organizational cyber risk management. In: Proceedings of the 8th International Conference on Computational Methods in Systems and Software, CoMeSySo'24. Lecture Notes in Networks and Systems. Springer Nature, Cham (2024)

Keijzer, S., Lochtenbergh, D., Marsman, T., Voorhoeve, S., Zwarts, N., Bouma, D., Treur, J., Roelofsma, P.H.M.P.: Managing Classified Information by a Third-Party Contractor: A Computational Cybersecurity Analysis. In: Plácido da Silva, H., Cipresso, P. (eds.) Computer-Human Interaction Research and Applications, CHIRA 2024. Communications in Computer and Information Science, vol. 2371, pp. 299–315. Springer, Cham. https://doi.org/10.1007/978-3-031-83845-3_18

Khan, M.A.: A survey of security issues for cloud computing. J Netw. Comput. Appl. **71**, 11–29 (2016)

Mauthe, A., Hutchison, D., Çetinkaya, E.K., Ganchev, I., Rak, J., Sterbenz, J.P., Gunkel, M., Smith, P., Gomes, T.: Disaster-resilient communication networks: principles and best practices. In: 2016 8th International Workshop on Resilient Networks Design and Modeling (RNDM), pp. 1–10 (2016)

Petermann, T., Bradke, H., Lüllmann, A., Poetzsch, M., Riem, U.: What Happens During a Blackout. Office of Technology Assessment at the German Bundestag (2011)

Rock, T.: 7 Real-Life Business Continuity Plan Examples You'll Want to Read. Invenio IT (2024). https://invenioit.com/continuity/4-real-life-business-continuity-examples/

Roelofsma, P.H.M.P., Jabeen, F., Taal, H.R., Treur, J. (eds.): Using Shared Mental Models and Organisational Learning to Support Safety and Security through Cyberspace: a Computational Analysis Approach. Springer Nature (2025)

Sjöstrand, K., Klingberg, J., Sedehi Zadeh, N., Haraldsson, M., Rosén, L., Lindhe, A.: The value of water—estimating water-disruption impacts on businesses. Water. **13**(11), 1565 (2021). https://doi.org/10.3390/w13111565

Tatineni, S.: Cloud-based business continuity and disaster recovery strategies. Int. Res. J. Modern. Eng. Technol. Sci. **5**(11) (2023). https://doi.org/10.56726/IRJMETS46236

Treur, J.: Network-oriented modeling and its conceptual foundations. In: Spiro, E., Ahn, Y. (eds.) Social Informatics. Proceedings of the 8th International Conference on Social Informatics, SocInfo'16. Lecture Notes in Computer Science, vol. 10047, pp. 157–175. Springer, Berlin (2016). https://doi.org/10.1007/978-3-319-47874-6_12

Treur, J.: Network-Oriented Modeling for Adaptive Networks: Designing Higher-Order Adaptive Biological, Mental and Social Network Models. Springer Nature (2020). https://doi.org/10.1007/978-3-030-31445-3

Treur, J.: On the dynamics and adaptivity of mental processes: relating adaptive dynamical systems and self-modeling network models by mathematical analysis. Cogn. Syst. Res. **70**, 93–100 (2021)

Treur, J., Van Ments, L. (eds.): Mental Models and Their Dynamics, Adaptation, and Control. Springer Nature, Cham (2022). https://doi.org/10.1007/978-3-030-85821-6

World Economic Forum: Internet shutdowns: What happens when the internet shuts down? World Economic Forum website (2022). https://www.weforum.org/agenda/2022/10/internet-shutdowns-explainer/. Accessed 10 Jun 2024

Adaptive Defence Mechanisms: An Adaptive Network Model to Protect HiX Systems Against Cyber Attacks

Tamás Börcsök[1], Ming Bos[1], Willemijn Buter[1], Daphne van der Hoeven[1], Atanas Velev[1], Niek Jan van den Hout[1], Debby Bouma[1,2], Jan Treur[1,2(✉)], and Peter H. M. P. Roelofsma[1]

[1] Risk Management and Cybersecurity Group, Cybersecurity Centre of Expertise, The Hague University of Applied Sciences, Hague, The Netherlands

[2] Social AI Group, Department of Computer Science, Vrije Universiteit Amsterdam, Amsterdam, The Netherlands

j.treur@vu.nl

Abstract. This paper analyses a sensitive part of the Cloud Supply Chain of a care provider that uses HiX through a hospital data server. The computational analysis approach used is based on adaptive network-oriented modelling. Models were created to simulate and analyse the process of a cyber-attack on a hospital data server. These simulations showed how by learning from experiences the ability to effectively counter an attack can increase per attack. In addition, it is analysed how the level of communication and the level of expertise amongst the Security Operations Centre are quite determinant in how effectively an attack is countered.

Keywords: Defence Mechanisms · Adaptive Network · HiX Systems · Cyber Attacks

1 Introduction

In an era where healthcare increasingly relies on digital infrastructure, ensuring the security of sensitive medical data is paramount. This paper focuses on mapping a critical segment of the Cloud Supply Chain for a hospital that utilizes HiX through a hospital data server. Since Erasmus Medical Center Rotterdam is one of the first hospitals to save its data at an external database (Bytesnet, 2024), it is used as an example of a hospital for this paper. In 2018, ChipSoft became market leader as supplier of Electronic Health Records [EHR] in the Netherlands (ChipSoft - Our Ambitions, n.d.). By utilizing an adaptive network-oriented modelling approach based on self-modeling networks (Treur, 2020) and relevant academic literature, models were developed to simulate the impact of a cyber-attack on a hospital data server.

This work is part of a research programme on computational analysis of safety and security in and through cyberspace. In (Abromaityte et al., 2024; Akers et al., 2024; Bouma et al., 2024; Capra et al., 2024; Hoffmans et al., 2024; Jeffery et al.,

R. Silhavy and P. Silhavy (Eds.): CoMeSySo 2024, LNNS 1489, pp. 376–401, 2025.
https://doi.org/10.1007/978-3-031-96798-6_29

2024; Keijzer et al., 2024) various other themes within cybersecurity were addressed. Moreover, in (Roelofsma et al., 2024) it was analysed how an AI coach can support safety and security for medical teams in hospitals.

This paper models a scenario in which an attacker attempts to access a hospital data server to collect sensitive data. The objective is to analyse the effects of this attack on the server's connections and to examine how the Security Operations Centre [SOC] and the Monitoring Application [MA] can learn from the attack to improve their defences against future threats.

To gain insight into the risks associated with the 'Cloud Supply Chain' of a hospital, the scenario mentioned in Sect. 2 has been processed into an adaptive network model. After this process, the adaptive network model was simulated in Matlab which is shown in Sect. 6. Two What-If analyses were performed and described in Sects. 7 and 8. Section 7 is the conceptual What-If analysis that tests performance in hypothetical scenarios, analyses the significance of various data features, and visualizes model behaviour across multiple models and data subsets. Section 8 is the probabilistic What-If analysis where the probabilities for each "if" option are examined and calculated for various simulation outcomes. A risk assessment is added in Sect. 8. The results and ideas for future research are discussed in Sect. 9.

2 Background Knowledge and Scenario Description

Many healthcare providers, such as hospitals, rely on IT services or cloud providers for their business applications. Such applications can include online portals for staff and patients, communication applications such as email, educational applications for medical staff in training, electronic prescription systems, and electronic health records such as Healthcare information eXchange [HiX]. HiX Digital Health Services is a service that, according to the manufacturer's website 'offers comprehensive support for transmural cooperation and operability in healthcare.' and "supports patient referrals and transfers, shared care, monitoring each other's patients, receiving them for consultations, and healthcare at a distance." (Chipsoft, n.d.). HiX is used by general practitioners, hospitals, pharmacies, and other types of healthcare providers.

However, many organizations do not have an overview of the network of providers the organization is dependent on and the consequences if one would be compromised, for example by the influence of a cyber-attack. Taking into consideration the scope of data which HiX provides access to, a compromise could cause the accuracy of schedules, the privacy of patients, and the legitimacy of prescriptions to be breached. To address this challenge, an approach called Cloud Supply Chain Cyber Risk Assessment [CSCCRA] was developed. This method involves visual modelling of the network of cloud service providers, known as Cloud Supply Chain Mapping [CSCM]. By creating a visual representation of the flow of information among cloud services, critical suppliers and Single Points of Failures can be identified.

In this paper, a scenario in which an attacker tries to gain access to a hospital data server with the goal to collect sensitive data is modelled. The objective is to research the effects of this attack on the connections of this data server and how the Security Operations Centre and the Monitoring Application will learn from the attack to counter future attacks.

When modelling, interactions such as between the MA and the SOC, mental processes must be considered. When an attack occurs, the first reaction can be stress. In the scenario, which will be described in more detail in Sect. 4, the SOC will experience the stress stimuli which will cause the reaction to counter the attack. Stress has a strong impact on both cognitive and affective processes. A positive effect of stress is a more adequate response to specific types of threats, as well as elevating surveillance, perception, and attention to threat-related stimuli. The positive effect stress has which is most relatable to the scenario is learning new decision-making behaviour which is better adapted to new situations encountered. Old decision-making is eliminated suppressing existing connections. As a result, more room is created for the ability to learn new connections, which is called the 'disconnect-reconnect' principle (Treur, 2020). The Scenario Simulations in Sect. 5 will show the process of the scenario and the manner in which the SOC and MA learn.

To determine the external Information Technology (IT) service providers of a hospital, the present study employed the Cloud Supply Chain Mapping (CSCM) method (Akinrolabu et al., 2019). This method allows for systematic identification and visualization of most stakeholders contributing to the cloud supply chain of any given firm, or as in our case, a hospital. Particularly, the Cloud Supply Chain Mapping model of the present research considers the Erasmus Medical Center Rotterdam as the foundation for the risk assessment model, therefore all nodes represented in the mapping tool are particularly appropriate to the Erasmus MC. However, the model is not exclusive to other applications outside of Erasmus MC, as it can be tailored to the needs of any clients. To gather data, Open-Source Intelligence [OSINT] was used to provide the most authentic representation possible concerning the Cloud Supply Chain. For illustrative purposes, Fig. 1 represents the applied Clouds Supply Chain Mapping method, as tailored to Erasmus Medical Center Rotterdam.

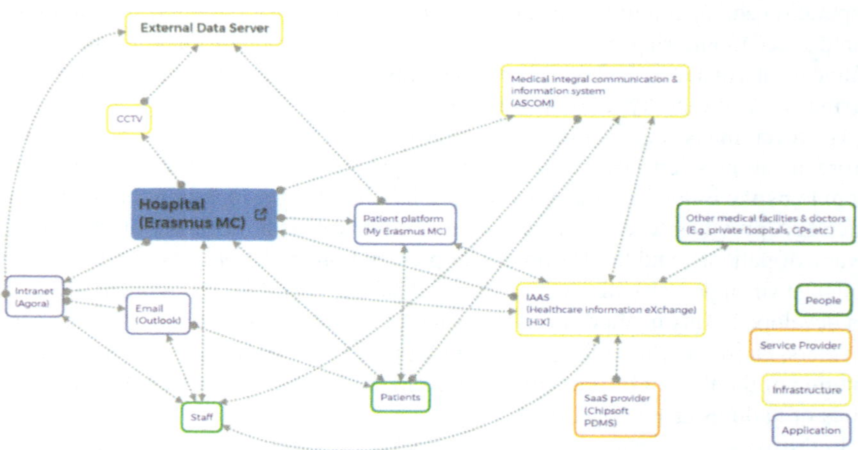

Fig. 1. The cloud supply chain map of Erasmus Medical Center Rotterdam

Firstly, the main service provider that supplies the hospital is the Dutch ChipSoft Patient Data Management System [PDMS]. This is a so-called Service as a Software provider (SaaS), meaning that it provisions and maintains the patient data management system of Erasmus MC. Secondly, the platform that is being provided in this manner is called the Healthcare information eXchange (HiX). Consequently, HiX is one of the platforms that serves as the backbone of data management within the hospital, therefore acts as an Infrastructure as a Service (IaaS). By being an Infrastructure provider, HiX ensures that the information exchange between the hospital staff and the patients is constantly up to date, hence data transfers with internal applications - such as the My Erasmus MC patient platform and the Agora intranet - are carried out with the help of HiX. Furthermore, other infrastructures within the cloud supply chain include the ASCOM Medical internal communication & information system, the CCTV network, and the external data server, all directly feeding, gaining, or exchanging data with Erasmus MC (Erasmus MC ICT Services, 2022). Finally, our hypothetical scenario assumes that a potential hacker breaches the external data server of the hospital, as well as manages to identify, access, and extract sensitive data. Consequently, the simulations strive to illustrate the effectiveness of the countermeasures employed within the hospital, namely the Security Operations Centre and its Monitoring Applications.

3 The Self-modeling Network Modelling Approach Used

To model dynamics in real-world processes, it is crucial to visualize the process by use of a Network-Oriented Modelling approach. An important note is that the word 'network' in this paper does not refer to a literal network such as a criminal network. Instead, the conceptualization of the term 'network' in this paper is a mental concept, a tool to conceptualize a (complex) process such as the process described in Sect. 2. Consequently, this paper does not model a network, it models by the use of networks (Treur, 2016).

The Network-Oriented Modelling approach incorporates a temporal dimension, which enables the interpretation of connections between nodes as temporal-causal connections. Social interactions, such as within the Security Operations Center, often involve multiple cyclic processes and adaptive elements. The temporal-causal dimension is therefore indispensable since it allows the network to show how nodes can have circular causal interactions, and how the timing of the interactions is (Treur, 2016). In temporal-causal networks, nodes Y in a network are interpreted as 'states' which have activation levels $Y(t)$ that vary over time, and connections $X \rightarrow Y$ between the states are interpreted as causal relations that define how a state X can affect a state Y over time. A temporal-causal network can be defined by three main types of characteristics:

(1) Connectivity characteristics
(2) Aggregation characteristics
(3) Timing characteristics

These characteristics are systematically represented in tables called role matrices, which provide a comprehensive specification of the model. The states in a Network-Oriented Modelling approach are therefore characterised by the connections to other states, and their *weight* (or *strength*) $\boldsymbol{\omega}_{X,Y}$ of the connection $X \rightarrow Y$. Consequently, in a role matrix called **mb** the incoming connection(s) to each specific node is indicated, and in a role matrix called **mcw** the weights of these connections.

In complex processes with elaborate interactions, multiple causal relations can affect a state. To combine multiple impacts, the combination function is used: $\mathbf{c}_Y(..)$. Different states and networks may require different combination functions. In the network model introduced here, two combination functions are used: the *advanced logistic sum function*: $\mathbf{alogistic}_{\sigma,\tau}(..)$ with parameters $\boldsymbol{\sigma}$ for steepness and $\boldsymbol{\tau}$ for threshold, as well as the $\mathbf{stepmod}_{\rho,\delta}(..)$ function is used, see Table 1.

The aggregation role matrix **mcfw** for combination function weights details how the effects from various nodes X on a specific node Y within the network combine to form an aggregated impact on node Y. Additionally, the aggregation role matrix **mcfp** for combination function parameters specifies the parameters of these functions. The speed factor role matrix **ms** defines speed factors $\boldsymbol{\eta}_Y$ that represent the rate at which nodes Y within the network change over time based on the aggregated impacts they receive, thereby representing the relative speeds at which different nodes respond. In addition, states have (activation) levels which may remain constant or vary over time. The factors that determine these activation levels at the start are the *initial values*, which are assigned as a numeric value for each node within the network. Subsequently, when the model is being simulated, the assigned initial value will determine the starting value of the respective node within the simulation. This value can be affected by the connections to other nodes and will therefore most likely change over the course of a simulation. For the application of the role matrices used in the present paper, see Sect. 4. Based on these network characteristics, dynamics of the network model uses for every state Y the difference equation

$$Y(t + \Delta t) = Y(t) + \eta_Y\left[\mathbf{c}_Y\left(\omega_{X_1,Y}X_1(t), \ldots, \omega_{X_k,Y}X_k(t)\right) - Y(t)\right] \Delta t$$

where X_i are the states from which state Y has incoming connections.

Table 1. Combination functions used.

Concept	Representation
Advanced logistic sum $\mathbf{alogistic}_{\sigma,\tau}(V_1, \ldots, V_k)$	$\left[\frac{1}{1+e^{-\sigma(V_1+\cdots+V_k-\tau)}} - \frac{1}{1+e^{\sigma\tau}}\right)](1 + e^{-\sigma\tau})$
Stepmod $\mathbf{stepmod}_{\rho,\delta}(V_1, \ldots, V_k)$	time t, repetition interval $\boldsymbol{\rho}$, tipping point $\boldsymbol{\delta}$ 0 if t mod $\rho < \delta$ else 1

It is important to note that the 'stepmod' function alternates between returning 0 (inactive) and 1 (active) at a consistent repetitive interval, which can be further explained by use of a metaphor. The function has an on/off characteristic which resembles a light switch. This light switch is turned on and off at regular times in a time interval which is indicated in the combination function parameters. The repetition parameter $\boldsymbol{\rho}$ specifies the duration of one complete cycle (comprising one inactive and one active period), while the duration parameter $\boldsymbol{\delta}$ determines the length of time before the value switches to 1.

Furthermore, the flexibility of states can differ. While some states can change within a relatively short period, other states lack this ability and are therefore slower. The speed factors $\boldsymbol{\eta}_Y$ are used for the timing of processes.

Note that in network modelling 'network characteristics' and 'network states' refer to two distinct aspects. Network characteristics describe the structure of the network by attributes like connection weights and excitability thresholds, while network states pertain to the dynamic conditions of the network. Self-modelling serves as a bridge between these two concepts by making the implicit characteristics explicit and representing them by a network state called a self-model state. Thus, a self-model involves transforming implicit network characteristics, such as connection weights (ω) and excitability thresholds (τ), into explicit states. This integration allows the network to maintain a self-representation of its structure. Like any state, a self-model state can evolve over time, thus enabling the network to adapt dynamically. This process can be extended to multiple levels, where characteristics at one level become states at the next. This hierarchical self-modelling approach can facilitate the creation of second-order or higher-order adaptive networks. For example, a network characteristic $\boldsymbol{\omega}_{X,Y}$ can be represented by a state $\mathbf{W}_{X,Y}$ that is added to the network, also called a **W**-state. This new state $\mathbf{W}_{X,Y}$ has a speed factor $\boldsymbol{\eta}_{W,X,Y}$ that can be made adaptive as well which can be achieved by adding a second-order self-model state $\mathbf{H}_{W,X,Y}$ to the network, also called an $\mathbf{H}_W$-state. This makes the network model second-order adaptive.

Any smooth dynamical system has a canonical representation as a network and any smooth (multi-order) adaptive dynamical system has a canonical representation as a (multi-order) self-modeling network, as is shown in (Treur, 2021; Hendrikse et al., 2023). Therefore, compared to adaptive dynamical systems in general, the network-oriented modeling approach used does not introduce any fundamental limitations concerning what it can model. This has also been confirmed by applications to complex adaptive processes in different disciplines, such as mental models in cognitive science, organisational learning in management science, synchronisation and bonding in social psychology, and teamwork in health science; e.g., (Treur and Van Ments, 2022; Canbaloğlu et al., 2023; Hendrikse, et al., 2024; Roelofsma et al., 2024).

4 The Introduced Adaptive Network Model

This section introduces the adaptive network model (see Fig. 2) tailored to the scenario as described in Sect. 2. The states of this adaptive network model are explained in Table 2 and the connections in Table 9 in the Appendix section.

In Table 3 the base connectivity of the states is indicated. Please note that this table only includes the incoming connections from the same level, or the levels below it. Consequently, downward connections (for example, a connection from the first order self-model level the base level), are not indicated in Table 3.

Table 4 provides a detailed overview of the weights of connections between states. As can be seen, most connection weights have a value of 1. However, for the connections regarding the two threshold states (T1 and T2), more tailored connection weights were appointed.

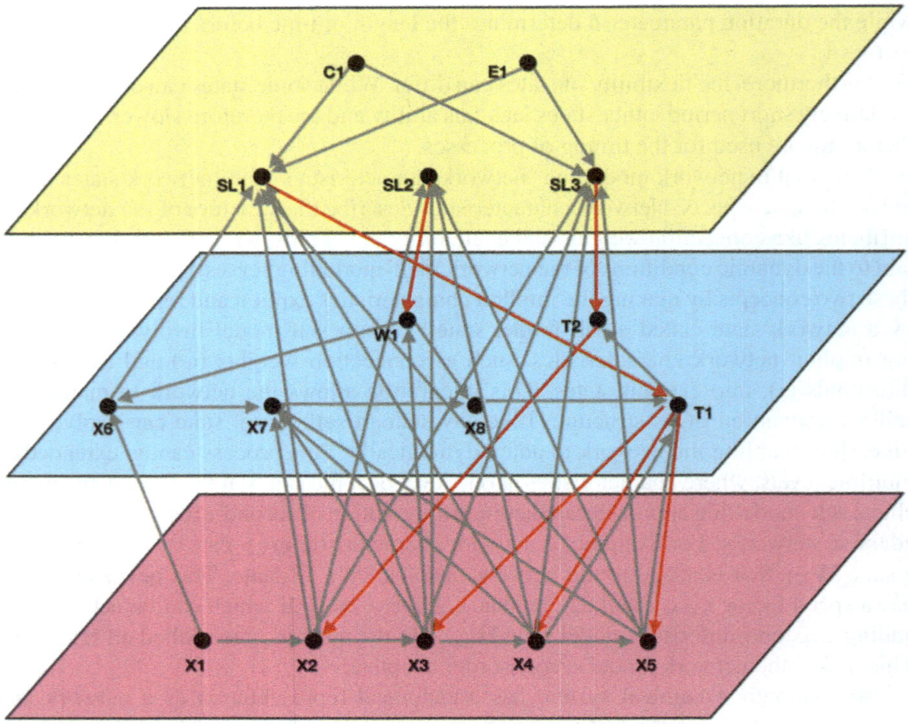

Fig. 2. Picture of the second-order adaptive network model

Table 2. States and their explanation

Nr	Name	Explanation of the node	Connection from
X1		The attacker tries to gain access to the hospital data server	X1
X2		The attacker breaches the security system and gains access	X1
X3		Sensitive data is identified by the attacker	X2
X4		Sensitive data is accessed by the attacker	X3
X5		Sensitive data is extracted by the attacker	X4
X6		Monitoring application MA notices the threat posed by the attacker	X1
X7		Monitoring Application MA notices the attack	X6, X2, X3, X4, X5
X8		Security Operations Centre SOC is notified by the Monitoring Application MA	X7, **W1**

(*continued*)

Table 2. (*continued*)

Nr	Name	Explanation of the node	Connection from
X9	**T1**	Security Operations Centre SOC actively counters the attack. First-order self-model **T**-state for excitability threshold of X3, X4, X5	X4, X5, X8
X10	**W1**	Monitoring Application [MA] early alerts the Security Operations Centre SOC for red flags. First-order self-model **W**-state for the weight of the connection X2 → X3	X3
X11	**T2**	Security Operations Centre SOC improves Monitoring Application MA rules. First-order self-model **T**-state for excitability threshold of X2	**T1**
X12	**SL1**	Speed of Security Operations Centre [SOC] actively countering the attack. Second-order self-model $\mathbf{H_W}$-state for X9	X6, C1, E1
X13	**SL2**	Speed of Monitoring Application [MA] early alerting Security Operations Centre [SOC] for red flags. Second-order self-model $\mathbf{H_W}$-state for X10	**SL2**
X14	**SL3**	Speed of Security Operations Center [SOC] improving Monitoring Application [MA] rules. Second-order self-model $\mathbf{H_W}$-state for X11	C1, E1
X15	C1	The level of communication	C1
X16	E1	The level of expertise	E1

Table 3. Base connectivity **mb**

mb	Base connectivity (Incoming arrows)							
	1	2	3	4	5	6	7	8
1. X1	X1							
2. X2	X1							
3. X3	X2							
4. X4	X3							
5. X5	X4							
6. X6	X1	X6						
7. X7	X2	X3	X4	X5	X6			
8. X8	X7	X10						
9. T1	X4	X5	X8	X9				

(*continued*)

Table 3. (*continued*)

mb	Base connectivity (Incoming arrows)							
	1	2	3	4	5	6	7	8
10. W1	X3							
11. T2	X9	X11						
12. SL1	X2	X3	X4	X5	X6	X12	X15	X16
13. SL2	X2	X3	X4	X5	X13			
14. SL3	X2	X3	X4	X5	X14	X15	X16	
15. C1	X15							
16. E1	X16							

Table 4. Connection weights **mcw**

mcw	Connection weights							
	1	2	3	4	5	6	7	8
1. X1	1							
2. X2	1							
3. X3	1							
4. X4	1							
5. X5	1							
6. X6	1	1						
7. X7	1	1	1	1	1			
8. X8	1	1						
9. T1	0.8	0.8	0.8	0.9				
10. W1	1							
11. T2	0.8	0.9						
12. SL1	1	1	1	1	1	1	1	1
13. SL2	1	1	1	1	1			
14. SL3	1	1	1	1	1	1	1	
15. C1	1							
16. E1	1							

As explained in Sect. 3, the functions $\mathbf{alogistic}_{\sigma,\tau}(..)$ and $\mathbf{stepmod}_{\rho,\delta}(..)$ were used during this research. Table 5 specifies the combination function parameters. For the $\mathbf{alogistic}_{\sigma,\tau}(..)$ function, the steepness σ and threshold τ are indicated. For the steepness, a value of 5 was chosen. For the threshold, it varied per state. For X1, the $\mathbf{stepmod}_{\rho,\delta}(..)$ function was used. Therefore, the right two columns were used to indicate the period of the repeated time interval, as well as the time point in which the **stepmod** becomes active.

Table 5. Combination function parameters for $\mathbf{alogistic}_{\sigma,\tau}$(..) and $\mathbf{stepmod}_{\rho,\delta}$(..)

mcfpv – combination function parameters	1 alogistic(..)		2 stepmod(p,v)	
	1 (steepness) σ	2 (threshold) τ	Duration of cycle ρ	Activation point δ
1. X1			45	40
2. X2	5	X11		
3. X3	5	X9		
4. X4	5	X9		
5. X5	5	X9		
6. X6	5	0.5		
7. X7	5	0.5		
8. X8	5	0.5		
9. T1	5	1		
10. W1	5	0.5		
11. T2	5	0.95		
12. SL1	5	3.5		
13. SL2	5	2		
14. SL3	5	3.5		
15. C1	5	0.5		
16. E1	5	0.5		

In other words, during every 45-time period, X1 is activated at time point 40. Note that here it is indicated as well that X9 (**T1**) and X11 (**T2**) are used as (first-order) self-model **T**-states for the thresholds of states X2 to X5, see the threshold column in the rows for the latter four states. By simply indicating these X9 and X11 instead of values in those cells, the network model becomes (first-order) adaptive according to the self-modeling principle.

Table 6 contains a detailed overview of the used speed factors and initial values. For X1, a speed factor with a value of 2 was used. This speed factors ensures the on-off switch-like characteristic of the **stepmod(..)** function (as described in Sect. 3) turns 'on' immediately, instead of gradually. The other speed factors depend on the state and its characteristics. As for the following nodes, beginning with X2 and up until X5a speed factor of 0.5 has been assigned. This is due to the fact that these nodes represent the 'attack chain' of a potential hacker, therefore a consistent speed factor has been assigned for all subsequent nodes to attain the most realistic simulation possible. Furthermore, the nodes from X6 to X8 represents the countermeasures in the scenario. These nodes act as a trigger mechanism against a potential attack, thus significantly lower values have been assigned for two reasons. At first, it must be slower than the speed of the attack chain to allow for triggering the countermeasures. Secondly, our simulations suggested that the minimal value of 0.1 allowed for the most authentic representation of the improvement of the learning curve for both the Monitoring Application and the Security Operations Centre alike. Note that here in the speed factors column it is indicated that X12 to X14 (**SL1** to **SL3**) are used as second-order self-model $\mathbf{H_W}$-states for the speed factors of states X9 to X11. By just indicating these X12 to X14 instead of values in those cells, the network model becomes a second-order adaptive network model according to the self-modeling principle.

Table 6. Speed factors and initial values

msv	Speed factors	Initial values
1. X1	2	0
2. X2	0.5	0
3. X3	0.5	0
4. X4	0.5	0
5. X5	0.5	0
6. X6	0.1	0
7. X7	0.1	0
8. X8	0.1	0
9. T1	X12	0
10. W1	X13	0
11. T2	X14	0
12. SL1	0.05	0
13. SL2	0.05	0
14. SL3	0.05	0
15. C1	0	0.6
16. E1	0	0.6

In the rightmost column in Table 6 the initial values are indicated. Due to the use of the **stepmod(..)** function that takes care of activation, most initial values started at 0. The only states differentiating from the others were the level of communication (C1) and the level of expertise (E1). The initial value for these states is 0.6, indicating that the Security Operations Centre [SOC] does have a basic level of communication and expertise but there is room for improvement (toward a value of 1).

5 Simulation Results

In this section, various simulations of the scenario as indicated in Sect. 2 and according to the values as indicated in the role matrices included above and in the Appendix section. See Fig. 3 for the overall plot; the Figs. 4, 5 and 6 each show a detailed plot of a single level of the model.

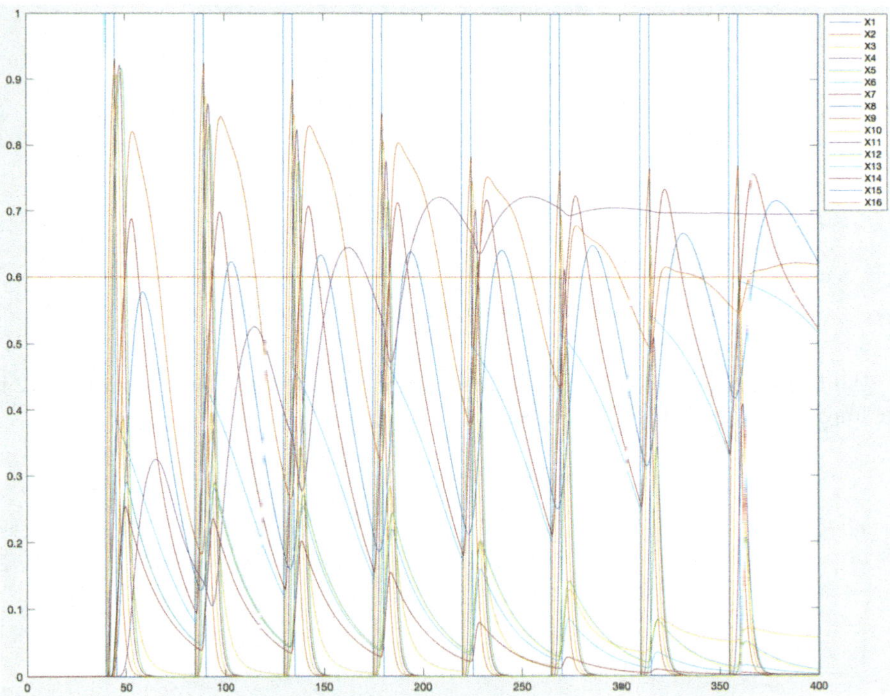

Fig. 3. The adaptive network model: Simulation 1

There are multiple peaks in the graph, indicating repeated attacks. These decrease well overtime due to the learning curve of the self-model **T**-states for learning and therefore lower the intensity of the attacks. For example, X11, where monitoring application rules are improved, becomes more secure further in time, resulting in a steadier tendency.

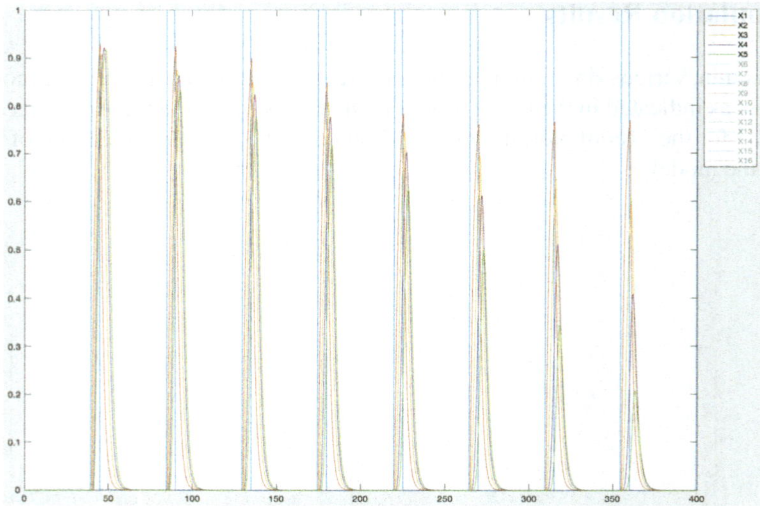

Fig. 4. A plot only showing the base level for Simulation 1

All five states tend to generally decrease proportionally with the exception of X2 in the long term, which slightly increases.

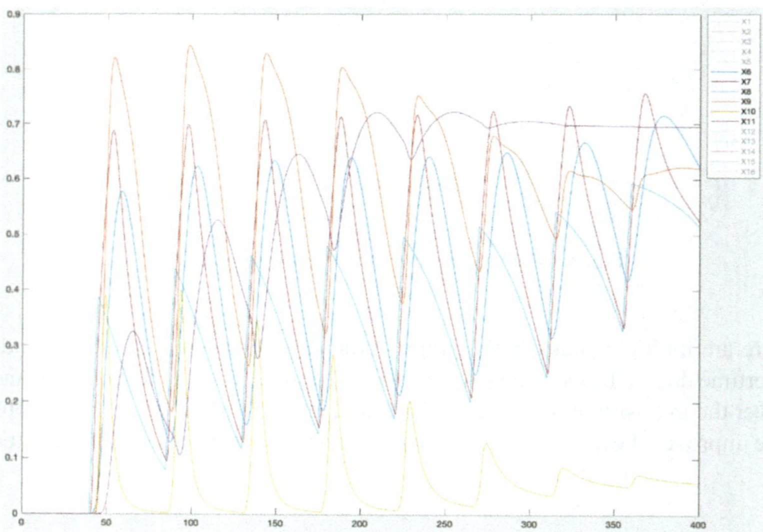

Fig. 5. A plot showing only the first-order adaptation level for Simulation 1

States X6, X7, X8, and X9 share the tendency of increasing slightly in general, except for X10 (the monitoring application early alerting the Security Operations Center for red flags) which in the long-term results in a decrease in attacks. Figure 5 focuses on the second level. In the second simulation, certain initial values and certain other values within the role matrices were adjusted.

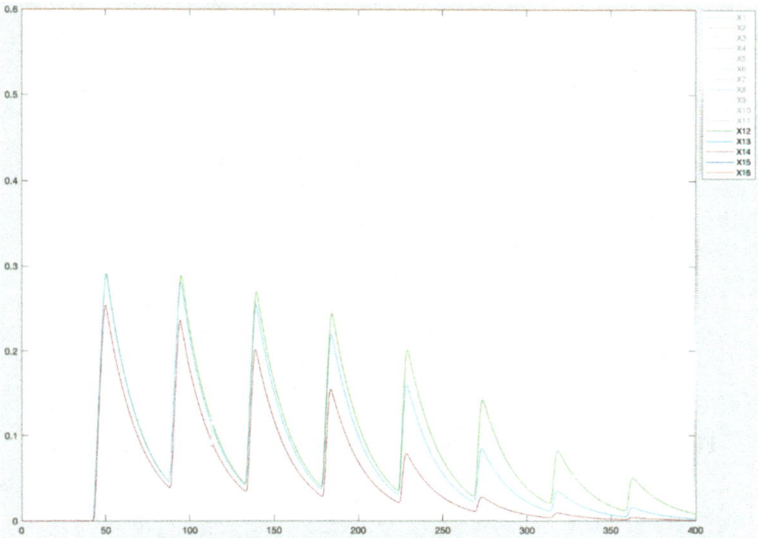

Fig. 6. A plot only showing the second-order adaptation level for Simulation 1

For running Simulation 2, the initial values of C1 and E1 were changed from 0.6 to 1.0. The simulation then ran as shown in Figs. 7 and 8. Overall, the quality is improved. This can be seen by the general trend of higher values on the vertical axis in comparison to Fig. 3. With a higher initial value in the level of expertise and communication, the process takes longer, but the quality is much higher. The most significant outcome is the change in X9 and X11, which is further described below. With a higher initial value in the level of expertise and communication, the process takes longer, but the quality is significantly higher. This is demonstrated in Fig. 7, where the peaks of X11 were considerably higher compared to the original values graph. The quality becomes higher, but it flattens in the end as the quality is already high. State X9 SOC actively countering each attack, results in the drops pictured in the graph that define the peaks after the attack is countered.

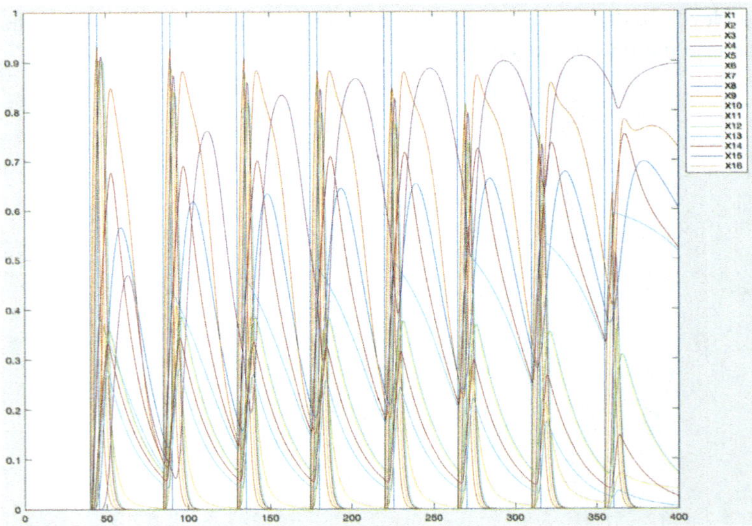

Fig. 7. Simulation 2 with initial values for C1 and E1 increased to 1.0

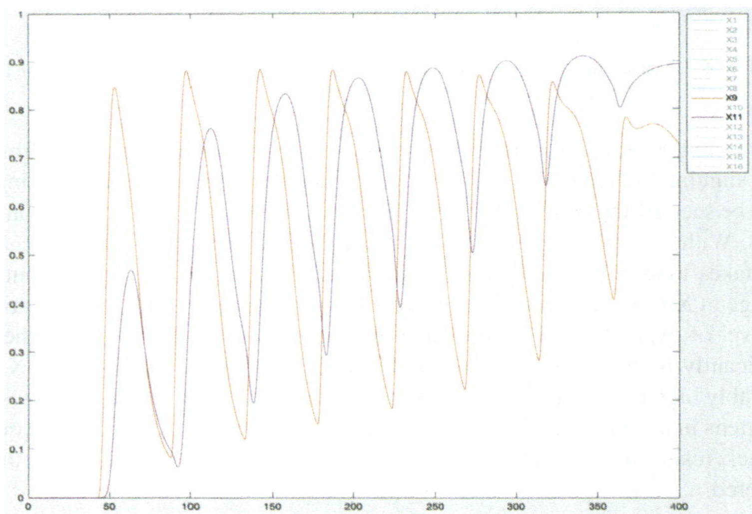

Fig. 8. Simulation 2 with initial values for C1 and E1 increased to 1.0 with focus on T1 and T2

6 What-If Analysis

A What-If analysis enables to test performance in hypothetical situations, analyse the importance of certain features of the situation, and visualise model behaviour across multiple variations (Wexler et al., 2019). In this paper, the hypothetical situation as explained in Sect. 2 is tested according to an already provided model. The what-if analysis tool is used to visualise the working of the self-adaptive model that is described in Sect. 4. In particular, it is analysed how development of measures depends on experiences with attacks. Since the impact of the cyber-attacks is analysed, this approach focuses on the network characteristics for X2, X3, X4, and X5. To measure their effect on the state X5, the If factor 1–4 are the weights of the connections to X1 → X2, X2 → X3, X3 → X4, and X4 → X5, so they were changed individually, resulting in the running of $2^4 = 16$ different scenarios that were simulated. For low weight, the If factor option value of 0.3 was used, while for high weight, the If factor option value of 0.8 was used. For an overview of these criteria see Table 7.

Table 7. Overview of If factor options and their labels, value intervals and probabilities.

If factors	If factor options	If factor option labels	If factor option value intervals	If factor option probabilities
1: Connection weight X1 → X2	If factor 1 option 1	low	0–0.5	0.3
	If factor 1 option 2	high	0.5–1	0.7
2: Connection weight X2 → X3	If factor 2 option 1	low	0–0.5	0.3
	If factor 2 option 2	high	0.5–1	0.7
3: Connection weight X3 → X4	If factor 3 option 1	low	0–0.5	0.3
	If factor 3 option 2	high	0.5–1	0.7
4: Connection weight X4 → X5	If factor 4 option 1	low	0–0.5	0.3
	If factor 4 option 2	high	0.5–1	0.7

When analysing the What factor X5 outcomes, a distinction was made between short-term and long-term effects. The reason for this, is that it is crucial for understanding the direct and sustained impact of the adaptive countermeasures. Specifically, for the SOC and MA it is essential to understanding their true effectiveness. To enable a probabilistic What-If analysis, probabilities for each of the If options were estimated, see Table 7, last column.

In addition, a risk assessment is introduced. This risk assessment considers both the probabilities as stated in the probabilistic What-If analysis and the impact levels. For the impact indicators for the categories 'Short-term What factor X5; low' and the 'Short-term What factor X5; high' the values 9 and 4 were estimated respectively. If the connection weight If factor 1 (X2) and If factor 2 (X3) is relatively high, the 'What factor outcome' (the peaks) are relatively low. This shows that the countermeasures become more effective. However, this process needs time. Therefore, the 'Short-term What factor 4; low' shows a relatively high What factor outcome, demonstrating that the countermeasures take time to become effective.

We chose only What factor X5 because it is the one with the most drastic and severe outcome as well as with the highest impact in case of a breach in the data server.

Short-term low indicator, equalling 9, is a high impact one due to the countermeasures threshold not being reached, as well as the internal process of the learning states has not been yet implemented due to the insufficient time. Furthermore, the high option is a lower impact indicator due to reaching the countermeasures threshold and therefore limiting the impact. The long-term indicators lead to a lower risk due to the sufficient time and experience necessary for their development and use. These indicators are further accounted for in the risk formula 'Impact * Probability = Risk' which is visible in Table 8. In the fifth column in Table 8 it is shown that the probabilities for the different If factors are assumed independent, so that the product rule p(A and B) = p(A) * p(B) can be applied to estimate the probabilities of the If factor option combinations. Some more info can be found as Linked Data at https://www.researchgate.net/publication/382739769.

Table 8. Risk assessment scheme

Risk Assessment: Probability x Impact Level					Short term What factor X5		Long term What factor X5	
If factor 1 options	If factor 2 options	If factor 3 options	If factor 4 options	If factor option combination probabilities	Low	High	Low	High
If factor 1 option 1: Low 0.3	If factor 2 option 1: low 0.3	If factor 3 option 1: low 0.3	If factor 4 option 1: low (0.3)	0.3 x 0.3 x 0.3 x 0.3 = 0.0081	0.0081		0.0081	
			If factor 4 option 2: high (0.7)	0.3 x 0.3 x 0.3 x 0.7 = 0.0189	0.0189		0.0189	
		If factor 3 option 2: high 0.7	If factor 4 option 1: low (0.3)	0.3 x 0.3 x 0.7 x 0.3 = 0.0189	0.0189		0.0189	
			If factor 4 option 2: high (0.7)	0.3 x 0.3 x 0.7 x 0.7 = 0.0441	0.0441		0.0441	
	If factor 2 option 2: high 0.7	If factor 3 option 1: low 0.3	If factor 4 option 1: low (0.3)	0.3 x 0.7 x 0.3 x 0.3 = 0.0189	0.0189		0.0189	
			If factor 4 option 2: high (0.7)	0.3 x 0.7 x 0.3 x 0.7 = 0.0189	0.0189		0.0189	
		If factor 3 option 2: high 0.7	If factor 4 option 1: low (0.3)	0.3 x 0.7 x 0.7 x 0.3 = 0.0441	0.0441		0.0441	

(continued)

Table 8. *(continued)*

Risk Assessment: Probability x Impact Level					Short term What factor X5		Long term What factor X5	
			If factor 4 option 2: high (0.7)	0.3 x 0.7 x 0.7 x 0.7 = 0.1029		0.1029	0.1029	
If factor 1 option 2: High 0.7	If factor 2 option 1: low 0.3	If factor 3 option 1: low 0.3	If factor 4 option 1: low (0.3)	0.7 x 0.3 x 0.3 x 0.3 = 0.0189	0.0189		0.0189	
			If factor 4 option 2: high (0.7)	0.7 x 0.3 x 0.3 x 0.7 = 0.0441	0.0441		0.0441	
		If factor 3 option 2: high 0.7	If factor 4 option 1: low (0.3)	0.7 x 0.3 x 0.7 x 0.3 = 0.0441	0.0441		0.0441	
			If factor 4 option 2: high (0.7)	0.7 x 0.3 x 0.7 x 0.7 = 0.1029		0.1029	0.1029	
	If factor 2 option 2: high 0.7	If factor 3 option 1: low 0.3	If factor 4 option 1: low (0.3)	0.7 x 0.7 x 0.3 x 0.3 = 0.0441	0.0441		0.0441	
			If factor 4 option 2: high (0.7)	0.7 x 0.7 x 0.3 x 0.7 = 0.1029	0.1029		0.1029	
		If factor 3 option 2: high 0.7	If factor 4 option 1: low (0.3)	0.7 x 0.7 x 0.7 x 0.3 = 0.1029	0.1029		0.1029	
			If factor 4 option 2: high (0.7)	0.7 x 0.7 x 0.7 x 0.7 = 0.2401		0.2401	0.2401	
Sum of probabilities					0.5289	0.4459	0.9748	0.0
Impact indicator					9	4	3	1
Risk assessment					0.53 x 9 = 4.76	0.45 x 4 = 1.78	0.97 x 3 = 2.92	0.0 x 1 = 0

7 Discussion

This paper aimed to map a sensitive part of the Cloud Supply Chain of a care provider that uses HiX through a hospital data server. By using the software an adaptive network model was created to simulate the process of a cyber-attack on a hospital data server. The simulations showed that the ability to effectively counter an attack increases per attack. In addition, the level of communication and the level of expertise amongst the Security Operations Centre are determinant in how effectively an attack is countered.

It should be taken into account however, that the model presented in this paper is a simplified model. It includes a limited number of variations and therefore, future research should aim to include more variations within the scenario to create a model which is even more realistic. Additionally, the ability of the level of expertise and the level of communication to learn should be taken into account in further research. In this paper, an initial value of 0.6 was chosen for the level of communication (C1) and the level of expertise (E1). Technically speaking, the level of communication and the level of expertise can improve over time, due to gained experience and/or knowledge. However, in this paper, this ability was not taken into account. For future research, it is therefore recommended to take the ability of C1 and E1 to improve into account to create a self-adaptive network model that is more extensive than this current version.

Lastly, future research is encouraged to specify the sort of cyber-attack in a more detailed manner. The current model focuses on a cyber-attack in general. However, there are multiple strategies or manners in which a hacker can try to gain access to the external data server. A ransomware attack may require different actions or strategies than a phishing link attack. Tailoring the model to different types of attacks would increase the effectiveness of the appliance of this model and would consequently increase the practicality of the model.

Appendix

See Tables 9 and 10.

Table 9. Explanations of the connections found in the adaptive network model

Connection	Connection in full	Connection explanation
X1 → X2	The attacker tries to gain access to the hospital data server → The attacker breaches the security system and gains access	If the attacker successfully breaches the hospital's data server, he can gain access to the data server
X1 → X6	The attacker tries to gain access to the hospital data server → Monitoring Application [MA] notices the threat posed by the attacker	The Monitoring Application [MA] detects the threat as the attacker tries to access the hospital data server
X2 → X3	The attacker breaches the security system and gains access → Sensitive data is identified by the attacker	If the attacker successfully breaches the hospital's security system and gains access, he can start identifying the data that is sensitive and therefore valuable
X2 → **SL1**	The attacker breaches the security system and gains access → Speed of Security Operations Centre [SOC] actively countering the attack	When the attacker breaches the security system and gains access to the hospital security system, the Security Operations Centre [SOC] quickly starts countering the attack

(*continued*)

Table 9. (*continued*)

Connection	Connection in full	Connection explanation
X2 → **SL2**	The attacker breaches the security system and gains access → Speed of Monitoring Application [MA] early alerting Security Operations Centre [SOC] for red flags	When the attacker breaches the hospital's security system and gains access, the Monitoring Application [MA] can quickly alert the Security Operations Centre [SOC] to the red flags
X2 → **SL3**	The attacker breaches the security system and gains access → Speed of Security Operations Centre [SOC] improving Monitoring Application [MA] rules	After the attacker breaches the security system and gains access, the Security Operations Centre [SOC] quickly starts improving the Monitoring Application [MA] rules
X3 → X4	Sensitive data is identified by the attacker → Sensitive data is accessed by the attacker	The attacker accesses the sensitive data that he has identified
X3 → X7	Sensitive data is identified by the attacker → Monitoring Application [MA] notices the attack	As the attacker identifies the sensitive data, the Monitoring Application notices the attack
X3 → W1	Sensitive data is identified by the attacker → Monitoring Application [MA] early alerts the Security Operations Centre [SOC] for red flags	As the attacker identifies the sensitive data, the Monitoring Application [MA] early alerts the Security Operations Centre [SOC] for red flags
X3 → **SL1**	Sensitive data is identified by the attacker → Speed of Security Operations Centre [SOC] actively countering the attack	After the sensitive data is identified by the attacker, the Security Operations Centre [SOC] quickly counters the attack
X3 → **SL2**	Sensitive data is identified by the attacker → Speed of Monitoring Application [MA] early alerting Security Operations Centre [SOC] for red flags	After the sensitive data is identified by the attacker, the Monitoring Application [MA] quickly early alerts the Security Operations Centre [SOC] for red flags
X3 → SL3	Sensitive data is identified by the attacker → Speed of Security Operations Centre [SOC] improving Monitoring Application [MA] rules	After the sensitive data is identified by the attacker, the Security Operations Centre [SOC] quickly improves the Monitoring Application [MA] rules
X4 → X5	Sensitive data is accessed by the attacker → Sensitive data is extracted by the attacker	After the sensitive data is accessed by the attacker, he extracts it
X4 → X7	Sensitive data is accessed by the attacker → Monitoring Application [MA] notices the attack	The Monitoring Application [M] notices that the sensitive data is accessed by the attacker

(*continued*)

Table 9. (*continued*)

Connection	Connection in full	Connection explanation
X4 → **T1**	Sensitive data is accessed by the attacker → Security Operations Centre [SOC] actively counters the attack	The Security Operations Centre [SOC] actively counters when the sensitive data is accessed by the attacker
X4 → **SL1**	Sensitive data is accessed by the attacker → Speed of Security Operations Centre [SOC] actively countering the attack	When the sensitive data is accessed by the attacker, the Security Operations Centre [SOC] quickly counters the attack
X4 → **SL2**	Sensitive data is accessed by the attacker → Speed of Monitoring Application [MA] early alerting Security Operations Centre [SOC] for red flags	When the sensitive data is accessed by the attacker, the Monitoring Application [MA] quickly alerts the Security Operations Centre [SOC] for red flags
X4 → **SL3**	Sensitive data is accessed by the attacker → Speed of Security Operations Centre [SOC] improving Monitoring Application [MA] rules	When the sensitive data is accessed by the attacker the Security Operations Centre [SOC] improves the Monitoring Application [MA] rules
X5 → X7	Sensitive data is extracted by the attacker → Monitoring Application [MA] notices the attack	The Monitoring Application [MA] notices that sensitive data is extracted by the attacker
X5 → **T1**	Sensitive data is extracted by the attacker → Security Operations Centre [SOC] actively counters the attack	The Security Operations Centre [SOC] actively counters the attack of sensitive data being extracted by the attacker
X5 → **SL1**	Sensitive data is extracted by the attacker → Speed of Security Operations Centre [SOC] actively countering the attack	The Security Operations Centre [SOC] rapidly counters the attack of sensitive data being extracted by the attacker
X5 → **SL2**	Sensitive data is extracted by the attacker → Speed of Monitoring Application [MA] early alerting Security Operations Centre [SOC] for red flags	The Monitoring Application [MA] quickly alerts the Security Operations Centre [SOC] for red flags
X5 → **SL3**	Sensitive data is extracted by the attacker → Speed of Security Operations Centre [SOC] improving Monitoring Application [MA] rules	After the sensitive data is extracted by the attacker, the Security Operations Centre [SOC] improves the Monitoring Application [MA] rules

(*continued*)

Table 9. (*continued*)

Connection	Connection in full	Connection explanation
X6 → X7	Monitoring application [MA] notices the threat posed by the attacker → Monitoring Application [MA] notices the attack	The Monitoring Application notices the threat and sees it as an attack
X6 → **SL1**	Monitoring Application [MA] notices the threat posed by the attacker → Speed of Security Operations Centre [SOC] actively countering the attack	As the Monitoring Application [MA] notices the threat, the Security Operation Centre [SOC] quickly counters the attack
X7 → X8	Monitoring Application [MA] notices the attack → Security Operations Centre [SOC] is notified by the Monitoring Application [MA]	The Security Operations Centre [SOC] is notified of the attack that was noticed by the Monitoring Application [MA]
X8 → **T1**	Security Operations Centre [SOC] is notified by the Monitoring Application [MA] → Security Operations Centre [SOC] actively counters the attack	The Security Operations Centre [SOC] actively counters the attack after being notified by the Monitoring Application [MA]
T1 → **T2**	Security Operations Centre [SOC] actively counters the attack → Security Operations Centre [SOC] improves Monitoring Application [MA] rules	The Security Operations Centre [SOC] improves the Monitoring Application [MA] rules after actively countering the attack
T1 → X3	Security Operations Centre [SOC] actively counters the attack → Sensitive data is identified by the attacker	The Security Operations Centre [SOC] actively countering the attack increases the threshold for the attacker to identify sensitive data
T1 → X4	Security Operations Centre [SOC] actively counters the attack → Sensitive data is accessed by the attacker	The Security Operations Centre [SOC] actively countering the attack increases the threshold for the attacker to access sensitive data
T1 → X5	Security Operations Centre [SOC] actively counters the attack → Sensitive data is extracted by the attacker	The Security Operations Centre [SOC] actively countering the attack increases the threshold for the attacker to extract sensitive data
W1 → X6	Monitoring Application [MA] early alerts the Security Operations Centre [SOC] for red flags → Monitoring application [MA] notices the threat posed by the attacker	The Monitoring Application [MA] early alerting the Security Operations Centre [SOC] for red flags increases the weight of the Monitoring Application [MA] noticing the threat

(*continued*)

Table 9. (*continued*)

Connection	Connection in full	Connection explanation
W1 → X8	Monitoring Application [MA] early alerts the Security Operations Centre [SOC] for red flags → Security Operations Centre [SOC] is notified by the Monitoring Application [MA]	The Monitoring Application [MA] early alerting the Security Operations Centre [SOC] for red flags increases the weight of the Security Operations centre [SOC] being notified by the Monitoring Application [MA]
T2 → X2	Security Operations Centre [SOC] improves Monitoring Application [MA] rules → The attacker breaches the security system and gains access	The Security Operations Centre [SOC] improving the Monitoring Application [MA] rules increases the threshold for the attacker to breach the security system and gaining access
SL1 → **T1**	Speed of Security Operations Centre [SOC] actively counters the attack → Security Operations Centre [SOC] actively counters the attack	The speed of the Security Operations Centre [SOC] actively countering the attack affects the effectiveness of the counterattack
SL2 → W1	Speed of Monitoring Application [MA] early alerting Security Operations Centre [SOC] for red flags → Monitoring Application [MA] early alerts the Security Operations Centre [SOC] for red flags	The speed of the Monitoring Application [MA] early alerting the Security Operations Centre [SOC] for red flags affects the timeliness of the Monitoring Application alerting the Security Operations Centre [SOC]
SL3 → **T2**	Speed of Security Operations Center [SOC] improving Monitoring Application [MA] rules → Security Operations Centre [SOC] improves Monitoring Application [MA] rules	The speed of the Security Operations Centre [SOC] improving the Monitoring affects the timeliness of the Security Operations Centre [SOC] to improve the Monitoring Application [MA] rules
C1 → **SL1**	The level of communication → Speed of Security Operations Centre [SOC] actively countering the attack	The level of communication can lower/increase the efficiency of the Security Operations Centre [SOC] actively countering the attack
C1 → **SL3**	The level of communication → Speed of Security Operations Centre [SOC] improving Monitoring Application [MA] rules	The level of communication can lower/increase the efficiency of the Security Operations Centre [SOC] improving the Monitoring Application [MA] rules

(*continued*)

Table 9. (*continued*)

Connection	Connection in full	Connection explanation
E1 → **SL1**	The level of expertise → Speed of Security Operations Centre [SOC] actively countering the attack	The level of expertise can lower/increase the speed of the Security Operations Centre [SOC] actively countering the attack
E1 → **SL3**	The level of expertise → Speed of Security Operations Centre [SOC] improving Monitoring Application [MA] rules	The level of expertise can lower/increase the speed of the Security Operations Centre [SOC] improving the Monitoring Application [MA] rules

Table 10. Combination functions

mcfwv	Combination functions	
	1 $\text{alogistic}_{s,t}(..)$	2 stepmod(..)
1. X1		1
2. X2	1	
3. X3	1	
4. X4	1	
5. X5	1	
6. X6	1	
7. X7	1	
8. X8	1	
9. T1	1	
10. W1	1	
11. T2	1	
12. SL1	1	
13. SL2	1	
14. SL3	1	
15. C1	1	
16. E1	1	

References

Abromaityte, M., et al.: Computational analysis of user experience of password-based authentication systems. In: Proceedings of the 16th International Conference on Intelligent Human Computer Interaction, IHCI'24. Lecture Notes in Computer Science. Springer Nature, Cham (2024)

Akers, D., et al.: Organizational response to APT attacks: computational analysis by behavioral network modelling. In: Proceedings of the 16th International Conference on Intelligent Human Computer Interaction, IHCI'24Volume: Lecture Notes in Computer Science. Springer Nature, Cham (2024)

Akinrolabu, O., New, S., Martin, A.: CSCCRA: a novel quantitative risk assessment model for SaaS Cloud service providers. Computers **8**(3), 66 (2019). https://doi.org/10.3390/computers8030066

Bouma, D., Hoffmans, C., Van den Hout, N.J., Zwarts-Wojtowicz, N., Treur, J., Roelofsma, P.H.M.P.: Cyber security in hospitals: a network-oriented model for behavioural learning of employees during phishing simulations. In: Proceedings of the 17th International Conference on Computational Intelligence in Security for Information Systems, CISIS'24. Lecture Notes in Networks and Systems, Springer Nature, Cham (2024)

Bytesnet: Erasmus MC vertrouwt voor alle datacenterdiensten en -verbindingen op Bytesnet – Bytesnet (2024). https://www.bytesnet.nl/cases/erasmus-mc/c

Canbaloğlu, G., Treur, J., Wiewiora, A. (eds.): Computational Modeling of Multilevel Organisational Learning and Its Control Using Self-Modeling Network Models. Springer Nature, Cham (2023). https://doi.org/10.1007/978-3-031-28735-0

Capra, A.M.B., et al.: Computational analysis of human factors in spear-phishing attacks: an adaptive network model. In: Proceedings of the 16th International Conference on Intelligent Human Computer Interaction, IHCI'24. Lecture Notes in Computer Science. Springer Nature, Cham (2024)

ChipSoft - Our ambitions: (n.d.). https://www.chipsoft.com/en/organization/ambition

Erasmus MC ICT Services: Aansluitvoorwaarden Erasmus MC-network. Erasmus MC (2022)

Hee, Q.L., Hogendorp, L., Warnaars, D., Treur, J.: The effect of sleep disturbances on the quality of sleep: an Adaptive Computational Network model. In: Proceedings of the Second International Conference on Innovations in Computing Research (ICR'23). Lecture Notes in Networks and Systems, vol. 721, pp. 105–117. Springer Nature, Cham (2023)

Hendrikse, S.C.F., Treur, J., Koole, S.: Modeling emerging interpersonal synchrony and its related adaptive short-term affiliation and long-term bonding: a second-order multi-adaptive neural agent model. Int. J. Neural Syst. **33**(7), 2350038 (2023)

Hendrikse, S.C.F., Treur, J., Koole, S.L. (eds.): New Analysis and Modeling Directions in Social Interaction Science: Emergent Multimodal Interpersonal Synchrony, Affiliation, and Bonding. Springer Nature, Cham (2024)

Hoffmans, C.C.A., et al.: Analysing the rise of biohacking in a Tech-Driven City: assessing risks through adaptive network modeling. In: Proceedings of the 16th International Conference on Intelligent Human Computer Interaction, IHCI'24. Lecture Notes in Computer Science. Springer Nature, Cham (2024)

Jeffery, E., et al.: Computational analysis of the effectiveness of the intelligence cycle for organizational cyber risk management. In: Proceedings of the 8th International Conference on Computational Methods in Systems and Software, CoMeSySo'24. Lecture Notes in Networks and Systems. Springer Nature, Cham (2024)

Keijzer, S., et al.: Managing Classified Information by a Third-Party Contractor: A Computational Cybersecurity Analysis (2024). Under submission

Köhler, M., Siepel, S., Treur, J.: The interplay between craving and impulse control: an adaptive network model for drug addiction treatment. Procedia Comput. Sci. **190**, 508–520 (2021)

Moulie, H., Van den Berg, R., Treur, J.: An adaptive network model for procrastination behaviour including self-regulation and emotion regulation. In: International Conference on Computational Science, ICCS'21. Lecture Notes in Computer Science, vol. 12742, pp. 540–554. Springer International Publishing, Cham (2021)

Roelofsma, P.H.M.P., Jabeen, F., Taal, H.R., Treur, J. (eds.): Using Shared Mental Models and Organisational Learning to Support Safety and Security through Cyberspace: a Computational Analysis Approach. Springer Nature, Cham (2024)

Treur, J.: Network-oriented modeling and its conceptual foundations. In: Spiro, E., Ahn, Y.Y. (eds.) Social Informatics. Proceedings of the 8th International Conference on Social Informatics, SocInfo'16. Lecture Notes in Computer Science, vol. 10047, pp 157–175. Springer, Cham (2016). https://doi.org/10.1007/978-3-319-47874-6_12

Treur, J.: Network-oriented modeling for adaptive networks: designing higher-order adaptive biological, mental and social network models. In: Studies in Systems, Decision and Control (2020). https://doi.org/10.1007/978-3-030-31445-3

Treur, J.: On the dynamics and adaptivity of mental processes: relating adaptive dynamical systems and self-modeling network models by mathematical analysis. Cogn. Syst. Res. **70**, 93–100 (2021)

Treur, J., Van Ments, L. (eds.): Mental Models and Their Dynamics, Adaptation, and Control. Springer Nature (2022). https://doi.org/10.1007/978-3-030-85821-6

Wexler, J., Pushkarna, M., Bolukbasi, T., Wattenberg, M., Viegas, F., Wilson, J.: The What-If Tool: interactive probing of machine learning models. IEEE Trans. Vis. Comput. Graph. **26**, 56–65 (2019). https://doi.org/10.1109/TVCG.2019.2934619

Transparency in Detecting Man-in-the-Middle Attacks on SS7 Networks Using SHAP Explainable AI

Fauzan Rizqi Muhammad[1], Parman Sukarno[1(✉)], and Aulia Arif Wardana[2]

[1] School of Computing, Telkom University, Bandung, Indonesia
zannrz@student.telkomuniversity.ac.id, psukarno@telkomuniversity.ac.id
[2] Wrocław University of Science and Technology, Wrocław, Poland
aulia.wardana@pwr.edu.pl

Abstract. This study addresses the security vulnerabilities inherent in the SS7 protocol, with a specific focus on mitigating Man-in-the-Middle (MITM) attacks. Despite technological advancements, persistent security issues in Signaling System No. 7 (SS7) underscore the need to enhance user protection and minimize risks. The primary objective is to leverage Explainable Artificial Intelligence (XAI) to make AI decisions in telecommunications more transparent and justifiable. This research employs advanced machine learning algorithms, including Random Forest, Autoencoders, and K-Means Clustering, integrated with SHapley Additive exPlanations (SHAP) to enhance the interpretability of AI models. Given the limited availability of specific SS7 datasets, the study extrapolates data from existing scholarly articles. The research methodology involves systematic data collection and preprocessing, followed by the implementation and optimization of algorithms to effectively detect and analyze vulnerabilities. The integration of XAI with robust analytical tools aims to make the machine learning detection process transparent, thus improving the security of the SS7 protocol. This approach is crucial to identify potential attack vectors and reduce associated risks. The results demonstrate high precision, with the Random Forest algorithm achieving 94% accuracy, the autoencoders showing a low loss around ±0.10, and K-Means Clustering achieving a high Silhouette Score of 0.999. Furthermore, SHAP values provide information on the distinctions and similarities between the algorithms.

Keywords: MITM Attacks · SS7 Protocol · XAI · Machine Learning · Intrusion Detection

1 Introduction

In today's cybersecurity landscape, MITM attacks remain a significant threat, as attackers exploit unsecured communication channels to intercept and manipulate sensitive data. With the increasing complexity of network environments and the

R. Silhavy and P. Silhavy (Eds.): CoMeSySo 2024, LNNS 1489, pp. 402–421, 2025.
https://doi.org/10.1007/978-3-031-96798-6_30

proliferation of remote work, organizations rely heavily on Intrusion Detection Systems (IDS) to monitor network traffic and detect any signs of unauthorized access or malicious activity, including MITM attempts. As cyber threats evolve, incorporating advanced MITM prevention techniques like end-to-end encryption and integrating robust IDS solutions has become essential for organizations to enhance their security posture and protect against the growing risks in today's digital world [1,2]. MITM attacks exploit vulnerabilities in communication protocols, such as SS7. Despite the introduction and increasing adoption of 5G and newer technologies, 3G remains in use. As illustrated in Fig. 1, the 3G phaseout is far from complete. Securing SS7 poses significant challenges due to limited access to open datasets, which are restricted by privacy concerns [3]. Although efforts to improve SS7 security using machine learning have been documented [4], the inherent transparency of these "black-box" models raises concerns about their trustworthiness [5].

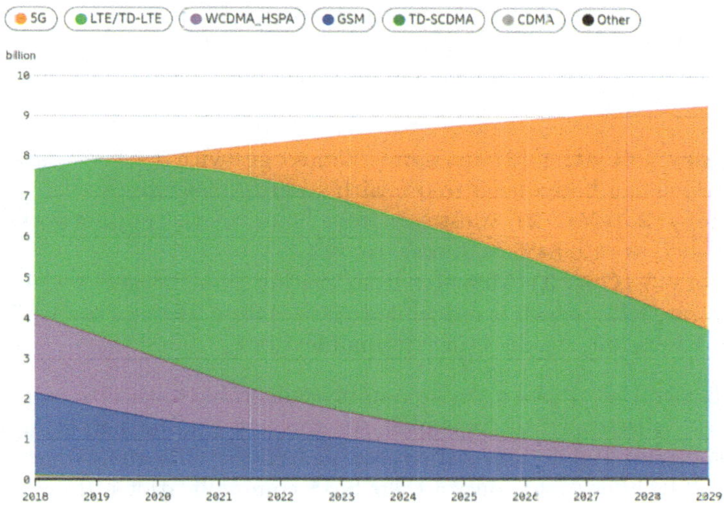

Fig. 1. Number of 3G, 4G, and 5G Subscribers as per Ericsson Study [6]

To enhance transparency, tools such as SHAP to explain AI decision-making processes, increasing trust in these technologies. This study employs interpretable algorithms, including Random Forest, Autoencoders, and K-Means, chosen for their balance of precision and interpretability. Random Forest improves detection accuracy and mitigates overfitting through the use of ensemble decision trees. Meanwhile, Autoencoders and K-Means are utilized for dimensionality reduction and data clustering tasks, respectively.

This research develops detection models that are interpretable, reliable, and robust. Detailed results and analysis are provided in the subsequent sections of this study.

2 Literature Review

2.1 SS7 Protocol and MITM Attacks

SS7 is essential for the functions of cellular networks, including call initiation, routing, and Short Message Service (SMS) service. It depends on three primary nodes:

1. **Service Switching Point (SSP)**: Manages call initiation and termination, and queries SCPs for routing information.
2. **Signal Transfer Point (STP)**: Routes messages between SSP and SCP, maintaining network performance.
3. **Service Control Point (SCP)**: Centralized databases providing vital call processing information.

However, the trust-based design of SS7 lacks robust authentication, making it vulnerable to data intercept, fraud, and unauthorized activities [7]. Advanced security protocols are needed to protect the network.

A MITM attack in SS7 involves intercepting and modifying SS7 messages. SS7 messages are categorized by communication path, each facing different security risks:

1. **Category 1 (Cat. 1)**: Home network messages with a success rate of attack 23%, which can be reduced to 0% with security measures.
2. **Category 2 (Cat. 2)**: Messages from home to visiting networks with an attack success rate 84%, reducible to 44%.
3. **Category 3 (Cat. 3)**: Messages from visiting home networks with an attack success rate 87%, which is reducible to 65%. Often involves the Update Location (UL) exploit, misdirecting communications.

Specifically, Cat. 3 messages are notably vulnerable to the UL call flow intercept attack, a type of MITM exploit. In this attack, the perpetrator utilizes the International Mobile Subscriber Identifier (IMSI) of a subscriber on a visiting network to hijack and control their SMS communications. If successful, the attacker can read, manipulate, and intercept messages, potentially gaining access to sensitive third-party applications such as online banking and social networks [8,9]. Research [10] provides a real-world example of such attacks, highlighting how attackers intercept SMS messages, including those used for two-factor authentication (2FA).

Figure 2 illustrates both legitimate and malicious uses of the Update Location call flow. Under normal circumstances, when a subscriber relocates geographically, their location data is updated across network databases via the Mobile Switching Center (MSC) and Visitor Location Register (VLR). However, in the MITM attack scenario, the attacker impersonates a legitimate MSC/VLR to initiate a fraudulent location update. This action creates a false representation of the subscriber's location, severely compromising their SMS security and exposing them to additional digital threats.

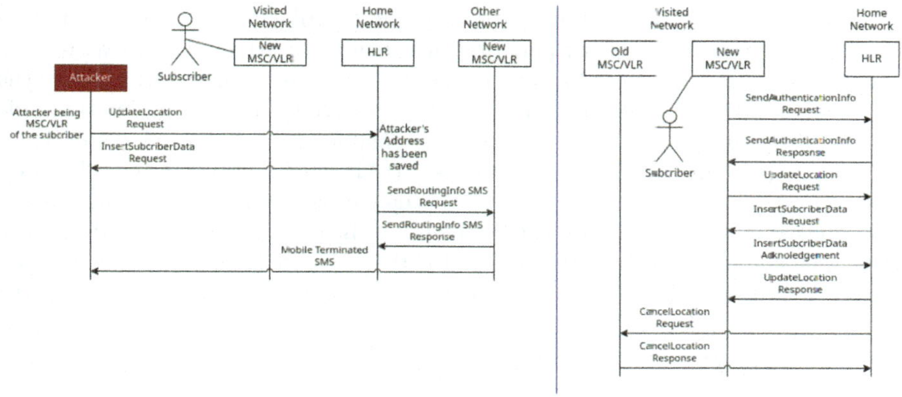

Fig. 2. Comparative Visualization of System Configurations: Attack Operation (Left) vs. Normal Scenario (Right) [11]

2.2 Explainable AI

Recent advances in artificial intelligence have led to a growing emphasis on the trustworthiness of AI systems, resulting in the emergence of XAI. This new domain seeks to make the decision-making processes of machine learning models transparent and understandable to humans. According to [12], XAI addresses the inherent opacity of modern AI, such as Deep Neural Networks (DNN). They provide a comprehensive taxonomy of methods, discuss goals such as trustworthiness, causality, transferability, and fairness, and highlight the challenges of developing metrics for explainability and making complex models interpretable.

Similarly, [13] emphasizes the practical importance of XAI, particularly in critical fields such as medical diagnosis. They explore various methods, including LIME, occlusion analysis, integrated gradients, and Layerwise Relevance Propagation (LRP), comparing them based on interpretability, faithfulness, and computational cost. These methods aim to elucidate the decision-making processes of deep neural networks, enhancing their accessibility and trustworthiness.

In research [14] investigate the practical application of explainability methods by organizations. Their findings reveal that most deployments are primarily used internally by machine learning engineers for debugging purposes, rather than for end-users. This study underscores a significant gap between current practices and the overarching goal of transparency, highlighting the need for frameworks that establish clear goals for explainability and address concerns such as spurious correlations and the feasibility of providing real-time explanations.

Furthermore, research [15], provides an extensive survey of the XAI landscape. They discuss the widespread adoption of AI across various sectors and the essential need for transparency to maintain trust in these systems. Their comprehensive review of existing XAI approaches, trends, and future research directions underscores the importance of explainability in the ongoing advancement of AI.

Collectively, these works underscore the critical role of XAI in improving the transparency, trustworthiness, and practical utility of AI models in real-world applications. They advocate for ongoing research and development to bridge the gap between AI's detection power and the necessity for human-understandable explanations, ultimately fostering more responsible and reliable AI systems.

Among the numerous XAI frameworks, SHAP [16,17] stands out for leveraging cooperative game theory to achieve a balance between accuracy and clarity, which makes it particularly advantageous in the fields of healthcare and finance. SHAP incorporates methodologies like LIME [18] and DeepLIFT [19] into an integrated framework. The SHAP value calculation is concisely captured by the Eq. 1:

$$\phi_i(f, x) = \sum_{z' \subseteq x'} \frac{|z'|!(M - |z'| - 1)!}{M!}[f_x(z') - f_x(z' \setminus i)] \tag{1}$$

Within this equation, $\phi_i(f, x)$ represents the SHAP value for feature i, indicating its contribution to the model output for input x. The function f denotes the model under evaluation, while x represents the complete input vector. The sum of $z' \subseteq x'$ accounts for all subsets of the perturbed input x, where $|z'|$ means the number of features in the subset, and M is the total number of features. The combinatorial factor $\frac{|z'|!(M-|z'|-1)!}{M!}$ compensates for all permutations, including features in z' preceding features i. The expressions $f_x(z')$ and $f_x(z' \setminus i)$ denote the model outputs with and without feature i, respectively, illustrating its marginal effect.

2.3 Related Works

In the field of telecommunications, machine learning techniques are increasingly employed to enhance the security of the SS7 protocol, as evidenced by numerous significant studies:

The research [11] utilized real and simulated datasets to train deep learning models for detecting SS7 attacks, with a specific focus on SMS interceptions within roaming environments. The study used advanced feature selection techniques and semi supervised learning, which demonstrated superior generalization by leveraging both labeled and unlabeled data. The findings indicated that semi-supervised models, particularly PReNet, performed exceptionally well across various metrics, underscoring the potential of deep learning to mitigate SS7 vulnerabilities in high-risk roaming scenarios.

Another study [20], created a dataset with an SS7 attack simulator to safely apply machine learning without exposing real network data. This research used the seasonal hybrid extreme studentized deviation algorithm (S-H-ESD) to identify deviations from normal patterns in SS7 traffic, indicating potential security problems. The study demonstrated a scalable and efficient framework using open-source tools for real-time data processing and anomaly detection, demonstrating its applicability in maintaining secure telecommunications networks.

The research [21] targeted the weaknesses of the SS7 network using a machine learning-based system developed through an SS7 Attack Simulator. This system used the S-H-ESD algorithm to detect anomalies, demonstrating the utility of machine learning to improve telecommunications security by identifying SS7 attacks and providing robust protection.

In the study [3], a dataset was generated using an open-source SS7 attack simulator. Various machine learning and rule-based filtering methods were compared for anomaly detection in SS7 traffic. The findings suggested that a hybrid approach, combining rule-based filtering with machine learning techniques, could offer a robust anomaly detection solution in SS7 networks.

The study [22] utilized a real-world dataset trained with a Convolutional Neural Network (CNN) algorithm, highlighting the efficiency of semi-supervised learning models in detecting SS7 attacks through specialized hyperparameter tuning. The results demonstrated the potential of CNN to identify anomalies in SS7 traffic, contributing to improved security measures in telecommunications.

Finally, the study [23] focused on using a dataset optimized through virtual machine configurations for SS7, employing various machine learning algorithms to detect anomalies with significant success in terms of speed and accuracy, particularly using decision trees and random forests.

These studies collectively emphasize the importance of machine learning in addressing SS7 vulnerabilities, particularly Category 3 messages such as SMS intercepts, which are notably less protected. Despite their promising results, these studies reveal a significant limitation: the lack of interpretability of machine learning models, which can hinder trust and regulatory compliance in critical telecommunication environments. The baseline for the metrics will be defined based on the results of the study [11] as this study uses the same dataset.

3 Methodology

3.1 Dataset

Due to the limited availability of labeled data, this study utilizes a dataset from [11], including simulated data generated using the JSS7 Attack Simulator, with a focus on Cat. 3 attacks. The dataset is composed of 85 labeled instances, where 57 represent normal traffic and 28 represent attack traffic, in addition to 4,557 unlabeled instances. This imbalance in the dataset highlights the challenge of working with limited labeled data, which is a common issue in the field of cybersecurity, particularly when detecting sophisticated attacks like those on the SS7 protocol.

The study focuses on the 85 labeled instances for both supervised and unsupervised learning methodologies. Each instance contains 43 distinct characteristics that capture the network traffic and signaling behaviors of the SS7 protocol.

However, utilizing all 43 features can lead to the problem of high dimensionality, which increases the risk of overfitting, making it harder for the model to generalize effectively. To mitigate this, feature selection techniques were employed, reducing the number of relevant features from 43 to 18. This reduction was

Table 1. Feature Description of the selected 18 feature

No.	Feature Name	Description
1	Label	Indicator of normal (0) or abnormal (1) instances.
2	f_count_ok_psi_between2lu	Number of successful Provide Subscriber Info (PSI) transactions between two consecutive location updates.
3	f_count_ok_fwsm_mt_between2lu	Number of successful SMS-MT (mobile terminated) forward events between two consecutive location updates.
4	f_count_ok_srism_between2lu	Number of successful SMS send routing info events (SRI-SM) between two location updates.
5	f_count_ok_sri_between2lu	Number of successful Send Routing Info (SRI) events between two location updates.
6	f_count_ok_purge_ms_between2lu	Number of successful Purge Mobile Subscriber (PurgeMS) transactions between two location updates.
7	f_c_ossn_others	Identifies whether the source of the signaling messages is HLR or SGSN.
8	f_count_ok_dsd_between2lu	Number of successful Delete Subscriber Data (DSD) events between two location updates.
9	c_cggt	Presumably a count or feature linked to a global title (CGT), details not provided.
10	f_count_ok_ul_between2lu	Number of successful UL events between two Location Updates.
11	f_frequent_ok_ul_between2lu	Frequency of successful UL events between two location updates.
12	f_count_ok_prn_between2lu	Number of successful Provide-Roaming-Number (PRN) transactions between two location updates.
13	f_count_ok_cl_between2lu	Number of successful Cancel-Location (CL) events between two location updates.
14	f_frequent_ok_psi_between2lu	Frequency of successful Provide Subscriber Info (PSI) events between two location updates.
15	f_count_gap_ok_sai_and_all_lu	Difference in the number of Send Authentication Info (SAI) and all update location messages.
16	f_same_cgtt_is_hlr_ossn	Checks if the same calling Global Title (GT) is identified as HLR based on SSN within the last set of minutes.
17	f_count_unloop_country_last_x_hours_ul	Counts the unique countries visited based on the update location messages from the same IMSI in the last X hours.
18	f_frequent_ok_cl_between2lu	Frequency of successful CL events between two location updates.

critical in enhancing the model's performance by eliminating redundant or irrelevant data, which could obscure the model's ability to detect malicious activities accurately.

The 18 selected features were specifically chosen due to their importance in distinguishing between normal and attack instances. These features provide crucial insights into the SS7 traffic patterns and signaling behaviors, making them invaluable for the model's ability to identify anomalies indicative of an attack. More comprehensive details about the feature selection process are elaborated in the Feature Selection section. Additionally, Table 1 offers a thorough explanation of the 18 selected features, underscoring their relevance in improving the model's detection accuracy for Category 3 attacks in SS7 traffic.

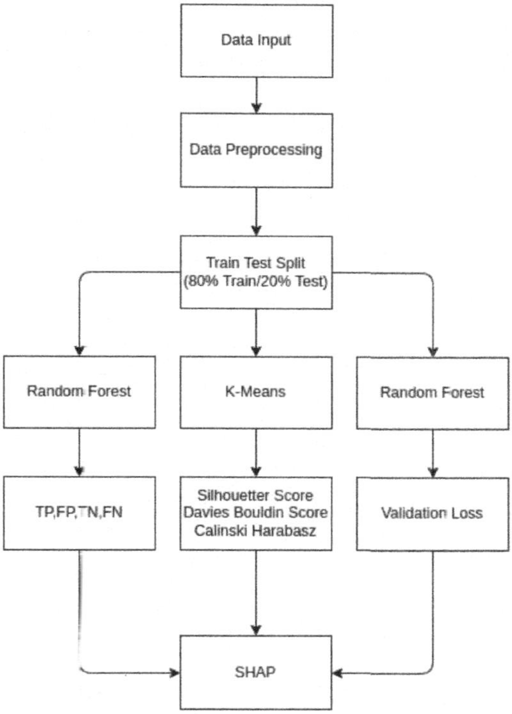

Fig. 3. Flowchart of the designed system

3.2 System

To address the research issue outlined in this document, a methodical strategy will be utilized, beginning with thorough data pre-processing, followed by the deployment of three different machine learning algorithms: Random Forest, Auto-Encoders, and K-Means Clustering. These algorithms were chosen for their diverse features and strengths in the field of machine learning, allowing for an extensive examination of the issue from multiple angles. This process is illustrated in Fig. 3.

3.3 Data Preprocessing

The preprocessing of the dataset is critical to ensure the quality and effectiveness of the machine learning models used. The preprocessing steps are detailed and illustrated in Fig. 4.

1. **Initial Cleanup:** Remove any rows with missing values to maintain data integrity. Although the dataset from the JSS7 simulator is generally clean, this step ensures that no incomplete records are processed.
2. **Feature Selection:** Remove features with no correlation or missing values (NaN) with the target variable. Ignore features with a correlation coefficient below 0.25, as they lack sufficient important value and could undermine model efficiency and precision.
3. **Final Dataset Formation:** The dataset, now streamlined with relevant characteristics, is split into 80% training and 20% testing sets to validate the effectiveness of modeling approaches.

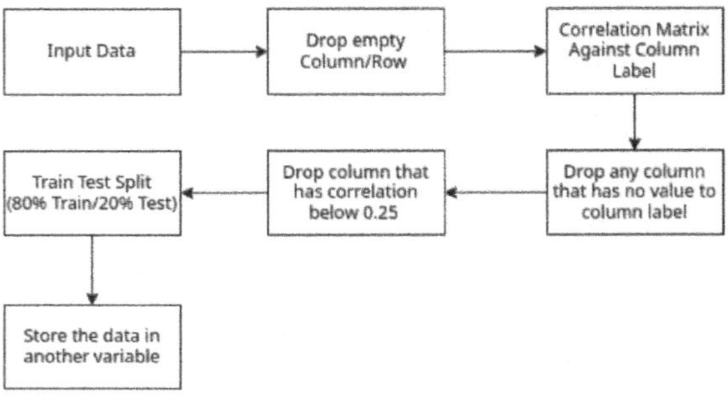

Fig. 4. Flowchart of the Data Preprocessing Method

These preprocessing steps are designed to refine the dataset into a robust form suitable for training supervised learning models, thus enhancing the accuracy and reliability of the outcomes. This process is illustrated in the figure below.

3.4 Learning Process and Testing Plan

The learning process involves applying three distinct algorithms, each tailored with specific parameters to optimize performance and address research objectives. The testing plan evaluates these algorithms based on specific metrics to ensure robustness and dependability, particularly in security applications where accuracy is crucial.

Random Forest Classifier. The Random Forest model [24], implemented through the `RandomForestClassifier` from scikit-learn, is configured with 25 trees (`n_estimators=25`) and a maximum depth of 30 (`max_depth=30`) to enhance performance while mitigating overfitting. In this study, the `random_state` parameter is set to 42 to ensure reproducibility (`random_state=42`). The effectiveness of the model is tested using True Positives (TP), False Positives (FP), True Negatives (TN), and False Negatives (FN) to evaluate precision and reliability in detecting anomalies [25].

Auto Encoders [26]. The dataset underwent normalization through the use of `MinMaxScaler`. Our autoencoder architecture includes an encoder that compresses the input down to eight dimensions through a dense layer activated by ReLU and a decoder that uses sigmoid activation to restore the original dimensions. In an attempt to have reproducibility, the model was initialized with fixed random seeds specified by `np.random.seed(42)` `atf.random.set_seed(42)`, on top of that there is parameter `shuffle` set to **False**. The model was compiled with the Adam optimizer [27] at a learning rate of 0.001 and trained for 1024 epochs using a batch size of 1024. A validation split of 20% was implemented to monitor performance. This configuration effectively minimizes mean squared error, promoting efficient data compression and reconstruction, which is crucial for anomaly detection. The Reconstruction Error will serve as an evaluation metric for this model. The accuracy can be calculated using the formula $Reconstruction\ Error - 1$

K-Means Clustering [28]. The optimal number of clusters was determined using the elbow method, resulting in 3 clusters (`n_clusters=3`) using `KMeans` from scikit-learn. After clustering, the labels were binarized and used as targets for training a secondary `RandomForestClassifier` to predict the membership of the cluster. The quality of the clustering was evaluated using the silhouette score [29].

All evaluation metrics prioritize accuracy to guarantee model robustness and dependability. The research aims to reduce error rates to maintain reliable performance, essential for security applications where detection accuracy is critical.

3.5 Analysis Using XAI

After obtaining the model output, SHAP will be used for XAI analysis. This includes:

- **Calculation of SHAP Values:** Determine the impact of individual input features on the model.
- **Feature Importance Visualization:** Visualize the overall importance of each feature across the dataset.
- **Decision Plot Analysis:** Trace the cumulative effect of each feature from the base value to the final model.
- **Impact Summary:** Show the distribution of feature impacts on the model output.

The summary plot will be depicted with the Y-axis displaying the features arranged by their significance, with the most influential feature at the top, and the X-axis indicating the SHAP values, where positive values suggest a rise in the predicted value and negative values suggest a decline. The colour gradient shows the feature values, ranging from blue for low values to pink/red for high values. The shape of the violin plot illustrates the density of data points at each SHAP value [30].

4 Result

4.1 Random Forest

Table 2. Overall Report for Random Forest Model

	Precision	Recall	F1-Score	Accuracy
Overall	0.95	0.94	0.94	0.94

Table 3. Detailed Report for Random Forest Model

	Precision	Recall	F1-Score	Support	TP	TN	FP	FN
Class 0	1.00	0.92	0.96	12	–	11	1	–
Class 1	0.83	1.00	0.91	5	5	–	–	0

The Random Forest model in this study demonstrates robust performance across multiple metrics, achieving high precision, recall, F1-scores, and overall accuracy. Table 2 shows that the model reaches an accuracy of 94%, with precision and recall values close to perfect, underscoring its reliability in correctly identifying both normal and attack instances in SS7 traffic. The precision of 95% indicates that most of the instances labeled as attacks are truly attack cases. At the same time, the recall of 94% suggests that the model successfully identifies the majority of actual attack instances. With an F1-score of 94%, the balance between precision and recall is maintained, making the Random Forest model a solid candidate for this classification task.

Looking deeper into the performance of the model across different classes, Table 3 breaks down the metrics for Class 0 (normal instances) and Class 1 (attack instances). For Class 0, the model achieves perfect precision of 100%, meaning it correctly identifies all normal instances without falsely labeling them as attacks. However, its recall is slightly lower at 92%, indicating that while most normal instances are identified, a few might still be misclassified as attacks. On the other hand, Class 1 (attack instances) has a perfect recall of 100%, meaning that every attack instance is detected, but a lower precision of 83%, showing that a small number of normal instances are incorrectly classified as attacks. The F1-score for Class 1 stands at 91%, indicating a strong balance between identifying all attacks and minimizing false positives.

The model's effectiveness is further illustrated through the detailed breakdown of TP, TN, FP, and FN. For Class 0, there are 11 true negatives, indicating that normal instances are correctly classified, and only 1 false positive, meaning the model incorrectly labeled a single normal instance as an attack. For Class 1, the model achieves 5 true positives, accurately classifying all attack instances without generating any false negatives. This balance between minimizing false

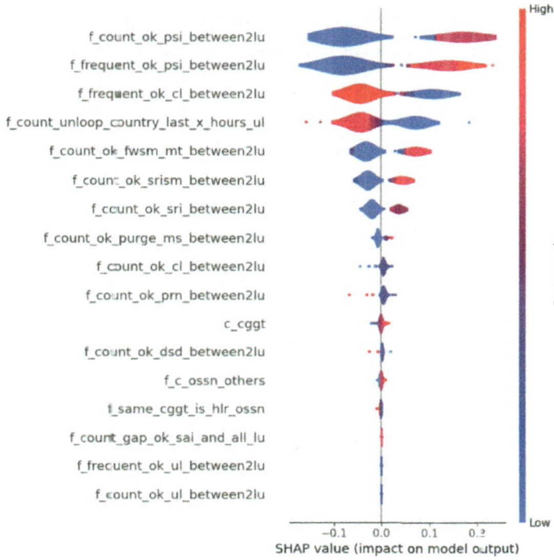

Fig. 5. SHAP Summary Plot for the Random Forest Model

positives and ensuring zero false negatives is crucial in a cybersecurity context, where misclassifying an attack could result in serious breaches, while too many false positives can overwhelm analysts with false alarms.

The importance of the features used in the Random Forest model is further explored through analysis. As illustrated in Fig. 5, the following features exhibit a broad spectrum of SHAP values:

- `f_count_ok_psi_between2lu`
- `f_frequent_ok_psi_between2lu`
- `f_frequent_ok_cl_between2lu`

This indicates that these features significantly impact the model's output and can either increase or decrease the probability of an instance being classified as an attack based on their specific values. For instance, higher values of `f_count_ok_psi_between2lu` may indicate abnormal signaling activity, tipping the model toward classifying the instance as an attack. In contrast, features with narrow SHAP value ranges contribute less to the model's decisions, suggesting they have less influence in distinguishing between normal and attack traffic.

This SHAP-based visualization provides an in-depth look at how individual features contribute to the model's predictions, offering transparency into the decision-making process of the Random Forest model. This is particularly valuable in cybersecurity, where understanding the reasons behind an attack classification can help analysts refine detection strategies and respond more effectively to threats. Overall, the Random Forest model's high performance across both classes, combined with insights from feature importance analysis, demonstrates its suitability for detecting Category 3 SS7 attacks while maintaining a balanced trade-off between sensitivity (recall) and specificity (precision).

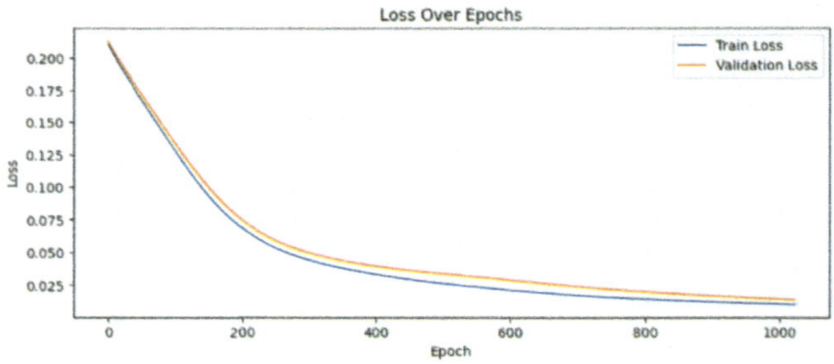

Fig. 6. Auto Encoder Result

4.2 Auto Encoder

Figure 6 illustrates the training and validation loss over 1000 epochs for the Auto Encoder model. The plot shows that both the training loss (blue line) and validation loss (orange line) steadily decrease over the epochs, indicating effective learning. The training loss decreases from 0.2108 to 0.0103, while the validation loss decreases from approximately 0.2133 to 0.0141. Although the validation loss is consistently higher than the training loss and the gap is increasing, suggesting that there is minor overfitting, the model maintains good generalizability. There is a diminishing return after Epoch 400 where it may lead to just longer time instead of actual improvement.

The performance results of the model show a reconstruction error of 0.01105 and an accuracy of 0.9889. The low reconstruction error indicates that the model has a high fidelity in reconstructing the input data, while the high accuracy demonstrates the model's effectiveness in correctly classifying the data.

Figure 7 illustrates the SHAP summary plot for the auto-encoder model, showing the impact of various features on the output of the model. The plot reveals that features such as `f_same_cgt_is_hlr_ossn` have a strong positive impact on the model's output with predominantly high SHAP values, indicating that higher feature values substantially affect the model. In contrast, `f_count_gap_ok_sai_and_all_u` exhibits a broad spread of both high and low SHAP values, suggesting that its impact can significantly vary, enhancing or reducing the model's output depending on its value. The following features predominantly display negative SHAP values:

- `f_frequent_ok_ul_between2lu`
- `f_count_ok_ul_between2lu`

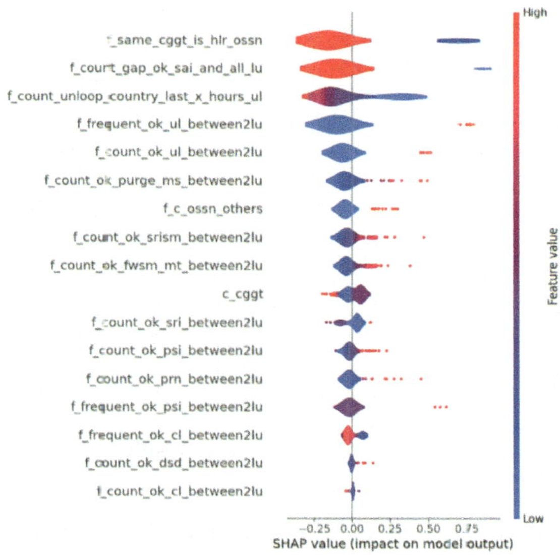

Fig. 7. SHAP Summary Plot for the Auto Encoder model

In these cases, larger feature values typically decrease the output, especially when the feature values are lower. With several reruns, the accuracy results of the Auto Encoder may change slightly with identical parameters, though not by much, while the SHAP values can vary greatly. Attempts to reduce its randomness seem to have been unsuccessful.

4.3 K-Means

In K-Means clustering, the Elbow method is used to identify the optimal number of clusters for the best result, and Fig. 8 illustrates the results obtained using the Elbow method. The plot presented illustrates the sum of squared errors (SSE) for different numbers of clusters, ranging from 1 to 10, with the x-axis representing the number of clusters and the y-axis showing the corresponding SSE values. As the number of clusters increases, the SSE generally decreases, reflecting that more clusters lead to smaller within-cluster variances. However, the critical observation here is the "elbow" point, which occurs at 3 clusters. This point marks where the SSE sharply declines and then starts to level off, suggesting that using 3 clusters provides an optimal balance between reducing within-cluster variance and maintaining model simplicity. In other words, adding more clusters beyond this point results in diminishing returns, as the improvement in SSE becomes less significant, while the complexity of the model increases. The "elbow" method is a well-known heuristic in clustering, often used to determine the ideal number of clusters by identifying this trade-off point.

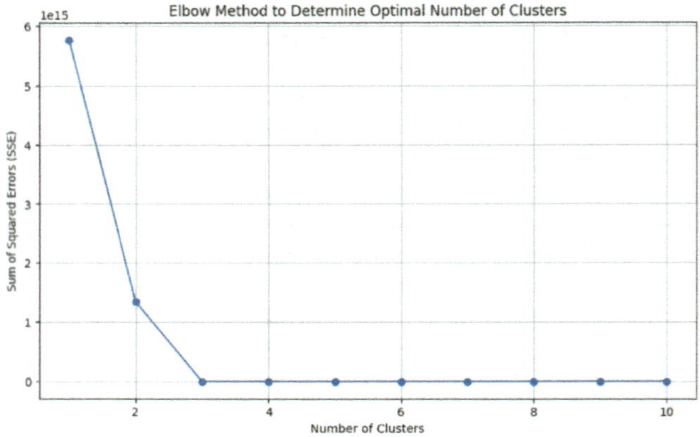

Fig. 8. K-Mean Elbow result

To further assess the effectiveness of the clustering, various essential evaluation metrics were utilized. One of the primary metrics used was the Silhouette Score, which achieved an exceptionally high value of 0.9999995118188754, indicating near-perfect clustering performance. The Silhouette Score measures how similar an instance is to its own cluster compared to other clusters, with values close to 1 indicating that instances are well matched to their clusters and poorly matched to others. The near-perfect score suggests that the clusters formed are highly cohesive and well separated, which is a strong indication of successful clustering. Additionally, other clustering evaluation metrics, such as the Davies-Bouldin Index and the Calinski-Harabasz Index, were considered. The extremely low Davies-Bouldin Index further confirms the compactness and separation of the clusters, as a lower score implies that clusters are tight and distinct from each other. Similarly, the high Calinski-Harabasz Index supports the observation that the clusters are dense and well separated, reinforcing the conclusion that 3 clusters are indeed optimal for this dataset.

Finally, Fig. 9 provides a SHAP summary plot for the K-means clustering model, offering insights into the importance of various characteristics in shaping the model's clustering output. SHAP values explain the contribution of each feature to the clustering decisions, highlighting which features most strongly influence whether an instance is assigned to a specific cluster. This plot offers interpretability, allowing a deeper understanding of the clustering results and how specific features drive the separation of clusters. Overall, the combination of SSE analysis, validation metrics, and SHAP interpretation confirms that the 3-cluster solution provides the most meaningful and effective partitioning of the dataset.

Among the primary features, `c_cggt` significantly impacts clustering results, possibly dominating the process due to its categorical attributes (values 1, 2, 3). This strong influence might overshadow other key features. Additionally, `f_c_ossn_others` exhibits substantial influence with both positive and negative contributions.

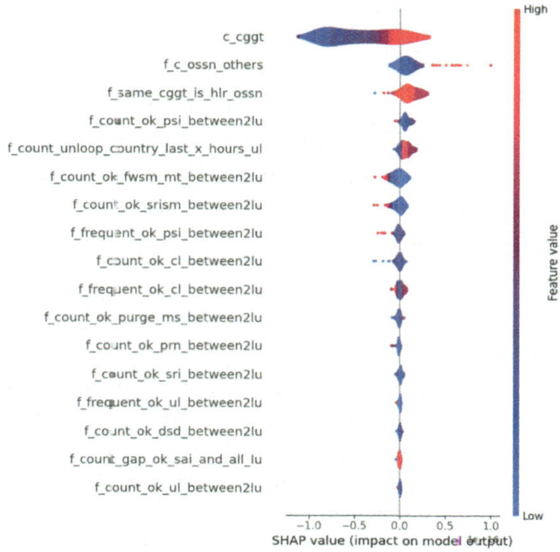

Fig. 9. SHAP Summary Plot for the K-means Clustering Model

Features such as `f_same_cggt_is_hlr_ossn`, `f_count_ok_psi_between2lu`, and `f_frequent_ok_cl_between2lu` also have notable impacts with varied contributions. On the other hand, `f_count_unloop_country_last_x_hours_ul` and `f_count_ok_ul_between2lu` show a balanced SHAP value distribution around zero, indicating mixed contributions. Features like `f_count_ok_prn_between2lu` and `f_count_gap_ok_sai_and_all_lu` have a minor impact on the model's output, as indicated by SHAP values near zero. SHAP analysis suggests that `c_cggt` strongly influences the output of the K-means clustering model, which could reflect an overreliance on this feature and highlights the need to evaluate the behavior of the model and the influence of additional features.

5 Discussion

The results were assessed against the baseline; however, the approach used to determine these results differs from that used for the baseline. This variation led to some empty columns and required adding 10 to all numerical values to align with the baseline format. The top three results, `NODE`, `PreNet`, and `AutoInt`, will be used as the baseline in this context.

All models in this study demonstrated high accuracy, some even exceeding baseline metrics. However, this may indicate a degree of overfitting as the algorithm's application differs from the standard or optimal approaches documented in previous studies. The results are comparable to or better than the baseline. Table 4 compares the performance of several algorithms, highlighting differences between the baseline methods (NODE, PreNet, AutoInt) and other techniques

Table 4. Comparison of *Baseline (average ± standard deviation %) and Results.

Algorithm	Accuracy	Precision	Recall	F1-Score
*NODE	93.35 ± 5.68	97.96 ± 6.55	83.56 ± 13.25	89.01 ± 11.14
*PreNet	93.18 ± 5.62	86.25 ± 11.47	97.28 ± 8.02	90.93 ± 8.16
*AutoInt	93.26 ± 10.74	95.39 ± 11.68	88.97 ± 16.68	90.89 ± 13.49
Random Forest	94	95	94	94
Auto Encoder	98.89	–	–	–
K-Means Clustering	99	–	–	–

(Random Forest, Autoencoder, K-Means Clustering) in terms of accuracy, precision, recall, and F1 score, with baseline metrics reported as average ± standard deviation. NODE achieved an accuracy of 93.35% ± 5.68%, PreNet 93.18% ± 5.62%, and AutoInt 93.26% ± 10.74%. In contrast, Random Forest and K-Means Clustering demonstrated higher accuracies of 94% and 99%, respectively, with Auto Encoder at 98.89%. For precision, NODE led with 97.96% ± 6.55%, followed by AutoInt at 95.39% ± 11.68%, and PreNet at 86.25% ± 11.47%. Random Forest maintained a consistent precision of 95%. The recall values were highest for PreNet at 97.28% ± 8.02%, compared to AutoInt at 88.97% ± 16.68% and *NODE at 83.56% ± 13.25%. Random Forest effectively balanced precision and recall, showing a recall of 94%. The F1 scores were notable for PreNet with 90.93% ± 8.16%, followed by AutoInt at 90.89% ± 13.49%, and NODE at 89.01% ± 11.14%. Random Forest achieved a robust F1 score of 94%. The baseline methods (NODE, PreNet, AutoInt) provided solid performance with noted variability, while Random Forest and K-Means Clustering offered superior accuracy. Autoencoder exhibited exceptionally high accuracy. This analysis underscores the trade-offs and strengths of each algorithm, guiding the selection based on desired performance metrics.

Despite this, our primary focus is on explainable AI, as reflected in the varying SHAP values for each algorithm. The Random Forest algorithm, using its ensemble technique, favors categorical features over binary components. This preference likely stems from the ability of categorical features to distinctly separate different groups, allowing precise splits that align with significant data distinctions, which is particularly advantageous in classification tasks. Autoencoders, leveraging a deep learning approach and the Adam Optimizer, can yield varying results due to the optimizer's adaptive learning rate. The model in this study emphasizes binary and categorical values, which may differ from those in other research. The K-Means algorithm exhibited notable variations due to its clustering nature. The feature `c_cggt`, with only three categorical values, may pose problems because it matches the number of clusters in K-Means, potentially introducing bias into decision making.

6 Conclusion

This study improves transparency in AI decision making by incorporating SHAP with algorithms such as Random Forest, Autoencoders, and K-Means. The application of SHAP values has significantly improved trust in these models by providing clear explanations of their decision-making processes. Furthermore, SHAP values have effectively extracted and elucidated the behavior and importance of each algorithm when applied to a dataset, thus deepening the understanding of the attack mechanisms of SS7.

To address the issues identified in the results and analysis, more detailed visualizations of the models are necessary to provide deeper insights into their mechanisms. A thorough investigation into the behavior of Auto Encoders and K-Means is essential to identify and mitigate potential biases. Furthermore, implementing more complex models could improve the robustness and accuracy of detections.

In general, the developed detection models are not only accurate, but also interpretable, reliable, and robust. These outcomes validate the study's objective of creating AI systems that are both precise and transparent. The detailed analyses provided demonstrate the effectiveness and practical applicability of the model in real-world scenarios. By addressing identified issues and exploring more sophisticated models, the study lays a solid foundation for the development of even more reliable and transparent AI systems in the future.

References

1. Yunizal, E., Wardana, A.A., Niarman A.: Simplification of workflow-oriented security assessment. IJAIT (Int. J. Appl. Inf. Technol.), 134–144 (2023)
2. Wardana, A.A., Sukarno, P., Salman, M.: Collaborative botnet detection in heterogeneous devices of internet of things using federated deep learning. In: Proceedings of the 2024 13th International Conference on Software and Computer Applications, ICSCA 2024, pp. 287 291. Association for Computing Machinery, New York (2024)
3. Ullah, K., Rashid, I., Afzal, H., Iqbal, M.M.W., Bangash, Y.A., Abbas, H.: SS7 vulnerabilities a survey and implementation of machine learning vs rule based filtering for detection of SS7 network attacks. IEEE Commun. Surv. Tutor. **22**(2), 1337–1371 (2020)
4. Rao, S.P., Chen, H.-Y., Aura, T.: Threat modeling framework for mobile communication systems. Comput. Secur. **125**(103047), 103047 (2023)
5. Gilpin, L.H., Bau, D., Yuan, B.Z., Bajwa, A., Specter, M., Kagal, L.: Explaining explanations: an overview of interpretability of machine learning. In: 2018 IEEE 5th International Conference on Data Science and Advanced Analytics (DSAA), pp. 80–89 (2018)
6. Ericsson: Mobile subscriptions outlook (n.d.). https://www.ericsson.com/en/reports-and-papers/mobility-report/dataforecasts/mobile-subscriptions-outlook. Accessed 02 July 2024
7. Dryburgh, L., Hewett, J.: Signaling System No. 7 (SS7/C7): Protocol, Architecture, and Services. Cisco Press (2005)

8. Rao, S.P., Kotte, B.T., Holtmanns, S.: Privacy in LTE networks. In: Proceedings of the 9th EAI International Conference on Mobile Multimedia Communications, MobiMedia 2016, Brussels, BEL. ICST (Institute for Computer Sciences, Social-Informatics and Telecommunications Engineering) (2016)
9. Kacer, M., Langlois, P.: SS7 attacker heaven turns into riot: How to make nation-state and intelligence attackers lives much harder on mobile networks. P1 Security (2017)
10. Peeters, C., Patton, C., Munyaka, I.N.S., Olszewski, D., Shrimpton, T., Traynor, P.: SMS OTP security (SOS): hardening SMS-based two factor authentication. In: Proceedings of the 2022 ACM on Asia Conference on Computer and Communications Security, ASIA CCS 2022, pp. 2–16. Association for Computing Machinery, New York (2022)
11. Guo, Y., Ermis, O., Tang, Q., Trang, H., De Oliveira, A.: An empirical study of deep learning-based SS7 attack detection. Information **14**, 509 (2023)
12. Arrieta, A.B., et al.: Explainable artificial intelligence (XAI): concepts, taxonomies, opportunities and challenges toward responsible AI. Inf. Fusion **58**, 82–115 (2020)
13. Samek, W., Montavon, G., Lapuschkin, S., Anders, C.J., Müller, K.-R.M.: Explaining deep neural networks and beyond: a review of methods and applications. Proc. IEEE **109**(3), 247–278 (2021)
14. Bhatt, U., et al.: Explainable machine learning in deployment. In: Proceedings of the 2020 Conference on Fairness, Accountability, and Transparency, FAT* 2020, pp. 648–657. Association for Computing Machinery, New York (2020)
15. Adadi, A., Berrada, M.: Peeking inside the black-box: a survey on explainable artificial intelligence (XAI). IEEE Access **6**, 52138–52160 (2018)
16. Lundberg, S.M., Lee, S.-I.: A unified approach to interpreting model predictions. In: 31st Conference on Neural Information Processing Systems (NIPS 2017), Long Beach, CA, USA (2017). https://arxiv.org/abs/1705.07874
17. Roshan, K., Zafar, A.: Using kernel SHAP XAI method to optimize the network anomaly detection model. In: 2022 9th International Conference on Computing for Sustainable Global Development (INDIACom), pp. 74–80 (2022)
18. Ribeiro, M.T., Singh, S., Guestrin, C.: "Why should i trust you?": explaining the predictions of any classifier. In: Proceedings of the 22nd ACM SIGKDD International Conference on Knowledge Discovery and Data Mining, KDD 2016, pp. 1135 1144. Association for Computing Machinery, New York (2016)
19. Shrikumar, A., Greenside, P., Kundaje, A.: Learning important features through propagating activation differences. In: Precup, D., Teh, Y.W. (eds.) Proceedings of the 34th International Conference on Machine Learning, volume 70 of Proceedings of Machine Learning Research, pp. 3145–3153. PMLR (2017)
20. K. Jensen, H.T. Nguyen, Van Do, T., Årnes, A.: A big data analytics approach to combat telecommunication vulnerabilities. Cluster Comput. **20**(3), 2363–2374 (2017)
21. Jensen, K., van Do, T., Nguyen, H.T., Årnes, A.: Better protection of SS7 networks with machine learning. In: Proceedings of the IEEE Conference on Network Security, Norwegian University of Science and Technology, Norway; Telenor ASA, Fornebu, Norway, pp. 1–8. IEEE (2016)
22. Ermis, O., et al.: A CNN-based semi-supervised learning approach for the detection of SS7 attacks. In: Information Security Practice and Experience. Lecture notes in computer science, pp. 345–363. Springer, Cham (2022)

23. Qasim, T., Durad, M.H., Khan, A., Nazir, F., Qasim, T.: Detection of signaling system 7 attack in network function virtualization using machine learning. In: 2018 15th International Bhurban Conference on Applied Sciences and Technology (IBCAST), pp. 484–488 (2018)
24. Breiman, L.: Mach. Learn. **45**(1), 5–32 (2001)
25. Muhammad Raihan Rawadi, H., Sukarno, P., Wardana, A.A.: Performance analysis of deep neural decision forest with extratrees feature selection for obfuscated malware detection. In: 2024 IEEE 3rd International Conference on Computing and Machine Intelligence (ICMI), pp. 1–5 (2024)
26. Bank, D., Koenigstein, N., Giryes, R.: Autoencoders. In: Machine Learning for Data Science Handbook, pp. 353–374. Springer, Cham (2023)
27. Zhang, Z.: Improved adam optimizer for deep neural networks. In: 2018 IEEE/ACM 26th International Symposium on Quality of Service (IWQoS), pp. 1–2 (2018)
28. Sinaga, K.P., Yang, M.-S.: Unsupervised k-means clustering algorithm. IEEE Access **8**, 80716–80727 (2020)
29. Shahapure, K.R., Nicholas, C.: Cluster quality analysis using silhouette score. In: 2020 IEEE 7th International Conference on Data Science and Advanced Analytics (DSAA), pp. 747–748 (2020)
30. Salih, A.M., et al.: A perspective on explainable artificial intelligence methods: SHAP and LIME. Adv. Intell. Syst. (2024)

Cyber Risk Management and Organisational Alignment Using Endley's Risk Management Model: Computational Analysis from an Adaptive Dynamical System Perspective

Debby Bouma[1,2], Thijmen Den Oudsten[1], Adam Naffakh[1], Manir Waliji[1], Jan Treur[1,2](✉), and Peter H. M. P. Roelofsma[1]

[1] Cybersecurity Expertise Centre, Risk Management and Cybersecurity Group, The Hague University of Applied Sciences, The Hague, Netherlands
j.treur@vu.nl, p.h.m.p.roelofsma@hhs.nl

[2] Department of Computer Science, Social AI Group, Vrije Universiteit Amsterdam, Amsterdam, Netherlands

Abstract. In this paper a novel approach is introduced to improve cyber risk management and organizational alignment. It is illustrated for a case study concerning the Amsterdam Airport at Schiphol. It applies Endsley's Risk Management Model to understand threats at different levels. An adaptive dynamical system model specifically designed for this cyber domain will be proposed, addressing the interconnectedness of vulnerabilities. This includes adaptation mechanisms to account for ocurring threats. A controlled experiment will then evaluate the model's effectiveness in enhancing situational awareness and its impact on risk management and alignment. The research also explores future risks using What-If analysis and probabilistic risk assessment.

Keywords: Adaptive Dynamical System Model · Endley Management Model · Cyber Risk

1 Introduction

This paper explores the potential of leveraging Endsley's Risk Management Model to enhance cyber risk management and organizational alignment at Schiphol Airport. Mica Endsley's Risk Management Model (Endsley, 1995, 2015, 2021; Salerno et al., 2004) provides a hierarchical framework for understanding dynamic environments through information processing. It defines situational awareness (SA) as a three-level construct encompassing:

- **Level 1: Perception:** Gathering information from the environment through sensory inputs.
- **Level 2: Comprehension:** Interpreting the gathered information and understanding its significance within the context of the situation.
- **Level 3: Projection:** Predicting future possibilities based on the current understanding of the situation.

R. Silhavy and P. Silhavy (Eds.): CoMeSySo 2024, LNNS 1489, pp. 422–444, 2025.
https://doi.org/10.1007/978-3-031-96798-6_31

After the above-mentioned situational awareness comes the decision making and the performance of action to complete the sequence of the model. The sequential nature of this model emphasizes the importance of successfully completing each level before progressing to the next. Endsley's model's strength lies in its generic nature, making it applicable across diverse domains, including aviation, military operations, and healthcare. It also acknowledges the influence of individual and environmental factors on SA (Scmo, 2021).

Research Objectives

1. **Establish a Theoretical Foundation:** Explore relevant areas such as cyber knowledge specific to Schiphol Airport and the psychology of human cognition and social processes within the airport environment.
2. **Develop a Network Model:** Design a network model tailored to the chosen scenario at Schiphol Airport. This includes defining network states, connections, weights, speed factors, initial values, self-adaptation mechanisms, and combination factors.
3. **Evaluate Model Effectiveness:** Conduct a controlled experiment to assess the network model's ability to enhance SA within the chosen scenario. Analyze the impact on organizational risk management and alignment.
4. **Explore Risk Landscape:** Utilize what-if analysis and probabilistic risk assessment to identify potential future developments and associated risks within the airport's cyber domain.
5. **Refine and Evaluate:** Based on the findings, refine the network model and conduct further evaluations. To will provide a comprehensive discussion of the research findings, highlighting the key takeaways, limitations of the proposed framework, and potential avenues for future research in this area. By employing Endsley's Situational Awareness Model and a network modelling approach, this research aims to contribute to improved cyber risk management and organizational alignment at Schiphol Airport, ultimately enhancing overall security and operational effectiveness.

This work is part of a research programme on computational analysis of safety and security in and through cyberspace. In (Abromaityte et al., 2024; Akers et al., 2024; Bouma et al., 2024; Capra et al., 2024; Hoffmans et al., 2024; Jeffery et al., 2024; Keijzer et al., 2024) various other themes within cybersecurity were addressed. Moreover, in (Roelofsma et al., 2024) it was analysed how an AI coach can support safety and security for medical teams in hospitals.

2 Background Information and Scenario Description

Schiphol Airport, a globally recognized aviation hub, manages a complex network of systems that are pivotal for the seamless operation of flights worldwide. This heavy reliance on technology not only underscores the airport's technological prowess but also creates a vulnerability to cyberattacks (Cybersecurity ACI World, n.d.). Furthermore, the potential for disruption caused by successful attacks on air traffic control systems makes airports, especially ones of Schiphol's stature, attractive targets for malicious actors (Humayed et al., 2017). Beyond technological vulnerabilities, human psychology also plays a significant role in cybersecurity breaches. Social engineering tactics can exploit human trust or trick employees into compromising systems (Lederer et al., 2020), while simple human error can expose systems (Liu et al., 2018).

Schiphol Airport detected anomalies in network traffic originating from their central data center during a routine security audit. This initial red flag prompted further investigation, which revealed unauthorized access to the flight management system, a critical component for air traffic control. This unauthorized access could have resulted in significant disruptions at the airport. Intelligence suggested the attack originated from a Chinese hacking group, adding a geopolitical dimension to the incident.

Schiphol's response to the cyberattack was a testament to the effectiveness of its predefined protocols and the expertise of its specialized teams. The airport's well-established Incident Response Plan served as a clear roadmap for containment and recovery efforts (National Institute of Standards and Technology (NIST), 2020). The swift activation of a temporary crisis team, the CERT/CSIRT, composed of highly skilled cybersecurity experts, underscored the value of having specialized personnel readily available to address cyber threats (Sans Institute, 2023). Equally important was the active involvement of senior management, whose leadership was instrumental in ensuring access to necessary resources and facilitating critical decision-making throughout the incident (Humayed et al., 2017).

Effective communication during a crisis is not just important, it's crucial. Schiphol adhered to its established communication protocols, promptly notifying relevant stakeholders, including department heads, regulatory bodies, international partners, and the public (Federal Emergency Management Agency (FEMA), 2023). This transparent communication not only fostered trust but also ensured a unified response network to address the attack, making all stakeholders feel informed and involved in the airport's response.

Maintaining trust with the public during a cyberattack is crucial. Schiphol's commitment to clear and regular communication throughout the incident reassured passengers, airline partners, and the public that the airport prioritized safety and was actively addressing the situation (FEMA, 2023). Furthermore, implementing cybersecurity training for all staff acknowledges the human element in cyberattacks. By educating staff on social engineering tactics and red flags for suspicious activity, Schiphol aims to minimize the potential for human error as a vulnerability (Lederer et al., 2020).

Following the threat's immediate containment, Schiphol Airport showed its commitment to robust recovery and prevention efforts. A thorough review of cybersecurity policies and procedures was conducted to identify and address any vulnerabilities exposed during the attack. Notably, Schiphol proactively employed an AI coach trained on vast cybersecurity data to assist in this review process, highlighting the airport's commitment to staying ahead of evolving cyber threats (Al-Riyami & Hajri, 2020). Furthermore, network monitoring capabilities were bolstered to detect anomalies and prevent future attacks, demonstrating Schiphol's dedication to maintaining a secure environment. These proactive measures underscore the airport's commitment to cybersecurity and its determination to prevent similar incidents in the future.

The case of Schiphol Airport demonstrates the value of a multifaceted response to a cyberattack. Pre-defined plans, skilled cybersecurity teams, active management involvement, clear communication protocols, staff training, and continuous improvement through AI and other means are essential components of a robust cybersecurity strategy

(Humayed et al., 2017). By implementing these measures, airports can strive to create a more secure and resilient environment for the aviation industry, ensuring the safety of passengers and the smooth operation of air travel.

3 The Adaptive Network Modeling Approach

The following passage outlines a new method for improving cyber risk management and organizational alignment at Schiphol Airport. It utilizes Endsley's Risk Management Model to analyze threats across different levels. A customized network model explicitly designed for Schiphol's cyber domain will be created to capture the interconnected vulnerabilities. A phased research plan will lay the groundwork and construct the network model, incorporating features such as self-adaptation mechanisms to address evolving threats. An experiment will then assess the model's effectiveness in improving situational awareness and its impact on risk management and alignment.

The Network-Oriented Modeling approach is established in (Treur, 2019, 2020) for this document. The basic architecture chosen is a temporal-causal network architecture defined by three types of network structure characteristics: the network's connectivity, the aggregation of multiple connections in the network, and timing.

A. Connectivity of the network: Connection weights $\boldsymbol{\omega}_{X,Y}$ for each connection from a state (or node) X to a state Y.
B. Aggregation of multiple connections in the network: A combination function $\mathbf{c}_Y(...)$ for each state Y to determine the aggregation of incoming causal impacts.
C. Timing in the network: A speed factor $\boldsymbol{\eta}_Y$ for each state Y.

A complete conceptual representation of a temporal-causal network structure includes the above labels for a network graph. These three notions, connection weight, combination function, and speed factor, make the graph of states and connections a labeled graph. This labeled graph forms the defining network structure of a temporal-causal network model (Treur, 2020). States Y have activation levels over time denoted by $Y(t)$ with t time. In Table 1 it is shown how based on these network characteristics dynamics of the states are defined. Table 2 shows the specific combination functions used in the introduced network model. The lower part of Table 1 shows the numerical representation describing the dynamics of a temporal-causal network by difference equations, defined based on the network structure as described. Thus, dynamic semantics are associated with any conceptual temporal-causal network specification in a numerical-mathematically defined manner. This provides a well-defined relation between network structure and network dynamics at the base level. The difference equations can be used both for simulation and mathematical analysis.

A network is adaptive if some of its characteristics change over time. To model adaptive networks, the self-modeling principle is used. For example, if connection weight $\boldsymbol{\omega}_{X,Y}$ is adaptive, a self-model state $\mathbf{W}_{X,Y}$ is added to the network that represents the value of $\boldsymbol{\omega}_{X,Y}$.

Any smooth dynamical system has a canonical representation as a network and any smooth (multi-order) adaptive dynamical system has a canonical representation as a (multi-order) self-modeling network, as is shown in (Treur, 2021; Hendrikse et al.,

Table 1. Dynamics defined in terms of the network structure characteristics; adopted from (Treur, 2020, Table 2.2)

Concept	Representation	Explanation
State values over time t	$Y(t)$	At each time point t each state Y in the model has a real number value in [0, 1]
Single causal impact	$\mathbf{impact}_{X,Y}(t)$ $= \boldsymbol{\omega}_{X,Y}\, X(t)$	At t state X with connection to state Y has an impact on Y, using connection weight $\boldsymbol{\omega}_{X,Y}$
Aggregating multiple impacts	$\mathbf{aggimpact}_Y(t)$ $= \mathbf{c}_Y(\mathbf{impact}_{X_1,Y}(t),\ldots, \mathbf{impact}_{X_k,Y}(t))$ $= \mathbf{c}_Y(\boldsymbol{\omega}_{X_1,Y}X_1(t), \ldots, \boldsymbol{\omega}_{X_k,Y}X_k(t))$	The aggregated causal impact of multiple states X_i on Y at t, is determined using a combination function $\mathbf{c}_Y(V_1, \ldots, V_k)$ and apply it to the k single causal impacts.
Timing of the causal effect	$Y(t+\Delta t) = Y(t) + \boldsymbol{\eta}_Y\, [\mathbf{aggimpact}_Y(t) - Y(t)]\, \Delta t$ $= Y(t) + \boldsymbol{\eta}_Y\, [\mathbf{c}_Y(\boldsymbol{\omega}_{X_1,Y}X_1(t), \ldots, \boldsymbol{\omega}_{X_k,Y}X_k(t)) - Y(t)]\, \Delta t$	The causal impact on Y is exerted over time gradually, using speed factor $\boldsymbol{\eta}_Y$; here the X_i are all states from which state Y has incoming connections.

Table 2. Combination functions used

Concept	Representation
Advanced logistic sum $\mathbf{alogistic}_{\sigma,\tau}(V_1, \ldots, V_k)$	$[\frac{1}{1+e^{-\sigma(V_1+\cdots+V_k-\tau)}} - \frac{1}{1+e^{\sigma\tau}}](1+e^{-\sigma\tau})$
Stepmodulo $\mathbf{stepmod}_{\rho,\delta}(V_1, \ldots, V_k)$	time t, repetition interval $\boldsymbol{\rho}$, tipping point $\boldsymbol{\delta}$ 0 if $t \bmod \rho < \delta$ else 1

2023). Therefore, compared to adaptive dynamical systems in general, the network-oriented modeling approach used does not introduce any fundamental limitations concerning what it can model. This has also been confirmed by applications to complex adaptive processes in different disciplines, such as mental models in cognitive science, organisational learning in management science, synchronisation and bonding in social psychology, and teamwork in health science; e.g., (Treur & Van Ments, 2022; Canbaloğlu et al., 2023; Hendrikse et al., 2025; Roelofsma et al., 2024).

4 Design of the Adaptive Network Model

The designed adaptive network model is depicted in Fig. 1. It was inspired by the approach to model organisational learning introduced in (Canbaloğlu et al., 2022) that was using the approach to modeling of (shared) mental models as shown in (van Ments et al., 2021) and (Treur & van Ments, 2022).

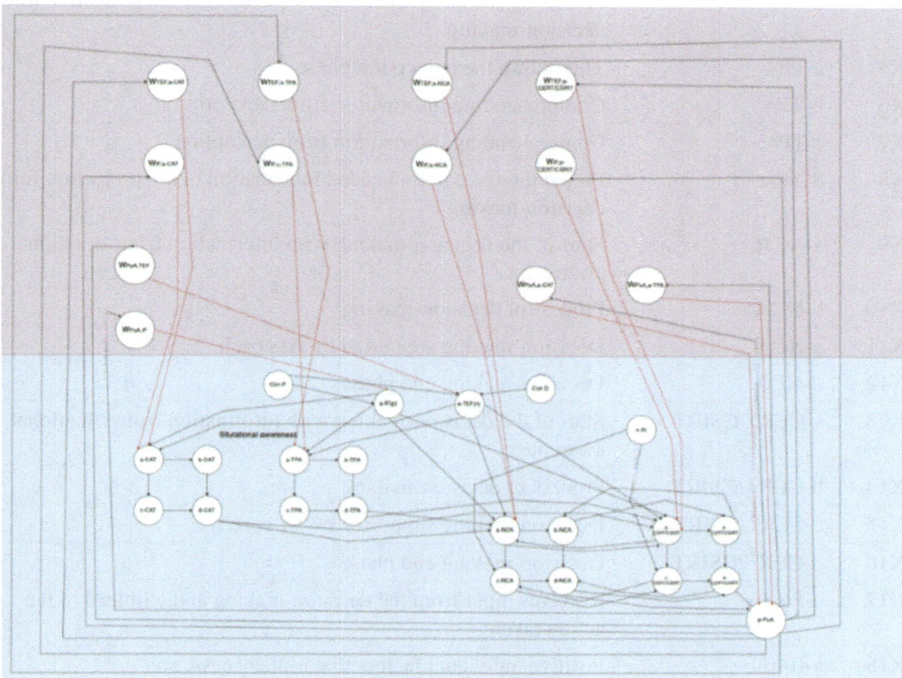

Fig. 1. Overview of the adaptive network model

The blue plane depicts the base level and the purple plane the adaptation level. In tables corresponding shading colours are used. In Table 3 an overview of all states and their explanation is provided. In this model the following abbreviations are used:

- CAT = Cyber Audit Team, is part of the situational awareness in Endsley's risk management model and will run audits to detect a possible breach.
- TPA = Third Party Audit, is part of the situational awareness in Endsley's risk management model, and will run audits to detect a possible breach.
- NCA = National Cyber Authorities, this is part of the decision-making part of Endsley's risk management model, and after the input from the situational awareness it will make a decision based on input on the ai coach and communicate with the incident response team.
- CERT/CSIRT = incident response team, this is part of the decision-making part of

Table 3. States and their explanation

States	States	Explanation
X1	a-CAT	This shows the perception phase
X2	b-CAT	Comprehending information from perception
X3	c-CAT	Comprehending information from perception
X4	d-CAT	Projecting the comprehended information to the projection for decision-making
X5	a-TPA	This shows the perception phase
X6	b-TPA	Comprehending information from perception
X7	c-TPA	Comprehending information from perception
X8	d-TPA	Projecting the comprehended information to the projection for decision-making
X9	a-NCA	Start of the decision-making with information from situational awareness
X10	b-NCA	Process of decision-making
X11	c-NCA	Decision-making supported by AI coach
X12	d-NCA	Decision-making end phase
X13	a-CERT/CSIRT	Start of the decision-making with information from situational awareness
X14	b-CERT/CSIRT	Process of decision-making
X15	c-CERT/CSIRT	Decision-making supported by AI coach
X16	d-CERT/CSIRT	Decision-making end phase
X17	a-PoA	Receives input from the decision-making and will lead to the action taken
X18	a-IF(p)	Positive individual factors that influence SA and decision-making
X19	a-TEF(n)	Negative Task and Environmental factors
X20	a-AI	AI coach that supports in the decision making
X21	Con P	Context factors personalize
X22	Con D	Context factor day
X23	$\mathbf{W}_{\text{PoA,TEF}}$	Allows the TEF to learn from the PoA
X24	$\mathbf{W}_{\text{PoA,IF}}$	Allows the IF to learn from the PoA
X25	$\mathbf{W}_{\text{PoA,a-CAT}}$	Allows to learn from the PoA
X26	$\mathbf{W}_{\text{PoA,a-TPA}}$	Allows to learn from the PoA
X27	$\mathbf{W}_{\text{IF,a-CAT}}$	Can learn from IF through PoA
X28	$\mathbf{W}_{\text{TEF,a-CAT}}$	Can learn from TEF through PoA

(*continued*)

Table 3. (*continued*)

States	States	Explanation
X29	$\mathbf{W}_{IF,a\text{-}TPA}$	Can learn from IF through PoA
X30	$\mathbf{W}_{TEF,a\text{-}TPA}$	Can learn from TEF through PoA
X31	$\mathbf{W}_{IF,a\text{-}NCA}$	Can learn from IF through PoA
X32	$\mathbf{W}_{TEF,a\text{-}NCA}$	Can learn from TEF through PoA
X33	$\mathbf{W}_{IF,a\text{-}CERT/CSRIT}$	Can learn from IF through PoA
X34	$\mathbf{W}_{TEF,a\text{-}CERT/CSRIT}$	Can learn from TEF through PoA
X35	Context	

Endsleys risk management model, and after the input from the situational awareness it will make a decision based on input on the AI coach and communicate with the National Cyber Authorities.

- PoA = performance of action, in this part the decisions taken by the decision making units will be transformed in actions.
- IF = individual factors, such as training knowledge, ability, these influence the above mentioned units in situational awareness and decision making. IF gets influence by the Con p factor. This is a positive influence.
- TEF = Task and Environmental Factors, such as stress, workload. This influences also the situational awareness and decision making. And it gets influenced by con D and con P. This factor is negative.

The design of a network model is defined by its network characteristics, specified by a number of tables in a standard format called role matrices. There are two tables for connectivity characteristics: **mb** for base connectivity and **mcw** for connection weights; see Tables 4 and 5. Moreover, in the Appendix section the other tables are included, **mcfw** and **mcfp** for aggregation characteristics and **ms** and **iv** for timing characteristics.

In role matrix **mb** per row for each of the states it is indicated from which state it receives incoming connections. In role matrix **mcw** (see Table 5) in the corresponding cells the connection weights are specified. In the green cells show the nonadaptive values and the light brown cells the self-model states that represent the adaptive values: X23 to X34.

Table 4. Role matrix **mb** for base connectivity

mb		1	2	3	4	5
X1	a-CAT	X1	X18	X19		
X2	b-CAT	X1				
X3	c-CAT	X1				
X4	d-CAT	X2	X3			

(*continued*)

Table 4. *(continued)*

mb		1	2	3	4	5
X5	a-TPA	X5	X18	X19		
X6	b-TPA	X5				
X7	c-TPA	X5				
X8	d-TPA	X6	X7			
X9	a-NCA	X4	X8	X13	X18	X19
X10	b-NCA	X9	X14			
X11	c-NCA	X9	X15	X20		
X12	d-NCA	X10	X11	X16		
X13	a-CERT/CSIRT	X4	X8	X9	X18	X19
X14	b-CERT/CSIRT	X10	X13			
X15	c-CERT/CSIRT	X11	X13	X20		
X16	d-CERT/CSIRT	X12	X14	X15		
X17	a-PoA	X12	X16			
X18	a-IF(p)	X18	X21			
X19	a-TEF(n)	X19	X21	X22		
X20	a-AI	X20				
X21	Con P	X21				
X22	Con D	X22				
X23	$\mathbf{W}_{\text{PoA,TEF}}$	X17	X23	X35	X19	
X24	$\mathbf{W}_{\text{PoA,IF}}$	X17	X24	X35	X18	
X25	$\mathbf{W}_{\text{PoA,a-CAT}}$	X17	X25	X35	X1	
X26	$\mathbf{W}_{\text{PoA,a-TPA}}$	X17	X26	X35	X5	
X27	$\mathbf{W}_{\text{IF,a-CAT}}$	X17	X27	X35	X18	
X28	$\mathbf{W}_{\text{TEF,a-CAT}}$	X17	X28	X35	X19	
X29	$\mathbf{W}_{\text{IF,a-TPA}}$	X17	X29	X35	X18	
X30	$\mathbf{W}_{\text{TEF,a-TPA}}$	X17	X30	X35	X19	
X31	$\mathbf{W}_{\text{IF,a-NCA}}$	X17	X31	X35	X18	
X32	$\mathbf{W}_{\text{TEF,a-NCA}}$	X17	X32	X35	X19	
X33	$\mathbf{W}_{\text{IF,a-CERT/CSRIT}}$	X17	X33	X35	X18	
X34	$\mathbf{W}_{\text{TEF,a-CERT/CSRIT}}$	X17	X34	X35	X19	
X35	Context	X35				

For the connectivity characteristics, the first table of the role matrices is **mb**, representing the base connections in Fig. 1 into a table that represents each state in a row

Table 5. Role matrix **mcw**

mcw						
X1	a-CAT	1	X27	X28		
X2	b-CAT	0.7				
X3	c-CAT	0.7				
X4	d-CAT	0.7	0.7			
X5	a-TPA	1	X29	X30		
X6	b-TPA	0.7				
X7	c-TPA	0.7				
X8	d-TPA	0.7				
X9	a-NCA	0.9	0.9	0.9	X31	X32
X10	b-NCA	0.9	0.9			
X11	c-NCA	0.9	0.9	0.9		
X12	d-NCA	0.9	0.9	0.9		
X13	a-CERT/CSIRT	0.9	0.9	0.9	X33	X34
X14	b-CERT/CSIRT	0.9	0.9			
X15	c-CERT/CSIRT	0.9	0.9	0.9		
X16	d-CERT/CSIRT	0.9	0.9	0.9		
X17	a-PoA	X25	X26			
X18	a-IF(p)	X24	0.7			
X19	a-TEF(n)	X23	1	0.7		
X20	a-AI	1				
X21	Con P	1				
X22	Con D	1				
X23	$\mathbf{W}_{\text{PoA,TEF}}$	1	1	1	1	
X24	$\mathbf{W}_{\text{PoA,IF}}$	1	1	1	1	
X25	$\mathbf{W}_{\text{PoA,a-CAT}}$	1	1	1	1	
X26	$\mathbf{W}_{\text{PoA,a-TPA}}$	1	1	1	1	
X27	$\mathbf{W}_{\text{IF,a-CAT}}$	1	1	1	1	
X28	$\mathbf{W}_{\text{TEF,a-CAT}}$	1	1	1	1	
X29	$\mathbf{W}_{\text{IF,a-TPA}}$	1	1	1	1	
X30	$\mathbf{W}_{\text{TEF,a-TPA}}$	1	1	1	1	
X31	$\mathbf{W}_{\text{IF,a-NCA}}$	1	1	1	1	
X32	$\mathbf{W}_{\text{TEF,a-NCA}}$	1	1	1	1	

(continued)

Table 5. (*continued*)

mcw						
X33	$\mathbf{W}_{IF,a\text{-}CERT/CSRIT}$	1	1	1	1	
X34	$\mathbf{W}_{PoA,TEF}$	1	1	1	1	
X35	Context	1				

alongside their incoming connections (see Table 4). At the **mcw** we assigned every state a weight value, in this process we both kept the scenario and the network model in mind. So, for example have certain states in the situational awareness a lower weight value than in the decision-making as the communication of the two parties in the decision-making should have a higher connectivity (see Table 5). Here, the light brown cells indicate the self-model states for adaptive connection weights.

5 Simulation Results

Scenario 1: No AI Used
In this scenario (see Fig. 2) the TPA and the CAT are running audits to scan for a breach, they do not interact they runs audits both, Furthermore, if a breach is found it will pursue to the comprehension and perception phase, TPA and CAT then interact with the NCA and the CSERT and CSIRT in the decision making on which they gets influenced by the ai coach supporting them to make a decision. TPA and CAT thus interact and interchange information with the NCA and CSERT/CSIRT. After the decision making it will lead to the performance of action. The performance of action receives input from decision making and transforms it into the actions taken. Furthermore, the IF factors such as knowledge and training have a positive influence on the situational awareness and the decision-making, the IF factor receives input from the contextual P factor as the ability of somethings differs per person. The TEF task and environmental factor is negative such as workload, and stress also influence the situational awareness and decision-making, this factor is influenced by con D as stress is more present during the day, and also the con P factor. These factors are able to learn to see how certain factors affected them and also reviewing the performance of action to learn from their actions.

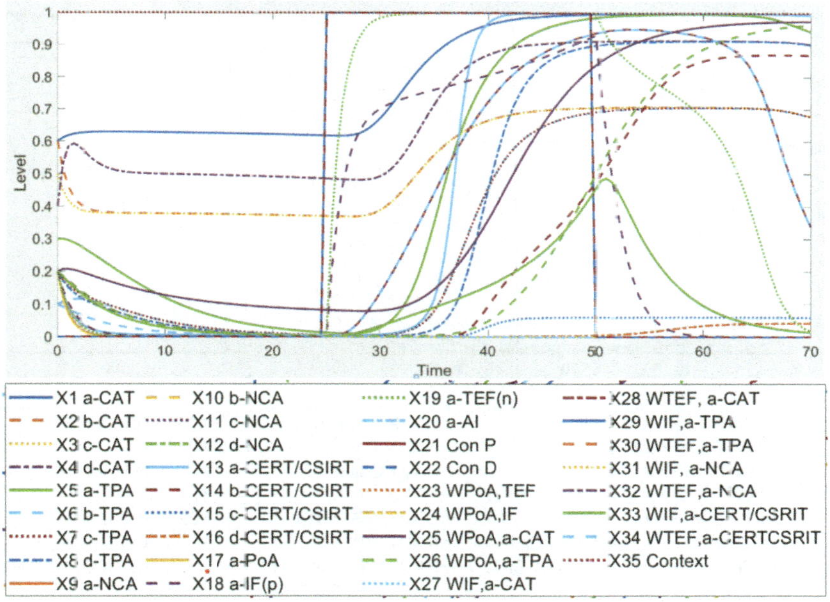

Fig. 2. Simulation 1: no AI used

As can be seen in the results, all states are relatively stable until the arrival of the context states, these ensure that everything happens. The situational awareness takes place, which influences the decision making phase, ensuring that the hacker group cannot take control and fails to create chaos at the airport. However, after X21 and X22 lose their value, all the other values drop as well. Meaning there should be a constant context to keep these states' values high.

Scenario 2: AI Is Used

In this scenario (see Fig. 3) simulation the TPA and CAT runs the situational awareness, but they fail to communicate the data breach to the decision-making units NCA and CSERT/CSIRT. Furthermore, as the two contextual factors con p and con d are not included in the scenario, will the two factors IF and TEF have a weaker influence on the situational awareness. As the two audit teams fails to communicate to the decision-making units, they fail to respond to the data breach, there will be then no performance of action as the decision makers are two late to respond. The states in this scenario will learn through the performance of action, as they can see what happened wrong in the sequence of the model, in this model thus the communication failed.

Similar to Scenario 1, the states are relatively stable until the arrival of the context states, these ensure that everything happens. The situational awareness takes place, which influences the decision making phase, ensuring that the hacker group cannot take control and fails to create chaos at the airport. However, with the constant presence of AI, most values stay relatively high, and state X17 (a-PoA) also increases.

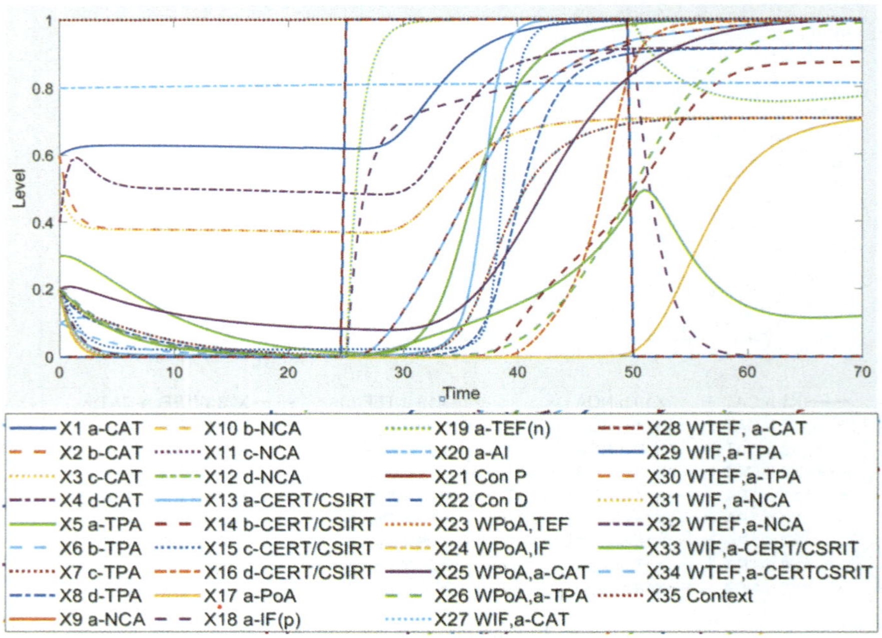

Fig. 3. Simulation 2: AI used

6 What-If Analysis

The focus is on which options for If factors connection weights, threshold and speed factor for X1 and X5 don't see data breach were can see effect in What factor X17 (PoA). During the What-If analysis, the matrices and initial values were used from the first simulation. The If factors and their options are shown in Table 6. In Table 7 the combinations of If factor options are shown and the simulation outcomes for the What factor X17 (PoA) indicated.

By assigning probabilities, a probabilistic What-If analysis and risk assessment was made, see Tables 8 and 9. First, for each of the If factor options a probability was estimated. See the last column in Table 8.

Based on these estimated probabilities for the If factor options, probabilities were determined for the If factor option combinations as seen in Table 9, columns 1 to 4. Here, it was assumed that the If factor option probabilities were independent so that for the combinations the product rule p(A and B and C) = p(A)*p(B)*p(C) can be applied. For each combination for which the What factor 1 X17 occurs in the corresponding simulation the related probabilities are shown in the last column of Table 9. By adding them up the overall probabilty is found that What factor 1 X17 occurs. The impact indicator for it was estimated as 9 (on a scale of 10) and the risk was determined by risk = probability*impact.

Table 6. If factors with their options and values

If factors	If factor options	If factor option labels	If factor option value intervals	Chosen If factors option values
If CW 1	If factor 1 option 1	low	0-0.35	0.17
	If factor 1 option 2	mediate	0.35-0.65	0.50
	If factor 1 option 3	high	0.65-1	0.82
If Tres 2	If factor 2 option 1	low	0-0.5	0.25
	If factor 2 option 2	high	0.5-1	0.75
If SP 3	If factor 3 option 1	low	0-0.5	0.25
	If factor 3 option 2	high	0.5-1	0.75

Table 7. If factor option combinations and the What factor X17 simulation outcomes

If CW 1 options	If Tres 2 options	If SP 3 options	What factor X17 outcome
If factor 1 option 1: low	If factor 2 option 1: low	If factor 3 option 1: low	X17 (PoA) has a shorter first increase, a longer lower decrease, and after the increase it stays high before decreasing.
		If factor 3 option 2: high	X17 (PoA) occurs faster and shorter overall
	If factor 2 option 2: high	If factor 3 option 1: low	The first peak is higher for the rest the same
		If factor 3 option 2: high	The first peak occurs faster and ends less high and decreases faster.
If factor 1 option 2: mediate	If factor 2 option 1: low	If factor 3 option 1: low	Overall, the lines are wider, first peak is higher, second peak is broader and decreases later.
		If factor 3 option 2: high	X17 occurs faster although still higher, the following decrease and increase also occur faster.

(continued)

Table 7. (*continued*)

If CW 1 options	If Tres 2 options	If SP 3 options	What factor 1 X17 outcome
	If factor 2 option 2: high	If factor 3 option 1: low	First decrease not much change, stays longer at bottom second increase the same
		If factor 3 option 2: high	Smaller first increase, overall line occurs earlier.
If factor 1 option 3: high	If factor 2 option 1: low	If factor 3 option 1: low	Significance change, much higher first wider peak, short decrease, increase before slowly decreasing.
		If factor 3 option 2: high	Still high first peak, second increase stays longer at top before decreasing faster.
	If factor 2 option 2: high	If factor 3 option 1: low	Much wider line stays longer at the second peak.
		If factor 3 option 2: high	Smaller first increase, longer at bottom, and longer also at the second peak.

Table 8. Overview of If factor options and their labels, value intervals and estimated probabilities

If factors	If factor options	If factor option labels	If factor option value intervals	If factor option probabilities
If factor 1	If factor 1 option 1	low	0-0.35	0.2
	If factor 1 option 2	mediate	0.35-0.65	0.3
	If factor 1 option 3	high	0.65-1	0.5
If factor 2	If factor 2 option 1	low	0-0.5	0.2
	If factor 2 option 2	high	0.5-1	0.8
If factor 3	If factor 3 option 1	low	0-0.5	0.3
	If factor 3 option 2	high	0.5-1	0.7

7 Discussion

This paper proposes a novel approach to improve cyber risk management and organizational alignment illustrated for the Amsterdam Airport Schiphol. It leverages Endsley's Risk Management Model (Endsley, 1995, 2015, 2021; Gini et al., 2011; Hoogendoorn et al., 2011) to understand threats at different levels. A network model specifically

Table 9. Overview of If factor option combination probabilities and What factor outcomes and risk assessment

If factor 1 options	If factor 2 options	If factor 3 options	If factor option combination probabilities		What factor 1 X17 category probabilities
If factor 1 option 1: Low 0.2	If factor 2 option 1: low 0.2	If factor 3 option 1: low 0.3	0.2*0.2*0.3=	0.012	X 0.012
		If factor 3 option 2: high 0.7	0.2*0.2*0.7=	0.028	
	If factor 2 option 2: high 0.8	If factor 3 option 1: low 0.3	0.2*0.8*0.3=	0.048	
		If factor 3 option 2: high 0.7	0.2*0.8*0.7=	0.112	X 0.112
If factor 1 option 2: Mediate 0.3	If factor 2 option 1: low 0.2	If factor 3 option 1: low 0.3	0.3*0.2*0.3=	0.018	
		If factor 3 option 2: high 0.7	0.3*0.2*0.7=	0.042	
	If factor 2 option 2: high 0.8	If factor 3 option 1: low 0.3	0.3*0.8*0.3=	0.072	X 0.072
		If factor 3 option 2: high 0.7	0.3*0.8*0.7=	0.168	X 0.168
If factor 1 option 3: High 0.5	If factor 2 option 1: low 0.2	If factor 3 option 1: low 0.3	0.5*0.2*0.3=	0.03	
		If factor 3 option 2: high 0.7	0.5*0.2*0.7=	0.07	X 0.07
	If factor 2 option 2: high 0.8	If factor 3 option 1: low 0.3	0.5*0.8*0.3=	0.12	

(continued)

Table 9. (*continued*)

If factor 1 options	If factor 2 options	If factor 3 options	If factor option combination probabilities		What factor 1 X17 category probabilities
		If factor 3 option 2: high 0.7	0.5*0.8*0.7=	0.28	X 0.28
			Sum of probabilities		0.714
			Impact indicator		9
			Risk assessment		**6.426**

designed for Schiphol's cyber domain has been developed, capturing the interconnectedness of vulnerabilities. A phased research plan established a theoretical foundation and meticulously designs the network model. This included features like self-adaptation mechanisms to account for evolving threats. A controlled experiment has evaluated the model's effectiveness in enhancing situational awareness and its impact on risk management and alignment.

The research has also explored future risks using What-If analysis and probabilistic assessment. The model will be continuously refined based on findings. This approach has the potential to significantly impact cyber risk management models by providing a more holistic view of the threat landscape and facilitating a proactive approach to cyber risk mitigation. Furthermore, the network model offers a novel assessment methodology. It can continuously assess risks by simulating attacks and incorporating real-time data for up-to-date assessments. Additionally, it could be integrated with existing frameworks for a more comprehensive approach. Future research directions include adapting the model for other airports, incorporating human factors, and exploring the potential of machine learning for enhanced threat prediction and risk assessment. This research can significantly advance cyber risk management strategies for critical infrastructure operators.

Appendix

The role matrices are standardized tables that specify the network characteristics for our adaptive network model in a systematic manner. The values within the matrices were properly chosen alongside our network model. Connectivity role matrices **mb** and **mcw** were already discussed in Sect. 4. For the timing characteristics, in **msv** (see Table 10) we assigned every state a speed factor according to their role and placement in the scenario and in the network model.

So, for example has the phase of perception a lower speed factor as they are doing checkups, while look at the comprehension phase that already is a faster state because in the comprehension, they convert the detected breach into a series of problems and their outcomes. In **iv** the initial values are shown.

Table 10. Role matrices **ms** and **iv** for timing characteristics

ms			iv		
X1	a-CAT	0.25	X1	a-CAT	0.6
X2	b-CAT	0.7	X2	b-CAT	0.6
X3	c-CAT	0.7	X3	c-CAT	0.5
X4	d-CAT	0.7	X4	d-CAT	0.4
X5	a-TPA	0.25	X5	a-TPA	0.3
X6	b-TPA	0.7	X6	b-TPA	0.1
X7	c-TPA	0.7	X7	c-TPA	0.2
X8	d-TPA	0.7	X8	d-TPA	0.2
X9	a-NCA	0.8	X9	a-NCA	0.2
X10	b-NCA	0.8	X10	b-NCA	0.2
X11	c-NCA	0.8	X11	c-NCA	0.2
X12	d-NCA	0.8	X12	d-NCA	0.2
X13	a-CERT/CSIRT	0.8	X13	a-CERT/CSIRT	0.2
X14	b-CERT/CSIRT	0.8	X14	b-CERT/CSIRT	0.2
X15	c-CERT/CSIRT	0.8	X15	c-CERT/CSIRT	0.2
X16	d-CERT/CSIRT	0.8	X16	d-CERT/CSIRT	0.2
X17	a-PoA	0.8	X17	a-PoA	0.2
X18	a-IF(p)	0.6	X18	a-IF(p)	0.2
X19	a-TEF(n)	0.7	X19	a-TEF(n)	0.2
X20	a-AI	0.2	X20	a-AI	0.8
X21	Con P	2	X21	Con P	0
X22	Con D	2	X22	Con D	0
X23	$\mathbf{W}_{PoA,TEF}$	0.2	X23	WPoA,TEF	0.2
X24	$\mathbf{W}_{PoA,IF}$	0.2	X24	WPoA,IF	0.2
X25	$\mathbf{W}_{PoA,a\text{-}CAT}$	0.2	X25	WPoA,a-CAT	0.2
X26	$\mathbf{W}_{PoA,a\text{-}TPA}$	0.2	X26	WPoA,a-TPA	0.2
X27	$\mathbf{W}_{IF,a\text{-}CAT}$	0.2	X27	WIF,a-CAT	0.2
X28	$\mathbf{W}_{TEF,a\text{-}CAT}$	0.2	X28	WTEF,a-CAT	0.2
X29	$\mathbf{W}_{IF,a\text{-}TPA}$	0.2	X29	WIF,a-TPA	0.2
X30	$\mathbf{W}_{TEF,a\text{-}TPA}$	0.2	X30	WTEF,**a**-TPA	0.2
X31	$\mathbf{W}_{IF,a\text{-}NCA}$	0.2	X31	WIF,a-NCA	0.2
X32	$\mathbf{W}_{TEF,a\text{-}NCA}$	0.2	X32	WTEF,a-NCA	0.2

(continued)

Table 10. (*continued*)

ms			iv		
X33	$\mathbf{W}_{\text{IF,a-CERT/CSRIT}}$	0.2	X33	WIF,a-CERT/CSRIT	0.2
X34	$\mathbf{W}_{\text{PoA,TEF}}$	0.2	X34	WTEF,a-CERT/CSRIT	0.1
X35	Context	0	X35	Context	1

Table 11. Aggregation characteristics: role matrix **mcfw** for choice of combination function

mcfw		alogistic	stepmod
X1	a-CAT	1	
X2	b-CAT	1	
X3	c-CAT	1	
X4	d-CAT	1	
X5	a-TPA	1	
X6	b-TPA	1	
X7	c-TPA	1	
X8	d-TPA	1	
X9	a-NCA	1	
X10	b-NCA	1	
X11	c-NCA	1	
X12	d-NCA	1	
X13	a-CERT/CSIRT	1	
X14	b-CERT/CSIRT	1	
X15	c-CERT/CSIRT	1	
X16	d-CERT/CSIRT	1	
X17	a-PoA	1	
X18	a-IF(p)	1	
X19	a-TEF(n)	1	
X20	a-AI	1	
X21	Con P		1
X22	Con D		1
X23	$\mathbf{W}_{\text{PoA,TEF}}$	1	
X24	$\mathbf{W}_{\text{PoA,IF}}$	1	
X25	$\mathbf{W}_{\text{PoA,a-CAT}}$	1	

(*continued*)

Table 11. (*continued*)

mcfw		alogistic	stepmod
X26	$\mathbf{W}_{PoA,a\text{-}TPA}$	1	
X27	$\mathbf{W}_{IF,a\text{-}CAT}$	1	
X28	$\mathbf{W}_{TEF,a\text{-}CAT}$	1	
X29	$\mathbf{W}_{IF,a\text{-}TPA}$	1	
X30	$\mathbf{W}_{TEF,a\text{-}TPA}$	1	
X31	$\mathbf{W}_{IF,a\text{-}NCA}$	1	
X32	$\mathbf{W}_{TEF,a\text{-}NCA}$	1	
X33	$\mathbf{W}_{IF,a\text{-}CERT/CSRIT}$	1	
X34	$\mathbf{W}_{PoA,TEF}$	1	
X35	Context		1

Table 12. Aggregation characteristics: role matrix **mcfp** for the parameters of chosen combination functions

mcfp		alogistic		stepmod	
		Steepness	Threshold	Repetition	Tipping point
X1	a-CAT	5	0.5		
X2	b-CAT	5	0.5		
X3	c-CAT	5	0.5		
X4	d-CAT	5	0.5		
X5	a-TPA	5	0.5		
X6	b-TPA	5	0.5		
X7	c-TPA	5	0.5		
X8	d-TPA	5	0.5		
X9	a-NCA	5	1.5		
X10	b-NCA	5	1.3		
X11	c-NCA	5	1.5		
X12	d-NCA	5	1.5		
X13	a-CERT/CSIRT	5	1.5		
X14	b-CERT/CSIRT	5	1.3		
X15	c-CERT/CSIRT	5	1.5		
X16	d-CERT/CSIRT	5	1.5		

(*continued*)

Table 12. (*continued*)

mcfp		alogistic		stepmod	
		Steepness	Threshold	Repetition	Tipping point
X17	a-PoA	5	1.8		
X18	a-IF(p)	5	0.5		
X19	a-TEF(n)	5	0.5		
X20	a-AI	5	0.5		
X21	Con P			50	25
X22	Con D			50	25
X23	$\mathbf{W}_{\text{PoA,TEF}}$	5	2.2		
X24	$\mathbf{W}_{\text{PoA,IF}}$	5	2.2		
X25	$\mathbf{W}_{\text{PoA,a-CAT}}$	5	2.2		
X26	$\mathbf{W}_{\text{PoA,a-TPA}}$	5	2.2		
X27	$\mathbf{W}_{\text{IF,a-CAT}}$	5	2.2		
X28	$\mathbf{W}_{\text{TEF,a-CAT}}$	5	2.2		
X29	$\mathbf{W}_{\text{IF,a-TPA}}$	5	2.2		
X30	$\mathbf{W}_{\text{TEF,a-TPA}}$	5	2.2		
X31	$\mathbf{W}_{\text{IF,a-NCA}}$	5	2.2		
X32	$\mathbf{W}_{\text{TEF,a-NCA}}$	5	2.2		
X33	$\mathbf{W}_{\text{IF,a-CERT/CSRIT}}$	5	2.2		
X34	$\mathbf{W}_{\text{PoA,TEF}}$	5	2.2		
X35	Context	5	0.1		

For the aggregation characteristics, the next role matrix, **mcfw** (see Table 11) shows which of the state has the **alogistic** function and which has the **stepmod function**. In our network model only two states have the **stepmod** function, X21 Con P and X22 Con D, as they depend on repetition and duration while the other states have a **alogistic** function that depends on the steepness and threshold parameters. In **mcfp** (see Table 12) the first two columns of the table represent the values of the **alogistic** function and the third and fourth column of the **stepmod** function. In the table we had to configure each state alongside our network model and adjust them alongside the sequences of events.

References

Abromaityte, M., Vos, T., Lucassen, B., Kruger, Q., Dubero, E., Van den Hout, N.J., Bouma, D., Treur, J., Roelofsma, P.H.M.P.: Computational analysis of user experience of password-based authentication systems. In: Proceedings of the 16th International Conference on Intelligent Human Computer Interaction, IHCI'24 Lecture Notes in Computer Science, vol. 15557, pp. 341–352. Springer Nature (2025)

Akers, D., Blomquist, T., Niedziela, J., Šebok, S., Van den Hout, N.J., Bouma, D., Treur, J., Roelofsma, P.H.M.P.: Organizational response to APT attacks: computational analysis by behavioral network modelling. In: Proceedings of the 16th International Conference on Intelligent Human Computer Interaction, IHCI'24 Lecture Notes in Computer Science, vol. 15557, pp. 411–425. Springer Nature (2024)

Al-Riyami, S., Hajri, M.A.: A survey of machine learning and deep learning applications in cybersecurity. Int. J. Adv. Comput. Sci. Appl. **11**(1), 482–491 (2020). https://doi.org/10.14569/IJACSA.2020.01.054

Bouma, D., Hoffmans, C., Van den Hout, N.J., Zwarts-Wojtowicz, N., Treur, J., Roelofsma, P.H.M.P. Cyber security in hospitals: a network-oriented model for behavioural learning of employees during phishing simulations. In: Proceedings of the 17th International Conference on Computational Intelligence in Security for Information Systems, CISIS'24 Lecture Notes in Networks and Systems, vol. 957, pp. 98–111. Springer Nature (2024)

Canbaloğlu, G., Treur, J., Roelofsma, P.H.M.P.: Computational modeling of organisational learning by self-modeling networks. Cogn. Syst. Res. **73**, 51–64 (2022)

Canbaloğlu, G., Treur, J., Wiewiora, A. (eds.): Computational Modeling of Multilevel Organisational Learning and Its Control Using Self-Modeling Network Models. Springer Nature (2023). https://doi.org/10.1007/978-3-031-28735-0

Capra, A.M.B., Dekker, E., Hoffmann, L., Hoffmans, C.C.A., Mueller, T., Van den Hout, N.J., Treur, J., Roelofsma, P.H.M.P.: Computational analysis of human factors in spear-phishing attacks: an adaptive network model. In: Proceedings of the 16th International Conference on Intelligent Human Computer Interaction, IHCI'24 Lecture Notes in Computer Science, vol. 15557, pp. 353–366. Springer Nature (2024)

Cybersecurity | ACI World: ACI World. (n.d.). https://aci.aero/airport-advocacy/airport-information-technology/cybersecurity/

Federal Emergency Management Agency (FEMA): Crisis Communication. Federal Emergency Management Agency (2023)

Endsley, M.R.: Toward a theory of situation awareness in dynamic systems. Hum. Factors. **37**(1), 32–64 (1995)

Endsley, M.R.: Final reflections: situation awareness models and measures. J. Cogn. Eng. Decis. Mak. **9**(1), 101–111 (2015)

Endsley, M.R.: Situation awareness. In: Handbook of Human Factors and Ergonomics, pp. 434–455. Wiley (2021)

Gini, M.L., Hoogendoorn, M., Van Lambalgen, R.M.: Learning belief connections in a model for situation awareness. In: Agents in Principle, Agents in Practice: 14th International Conference, PRIMA 2011, Wollongong, Australia, November 16–18, 2011. Proceedings 14, pp. 373–384. Springer, Berlin Heidelberg (2011)

Hendrikse, S.C.F., Treur, J., Koole, S.: Modeling emerging interpersonal synchrony and its related adaptive short-term affiliation and long-term bonding: a second-order multi-adaptive neural agent model. Int. J. Neural Syst. **33**(7), 2350038 (2023)

Hendrikse, S.C.F., Treur, J., Koole, S.L. (eds.): New Analysis and Modeling Directions in Social Interaction Science: Emergent Multimodal Interpersonal Synchrony, Affiliation, and Bonding. Springer Nature (2025)

Hoffmans, C.C.A., Van der Does, L., Van der Kolk, M., Van Rooyen, T., Barelds, N., Treur, J., Roelofsma, P.H.M.P.: Analysing the rise of biohacking in a tech-driven city: assessing risks through adaptive network modeling. In: Proceedings of the 16th International Conference on Intelligent Human Computer Interaction, IHCI'24 Lecture Notes in Computer Science. Springer Nature (2024)

Hoogendoorn, M., van Lambalgen, R.M., Treur, J.: Modeling situation awareness in human-like agents using mental models. In: Proceedings of the Twenty-Second International Joint Conference on Artificial Intelligence, pp. 1697–1704 (2011)

Humayed, H., Lin, J., Li, F., Luo, B.: Cyber-Physical Systems Security – A Survey. IEEE (2017). https://doi.org/10.1109/JIOT.2017.2703172

Jeffery, E., Malsagov, U., Tijssen, N., Zeisig, A., Hoffmans, C.C.A., Treur, J., Roelofsma, P.H.M.P.: Computational analysis of the effectiveness of the intelligence cycle for organizational cyber risk management. In: Proceedings of the 8th International Conference on Computational Methods in Systems and Software, CoMeSySo'24 Lecture Notes in Networks and Systems. Springer Nature (2024)

Keijzer, S., Lochtenbergh, D., Marsman, T., Voorhoeve, S., Zwarts, N., Bouma, D., Treur, J., Roelofsma, P.H.M.P., Managing Classified Information by a Third-Party Contractor: A Computational Cybersecurity Analysis. In: Proceedings of the 8th International Conference on Computer-Human Interaction Research and Applications, CHIRA'24. Communications in Computer and Information Science, vol. 2371, pp. 299–315 (2025)

Lederer, A.L., Carbone, A., Kannan, R.K.: Social engineering: a review of the literature and recommendations for mitigation. Comput. Secur. **91**, 101724 (2020). https://doi.org/10.1016/j.cose.2019.101724

Liu, L., De Vel, O., Han, Q.-L.H., Zhang, J., Iang, Y.X.: Detecting and Preventing Cyber Insider Threats: A Survey. IEEE (2018). https://doi.org/10.1109/COMST.2018.2800740

Roelofsma, P.H.M.P., Jabeen, F., Taal, H.R., Treur, J. (eds.): Using Shared Mental Models and Organisational Learning to Support Safety and Security through Cyberspace: a Computational Analysis Approach. Springer Nature (2025)

Salerno, J., Hinman, M., Boulware, D.: Building a framework for situation awareness. In: Proceedings of the 7th International Conference on Information Fusion, vol. 89, p. 90 (2004)

Scmo, C. B.: Weather and Situation Awareness in Aviation - CMOS BULLETIN SCMO. CMOS BULLETIN SCMO. https://bulletin.cmos.ca/weather-and-situation-awareness-in-aviation/ (2021)

Treur, J.: The ins and outs of network-oriented modeling: from biological and mental networks to social networks. In: Transactions on Computational Collective Intelligence Contents of Keynote Lecture at ICCCI'18, vol. 32, pp. 120–139. Springer Publishers (2019)

Treur, J.: Network-Oriented Modeling for Adaptive Networks: Designing Higher-Order Adaptive Biological, Mental, and Social Network Models. Springer Nature (2020)

Treur, J.: On the dynamics and adaptivity of mental processes: relating adaptive dynamical systems and self-modeling network models by mathematical analysis. Cogn. Syst. Res. **70**, 93–100 (2021)

Treur, J., Van Ments, L. (eds.): Mental Models and Their Dynamics, Adaptation, and Control. Springer Nature (2022). https://doi.org/10.1007/978-3-030-85821-6

van Ments, L., Treur, J., Klein, J., Roelofsma, P.H.M.P.: A second-order adaptive network model for shared mental models in hospital teamwork. In: Computational Collective Intelligence: Proceedings of the 13th International Conference, ICCCI 2021 Lecture Notes in Computer Science, vol. 12876, pp. 126–140. Springer (2021)

The Audit of Information Systems is a Crucial Aspect of Ensuring the Economic Security of a Business

A. D. Khristenko[1], F. S. Fokin[2], I. V. Osliakova[3](✉), O. S. Abaydullina[3], N. E. Ioffe[3], S. S. Katakhova[3], and L. N. Svetova[3]

[1] Financial University under the Government of the Russian Federation, Moscow, Russia
[2] Federal State Budget Educational Institution of Higher Education «MIREA - Russian Technological University», Moscow, Russia
[3] Foreign Languages Chair, Russian Technological University 78, Moscow, Russia
oslyakova@mirea.ru

Abstract. The article discusses the importance of information systems auditing as an integral part of business economic security in the context of digital transformation. It examines the history of information systems auditing, the legislative framework for information systems audit in Europe and Russia, and the key risks associated with information systems use. Additionally, it explores the impact of emerging technologies such as blockchain and artificial intelligence on economic security and their implications for information systems audits.

Keywords: Audit of Information Systems · Economic Security · Digtal Transformation · Risk Management · Compliance with Information Security Legislation · Blockchain · Artificial Intelligence · COSO and COBIT Frameworks

1 Introduction

Today, the significance of information technology for a business cannot be underestimated. All aspects of a company's operations, from data management to digital security, rely on information technology. Therefore, the review of information systems is crucial for ensuring the economic stability of an organization. In the context of widespread digitalization, information resource management has become even more crucial, influencing the financial performance of a modern business.

It is worth noting that the process of digitalization and computerization of society and the economy did not begin in the twenty-first century as is commonly believed, but rather in the twentieth century. This is indicative of how all existing information systems have been modified. The paper will examine the stage of development of information audit within the European context, which will highlight its relevance and significance both in the past century and in the present day.

R. Silhavy and P. Silhavy (Eds.): CoMeSySo 2024, LNNS 1489, pp. 445–449, 2025.
https://doi.org/10.1007/978-3-031-96798-6_32

The origins of the emergence of the phenomenon of information system auditing can be traced back to its most prominent manifestation – the emergence of this academic discipline in several developed countries at the end of the twentieth century. This development is confirmed by the increasing scientific discussion on the topic and the establishment of educational curricula, indicating the growing importance of this field as a component of economic security for both businesses and states.

In Russian economic terms, the concept of auditing is defined in accordance with Federal Law № 307-FZ dated December 30, 2008, which defines it as an independent assessment of the financial statements of a company in order to provide an opinion on their accuracy and reliability. However, within the context of this study, the term "audit" is used to refer to a process that differs from the traditional understanding of an audit of financial statements and associated business processes. And since the verification of certain aspects of an organization's financial reporting is directly linked to the economic viability of the company, information auditing in its broadest sense is also relevant.

The main aim of the research is to update knowledge on the challenges of information audit development in both European and Russian contexts. It is also necessary to monitor the variability in the opportunities for the formation of this field and the impact of this development on the economy.

2 Methods

Currently, in most developed countries, regulatory legal frameworks have been established and are being refined to control the development and implementation of computer information systems. Therefore, we can move on to the stage of establishing information system auditing and note laws on information security protection, which directly impact the economic security of businesses and states. In the Netherlands, for example, such regulations were established in the past century and have since then been updated. Since 2019, a new law – Computer Crime Act (WCC) III has come into effect that not only strengthens the cybersecurity of the country but also links the work of information system auditors with ensuring the economic security of organizations. This phenomenon is reflected in the Dutch Civil Code, which requires an auditor's assurance on the continuity and accuracy of data processing, particularly those related to a company's financial records.

In Russia, there is a law that ensures cybersecurity and protection of information and information systems – the Federal Law "On Information, Information Technologies, and Information Protection" dated July 27, 2006, № 149-FZ. Similar to other developed countries, the legislation in the area of security continues to grow. In particular, cooperation systems in the field of information system auditing have emerged in the European Union and continue to be developed.

All these legal documents form the basis for the development of audit expertise in the field of information systems. Therefore, there is a need for training in this area and educational and methodological materials are being developed for work in this direction. Some of these materials were studied during the collection and analysis of information for the research. Based on the teaching methods and topics covered in these educational programs, a clear understanding of the importance of audit of information systems to the economic security of a business has been developed.

3 Results

Based on the information provided, we can trace the development of auditing information systems as a natural consequence of digitalization. In the twentieth century, with the emergence of advanced computers and information analysis tools, the risks associated with their use began to emerge, which necessitated their management. As the concept of auditing is multifaceted and encompasses internal control, managing risks within this context is also important. The implementation of the relevant regulatory legal acts signified the official recognition of the need to validate data processing and analysis processes, as well as information transfer.

The development of academic fields directly or indirectly related to education in the field of information systems at universities in Europe and Russia has highlighted the demand for such professionals in the labor market. If we examine the training programs offered by technical and economic universities in Moscow for the formation of knowledge about the audit of information systems, economic security and their relationship, you can find that in such programs audit as an academic discipline is taught simultaneously with data analysis, the basics of working with information technology, programming and economic security as a scientific discipline. These training programs are offered by numerous universities, with specific examples being provided by RTU MIREA and the Financial University under the government of the Russian Federation in their programs on economic security and internal auditing.

Let's consider the auditing of information systems from a more practical perspective: as a process within the organization's framework, it is the inspection of the company's information system, including the software and other components of the internal control system. Special attention is paid to the application of internal control standards, such as the COSO and COBIT models. In practice, the auditor also needs to verify the efficiency of the company's information system. In Russia, this includes 1C accounting systems such as "Accounting", "Sail", and "Info-Accountant". In international practice, SAP and other systems are used.

An information systems audit is essential for ensuring the economic security of a company for several reasons:

- Minimizing risks: By identifying and addressing vulnerabilities in information systems, businesses can reduce the risk of cyberattacks, data breaches, and disruptions to business processes, thereby preventing significant financial losses.
- Cost optimization: Through an audit, companies can identify underutilized resources and optimize their IT infrastructure, software, and personnel costs.
- Improved efficiency: Implementing the recommendations of auditors can lead to increased productivity, reliability, and security of information systems, ultimately enhancing the overall efficiency of the organization.
- Compliance: Audits ensure that information systems adhere to legal requirements and industry standards for information security, particularly crucial for companies handling sensitive data.

And since the connection between economic security and information systems auditing has been proven not only theoretically, but also practically, it is essential to pay attention to specific issues addressed by this type of audit. These issues include dealing with the aforementioned risks, which are important to consider in greater detail:

- Risks directly related to the source data used in the preparation of financial statements
- Risk factors that may affect the prepared financial statements and the information presented therein.

In the area of risks associated with source data, some of the main areas of concern for information systems auditors include:

- Risks related to providing access to information systems (including potential threats of unauthorised entry into the system);
- Risk of erroneous recording of financial information by the company;
- Risk of recording financial information without proper identification;
- Potential loss or corruption of information due to software failures, computer hardware damage, or other technical issues;
- The risk of losing original primary documentation prior to its entry into the database;
- Possibility of automated generation of reporting data leading to potential distortion.

At present, the relatively new field of auditing involves the examination of information systems in classic scenarios involving the use of software and hardware by businesses for the collection and analysis of financial data, with subsequent reporting. As information technology advances, it is possible to observe trends in the evolution of risks associated with information systems that may pose a threat to an organization's economic security. This often includes discussions about blockchain technology and artificial intelligence applications. However, even in a changing landscape, established models for assessing internal control processes remain relevant, albeit with potential for further refinement.

4 Discussion

The aspects of information systems auditing discussed in the study form the basis of discussions at various academic institutions around the world. In Russia, these issues are often discussed at scientific conferences within the context of the wider discussion on the development of digitalization in the economy.

There is a growing trend towards increased risks to the economic stability and security of companies, which is associated with the development of blockchain, digital currencies, and the introduction of artificial intelligence in database management. These developments have the potential to produce both positive and negative effects, which may be influenced by human factors. Consequently, research in this area has become increasingly important.

5 Conclusion

In today's rapidly digitalizing world and with businesses increasingly relying on information technology, auditing information systems has become not only relevant but also a critical tool for ensuring the economic security of enterprises. The analysis has clearly demonstrated that a well-structured and regularly conducted information system audit can not only detect vulnerabilities and prevent financial loss but also form the basis for long-term sustainable business growth.

The study emphasizes the key benefits of information system audits in terms of economic security, at least in relation to financial reporting.

However, the audit of information systems is a field that is impacted by global developments. In the context of ongoing technological advancements and evolving cyber threats, a holistic and systematic approach is required. Regular audits conducted by skilled professionals will enable you to swiftly respond to new challenges and ensure the security and sustainable growth of your business in the digital era.

References

1. Baranova, O.V.: Audit of information systems. Finan. Theor. Pract. **1**, 58–60 (2009)
2. Dvoinishnikov, N.E., Islamutdinova, D.F.: The concept and essence of information systems security audit. Moscow Econ. J. **10** (2019) https://cyberleninka.ru/article/n/ponyatie-i-suschnost-audita-bezopasnosti-informatsionnyh-sistem, last 8 Aug 2024
3. Everson, M.E.A., Chesley, D.L., Martens, F.J., Bagin, M., Katz, H., Sylvius, K.T., Perraglia, S.J., Zelnik, K.C., & Grimshaw, M.: Enterprise risk management: integrating with strategy and performance. Committee of Sponsoring Organizations of the Treadway Commission (2017)
4. Federal Law of Russian Federation No. 149-FZ. On Information, Information Technologies and Information Protection. dated 27 July 2006
5. Federal Law of Russian Federation No. 307-FZ. On Auditing Activities. dated 30 Dec 12 2008
6. Sitnov, A.A.: Audit of information systems - a risk-oriented approach. Prob. Econ. Legal Pract. **5**, 94–97 (2019)
7. Berghout, E., Fijneman, R., Hendriks, L., de Boer, M., Butijn, B.J.: Auditing advanced information systems and technologies in a modern digital world. In: Advanced Digital Auditing. Progress in IS. Springer, Cham (2023). https://doi.org/10.1007/978-3-031-11089-4_1

Implementation of a Cross-Platform Simulator Application in the Process of Learning a Foreign Language for Students of Technical Specialties

V. S. Postolenko[1], V. S. Postolenko[1], A. S. Fokin[1], I. V. Osliakova[2(✉)], E. E. Rybakova[2], A. R. Prokopchuk[2], and O. A. Frolova[2]

[1] Federal State Budget Educational Institution of Higher Education, MIREA - Russian Technological University, Vernadsky Prospect, 78, Moscow, Russia
[2] Foreign Languages Chair, Russian Technological University, Vernadsky Avenue, Moscow 119454, Russia
oslyakova@mirea.ru

Abstract. The article examines the role of cross-platform simulator applications in the process of learning a foreign language and reviews the practical application of the interactive simulator in order to develop and improve the level of language proficiency in the field of computer science. #COMESYSO1120.

Keywords: Cross-Platform Application · Simulator · Interactivity · Personalization · Motivation · Learning Effectiveness · Interface

1 Introduction

In the context of globalization and the development of information technologies, the study of foreign languages is becoming increasingly relevant. Traditional methods of teaching foreign languages, based on textbooks and instructor-led classes, do not always provide sufficient effectiveness and motivation. Consequently, the role of mobile technologies in education, particularly cross-platform training applications, is growing. The combination of traditional teaching methods and modern interactive techniques is a crucial condition for effectively teaching a foreign language at a university, especially for students in technical fields. They typically encounter significant challenges and require narrowly focused language knowledge applicable to their specific field. Currently, there are no applications available that specifically help train technical language skills. While numerous applications assist in starting from scratch, much of the content may not be relevant for future professional use, and specialized terminology is rarely found in these applications. Therefore, the implementation of a cross-platform language training application tailored to a narrowly specialized technical context represents a pertinent task, contributing to the effective education of students in technical fields.

R. Silhavy and P. Silhavy (Eds.): CoMeSySo 2024, LNNS 1489, pp. 450–457, 2025.
https://doi.org/10.1007/978-3-031-96798-6_33

2 Advantages and Challenges of Implementation

The advantages of cross-platform training applications include accessibility, interactivity, personalization, feedback, diverse content, and motivation. The accessibility of the application on various devices allows users to study at any convenient time and place, without being tied to a specific location or time. This is especially relevant for students who can do their homework even while on the way home. Interactive exercises, games, and tasks make the learning process more engaging and contribute to active material absorption. Additionally, the application can adapt to the user's level, needs, and goals, offering relevant content and an individualized learning trajectory. Instant feedback enables progress tracking, error identification, and adjustment of the learning strategy. The application can include various materials such as audio, video, texts, grammar exercises, vocabulary drills, pronunciation practice, and speech comprehension, promoting comprehensive language skill development. Achievement systems, rankings, competitive elements, and the ability to communicate with other users sustain motivation and interest in language learning.

Furthermore, it is important to consider potential challenges that may arise during the integration process: staff training, individualized instruction, assessment and monitoring, language integration barriers, personal preferences. University instructors may require training on using the application, necessitating additional time and resources. Adapting the application to the needs of different student groups based on their knowledge levels and learning styles is crucial. Ensuring the ability to track students' learning progress and evaluate their achievements through the application is vital. If the application is available only in French, additional support or training for students may be necessary, especially if they do not have a proficient command of the language. Some students may prefer traditional teaching methods and face difficulties adapting to new technology.

3 Methods

The following methods were used for the research:

1. Literature review
   A review of modern foreign language learning techniques for students in technical specialties was conducted, along with an analysis of existing cross-platform training applications.
2. Student survey to identify their needs and expectations from a training application. During the survey, respondents answered 3 questions:
   - "Do you use any application for learning a foreign language?"
   - If yes, please write the name of the application.
   - What functionality in a foreign language learning application do you consider most essential? (select multiple options)
3. Mathematical statistics method
   It is designed for processing and analyzing data obtained during experimental research, as well as for interpreting results based on this data.

4. Student survey to analyze the developed application
   After developing the application concept based on the data from the first survey, a second survey was conducted with 2 questions:
   - How do you rate the convenience of the presented interface?
   - Would you use the application for learning a foreign language specifically in the technical field?

4 Results

The research revealed that, firstly, a significant 40% do not use any application for learning foreign languages (Fig. 1).

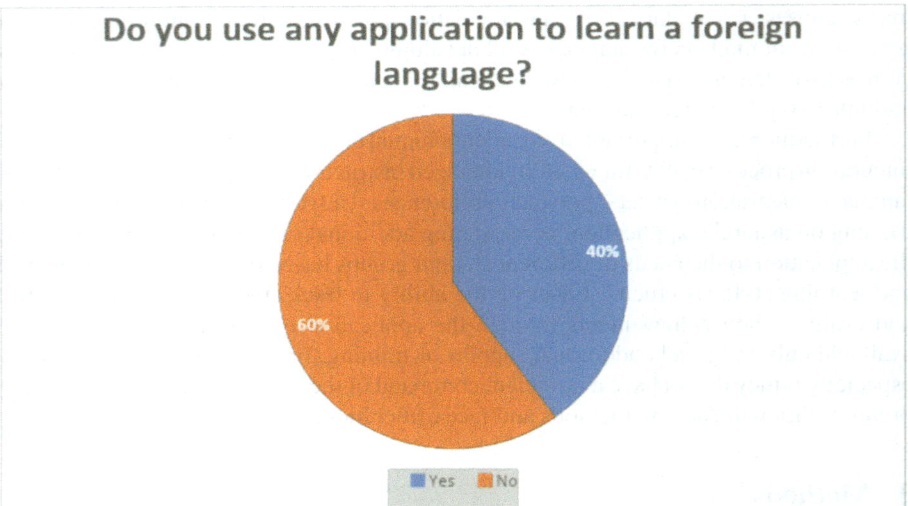

Fig. 1. Comparative analysis of using applications for learning foreign languages, %.

Secondly, it was determined that the most popular application is Duolingo (Fig. 2), so it can be used as a basis for developing the concept of our educational application.

Additionally, based on the respondents' answers, shortcomings of existing applications were identified, and it became clear that the application should definitely include additional audio and video lessons, articles, books, etc. (84% of respondents indicated this point). Also, 60% of the respondents would like the ability to add to favorites, and 48% would like a chat for discussing questions (Fig. 3).

These results helped us to fill our application with only useful functionality.

On the main screen, the progress in completing 4 sections will be displayed, which will be dedicated to vocabulary, grammar, listening, and reading tasks. Specifically, there will be various exercises to expand the technical vocabulary:

- Different types of exercises such as flashcards with terms and their translations, exercises for memorizing new words using associations, crosswords, and puzzles to effectively learn new vocabulary.

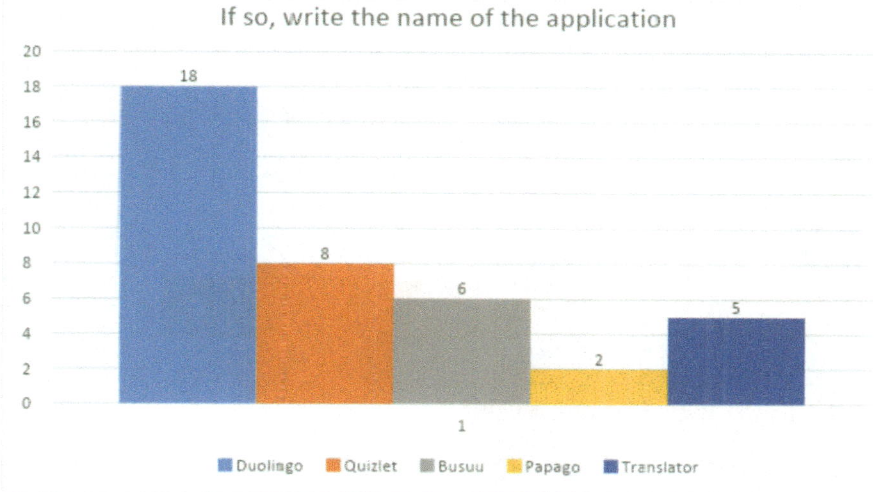

Fig. 2. Analysis of using applications for learning foreign languages.

- Interactive exercises with audio support to practice pronunciation and correct stress.
- There will also be a panel to navigate to the task completion. An example interface of the main screen can be seen in Fig. 4.
- The app will include interactive exercises for reinforcing the material, such as fill in the blanks, matching, and multiple-choice questions, which will contribute to language skill development and automation of grammar rules. An example interface for a fill in the blanks exercise is shown in Fig. 5.
- Language games like crosswords, quizzes, and puzzles will make the learning process more engaging and aid in memorizing new information.
- Teachers will be able to assess students' progress by viewing it on the main screen, facilitated by a user-friendly interface. This monitoring will help evaluate progress, identify weak areas, and adjust the teaching strategy.

The ability to communicate with other students and teachers to discuss questions and improve language skills will be provided through a "Chat" tab in the menu panel. Access to additional materials such as audio and video lessons, articles, and books will be available through the "Extra Materials" tab. Students will have the option to revisit challenging assignments by adding them to their "Favorites" tab, indicated by a heart icon on each task screen.

Teachers will be able to assess student performance by viewing progress on the main screen and comparing results with other students in the "Leaderboard" tab within the menu panel, facilitated by a user-friendly interface. This monitoring will help evaluate progress, identify weak areas, and adjust the teaching strategy.

In another survey, we asked participants to rate the developed interface on a scale of 1 to 5. The results showed that the proposed interface was sufficiently understandable, with 62% rating it as 5 and 36% rating it as 4 (see Fig. 6).

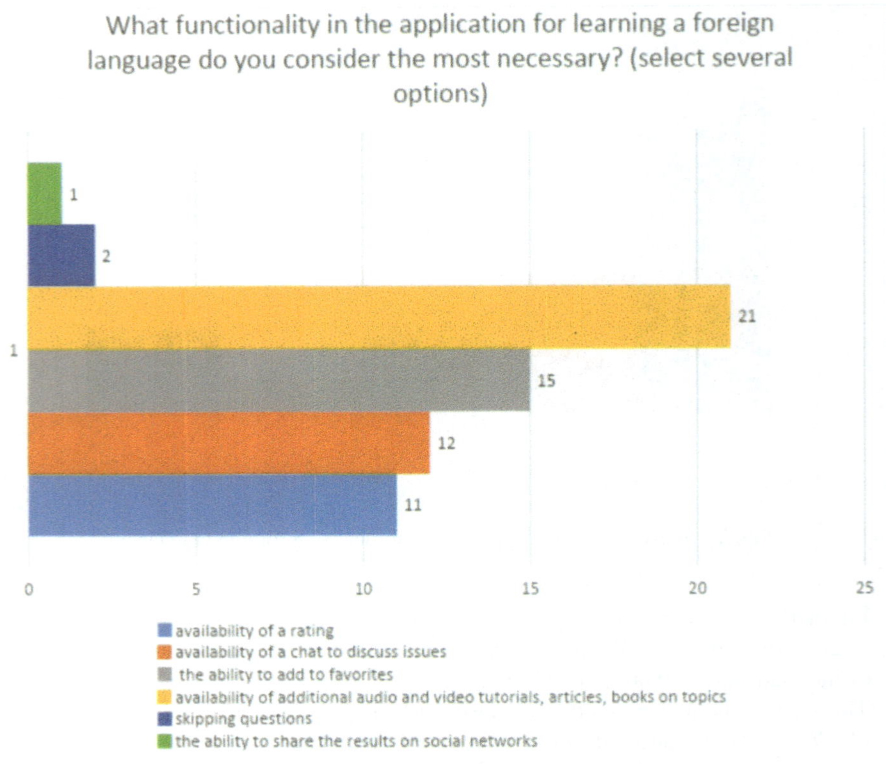

Fig. 3. Comparative analysis of using applications for learning foreign languages

Fig. 4. Example of the main screen interface.

Fig. 5. Example of the lesson interface.

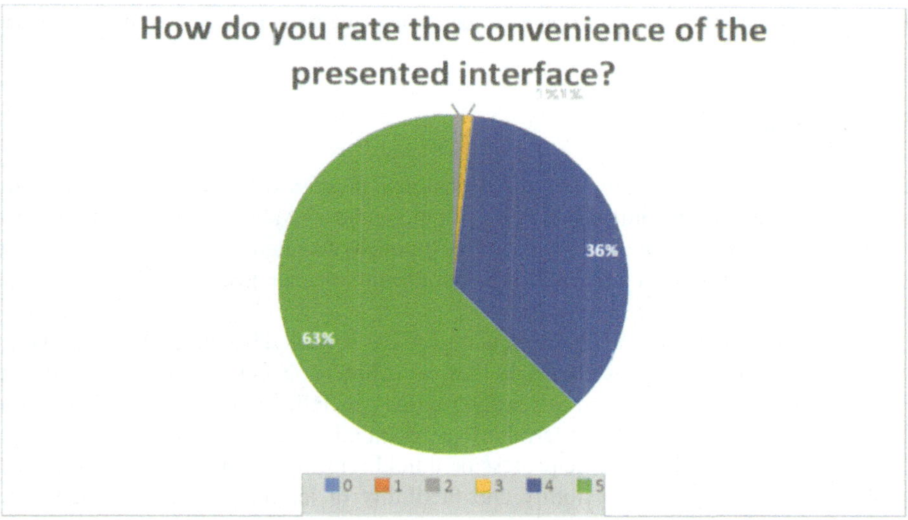

Fig. 6. Comparative analysis of rating the convenience of the presented interface.

Furthermore, it was determined that 90% of surveyed students from technical universities would use an application aimed at learning a foreign language in the technical field (Fig. 7).

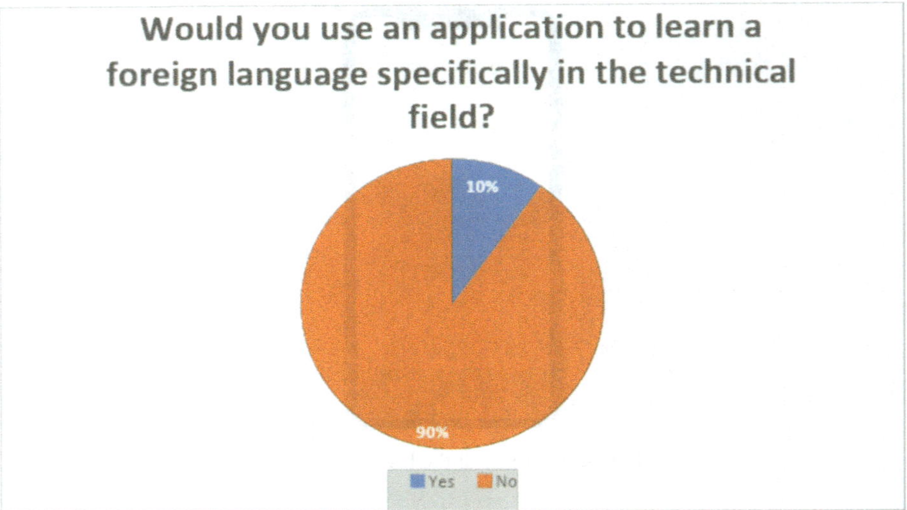

Fig. 7. The probability of using the application for learning foreign languages in the technical field.

5 Discussions

Integrating such an application into the educational process will allow students to study the language for their technical specialization more narrowly, which in turn will increase their competitiveness in the labor market. The interest of students in such tools is already evident, as the application will provide only relevant material necessary for their future profession.

This approach to learning a foreign language has several advantages. Firstly, it helps to increase the motivation of students, as they see a direct connection between the studied material and their future career. Secondly, narrow specialization allows optimizing the learning process, focusing on the terminology, grammatical structures, and communication skills demanded in a specific professional field. Thirdly, the use of a cross-platform application ensures the accessibility and flexibility of learning, allowing students to study at any time and in any place, individually adjusting the pace and intensity of their studies.

It is important to note that the developed application is not intended to replace traditional foreign language teaching methods, but to serve as an additional tool that expands the possibilities for both students and teachers. Integrating the application into the educational process will allow the teacher to devote more time to developing students' speaking, listening, and writing skills, as well as conducting interactive classes and discussions on professional topics.

The introduction of a cross-platform training application in the process of learning a foreign language for students of technical specialties is a promising direction in the development of the educational process. This tool can significantly improve the efficiency and quality of education, as well as prepare graduates for successful careers in the context of globalization and high competition. Further research in this area will be aimed at developing specific methods and approaches to integrating the application into the educational process, as well as evaluating its effectiveness in practice.

6 Conclusion

Thus, the introduction of a cross-platform simulator application will facilitate the process of learning a foreign language. Thanks to the ease of use, the application will increase the motivation of students, improve their knowledge in the field of technical language, and help students of non-humanitarian specialties to memorize the necessary terms more easily.

The use of such innovative technologies in the educational process opens up new opportunities to improve the effectiveness of teaching foreign languages. Such an application can become an important tool in the arsenal of a modern teacher, helping to make classes more interactive, exciting and effective for students of technical fields. Further development and improvement of such mobile simulators will contribute to a more successful development of professionally oriented language material.

References

1. Arzumanova, N.V.: The use of modern information technologies in the educational process. News of the Russian State Pedagogical University named after. A. I. Herzen 113 (2009)
2. Zakharova, I.G.: Information Technologies in Education: Study. Academy, Moscow (2003)
3. Gordeeva, N.O.: The practice of using virtual simulators for mastering algorithms for the professional activities of future economists. Modern Prob. Sci. Educ. **1**(6) (2022). https://doi.org/10.17513/spno.32257
4. Veksper, V.A., Reidel, P.B.: Interactive simulators and their importance in the educational process. Ther. Nova. **41**, 205–211 (2016)
5. https://frenchblogs.ru/learning/for-beginners/apps-for-french-learning.html, last accessed 8 Aug 2024
6. Godwin-Jones, R.: Smartphones and language learning. Lang. Learn. Technol. **21**(2), 3–17 (2017)
7. Kukulska-Hulme, A.: Mobile-assisted language learning. In: Chapelle, C.A. (ed.) The Concise Encyclopedia of Applied Linguistics. Wiley (2018)
8. Burston, J.: Mobile-assisted language learning: a selected annotated bibliography of implementation studies 1994–2012. Lang. Learn. Technol. **17**(3), 157–224 (2013)

A Set of Scientific and Technical Marketing Methods Based on the Use of Social Networks

Igor Janiszewski(✉)

Federal Research Center «Computer Science and Control» of the Russian Academy of Sciences, 44 building 2 Vavylova Street, Moscow 119333, Russia
igoryster@gmail.com

Abstract. A set of methods of scientific and technical marketing based on the use of social networks is presented, which provides a solution to the problems of establishing connections between the profiles of subjects-creators and subjects-consumers of scientific and technical content; filtering scientific and technical content; measuring public attention and influence; ranking manufacturers and scientific and technical content with the formation of recommendations to consumers; thematic targeting of scientific and technical content; collaborative filtering with topic modeling and group recommendation generation. The developed set of methods makes it possible to effectively use the potential of network interaction between group members, defined on the basis of paradigms of communities of scientific, technical and economic entities in the interests of creating and implementing competitive technologies that are in demand by the economy.

Keywords: scientific and technical marketing · social networks · scientific and technical content · group recommendations · topic modeling · public opinion

1 Introduction

With the development of information and communication technologies, numerous formal and informal social networks of organizations and researchers began to form [1]. For example, to implement joint projects or apply for grants for scientific research, associations are formed among universities, companies and research institutes. In this regard, the task of choosing strategic partners becomes urgent. The study examined a model of the importance of partners in scientific cooperation networks based on reputation measurement and developed a concept that includes a set of scientific and technical marketing methods based on the use of social networks.

A new and original element of the developed concept is the use of modern advances in the field of group recommendations, previously of interest to researchers in connection with industries such as tourism or entertainment, where some types of products are specifically intended for group consumption (excursions, performances). Although in relation to scientific and technical content it is not customary to say that it involves group consumption, it is known that groups can facilitate the perception and dissemination of new knowledge [3–5].

R. Silhavy and P. Silhavy (Eds.): CoMeSySo 2024, LNNS 1489, pp. 458–470, 2025.
https://doi.org/10.1007/978-3-031-96798-6_34

Within the framework of the developed concept, it is proposed to harmonize the recommendations of new resources to user groups. At the same time, the assumption is made that if a scientific and technical resource is recommended to an explicit or implicit group of users in accordance with the interests of the group, then they may begin to discuss the proposed material. Such discussion increases the chances that group members will assimilate and constructively use scientific knowledge and technology. It is proposed to take into account not only explicitly registered user groups, but also implicit groups, whose members may only be known to have viewed a certain resource on the Internet in the past. At first glance, this approach to grouping users may seem formal, but when we talk about implicit groups, we take into account the possibility of the existence or emergence in the future of connections between readers of the resource outside our focus on the Internet or beyond, in real life. At the same time, readers who prefer to rely on their own assessments are also taken into account.

The developed concept implements an integrated approach to scientific and technical marketing in social networks and includes the following modules:

module for establishing connections between the profiles of creator subjects (SS) and consumer subjects (SP) of scientific and technical content (STC);
scientific and technical content filtration module;
module for measuring public attention and influence;
module for ranking manufacturers and scientific and technological complexes and developing recommendations for consumers;
STC thematic targeting module;
collaborative filtering module with topic modeling;
group recommendation module.

2 Methods

To conduct the study, systems analysis, probabilistic methods, topic modeling methods, mathematical modeling methods and collective examination methods were used. Collaborative filtering is used to build predictions for system users that are based on information about the preferences of other similar users.

3 Module for Establishing Connections Between Subject Profiles

The task of this module is to identify profiles of the same subject in various digital databases and in various social systems [6, 7].

The problem is formalized as follows. Given a set $U = \{u_1, u_2, \ldots u_n\}$ of usernames I in various systems, as well as a possible name c of this subject on another platform. It is required (using machine learning) to construct an identification function $f(U,c)$, which takes the value 1 if c belongs to subject I, and otherwise the value 0. The authors R. Zafarani and H. Liu construct behavioral characteristics using a supervised approach. Authors O.Goga and H.Lei use geolocation, timestamps, and social media texting style to link user profiles on social networks such as Yelp, Flickr, and Twitter. Authors J.Tang, D.Zhang and L.Yao extract information from user profiles about Internet home pages, organizational affiliations, address, email, publications, etc. Then using a supervised learning method known as conditional random fields» (CRF), users' home pages are identified. Finally, a constrained probabilistic model is used to combine multiple subject publications in different bibliographic databases and eliminate ambiguous author names.

Based on these approaches, we make two significant improvements.

First, while recognizing that names are the most powerful identifiers of subjects, we implicitly take into account additional properties such as language and the author's skill level. These properties can be obtained, for example, from the conferences mentioned by the authors.

Second, we take into account the structural connections between actors in different networks. We hypothesize that co-authors of publications or participants at a particular conference are more likely to be connected in a social network. These considerations are taken into account in the machine learning algorithm in this module.

4 Scientific and Technical Content Filtering Module

This module solves the problem of identifying essential and discarding non-essential scientific and technical content in a social network. Authors K. Weller, E. Droge, C. Puschmann list several aspects of scientific and technical messages on social networks (tweets): (1) the presence of scientific and technical information or links to articles in scientific journals, (2) tweets authored by scientists, (3) tweets with hashtags related to science. With regard to text analysis, many approaches to unsupervised learning have been developed without using labeled samples, in particular those based on topic modeling [8–10]. The main difficulty here is to identify keywords and hashtags related to STC.

We build an academic lexicon from a corpus of abstracts of articles in a particular scientific field. Another approach involves analyzing combinations of terms that appear together in scientific texts. Authors M. Conover, J. Ratkiewicz, and M. Francisco start with a small set of hashtags and then expand the vocabulary based on how often hashtags are co-used in tweets. In our case, we can start with conference keywords and tags. In order to automatically determine whether a link points to a scientific Internet resource (blog, publication, etc.), we solve the classification problem. Authors S. D. Gollapalli, C. Caragea, P. Mitra, C. L. Giles analyze the structure of Internet addresses (URL) to classify the home pages of researchers.

In addition to these approaches, we build a predefined list of Internet pages, including digital libraries, slide sharing sites, etc. By aggregating the signals obtained using these approaches, we obtain a reliable model for identifying scientific tweets.

5 Module for Measuring Public Attention and Influence

This module solves the problem of measuring public attention and influence of scientific and technical content at a significant scale of source data. Ultimately, scientific discoveries can gain public recognition, and social media is one way to achieve this. Therefore, measuring public attention can provide feedback in addition to traditional metrics of scientific relevance. Authors of previous studies have examined the correlation of publication citations with mentions on Wikipedia, blogs, Twitter and various social networks. Based on these approaches, we have supplemented them in the following respects.

First, the broader reach of researchers, conferences, and scientific journals required the inclusion of the above-mentioned automatic methods for identifying and linking scientific content on social networks.
Second, to more comprehensively understand the impact of scientific content, we consider the following factors:

user reputation factor. For example, Internet links in a message from a computer specialist are more important than Internet links on behalf of a conference chatbot;
link source weight factor. For example, a link to a scientific publication in a Wikipedia article carries more weight than a mention of the same scientific publication in a personal blog or tweet.

6 Ranking and Recommendation Module

In this module, based on signals from social networks, recommendations are developed for consumers of such elements of scientific and technical content as blogs, tweets, articles, and the authors of STC themselves. To develop recommendations, the results of the work of the modules of our system described above are used: understanding of STC manufacturers; understanding of STC consumers; STC identification mechanism.

The functioning of the module is based on the following results.

The work of S. M. Kywe, E.-P. Lim and F. Zhu proposes a taxonomy of Twitter recommender systems based on the question they answer: (1) who to follow on Twitter? (2) what to write about? What link or hashtag should I include in my tweet? (3) who to mention, whose message to forward?

In the works of C. Wang, D. M. Blei, a regression model is proposed, in which collaborative filtering is combined with probabilistic topic modeling to develop STC recommendations.

The work of A. Younus, M. A. Qureshi and P. Manchanda applies topic modeling to tweets to generate recommendations for scientific articles.

When developing recommendations in this module, we take into account the experience and hierarchy of academic scientists (graduate student, candidate of sciences, professor) and representatives of industry science (researcher, leading scientist), as well as their area of competence in accordance with their publications. We model connections between scientific, technical and economic entities, taking into account co-authorship, citation and affiliation networks, as well as interactions on Twitter. This module includes a mechanism similar to cross-review for ranking STC and its authors, taking into account which articles and mentions are sent on Twitter, receive approving likes, etc., within their own research communities, in contrast to global counters.

7 Module for Thematic Targeting of Scientific and Technical Content

The task of this module is to analyze the consumption of scientific and technical content by readers, in contrast to the task of analyzing the production of scientific and technical content by authors. To analyze the consumption of scientific and technical content, we use a topic modeling (TM) approach in combination with evaluations of materials by readers operating within various projects [11–14]. Based on this analysis, we create news feeds that offer readers scientific and technical content within the framework of a specific project. In our approach, each user project is processed in isolation from other projects. Although we use the term "recommendation," our approach should not be confused with recommender systems that rely on similarities between users, since we do not assume similarities between different projects.

A project may include one or more participants evaluating scientific and technical content. Each participant can either rely on their own ratings, or receive materials recommended to other participants based on their ratings, or a combination of both approaches [15–18]. However, in this module we are targeting only those readers who rely on their own assessments, without the influence of other project participants. In other words, we will assume that each project is individual, and the transition from individual projects to group ones is possible by merging news flows generated by the leading participants in the project. Therefore, below we do not distinguish between the project and the reader, the consumer of scientific and technical content.

Each proposed scientific and technical material can be rated by the consumer on a scale from (+2) to (−2):

Rating (+2) – Yes, I'm interested in this, I want to receive more such scientific and technical content;
Rating (+1) – Yes, it is possible that such content may interest me, under certain conditions, but I am not ready to view all such materials in a row - additional filtering is required;
0 – refusal to rate: forget that I saw this material, it will neither help you select interesting content nor discard uninteresting content;
Rating (−1) – Rather "no": such content is unlikely to interest me;
Rating (−2) – A definite "no": not interesting, please do not offer me such content in the future.

Based on the assessment obtained, each document can be assigned to one of four classes of documents for a specific consumer. Each class C_m includes projects that received the same score $m \in \{+2, +1, -1, -2\}$. Materials rated 0 are not included in any document class. The combination of four classes forms a corpus of documents *D*. To analyze *D*, the topic modeling (TM) apparatus is used.

In [19], the concept of "thematic divergence" *TD*(*i*, *DD*) of a document $d_i \in D$ from a subset of documents $DD \subset D$ was introduced, defined as the Hellinger distance between the document topic probability vector θ_i and the topic probability vector averaged over the DD subset:

$$TD(i, DD) = H\left(\theta_i, \theta'(DD)\right).$$

The value of thematic divergence ranges from 0 to 1 and shows how much the topic of a document differs from the topic of a subset of documents: 0 - the topic is the same, 1 - has nothing in common.

In our case, we calculate the thematic divergence between each new scientific and technical document d received into the information system and four classes of documents

C_m, in order to generate a news flow for a specific consumer. For each document, the values dp2, dp1, dn1, dn2 are calculated – thematic divergence of the document from document classes C_m with user ratings $m \in \{+2, +1, -1, -2\}$, respectively.

To develop recommendations, we will introduce threshold levels of positive and negative ratings ($0 \leq pp1 \leq pp2 \leq 1$), ($0 \leq pn1 \leq pn2 \leq 1$). A positive recommendation for inclusion of a document in the news flow for a specific consumer is issued under the following conditions:

```
(dp2 >= pp2) // High interest
OR
(dp2 >= pp1 AND // Perhaps this is what we need
dn2 < pn2) // No good reason for refusal
OR
(dp1 >= pp2 AND // Interesting enough
dn1 < pn2 AND // Not enough reasons to refuse
dn2 < pn1) // Not enough reasons to refuse
OR
(dp1 >= pp1 AND // There is cause for interest
dn1 < pn1 AND // No reason to refuse
dn2 < pn1) // No reason to refuse
```

Recommended documents are sorted in descending order of weight W, which is calculated using the following pseudocode:

```
W = {
// High interest
if (dp2 >= pp2) return 3 + (dp2 - pp2);
// Quite interesting
if (dp2 >= pp1) return 2 + (dp2 - pp1);
if (dp1 >= pp2) return 2 + (dp1 - pp2);
// Interest is possible
return 1 + (dp1 - pp1);
}
```

At first glance, it may seem that the idea of using the TM apparatus for negative filtering of content that is uninteresting to the consumer is illegitimate due to the potentially infinitely wide variety of content of documents rejected by the consumer. However, in our approach, negative filtering is applied only to documents that have received a strong negative rating from the consumer. The wider the topic of the document corpus, the larger the topics calculated by the TM apparatus will be, the rougher the proposed content filtering algorithm will work, the more difficult it will be for the consumer to notice the materials of interest to him in the STC news stream, and the less useful our program will be.

Let us ask ourselves: is it possible to give practical advice to the consumer regarding the negative assessment of content in a way that avoids applying the TM apparatus to an excessively wide variety of topics? We believe that yes, such advice can be useful, and partly it can be supported at the user interface level of our information system.

We recommend that consumers carefully apply negative ratings to documents of incomprehensible, unfamiliar, or dubious content. Do not rush into a negative assessment of documents whose topics are far from the subject area of the project. It is better not to evaluate such documents, but to leave them without attention, than to bother giving a negative assessment. Most of the unnecessary documents will gradually disappear from the news feed due to the consumer's lack of interest in them. An analogy with a person who prudently ignores what does not concern him is appropriate here [20–22].

On the other hand, at the initial stage of each new project, while there are few documents with evaluations, it may be useful to completely exclude the possibility of negative evaluations in the user interface of our system. After accumulating a corpus of documents with positive ratings, you can begin to apply negative ratings to fine-tune thematic targeting. The same considerations speak in favor of initializing, "seeding" a new project with selected documents with positive assessments from another project on related topics [23, 24]. This feature can be supported in the user interface.

8 Collaborative Filtering Module with Topic Modeling

Classic recommender systems, like collaborative filtering, treat users as abstract entities, neglecting nonlinear processes such as attention and interpretation. In our study, we rely on the results of B. C. Love, D. L. Medin, T. M. Gureckis, who describe the model of category learning in people. The SUSTAIN (Supervised and Unsupervised STratified Adaptive Incremental Network) model takes into account the dynamics of category learning. Recently used categories in this model have a stronger impact on user navigation than categories used in the distant past. The key features of the model are flexibility and a focus on simplicity due to the fact that the number of hidden elements (clusters) is not selected in advance, but is gradually increased along the learning trajectory. The SUSTAIN model includes three layers: an input layer, an intermediate layer that contains clusters, and an output layer. At the beginning of training, the first cluster is created for the first sample from the training set. The model allows for both supervised learning (with feedback) and unsupervised learning. Supervised learning requires an external mechanism to verify the categorization of input samples. Incorrect categorization is interpreted as a learning error and causes the creation of a new cluster. In contrast, with unsupervised learning, no feedback is required, and the criterion for successful learning is the proximity of the training sample to the set of clusters. Each new cluster is created under the condition that the existing clusters are not enough to accommodate the next sample with a given permissible deviation from the cluster center.

Two additional parameters of the SUSTAIN model are the learning rate η and the attentional focusing coefficient r. The learning rate η determines the influence of the input stimulus on the displacement of the cluster in which this stimulus is included. The attention indicator r is a constant representing the ability of this person to focus on the information aspects of the problem being solved.

In our work, we use a variant with unsupervised training of the SUSTAIN model: the history of Internet resources viewed by the user is used for training. The trained model is then used to predict new resources from a given set of candidates. The inputs to the SUSTAIN model are topics extracted from resources using the LDA engine. For each resource, 500 topics are calculated. The activation of each input feature is controlled by the importance of the feature. The importance of features for a particular user is also determined during the training phase. The hidden layer consists of a set of clusters that group together similar resources encountered in the past. Each cluster represents one user's area of interest. During the testing phase, many clusters and their structure do not change. The decision to classify (select or not select for viewing) test Internet resources depends on the level of reaction of the most active (winning) cluster.

The SUSTAIN-based recommendation system assumes that, first of all, we analyze the tags assigned to the Internet resources in question and calculate 500 topics using the LDA method. These topics represent the n input features of our model. Then, based on the training set of resources that the user has marked for himself in the past, we calculate the user's personal attention quotient settings and a set of clusters. At the same time, we slightly adapted the unsupervised learning mechanism of the SUSTAIN model. Given the large number of input dimensions (500 topics), we limit the training focus to only topics that are related to this learning resource.

The LDA mechanism was implemented using the Java Mallet library with Gibbs sample selection and $l = 2000$ iterations. To reduce noise, we set the number of topics to 500 and limit topics whose probability in this resource is at least 0.01.

The term "candidate resources" refers to a set of resources from which the most suitable resource should be selected to be recommended to the user. We select 100 candidates using the user-based collaborative filtering CF_U method [20]. The CF_U method includes two steps: first, the k most similar neighbor users in a certain metric are selected. Then, from the resources of these neighbors, those that the target user has not yet seen are selected, and such resources are offered to him. This method is based on the assumption that if two users' tastes were similar in the past, then they are likely to be similar in the future. The similarity of user tastes is calculated using a binary user-resource matrix and using a cosine measure of vector similarity. Neighborhood size $k = 20$. More formally, the prediction $CF_U(u, r)$ for target user u and resource r is given by the following equation:

$$CF_U(u, r) = \sum_{v \in V_{u,r}} sim(\mathrm{u}, v),$$

where $V_{u,r}$ is the set of neighbors of user u who have tagged resource r, and $sim(u,v)$ is the cosine proximity between u and v.

9 Group Recommendation Module

Interest in STC group recommendations is based on the assumption that if a scientific and technical resource is recommended to a group of users in accordance with the interests of the group, they will begin to discuss the proposed material. If such a discussion takes place, it increases the chances of the assimilation and constructive use of scientific knowledge and technology by group members.

The quality of group recommendations can be assessed using various types of metrics:

classification metrics: accuracy - the proportion of correct recommendations in relation to all recommended resources and coverage - the proportion of recommended relevant resources in relation to all relevant resources that should have been recommended;
error metric – the average error of the recommendation system's prediction of the value of ratings for a set of resources;
the ranking metric takes into account the position of elements in the list of recommendations and characterizes the deviation from the "true" order of elements;
coverage and randomness: "coverage" is understood as the proportion of users or groups for which it was possible to generate at least one recommendation. Or alternatively, as the share of resources available (for example, in a product catalog) that were recommended to at least one user or group; "accidentality" characterizes the deviation of the number of (useful) recommendations from a primitive predictive model based on the popularity of resources;
Consensus and fairness: consensus characterizes the closeness of ratings given by group members, and fairness is measured by the proportion of group members rating (recommended to them) resources above a given threshold;

To recommend scientific and technical content on social networks, we propose to use deep learning methods based on the history of user ratings of resources. Our model relies on a tracking mechanism or attention mechanism. The attention mechanism in machine learning problems takes as input n vector arguments $x_1, x_2, \ldots, x_n$ and context c and returns vector y, which is a linear combination of vectors x_i with weights a_i depending on the "relevance" of the vector x_i in a given context. In our model, we apply the attention mechanism to users u_i and talk about the "weights" of users in the group according to the influence of the – user on the final choice of the group. It is important to note that we are studying groups that may be, so to speak, "ad-hoc" groups whose members do not have permanent relationships. Such groups can be created for a short time, to solve one or several specific tasks. Such "random" groups cannot be studied as "pseudo-users", in contrast to, for example, married couples.

Let $U = \{u_1, u_2, \ldots, u_M\}$ be a set of M users, and $I = \{i_1, i_2, \ldots, i_N\}$ be a set of N resources. Let us denote the browsing history log as $H = \{\langle g_1, s_1\rangle, \langle g_2, s_2\rangle, \ldots, \langle g_{N1}, s_N\rangle\}$, where $g_n \subset U$ denotes the group, and $s_n = i_k$ is the resource selected by the group. For this target group g_t, we need to generate a list of resource recommendations that members of this group $u_{g_t,i}$ might be interested in. In the group recommendation problem, given are users U, resources I, journal H, and target group g_t. It is required to construct a function for assessing resources by the target group $f_{g_t} : I \to \mathbb{R}$.

The developed model of group recommendations assumes that the assessment of resources by a group of users is built in two stages. First, group members "vote" one another in order to select the most competent group members for the task at hand. Each vote of user l in favor of user m in group i is expressed by the weight $\alpha(i,l,m)$. The most competent group members with the most votes then make a choice of resources to be offered to all group members. The use of the attention mechanism in the task of determining weights implies that each user l does not look at all members of the group, but focuses on a few experts, ignoring the rest. This motivates our proposal to use an attention mechanism to train a neural network capable of finding the weights $\alpha(i,l,m)$.

Formally, we model each user l by two vectors: a latent vector $\boldsymbol{u}_l$ and a context vector $\boldsymbol{c}_l$. The attention mechanism uses the context vector $\boldsymbol{c}_l$ to estimate the influence of other users $\{\boldsymbol{u}_1, \boldsymbol{u}_2, \ldots, \boldsymbol{u}_{l-1}, \boldsymbol{u}_{l+1}, \ldots, \boldsymbol{u}_n\}$ on user l. First, we split the group g_i into n subgroups, where each subgroup $g_{i,l}$ includes all users of the group g_i except user l. We apply a network with an attention mechanism to each of the subgroups and obtain a representation of the subgroup that reflects the views of user i on other users:

$$g_{i,l} = \sum_{m \neq l} \alpha(i, l, m)\boldsymbol{u}_m.$$

Further, the representation of group i is $g_i = \sum_l g_{i,l}$, that is, the votes of all group members are taken into account equally.

Next, we use Bayesian personality ranking to model implicit group preferences. Although this method has previously been used for personal ranking, we modify it for group recommendations. To do this, we form triplets that include group i and two resources: a positive resource j, which the group has previously seen, and a negative resource k, which the group has not previously seen. We assume that the group prefers all (previously seen by the group) positive resources over all negative (which it has not yet seen). Therefore, as the objective function we choose:

$$\arg\min_{\Theta} \sum_{(i,j,k) \in D_s} \left(-\ln \sigma\left(\hat{R}_{ij}(\Theta) - \hat{R}_{ik}(\Theta)\right) + \lambda \|\Theta\|^2\right),$$

where the set D_s contains all pairs of positive and negative resources for each group; Θ are the model parameters, $\hat{R}_{ij}(\Theta)$ is the predicted estimate by group i of resource j, σ is the logistic sigmoid function, and λ is the regularization parameter.

10 Results and Discussion

Group modeling and group recommendations have been studied for various industries such as social media, tourism, and entertainment. But compared to widely developed recommendation methods aimed at individual consumers, group recommendations (GRs) have been less studied [25–27].

GR methods can be divided into the following categories: aggregation of preferences, aggregation of assessments, and methods for modeling the decision-making process in a group.

The first approach combines group members' preferences to create a group profile and then makes recommendations based on the group profile.

The second approach first computes a list of recommended resources for each group member and then harmonizes the individual lists in various ways, such as averaging, or minimizing dissatisfaction or maximizing satisfaction within the group. At the same time, approaches associated with the aggregation of either preferences or the aggregation of assessments cannot reflect the interaction between group members in the decision-making process.

A number of researchers have developed model-based approaches to group recommendations. Thus, in the works of S. Seko, T. Yagi, M. Motegi, S. Muto, a model is proposed that is aimed at stable groups such as friends, couples or families who know each other well. However, in our task of distributing scientific and technical content, we also have to target "random" (ad-hoc) groups of unfamiliar users whose preferences are not so stable. In the works of L. Carvalho and H. Macedo, each group action is considered as a non-cooperative game in which the interests of players compete, and it is proposed to use a Nash game equilibrium for group recommendation. However, the conditions of such equilibrium may be satisfied not by one, but by several candidates, among which this approach does not help to choose a single option.

Probabilistic models have also been used more than once for group recommendations. The works of X. Liu, Y. Tian, M. Ye, W.-C. Lee proposed a model of "personal impact topic" (PIT), in which the most influential member of the group has the greatest influence on the choice of the group. However, this model does not take into account the different competencies of group members: one may be an expert in cinema, and another in restaurants. When solving various issues, the opinions of various group members may carry the greatest weight. The works of Q. Yuan, G. Cong, C.-Y. Lin propose a consensus model (COM) of group recommendations. It takes into account the dependence of the influence of group members on the subject area and takes into account both group interests and the interests of individual group members. However, this model also assumes that the user has the same probability of agreeing with the group decision of any group in which he could be a member, which is essentially equivalent to the disadvantage of the PIT model noted above.

Alternatively, in the works of J. Gorla, N. Lathia, S. Robertson and J. Wang, the evaluation of a candidate resource depends not only on the interests of each member of the group, but also on the interests of the entire group. This work proposes a model based on the theory of information correspondence, but, unfortunately, has a high computational complexity of the order of $O(|U|^2|I|^2)$, where $|U|$ is the number of users, and $|I|$ there is a number of groups. The work of L.Hu, J. Cao, G. Xu, L. Cao, Z. Gu and W. Cao uses a deep neural network to study the preferences of stable groups. But as noted above, we are interested not only in stable groups, but also in blurred and random ones.

We propose a new approach to group recommendations based on the assumption that group members tend to follow the opinions of a small number of the most influential members (leaders or experts) rather than evaluate the opinions of all group members. To model this pattern of group interaction, we propose a model using an attention mechanism combined with a collaborative filtering mechanism. Using the attention mechanism, our model is trained not only to reflect the interaction within a group, but also to dynamically calculate the user's weight in different groups, which previously proposed models cannot do.

11 Conclusion

The developed set of scientific and technical marketing methods based on the use of social networks makes it possible to effectively use the potential of network interaction between group members, defined on the basis of the paradigms of communities of scientific, technical and economic entities in the interests of creating and implementing competitive technologies that are in demand by the economy.

References

1. Tobin, D., Bogomolov, A., Golosovskiy, M.: Model of organization of software testing for cyberphysical systems. Stud. Syst. Decis. Contr. **418**, 51–60 (2022)
2. Lynch, R.D., Biederman, D.J., Silva, S., Demasi, K.: A syringe service program within a federal system: foundations for implementation. J. Addict. Nurs. **32**(2), 152–158 (2021)
3. Larkin, E.V., Bogomolov, A.V., Privalov, A.N., Dobrovolsky, N.N.: Discrete model of paired relay-race. Bull. South Ural State Univ. Series Math. Model. Program. Comput. Soft. **11**(3), 72–84 (2018)
4. Larkin, E.V., Bogomolov, A.V., Privalov, A.N., Dobrovolsky, N.N.: Relay races along a pair of selectable routes. Bull. South Ural State Univ. Series Math. Model. Program. Comput. Soft. **11**(1), 15–26 (2018)
5. Slavin, O.A., Janiszewski, I.M.: Effective practices of using spatial models in document image classification. Bull. South Ural State Univ. Series Math. Model. Program. Comput. Soft. **16**(4), 61–70 (2023)
6. Zakharchenko, O.V., Bakulich, O.O., Potapenko, T.P., Voloshenko, M.O., Kharuta, V.S.: Fundamentals of the system simulation methodology "personmachine" in project and program management. Int. J. Manag. **11**(3), 133–142 (2020)
7. Akkaya, A., Ayyıldız, E.: Automation software for semiconductor research laboratories: measurement system and instrument control program. Mapan. **35**(3), 343–350 (2020)
8. Larkin, E., Kotov, V., Bogomolov, A., Privalov, A.: Multiple swarm relay-races with alternative routes. Lect. Notes Comput. Sci. **10941**, 361–373 (2018)
9. Golosovskiy, M.S., Bogomolov, A.V., Tobin, D.S.: Algorithm for setting fuzzy logical inclusion systems based on statistical data. Auto. Document. Math. Ling. **57**(1), 1–9 (2023)
10. Janiszewski, I.: Software package for collecting and analyzing information for qualimetry of scientific projects. Lect. Notes Comput. Sci. **934**, 117–126 (2024)
11. Golosovskiy, M.S., Bogomolov, A.V., Balandov, M.E.: Optimized fuzzy inference for Sugeno-type systems. Auto. Document. Math. Ling. **56**(5), 237–244 (2022)
12. Larkin, E., Privalov, A., Bogomolov, A., Akimenko, T.: Model of digital control system by complex multi-loop objects. AIP Conf. Proc. **2700**, 030009 (2023)
13. Slavin, O.A., Arlazarov, V.L.: Algorithms of the tiger and cuneiform optical character recognition software. Patt. Recog. Image Anal. **33**(4), 669–684 (2023)
14. Ivanov, A., Bogomolov, A.: Heterogeneous information system for the integration of departmental databases on the state and development of human capital. Stud. Syst. Decis. Contr. **437**, 117–127 (2023)
15. Radionov, T., Mikhalev, A., Kukartsev, V., Tynchenko, V.: Program model of the interacting adaptive traffic control system. Commun. Comput. Info. Sci. **1304**, 218–234 (2020)
16. Slavin, O., Farsobina, V., Myshev, A.: Analyzing the content of business documents recognized with a large number of errors using modified Levenshtein distance. Stud. Syst. Deci. Contr. **417**, 267–279 (2022)
17. Larkin, E., Akimenko, T., Bogomolov, A., Sharov, V.: Reliability of robot's controller software. Lect. Notes Comput. Sci. **14214**, 289–299 (2023)
18. Mikhail, G., Aleksey, B., Mikhail, B.: A model of continuous integration and deployment of engineering software. Lect. Notes Comput. Sci. **2**, 789–796 (2021)
19. Sheshkus, A.V., Kondrashova, A.N., Nikolaev, D.P.: Development of the neural network based recognition methods at V.l. Arlazarov's scientific school. Patt. Recog. Image Anal. **33**(4), 717–729 (2023)
20. Pando, B., Dávila, A.A.: A systematic mapping study on software testing in the devops context. Proc. Inst. Syst. Prog. RAS. **35**(1), 163–188 (2023)

21. Bychkov, E.V., Bogomolov, A.V., Kotlovanov, K.Y.: Stochastic mathematical model of internal waves. Bull. South Ural State Univ. Series Math. Model. Program. Comput. Soft. **13**(2), 33–42 (2020)
22. Slavin, O.A., Janiszewski, I.M.: Extraction of information fields in administrative documents using constellations of special text points. Stud. Syst. Decis. Contr. **477**, 31–45 (2023)
23. Akimenko, T., Bogomolov, A., Larkin, E., Privalov, A.: A model of scene observing from mobile robot, moving on rough terrain. AIP Conf. Proc. **3021**(1), 090005 (2024)
24. Slavin, O.A., Pliskin, E.L.: Method for analyzing the structure of noisy images of administrative documents. Bull. South Ural State Univ. Series Math. Model. Program. Comput. Soft. **15**(4), 80–89 (2022)
25. Golosovskiy, M., Bogomolov, A.: Fuzzy inference algorithm using databases. Lect. Notes Comput. Sci. **724**, 444–451 (2023)
26. Novikov, D.: Models of strategic decision-making under informational control. Mathematics. **9**(16), 1889 (2021)
27. Bogomolov, A.V., Klimov, R.S.: Automation of information processing when conducting collective network examinations. Auto. Mod. Technol. **71**(11), 509–512 (2017)

Application of ResNet-18 and VGG-16 Algorithms in the Diagnosis of Breast Cancer

Hussein Alsajer[1(✉)], Yuriy N. Philippovich[1], Ekaterina A. Pukhova[1], and Tatiana V. Beketova[2]

[1] Moscow Polytechnic University, Moscow 107023, Russian Federation
h.sajerov@gmail.com
[2] V.A. Nasonova Research Institute of Rheumatology, Moscow 115522, Russian Federation

Abstract. Breast cancer is one of the most dangerous diseases and the second largest cause of female cancer death. Breast cancer starts when malignant, cancerous lumps start to grow from the breast cells. Self-tests and Periodic clinical checks help to early diagnosis and thereby improve the survival chances significantly. The breast cancer classification is a medical method that provides researchers and scientists with a great challenge. Neural networks have recently become a popular tool in cancer data classification. Deep learning (DL) technologies are becoming a buzzword these days, especially for breast histopathology image tasks, such as diagnosing, due to the high performance obtained in image classification. Among deep learning types, Convolutional Neural Networks (CNN) are the most common types of DL models utilized for medical image diagnosis and analysis. However, CNN suffers from high computation cost to be implemented and may require to adapt huge number of parameters. Thus, and in order to address this issue; several pre-trained models have been established with the predefined network architecture. In this study, we trained VGG-16 ResNet-18 on Breast images. a transfer learning model based on Visual Geometry Group with 16-layer deep model architecture (VGG16) is utilized to extract high-level features and ResNet-18 with 18 layers. The empirical outcomes confirm that the application of ResNet-18 provides the most reliable performance for accuracy where accuracy with ResNet-18 was 98.89% while accuracy with VGG-16 was 94.44%.

Keywords: Breast Cancer · Deep Learning · Convolutional Neural Networks · CNN · Classification · Visual Geometry Group

1 Introduction

Breast cancer is one of the most common and complicated diseases: it is a form of uncontrolled growth of the breast cells, and while it is possible for anyone to be diagnosed, it is much more common in women. Renowned as one of the most common cancer killers worldwide, in 2020, over 2.3 million new cases of the disease were diagnosed globally, emphasizing its burden of public health [1, 2]. The disease present itself in a few different varieties, which can generally be classified as either invasive and non-invasive types, including invasive ductal carcinoma and ductal carcinoma in situ [3]. Multiple

R. Silhavy and P. Silhavy (Eds.): CoMeSySo 2024, LNNS 1489, pp. 471–478, 2025.
https://doi.org/10.1007/978-3-031-96798-6_35

risk factors have been demonstrated, from genetic predisposition—e.g., mutations in the BRCA1 and BRCA2 genes—to lifestyle and environmental factors [4, 5].

Breast cancer is diagnosed with several methods and procedures to detect and confirm the presence of cancerous cells in breast tissue within Mammography, ultrasound and Why MRI, where used to look inside the human body as a non-invasive method for helping doctors for diagnose and treat [6, 7].

Treatment outcomes and survival rates are greatly improved with early detection. Nevertheless, traditional manual diagnosis is time-consuming and necessitates the proficiency of exceptionally trained pathologists, who may still be susceptible to diagnostic inaccuracies as a result of human constraints and variations in expertise.

In order to tackle these difficulties, Computer-Aided Diagnostic (CAD) systems have been created and proven to greatly assist in the diagnostic procedure [8].

Breast cancer detection using deep learning techniques has emerged as a revolutionary approach for the medical diagnostic field, significantly improving the rate of accurate breast cancer detection. This approach mainly employs Convolutional Neural Networks (CNNs) for histopathological image analysis to distinguish between different types of breast cancer [9, 10].

Convolutional Neural Networks (CNNs) have recently shown success in several areas of image classification, and CNNs are increasingly being used in medical image processing, particularly in the classification of histopathological images [9]. CNNs have the ability to detect and extract hierarchical features from images, making them well suited to identifying complex patterns associated with malignancies. The classification task requires that each image or image section be assigned to a corresponding category, such as benign or malignant [10, 11].

The process of classifying breast cancer consists of several basic steps. The first is Data Acquisition and Collection (collecting medical images), the second is Data Preprocessing, this stage prepares the data for model training by addressing any issues or enhancing features. Then data splitting (Training, Validation, Testing), the preprocessed data is divided into three sets to avoid overfitting and ensure a proper evaluation of the trained model. The fourth is building the deep learning model, choosing an appropriate deep learning architecture based on task requirements and data characteristics and then training the selected deep learning model using the training data. The last, Assessing the trained model's performance on the held-out test set using appropriate evaluation metrics as shown in Fig. 1.

2 Methods

In this study, we use the CNN method for data processing and build the program and use VGG-16 and ResNet-18 Architecture to build the model, we ran an experiment to examine the behaviors of VGG16 and ResNet, with the goal of determining both the effectiveness and efficiency of the algorithms for classification breast cancer.

VGG-16 and ResNet-18 are prominent convolutional neural network architectures that have significantly influenced the field of deep learning, particularly in image classification tasks [13, 14].

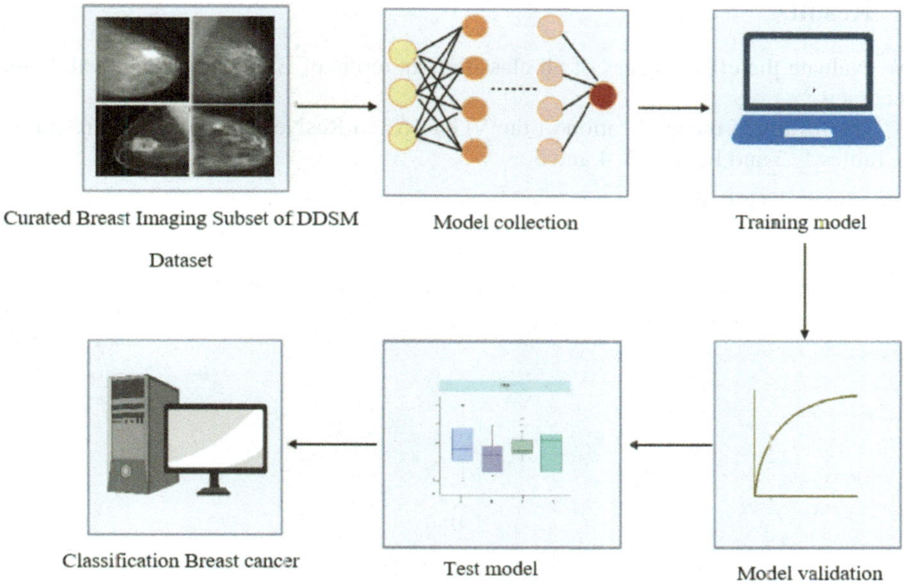

Fig. 1. General diagram of Deep learning architecture for classification Breast cancer [12].

ResNet-18 is a significant architecture in the domain of deep learning, specifically designed to improve image classification tasks. It is part of the Residual Network (ResNet) family, which was introduced to combat the vanishing gradient problem commonly encountered in deeper neural networks. The architecture allows for the creation of networks that are deeper and more capable of learning complex patterns in data, which enhances their performance on various tasks [15–17].

All experiments on the classifiers described in this paper were conducted using libraries from MATLAB. For data preprocessing, classification, regression, clustering, and association rules, MATLAB has a set of Deep learning algorithms. Deep learning approaches are used in MATLAB to solve a number of real-world challenges. Experimenters and developers can design and analyze their models using the program's well-defined structure.

We have used CBIS-DDSM dataset (Curated Breast Imaging Subset of DDSM) is an updated and standardized version of the Digital Database for Screening Mammography (DDSM). The DDSM is a database of 2,620 scanned film mammography studies.

It contains normal, benign, and malignant cases with verified pathology information, but we used a part of this Dataset because we wanted training time quicker so we toke 181 images from Benign Masses and 88 images from Malignant Masses.

3 Results

We evaluate the effectiveness of all classifiers in terms of Elapsed time and validation accuracy.

The results of the application of the VGG-16 and ResNet-18 algorithms are shown in Tables 1, 2 and Figs. 2, 3, 4 and 5.

Experiment VGG-16:

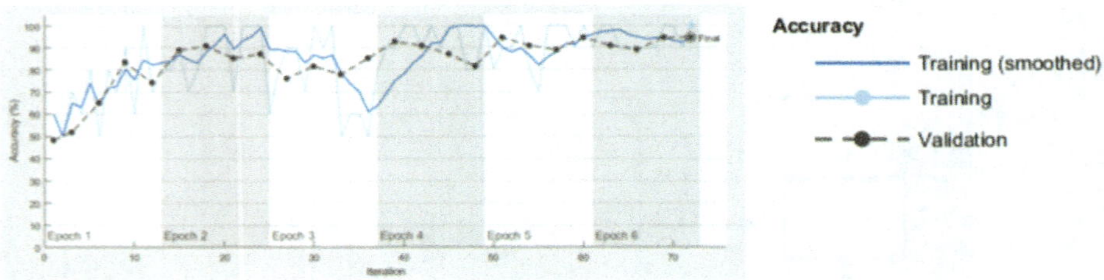

a)

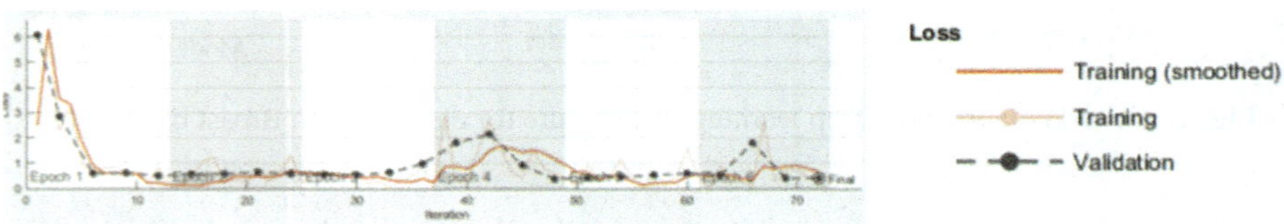

b)

Fig. 2. Results of VGG-16 experiments (a- accuracy of training using VGG-16 algorithm, b- loss of training using VGG-16 Algorithm).

Table 1. Accuracy and elapsed time of VGG-16 algorithm.

Number of Experiment	Algorithm	Validation Accuracy	Elapsed Time
1	VGG-16	94.44%	20 min 56 s
2	VGG-16	89.76%	19 min 17 s

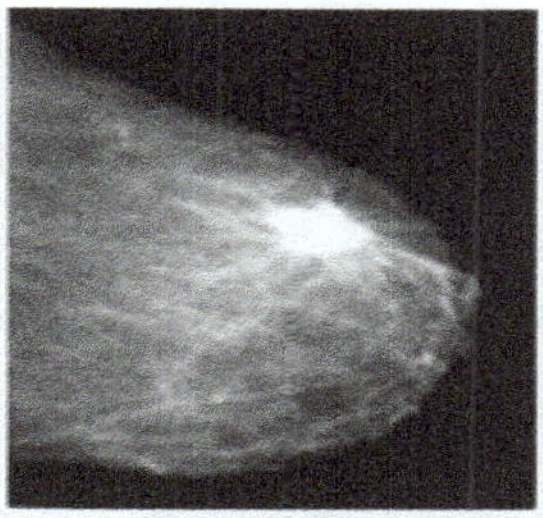

a)

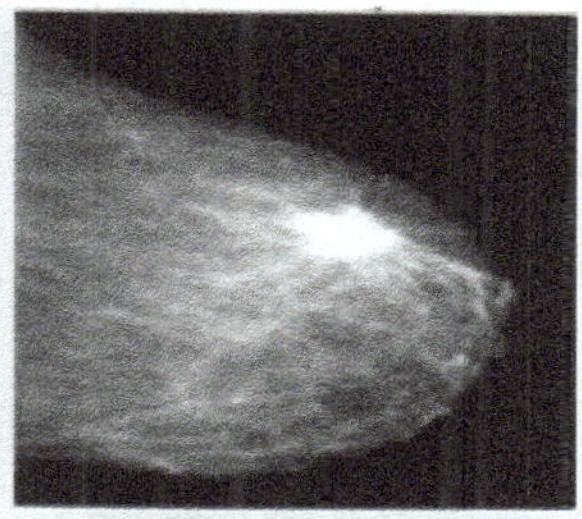

b)

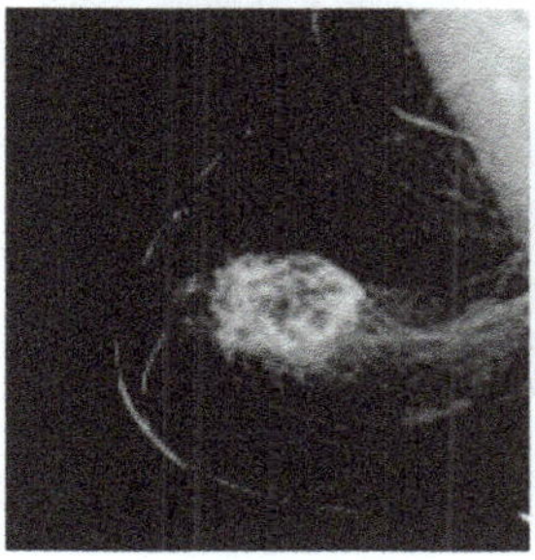

c)

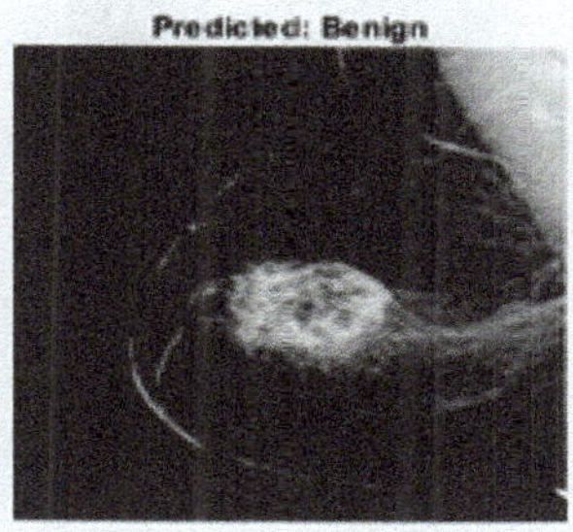

d)

Fig. 3. Results of first and second experiment of VGG-16 algorithms (a and c input images, b and d output image with results).

Experiment ResNet-18:

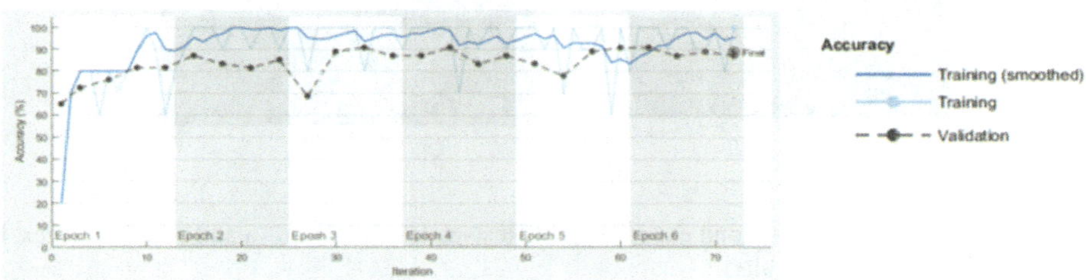

a)

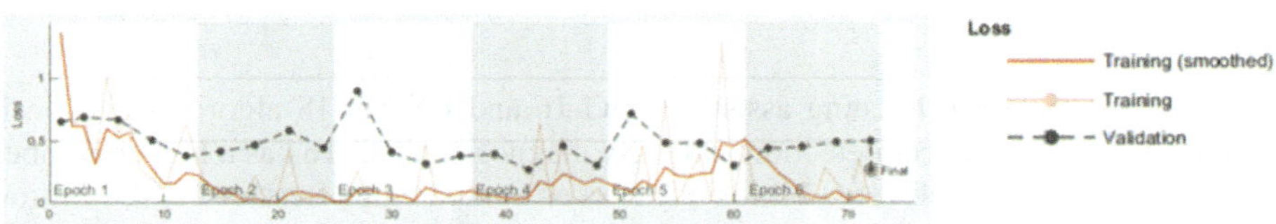

b)

Fig. 4. Results of ResNet-18 experiments (a- accuracy of training using ResNet-18 algorithm, b- loss of training using ResNet-18 Algorithm).

Table 2. Accuracy and elapsed time of ResNet-18 algorithm.

Number of Experiment	Algorithm	Validation Accuracy	Elapsed Time
1	ResNet-18	98.89%	3 min 50 s
2	ResNet-18	95.33%	3 min 20 s

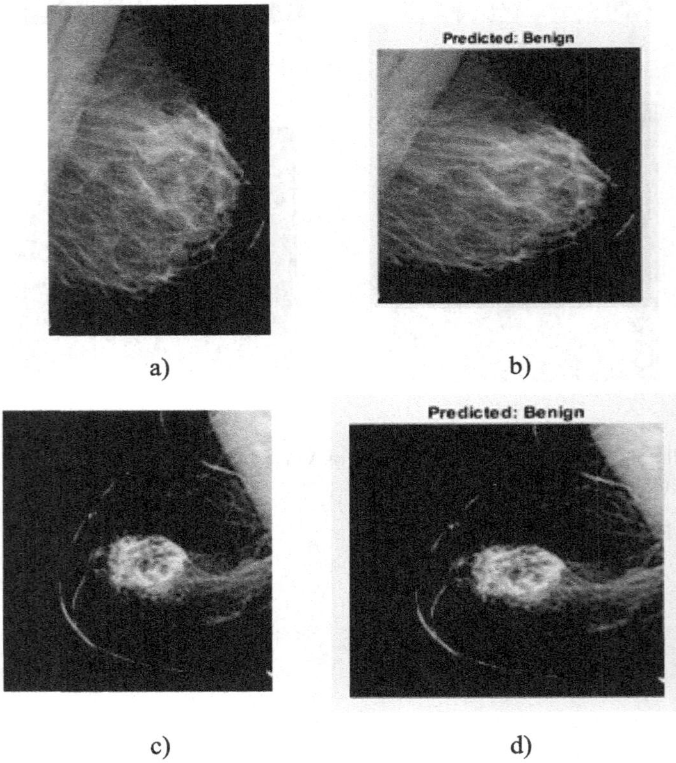

Fig. 5. Results of first and second experiment of ResNet-18 algorithms (a and c input images, b and d output image with results).

4 Conclusion

In this paper, the deep learning assisted VGG-16 and ResNet-18 algorithm has been proposed for breast cancer detection and early diagnosis. VGG-16 has a higher number of parameters and FLOP (15.3 billion FLOPs) as compared to ResNet-18 (1.8 billion FLOPs) and this elucidate clearly reason difference in Elapsed time between ResNet-18 and VGG-16, where VGG-16 toke 20 min and 56 s (Elapsed time), while ResNet-18 toke 3 min and 50 s (elapsed time) as we see in Table (1, 2).

Regarding the accuracy, we can say that the increase in the number of layers affects the accuracy, and from it we will get a higher accuracy by using ResNet-18 because it contains 18 hidden layers while VGG-16 it has a lower number of layers and this explains the difference in accuracy.

References

1. Ara, S., Das, A., Dey, A.: Malignant and benign breast cancer classification using machine learning algorithms. In: Proceedings of the 2021 International Conference on Artificial Intelligence (ICAI), pp. 97–101. IEEE, Islamabad, Pakistan (2021). https://doi.org/10.1109/ICAI52203.2021.9445249
2. Giaquinto, A.N., et al.: Breast Cancer Statistics. A Cancer J. Clinicians **72**, 524–541 (2022). https://doi.org/10.3322/caac.21754
3. Francies, F.Z., et al.: BRCA1, BRCA2 and PALB2 mutations and CHEK2 c.1100delC in different South African ethnic groups diagnosed with premenopausal and/or triple negative breast cancer. BMC Cancer **15**, 912 (2015). https://doi.org/10.1186/s12885-015-1913-6
4. Ayoub, N.M., Al-Shami, K.M., Yaghan, R.J.: Immunotherapy for HER2-positive breast cancer: recent advances and combination therapeutic approaches. Breast Cancer Targets Ther. **11**, 53–69 (2019). https://doi.org/10.2147/BCTT.S175360
5. Mertens, E., Barrenechea-Pulache, A., Sagastume, D., Vasquez, M.S., Vandevijvere, S., Peñalvo, J.L.: Understanding the contribution of lifestyle in breast cancer risk prediction: a systematic review of models applicable to Europe. BMC Cancer **23**, 687 (2023). https://doi.org/10.1186/s12885-023-11174-w
6. Cruz-Roa, A., et al.: 'Accurate and reproducible invasive breast cancer detection in whole-slide images: a deep learning approach for quantifying tumor extent. Sci. Rep. **7**(1), 46450 (2017)
7. Mambou, S., Maresova, P., Krejcar, O., Selamat, A., Kuca, K.: Breast cancer detection using infrared thermal imaging and a deep learning model. Sensors **18**(9), 2799 (2018)
8. Araújo, T., et al.: Classification of breast cancer histology images using convolutional neural networks. PLoS ONE **12**(6), e0177544 (2017)
9. Spanhol, F.A., Oliveira, L.S., Petitjean, C., Heutte, L.: Breast cancer histopathological image classification using convolutional neural networks. In: Proceedings of the 2016 International Joint Conference on Neural Networks (IJCNN), pp. 2560–2567. IEEE, Vancouver, BC, Canada (2016). https://doi.org/10.1109/IJCNN.2016.7727519
10. Amrane, M. Oukid, S. Gagaoua, I., Ensari, T.: Breast cancer classification using machine learning. In: Proceedings of the 2018 Electric Electronics, Computer Science, Biomedical Engineerings' Meeting (EBBT), pp. 1–4. IEEE, Istanbul, Turkey (2018). https://doi.org/10.1109/EBBT.2018.8391453.
11. Tran, K.A., Kondrashova, O., Bradley, A., Williams, E.D., Pearson, J.V., Waddell, N.: Deep learning in cancer diagnosis, prognosis and treatment selection. Genome Med **13**, 152 (2021). https://doi.org/10.1186/s13073-021-00968-x
12. Jiang, B., Bao, L., He, S., Chen, X., Jin, Z., Ye, Y.: Deep learning applications in breast cancer histopathological imaging: diagnosis, treatment, and prognosis. Breast Cancer Res. **26**, 137 (2024). https://doi.org/10.1186/s13058-024-01895-6
13. Hapsari, R.A., Purwinarko, A.: Implementation of convolutional neural network algorithm using Vgg-16 architecture for image classification in facial images. Recursive J. Inform. **1**, 83–92 (2023). https://doi.org/10.15294/rji.v1i2.68059
14. Albashish, D., Al-Sayyed, R., Abdullah, A., Ryalat, M.H., Ahmad Almansour, N.: Deep CNN Model based on VGG16 for breast cancer classification. In: Proceedings of the 2021 International Conference on Information Technology (ICIT), pp. 805–810. IEEE, Amman, Jordan (2021). https://doi.org/10.1109/ICIT52682.2021.9491631

15. Subaar, C., et al.: Investigating the detection of breast cancer with deep transfer learning using ResNet18 and ResNet34. Biomed. Phys. Eng. Express **10**, 035029 (2024). https://doi.org/10.1088/2057-1976/ad3cdf
16. He, K., Zhang, X., Ren, S., Sun, J.: Deep residual learning for image recognition. In: Proceedings of the 2016 IEEE Conference on Computer Vision and Pattern Recognition (CVPR), pp. 770–778. IEEE, Las Vegas, NV, USA (2016). https://doi.org/10.1109/CVPR.2016.90
17. Szegedy, C., Liu, W., Jia, Y., Sermanet, P., Reed, S., Anguelov, D., Rabinovich, A.: Going deeper with convolutions. In: Proceedings of the IEEE conference on computer vision and pattern recognition (CVPR), pp. 1–9. IEEE, Boston, MA, USA (2015). https://doi.org/10.1109/CVPR.2015.7298594

Identification of Key Meteorological Drought Factors Using Machine Learning

Kirill I. Kravtsov[1,2], Vadim S. Tynchenko[1,2,3](✉), and Svetlana V. Kukartseva[3]

[1] Reshetnev Siberian State University of Science and Technology, Krasnoyarsk 660037, Russia
vadimond@mail.ru

[2] Bauman Moscow State Technical University, Moscow 105005, Russia

[3] RSAU-MAA Named after K.A. Timiryazev, Moscow 127434, Russia

Abstract. The research aims to develop and evaluate machine learning models to predict the intensity of drought in the United States using meteorological data collected from satellites. It is assumed that machine learning models trained on meteorological data can predict the intensity of drought with high accuracy. Two models were used to predict drought: CART (Classification and Regression Trees) and random forest. The data was preprocessed, including filling in the missing values with the median method and normalization. Both models showed good classification accuracy. The random forest model achieved 94.2% accuracy, and the CART model achieved 93% accuracy. The analysis of the importance of the signs revealed that the pressure on the surface, the temperature range at an altitude of 2 m and the temperature of the earth's surface are the key meteorological parameters affecting the intensity of drought. This study demonstrates that machine learning models can be used to accurately predict drought, which makes it possible to develop more effective strategies for managing water resources and adapting to climate change, especially in agriculture.

Keywords: Sustainable Growth · Agricultural Development · Land Management

1 Introduction

Drought is a complex and multifaceted natural phenomenon characterized by a long period of precipitation deficiency, leading to a shortage of water in ecosystems, for agriculture, industry and the population. In the modern world, marked by an increase in anthropogenic pressure on the environment and climate change, drought is becoming an increasingly frequent and intense phenomenon, having a significant impact on the lives of people and nature [1–5].

Drought is one of the most widespread and devastating natural disasters in the world. Its impact is manifested in various spheres of life:

- Agricultural industry: Drought leads to lower yields and increased losses in the agricultural sector, which can lead to food shortages and higher food prices [6].
- Water resources: Lack of precipitation reduces the volume of water resources, which can lead to a shortage of drinking water for the population and to restrictions in the use of water in industry and energy [7].

R. Silhavy and P. Silhavy (Eds.): CoMeSySo 2024, LNNS 1489, pp. 479–489, 2025.
https://doi.org/10.1007/978-3-031-96798-6_36

- Ecosystems: Drought can lead to deterioration of forests, meadows, swamps and other ecosystems, threatening biodiversity and disrupting the ecological balance [8].
- Economy: Drought can cause economic losses in various sectors of the economy, including agriculture, tourism, energy and transport [9].

Climate change has a significant impact on the increase in the frequency and intensity of droughts. Rising temperatures increase water evaporation, which exacerbates precipitation shortages and leads to drier conditions. Changes in atmospheric circulation can lead to the displacement of precipitation zones and to an increase in the duration of dry periods [10–13].

Accurate drought forecasting is key to developing strategies for drought prevention and mitigation. To predict drought, it is necessary to study the relationship between meteorological parameters and its intensity, and use modern methods of modeling and data analysis [14–16]. Understanding the factors influencing the development and intensity of drought will help us develop more effective strategies for adaptation to climate change and ensure more sustainable use of water resources [17–20].

The research aims to develop and evaluate machine learning models to predict the intensity of drought in the United States using meteorological data.

2 Materials and Methods

A data set of meteorological observations over the territory of the United States was used for the study, the data were collected using satellite observations and include meteorological parameters for the period 2017–2018.

Drought monitoring in the United States involves measurements of drought created manually by experts using a wide range of data. The goal of this dataset is to explore the possibility of predicting and classifying droughts using only meteorological data, which could potentially lead to the generalization of drought forecasts from the US to other regions of the world (Fig. 1).

In the dataset, the intensity of drought is represented as a categorical variable taking values from 0 to 4, where 0 is the absence of drought and 4 is severe drought. Each entry in the dataset (a total of 543,357 records) represents a drought level at a specific moment in time in a specific US county, accompanied by data on 18 meteorological indicators:

- WS10M_MIN: Minimum wind speed at 10 m (m/s)
- QV2M: Specific humidity at 2 m (g/kg)
- T2M_RANGE: Temperature range at 2 m (°C)
- WS10M: Wind speed at 10 m (m/s)
- T2M: Temperature at 2 m (°C)
- WS50M_MIN: Minimum wind speed at 50 m (m/s)
- T2M_MAX: Maximum temperature at 2 m (°C)
- WS50M: Wind speed at 50 m (m/s)
- TS: Earth skin temperature (°C)
- WS50M_RANGE: Wind speed range at 50 m (m/s)
- WS50M_MAX: Maximum wind speed at 50 m (m/s)
- WS10M_MAX: Maximum wind speed at 10 m (m/s)

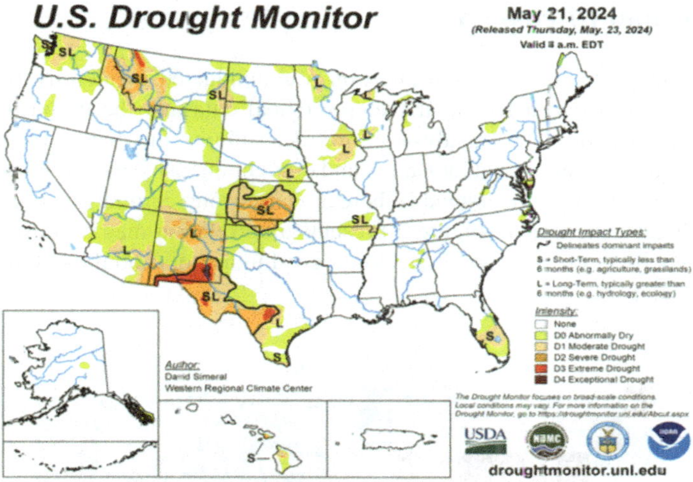

Fig. 1. U.S. Drought Monitoring Map for May 21, 2024

- WS10M_RANGE: Wind speed range at 10 m (m/s)
- PS: Surface pressure (kPa)
- T2MDEW: Dew/Frost point at 2 m (°C)
- T2M_MIN: Minimum temperature at 2 m (°C)
- T2MWET: Wet bulb temperature at 2 m (°C)
- Output Parameter: Drought intensity classes (D0–D4)

Missing values were found in the dataset, which were filled in using the median method. This method consists of replacing the missing values with the median value of the corresponding column. The data was then divided into training and test sets in a ratio of 80% to 20% [21–23].

The following machine learning models were used to predict the intensity of drought:

1. CART (Classification and Regression Trees): Decision trees are nonparametric models that divide data into subsets based on predictor values [24].
2. Random forest: An ensemble model consisting of a set of decision trees, each of which is trained on a random subset of data. The prediction is made by voting all the trees [25].

The following metrics were used to evaluate the performance of the models:

- Accuracy: The proportion of correctly classified samples [26].
- Completeness: The proportion of correctly classified positive samples [27].
- F1 is a measure: The harmonic mean of accuracy and completeness [28].
- Error Matrix: A matrix showing the number of correctly and incorrectly classified samples for each class

3 Results and Discussion

As part of the study, a correlation analysis was conducted to identify possible linear dependencies between the input meteorological data and the output parameter - drought intensity. Correlation analysis allows determining the degree of interrelation between various meteorological indicators and drought intensity, measured through D0–D4 classes. This helps to understand if there are any strong correlation links that might indicate a direct dependence. Figure 2 presents a correlation matrix showing Pearson correlation coefficients between all meteorological indicators and the output parameter (drought intensity).

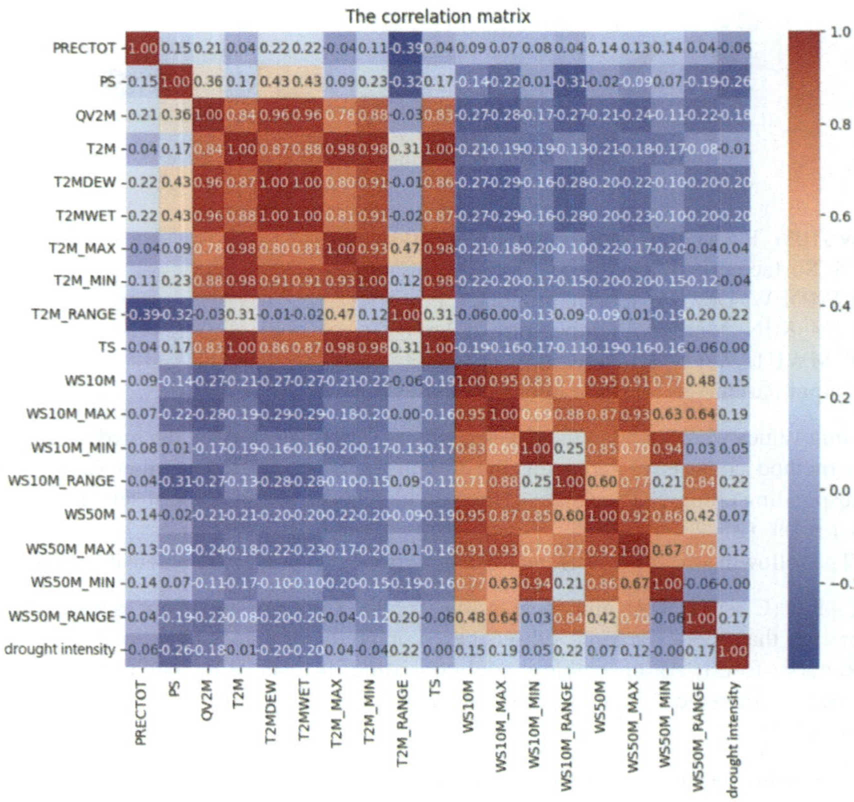

Fig. 2. Correlation matrix of meteorological indicators and drought intensity

In this study, a random forest model was used to predict the intensity of drought. The model was trained on a training dataset and evaluated on a test dataset. The results of the evaluation of the model showed good performance [29, 30].

Figure 3 is a histogram of the importance of features in a random forest model. As can be seen from the figure, the most important signs for predicting the intensity of drought are PS (surface pressure), T2M_RANGE (temperature range at an altitude of 2 m) and TS (temperature of the earth's surface). These features have the greatest contribution to solving the random forest model.

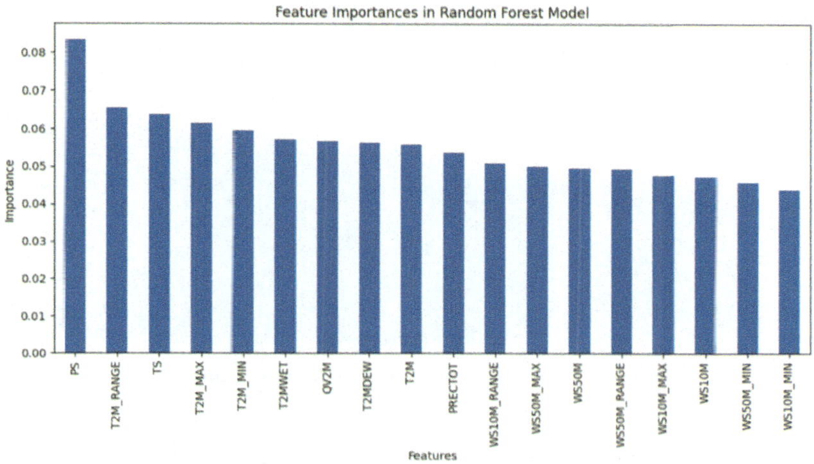

Fig. 3. A histogram of the importance of features in a random forest model

Figure 4 is a matrix of errors of the random forest model. The matrix shows the number of correctly and incorrectly classified samples for each class of drought intensity. As can be seen from the matrix, the model has high classification accuracy for classes 0 (no drought), 1 (mild drought) and 2 (moderate drought), but the classification accuracy for classes 3 (severe drought) and 4 (very severe drought) is slightly lower.

Figure 5 is a histogram of the importance of features in the CART model. As can be seen from the figure, the most important signs for predicting the intensity of drought are PS (surface pressure), T2M_RANGE (temperature range at an altitude of 2 m) and TS (temperature of the earth's surface). These features have the greatest contribution to the solution of the CART model.

Figure 6 is a CART model classification report that includes accuracy, completeness, F1 measure and support for each class of drought intensity. As can be seen from the figure, the CART model has high classification accuracy for all classes of drought intensity, with the exception of Class 4 (very severe drought).

The random forest model demonstrated a classification accuracy of 94.2%, which is slightly higher than that of the CART model (93%). The simulation results confirm the importance of surface pressure (PS), temperature range at an altitude of 2 m (T2M_RANGE) and Earth surface temperature (TS) for predicting the intensity of drought. These parameters are key indicators of the water balance in the region and have a significant impact on the development and intensity of drought.

For a deeper analysis of the effect of temperature on the intensity of drought, a random forest model was built using only the temperature parameters: 'T2M', 'QV2M', 'T2M_MAX', 'T2M_MIN', 'T2M_RANGE', 'T2MDEW', 'T2MWET'.

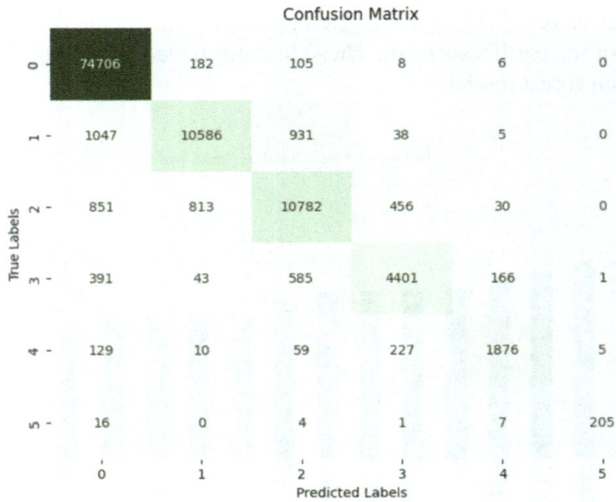

Fig. 4. The error matrix of the random forest model.

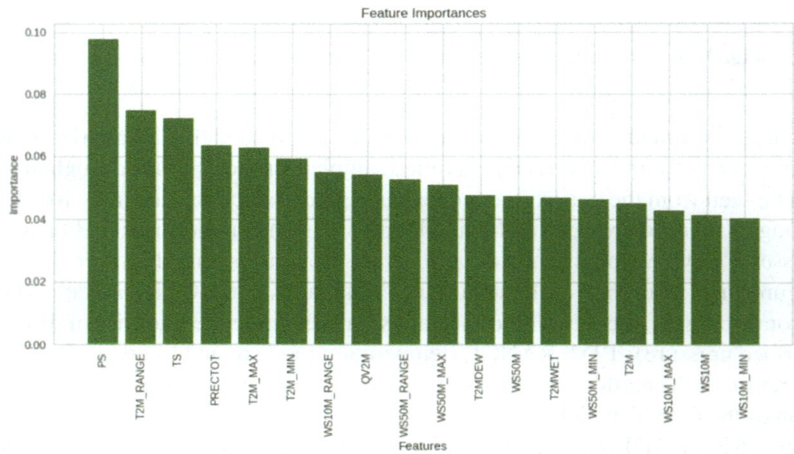

Fig. 5. A histogram of the importance of features in the CART model.

Figure 7 is a histogram of the importance of features in a random forest model based only on temperature parameters. As can be seen from the figure, the most important feature is PS (surface pressure), followed by T2M_RANGE (temperature range at a height of 2 m) and TS (temperature of the earth's surface).

Figure 8 is an error matrix of a random forest model based only on temperature parameters. As can be seen from the matrix, the model has high classification accuracy for classes 0 (no drought), 1 (mild drought) and 2 (moderate drought), but the classification accuracy for classes 3 (severe drought) and 4 (very severe drought) is slightly lower. Figure 9 is a report on the classification of the CART model based only on temperature

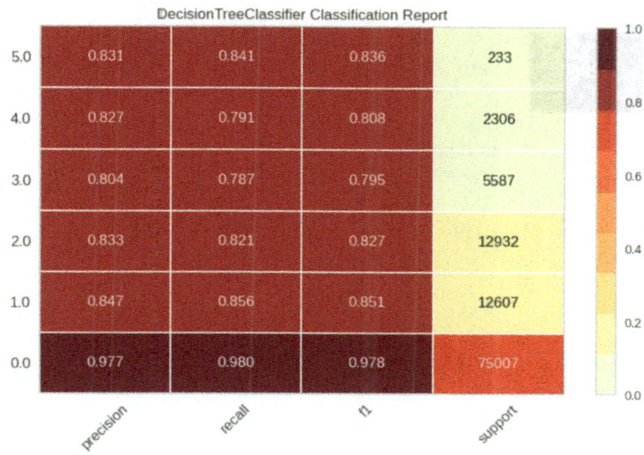

Fig. 6. A histogram of the importance of features in the CART model.

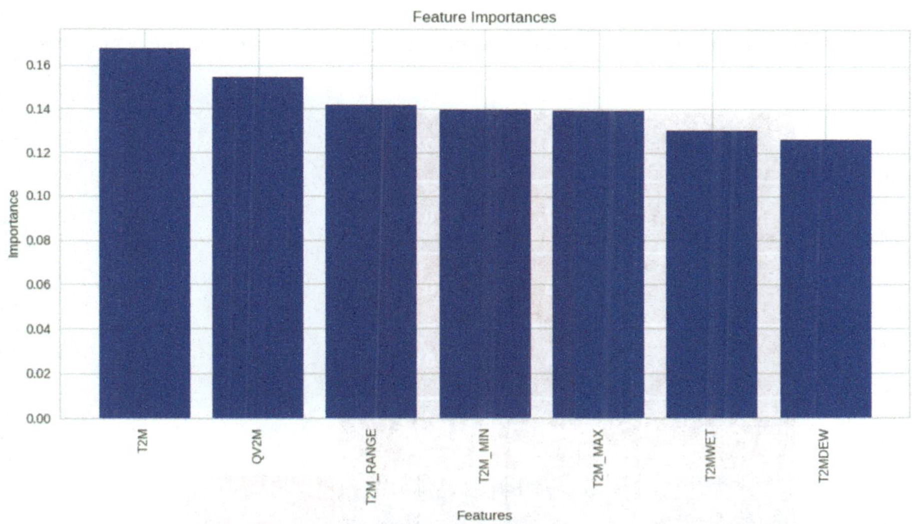

Fig. 7. A histogram of the importance of features in a random forest model based only on temperature parameters.

parameters. As can be seen from the figure, the CART model has high classification accuracy for all classes of drought intensity, with the exception of Class 4 (very severe drought).

An analysis of the simulation results using only temperature parameters shows that temperature has a significant effect on the intensity of drought, however, the inclusion of other meteorological parameters, such as surface pressure, increases the accuracy of forecasting.

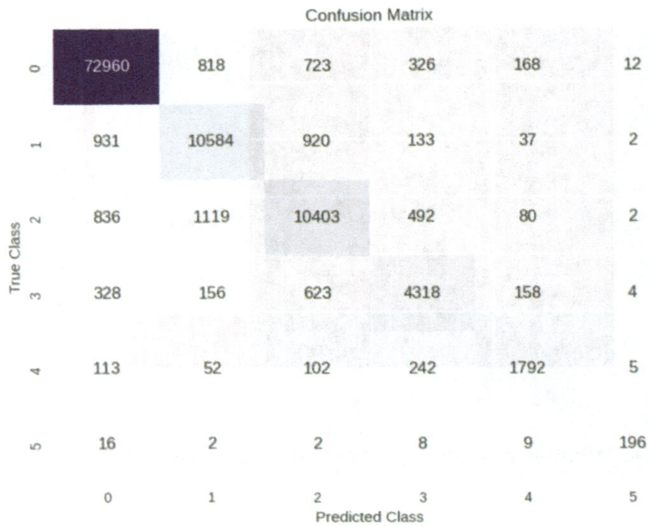

Fig. 8. The error matrix of a random forest model based only on temperature parameters.

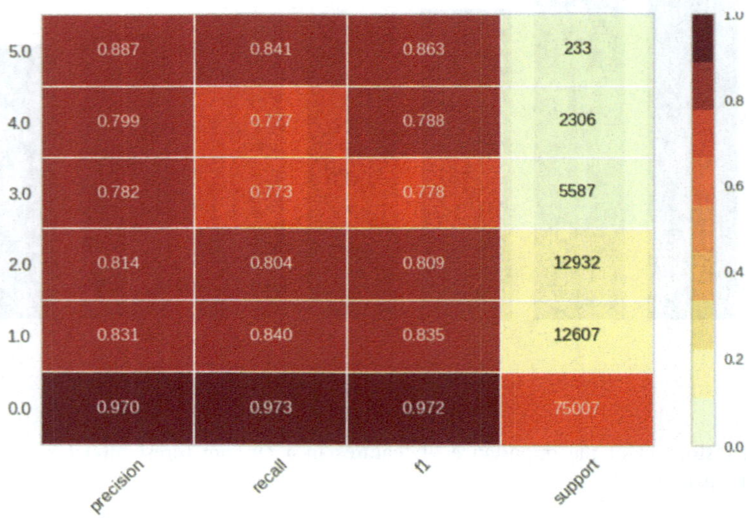

Fig. 9. A report on the classification of the CART model based only on temperature parameters.

4 Conclusion

In this study, CART and random forest models were evaluated to predict the intensity of drought in the United States using meteorological data collected through satellite observations. The results showed that both models have good classification accuracy and can be used to predict drought intensity. It has been revealed that surface pressure, temperature range, and ground surface temperature are key meteorological parameters that affect the intensity of drought.

The results of this study are important for the development of strategies for water management, adaptation to climate change and reducing the risk of drought. It is especially important to note the contribution of this study to agro-climatic science. Accurate forecasting of drought intensity can help farmers and agronomists make better informed decisions on agricultural water management, such as choosing more drought-tolerant crop varieties, optimizing irrigation systems, and crop planning.

This study opens up broad prospects for further research in the field of drought forecasting. In the future, it will be possible to conduct research aimed at using more extensive and detailed data sets, applying more complex machine learning models, studying the impact of anthropogenic factors on drought intensity and developing interactive drought forecasting systems.

This research contributes to the development and evaluation of machine learning models for predicting drought intensity, identifying key meteorological parameters that affect drought intensity, and providing valuable knowledge for the development of drought mitigation strategies, including for agro-climatic science. The results of this study can be used to improve water management systems, adapt to climate change and reduce the risk of drought in agriculture and other sectors of the economy.

References

1. Kuwayama, Y., Thompson, A., Bernknopf, R., Zaitchik, B., Vail, P.: Estimating the impact of drought on agriculture using the U.S. drought monitor. Am. J. Agric. Econ. **101**, 193–210 (2018). https://doi.org/10.1093/ajae/aay037
2. Ray, R., Ampim, P., Gao, M.: Crop protection under drought stress. In: Crop Protection Under Changing Climate, pp. 145–170. Springer, Cham (2020). https://doi.org/10.1007/978-3-030-46111-9_6
3. Madadgar, S., Aghakouchak, A., Farahmand, A., Davis, S.: Probabilistic estimates of drought impacts on agricultural production. Geophys. Res. Lett. **44**, 7799–7807 (2017). https://doi.org/10.1002/2017GL073606
4. Begna, T., et al.: Effects of drought stress on crop production and productivity. Int. J. Res. Stud. Agri. Sci. **6**(9), 2454–6224 (2020). https://doi.org/10.20431/2454-6224.0609005
5. Bosikov, I.I., et al.: Modeling and complex analysis of the topology parameters of ventilation networks when ensuring fire safety while developing coal and gas deposits. Fire. **6**(3), C. 95 (2023)
6. Chen, T., Xia, G., Liu, T., Chen, W., Chi, D.: Assessment of drought impact on main cereal crops using a standardized precipitation evapotranspiration index in Liaoning Province, China. Sustain. For. **8**, 1–16 (2016). https://doi.org/10.3390/SU8101069
7. Elbeltagi, A., Pande, C.B., Kumar, M., et al.: Prediction of meteorological drought and standardized precipitation index based on the random forest (RF), random tree (RT), and Gaussian process regression (GPR) models. Environ. Sci. Pollut. Res. **30**, 43183–43202 (2023). https://doi.org/10.1007/s11356-023-25221-3
8. Yang, X., Lu, M., Wang, Y., Wang, Y., Liu, Z., Chen, S.: Response mechanism of plants to drought stress. Horticulturae. **7** (2021). https://doi.org/10.3390/HORTICULTURAE7030050
9. Ribeiro, A., Russo, A., Gouveia, C., Páscoa, P., Pires, C.: Probabilistic modelling of the dependence between rainfed crops and drought hazard. Nat. Hazards Earth Syst. Sci. **19** 2795–2809 (2019). https://doi.org/10.5194/NHESS-19-2795-2019

10. Malozyomov, B.V., et al.: Study of supercapacitors built in the start-up system of the main diesel locomotive. Energies. **16**(9), C. 3909 (2023)
11. Fung, K.F., Huang, Y.F., Koo, C.H.: Coupling fuzzy–SVR and boosting–SVR models with wavelet decomposition for meteorological drought prediction. Environ. Earth Sci. **78**, 693 (2019). https://doi.org/10.1007/s12665-019-8700-7
12. Kukartsev, V., et al.: Using machine learning techniques to simulate network intrusion detection. In: 2024 International Conference on Intelligent Systems for Cybersecurity (ISCS), pp. 1–4. IEEE
13. Achite, M., Jehanzaib, M., Elshaboury, N., Kim, T.-W.: Evaluation of machine learning techniques for hydrological drought modeling: A case study of the Wadi Ouahrane Basin in Algeria. Water. **14**, 431 (2022). https://doi.org/10.3390/w14030431
14. Dehghani, M., et al.: Uncertainty analysis of streamflow drought forecast using artificial neural networks and Monte-Carlo simulation. Int. J. Climatol. **34**(4) (2014)
15. Brigida, V., et al.: Technogenic reservoirs resources of mine methane when implementing the circular waste management concept. Resources. **13**(2), 33 (2024)
16. Leng, G., Hall, J.: Crop yield sensitivity of global major agricultural countries to droughts and the projected changes in the future. Sci. Total Environ. **654**, 811–821 (2019). https://doi.org/10.1016/j.scitotenv.2018.10.434
17. Strateichuk, D.M., et al.: Morphological features of polycrystalline CdS1− xSex films obtained by screen-printing method. Crystals. **13**(5), C. 825 (2023)
18. Elbeltagi, A., et al.: Combination of limited meteorological data for predicting reference crop evapotranspiration using artificial neural network method. Agronomy. **12**(2), C. 516 (2022)
19. Zhang, P., Zhang, J., Chen, M.: Economic impacts of climate change on agriculture: the importance of additional climatic variables other than temperature and precipitation. J. Environ. Econ. Manag. **83**, C. 8-31 (2017)
20. Tynchenko, Y., Kukartsev, V., Tynchenko, V., Kukartseva, O., Panfilova, T., Gladkov, A., Nguyen, V., Malashin, I.: Landslide assessment classification using deep neural networks based on climate and geospatial data. Sustain. For. **16**, 7063 (2024). https://doi.org/10.3390/su16167063
21. Tynchenko, Y.A., Kukartsev, V.V., Xiaogang, W., Kravtsov, K.I.: Modeling the drought intensity in mountainous areas using meteorological parameters. Sustain. Dev. Mountain Territ. **16**(2), 655–668 (2024). https://doi.org/10.21177/1998-4502-2024-16-2-655-668
22. Seidel, D.J., et al.: Global radiosonde balloon drift statistics. J. Geophys. Res. Atmosph. **116**(D7) (2011)
23. Shutaleva, A., et al.: Sustainability of inclusive education in schools and higher education: teachers and students with special educational needs. Sustainability. **15**(4), C. 3011 (2023)
24. Tynchenko Ya, A., et al.: Probabilistic analysis of pump reliability indicators using a neural network. MIAB. Mining Inf. Anal. Bull. **2024**(7-1), 126–136 (2024). https://doi.org/10.25018/0236_1493_2024_71_0_126
25. Tynchenko, Y., Tynchenko, V., Kukartsev, V., Panfilova, T., Kukartseva, O., Degtyareva, K., Nguyen, V., Malashin, I.: Soil properties classification in sustainable agriculture using genetic algorithm-optimized and deep neural networks. Sustain. For. **16**, 8598 (2024). https://doi.org/10.3390/su16198598
26. Park, S., Im, J., Jang, E., Rhee, J.: Drought assessment and monitoring through blending of multi-sensor indices using machine learning approaches for different climate regions. Agric. For. Meteorol. **216**, 157–169 (2016). https://doi.org/10.1016/J.AGRFORMET.2015.10.011
27. Kukartsev, V.A., et al.: Study of the influence of the thermal capacity of the lining of acid melting furnaces on their efficiency. Metals. **13**(2), C. 337 (2023)
28. Malozyomov, B.V., et al.: Determination of the performance characteristics of a traction battery in an electric vehicle. World Electr. Vehic. J. **15**(2), 64

29. Panfilova, T., Kukartsev, V., Tynchenko, V., Tynchenko, Y., Kukartseva, O., Kleshko, I., Wu, X., Malashin, I.: Flood susceptibility assessment in urban areas via deep neural network approach. Sustain. For. **16**, 7489 (2024). https://doi.org/10.3390/su16177489
30. Kaung, P.A., Isakov, A.E., Panfilov, I.A., Tynchenko, V.V., Stupina, A.A.: Principles for forming environmentally safe and economically effective sustainable development of geo resources. MIAB. Mining Inf. Anal. Bull. **2024**(7-1), 159–175 (2024). https://doi.org/10.25018/0236_1493_2024_71_0_159

Robust Breast Cancer Detection Model Based on Machine Learning

Rubén Ruiz[1], Glem Ramos[1], and Wilfredo Ticona[1,2](✉)

[1] Universidad ESAN, Lima, Peru
{21100100,21100309}@ue.edu.pe, wmamani@esan.edu.pe
[2] Universidad Tecnológica del Perú, Lima, Peru

Abstract. Breast cancer is one of the most prevalent oncological diseases worldwide, and it has a significant impact on public health and the quality of life of those affected. The objective of this study was to identify the stages of breast cancer using Machine Learning algorithms. The proposed methodology consists of 5 phases: obtaining dataset; preprocessing (One-hot Encoding; Label Encoding; Data Scaling; Data Imputation, Data Balancing and Feature Engineering) and model implementation (SVM, RF, DT, KNN, RL and XGBoost); evaluation (Accuracy, Recall, Precision, F1-Score, Specificity, ROC and Confusion Matrix) and validation. Superior results were obtained with the SVM model, with an Accuracy of 82.72%, Specificity of 85.11%, and the following hyperparameters: C (1), gamma (scale), and kernel (linear). The results demonstrate that it is possible to identify breast cancer stages using Machine Learning models.

Keywords: Breast cancer · Hyperparameters · Machine Learning · Preprocessing · XGBoost · SVM

1 Introduction

Breast cancer is a disease in which altered breast cells multiply uncontrollably, forming tumors that, if left untreated, can spread throughout the body and prove fatal. These cancerous cells begin to develop in the milk ducts or milk-producing lobules of the breast. In stage 0, the cancer does not pose a lethal threat and can be detected in its early stages. However, cancer cells can invade the surrounding breast tissue, resulting in lumps or thickening. Invasive cancers can spread to nearby lymph nodes or other organs (metastasis), which can be life-threatening and fatal. Treatment depends on the characteristics of the patient, the type of cancer and its level of spread. Generally, a combination of surgery, radiotherapy and medication is used. Being one of the most prevalent oncological diseases worldwide, and it has a significant impact on public health and the quality of life of those affected. According to WHO [1], it is estimated that by 2024 there will be 310720 new cases of breast cancer in women and 2790 in men. With a mortality rate of 42250 women and 530 men will lose their lives due to breast cancer.

R. Silhavy and P. Silhavy (Eds.): CoMeSySo 2024, LNNS 1489, pp. 490–504, 2025.
https://doi.org/10.1007/978-3-031-96798-6_37

In Latin America, breast cancer is also a significant cause of mortality. According to PAHO data [2], approximately 187,000 new cases of breast cancer are diagnosed each year, of which approximately 63,000 are due to the disease. These figures highlight the urgent need for effective strategies for the early detection and timely treatment of breast cancer. In addition, it is estimated that approximately 685,000 women died from this disease.

In the national context, according to the Peruvian Ministry of Health (MINSA) [3], breast cancer poses a significant public health challenge. Although precise data may vary, 8000 new cases of breast cancer will be reported in 2023, and the incidence rate will continue to increase annually.

However, despite these challenges, artificial intelligence and machine learning are emerging as promising tools to fight breast cancer. Advances in machine learning are revolutionizing the diagnosis and treatment of diseases, enabling early and accurate detection of abnormalities in medical images such as mammograms and ultrasounds. In this study, machine learning models are used to enhance the early detection of breast cancer. Ultimately, improving clinical outcomes, reducing mortality, and bringing hope to millions of people around the world.

2 Related Work

In Li S [5], Li focused on the early detection of breast cancer using machine learning algorithms. The scope of this study was limited to the analysis of data provided by the Wisconsin Breast Cancer Data Set (WBCD) obtained from the UCI database. The research design was quantitative, employing predictive modeling techniques to evaluate the effectiveness of different machine learning algorithms in classifying breast tumors as benign or malignant. The study population comprised patients with breast cancer in the WBCD dataset. The sample included 683 cases after eliminating records with missing values. The sampling method was non-probability because the full sample was available in the dataset. Data collection methods were not applied in this study because it was based on a pre-existing data set. However, data analysis involves data preparation, machine learning algorithm comparison, and systematic elimination of variables to assess their impact on prediction accuracy. The theoretical results include a review of the underlying principles of the machine learning algorithms used, such as KNN, LR, DT, RF, NB, SVM, and NN. The empirical results demonstrate that the KNN, RF, and SVM classifiers were the most accurate with accuracy values of 98%, 95%, and 96%, respectively. In addition, "Clump Thickness" and "Bare Nuclei" were identified as the most influential variables in the prediction of breast cancer. In summary, the usefulness of machine learning algorithms for breast cancer detection was demonstrated and highlighted the importance of certain tumor characteristics in the diagnostic process.

In a study by Lou J [6], the Breast Cancer dataset, consisting of 570 records with 32 features, was used to develop predictive models for breast cancer. The focus of this study was to compare the effectiveness of different machine learning algorithms, including RF, SVM, and the combination of Principal Component Analysis (PCA) with SVM (PCA-SVM). The scope of this study focuses on evaluating the accuracy, F1 score, and training time of the models. The target population was women diagnosed with breast cancer, and the sample was randomly selected from the dataset. Sampling and data collection methods involved extracting historical patient records. Data analysis included the selection of the 10 most important features using SVM and the application of dimensionality reduction techniques with PCA. The results were promising reaching high accuracy rates for breast cancer prediction, and the PCA-SVM model proved to be the most efficient, with an accuracy of 96%, an F1 score of 0.95, and a training time of 0.001999 seconds.

Similarly, Pérez Núñez J [7] analyzed the effectiveness of multiple machine learning models in breast cancer prediction. The scope of the investigation was limited to two Wisconsin breast cancer datasets provided by Dr. William H. Solberg, one from 1995 and the other from 1992. These datasets were obtained from the Kaggle and UC Irvine repositories. The research design was quantitative, employing data mining and machine learning techniques to evaluate the predictive performance of the models. The study population comprised the clinical records of patients with breast cancer who were included in the Wisconsin database. The sample comprised all available cases in the datasets with no specified sample size. Data collection methods consisted of direct download of datasets from online repositories. Data analysis involves data preprocessing, building machine learning models using algorithms such as DT, K-NN, Naive Bayes (NB), LR, SVM, MLP, and RF, and evaluating the performance of these models using metrics such as the confusion matrix and receiver operating characteristic (ROC) curve. The theoretical results suggest that machine learning models are valuable tools for breast cancer prediction. Empirically, the MLP algorithm achieved the highest accuracy (97.20%) on the 1995 database, whereas the SVM algorithm achieved accuracy (97.20%) on the 1992 database. These findings highlight the potential of CAD (Computer-Aided Diagnosis) technologies and machine learning methodologies for the early detection and treatment of breast cancer.

Zhou et al. [8] conducted a retrospective study to develop a high-precision classification algorithm for breast cancer using the Wisconsin breast cancer dataset. The study aimed to analyze the dataset and create a machine learning model to accurately distinguish between benign and malignant breast tumors. The researchers employed standard data preprocessing techniques, including feature scaling and handling missing values, and performed statistical analyses such as Spearman correlation and the Wilcoxon rank sum test to select relevant features. They trained seven machine learning algorithms: decision tree, stochastic gradient descent, random forest, support vector machine, logistic regression, and AdaBoost. The AdaBoost-Logistic algorithm emerged as the most accurate, with a classification accuracy of 99.12%, outperforming other models and previous techniques. This study demonstrates the feasibility and effectiveness of using machine learning for breast cancer classification, potentially aiding in the early detection and treatment of the disease.

Anurag Dutta et al. [9] conducted research on the prognosis of breast cancer using machine learning classifiers. The study utilized a dataset of mammography images, which included both healthy and carcinogenic cases, and made the data publicly available on GitHub. The research design employed supervised learning algorithms such as Logistic Regression, Support Vector Machine, k-Nearest Neighbors, Random Forest Classifier, Gaussian Naïve Bayes, and Decision Tree to classify the images as benign or malignant. The performance of these classifiers was evaluated using metrics like precision, recall, and F1 score. The k-Nearest Neighbors classifier emerged as the most efficient, with an F1 score of 0.9537, precision of 0.9324, and recall of 0.9760. This study demonstrates the potential of machine learning in aiding the diagnosis of breast cancer, suggesting that such technologies could improve early detection and treatment outcomes.Finally, Chen H, Wang N, Du X et al. [10] used the WDBC breast cancer prediction database. This dataset was obtained from the Machine Learning repository of the University of California, Irvine (UCI). Several algorithms were employed, including XGBoost, K-NN, LR, and RF. The variables used in the study included features such as radius, perimeter, area, compactness, concavity, and concave points, among others, both in their mean values and their "worst" or largest values. These variables were used to train and evaluate the breast cancer prediction models. The best algorithm used in the breast cancer prediction study was XGBoost, with an accuracy of 97.40%.

3 Methodology

The methodology used in this study was based on previous studies [7, 9–11]. It consists of 5 phases: Data acquisition, Data preprocessing, Model implementation, Evaluation and Validation. A breast cancer dataset was used. In data preprocessing, techniques such as One-Hot coding, data scaling, data imputation, feature balancing, and feature engineering were applied. Subsequently, several machine learning models were implemented: SVM, DT, RL, XGBoost, KNN, and RF. The evaluation of these models was performed using metrics such as accuracy, precision, recall, specificity, and F1 score, in addition to ROC-AUC curves and confusion matrices. The final validation consists of selecting the most efficient model based on the evaluation results. This process is fundamental to developing accurate and reliable medical diagnostic systems, and a representative graph is presented in Fig. 1.

3.1 Obtaining Dataset

In this phase, the malignant and non-malignant breast cancer datasets were identified. This search was performed in official oncology research repositories and oncology scientific articles. The tasks performed to select the database are described in the following section.

- Examine and select the database containing malignant and non-malignant breast cancer data.
- Corroborate that the database source is previously labeled by oncology experts and has been evaluated in a laboratory.
- Download the database.

Finally, the dataset "Breast Cancer Dataset" [12] from the University Medical Center, Institute of Oncology located in Ljubljana, Yugoslavia, which shared the present dataset through the Kaggle platform, was selected: https://www.kaggle.com/datasets/faysalmiah1721758/breast-cancer-data. Table 1 lists the attributes and their descriptions.

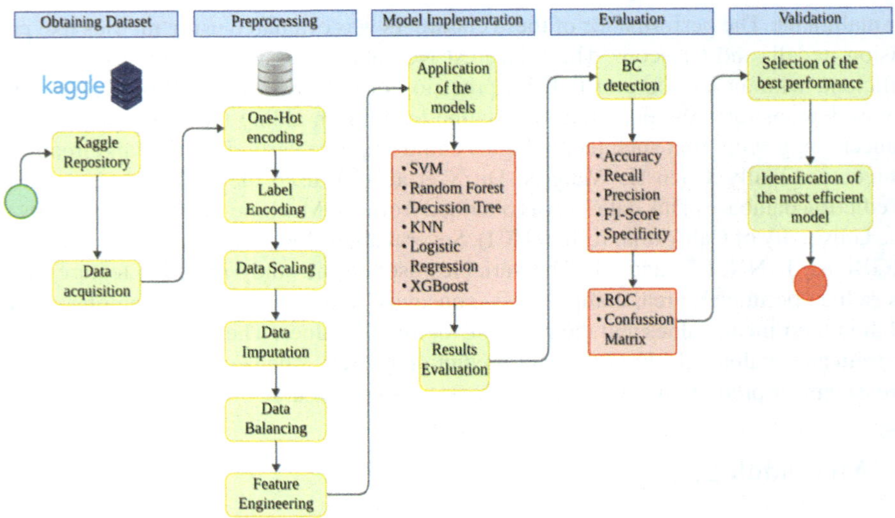

Fig. 1. Proposed methodology.

Table 1. Attributes and descriptions

Attributes	Description
class	This attribute classifies cancer recurrence (or no recurrence after the initial treatment.
age	Range of patient's age.
menopause	Menopause is a natural biological process that marks the end of a woman's menstrual cycle.
tumor-size	Tumor size refers to the dimensions or measurements of the tumor. In cancer diagnosis and treatment, tumor size.
inv-nodes	INV Nodes were the number (range 0 - 39) of axillary lymph nodes containing visible metastatic breast cancer.
node-caps	If the cancer metastasizes to a lymph node, it may remain outside the original site of the tumor.
deg-malig	One of the main causes of death is cancer. The most common cancer in women is breast cancer.
breast	Position of breast.
breast-quad	The breasts are divided into 4 quadrants with the nipple at the center.
irradiate	Irradiation is a common treatment for breast cancer.

3.2 Preprocessing

Data preprocessing is critical to ensure that Machine Learning algorithms work efficiently and deliver optimal results. This involves eliminating missing data, categorical variable coding, and data scaling. Therefore, through these operations, we aim to maximize the performance of the models and improve the accuracy of the prediction of breast cancer recurrence in patients.

Table 2 provides a structured and clear view of the preprocessing steps, ensuring that the data are in a format suitable for use in Machine Learning algorithms. Each activity was designed to improve data quality and maximize the accuracy of the predictive models, thereby contributing to a better understanding and prediction of breast cancer recurrence in patients.

- **One-Hot encoding:** The application of One-Hot Encoding to the variables "age", "tumer-size" and "inv-nodes" plays an essential role in data preparation. This strategy consists of creating a binary column (which can only contain the values 0 or 1) for each unique value that exists in the categorical variable to which we encode and mark a 1, the column corresponding to the value present in each record and leaving the other columns with a value of 0 [13].
- **Coding of categorical variables:** This transformation is crucial for data preparation and data analysis using machine learning algorithms. The function encodes the labels of a categorical feature into numerical values between 0 and 1 [14].
- **Missing values management with special label:** A special label, in this case "missing", was assigned to the null values in the "node caps" and "breast-quad" columns. This action was performed to identify and treat missing values in these columns. Subsequently, the missing values in both columns were counted to determine the amount of missing data. These missing values were then converted back to NaN (null values) to maintain consistency in the handling of missing values in the dataset. This approach allows for a clear and systematic treatment of missing values, which is essential for maintaining data integrity and quality during subsequent analysis and modeling.
- **Determination of the best value of k for KNN by cross-validation:** This process evaluates the performance of the KNN classifier with different values of k using the cross-validation technique. The k-iteration or k-fold cross-validation process divides the original data into k subsets. At the time of training, each k subset was taken as the model test set, and the remaining subset (k-1) was taken as the training set [15]. In this case, it was tested in the range of 1–15. Table 3 presents the values obtained by k, with 0.744767090139141 as the highest mean and a triple tie between $k = 12$, 13, and 15. However, we chose to use $k = 13$ because an odd value for k is preferred to avoid ties in classification, and a lower value makes the model slightly less complex and faster.
- **Missing value imputation with KNN:** A simple ranking algorithm that stores all available observations and ranks new cases based on a similar measure [16]. Thus, when a KNN is used, a new observation is ranked according to the votes of its K nearest neighbors. As a pro, it is simple to learn and implement. The disadvantage of this method is that it uses the entire dataset to train "each point" and therefore requires the use of a large amount of memory and processing resources [17]. For

these reasons, KNN tends to work better on small datasets without many features [17].

- **Data balancing with SMOTE:** The dataset is unbalanced. Class "0" (non-recurrence-events') has 70% and class "1" (recurrence-events') has 30%. This imbalance could negatively affect the performance of ML models, particularly in terms of accuracy and recall for the minority class. To mitigate the impact of imbalance, the SMOTE (Synthetic Minority Over-sampling Technique) technique will be used. SMOTE is an oversampling-based technique, which generates "synthetic" or artificial instances in the attribute space with the objective of balancing the data sample based on the Nearest Neighbor rule [18]. This rule assumes that instances close to each other have a higher probability of belonging to the same class. The generation process is performed by interpolating new instances rather than duplicating them, as in resampling-type algorithms do [18].
- **Feature Engineering:** Feature engineering is the act of extracting features from raw data and transforming them into formats that are suitable for the machine learning model. It is a crucial step in the machine learning pipeline because appropriate features can ease modeling difficulty, which enables the pipeline to output results of higher quality [19, 20].

3.3 Model Implementation

In this phase, 6 ML techniques were used: SVM, RF, KNN, DT, LR, and XGBoost. Table 4. shows the implementation activities.

- **SVM:** The goal of the SVM algorithm is to generate a line that distinguishes "as best as possible" between the two categories of points so that it can continue to classify points efficiently even if more points are later created in the queue. Any hyperplane has a minimum vertical separation between the data points on either side, and the margin is the sum of the two minimum separations [5].

$$f(x) = \text{sign}(\langle w, x \rangle + b) \tag{1}$$

- **KNN:** The KNN algorithm judges the category of unknown samples and performs pattern recognition to facilitate its wide use in prediction analysis. To determine the distance between known and unknown distributions, all samples from categories with known distributions were used as references. In addition, k known samples that are most like the unknown samples are placed into one category by majority vote [5].
- **DT:** The classification mechanism using a decision tree is as follows: start with the root node is started, an instance feature is tested, and then distribute the instance to its child nodes based on the test results [5].

$$H(S) = \sum_{i=1}^{c} p_i \log_2 p_i \tag{2}$$

- **RF:** The RF algorithm comprises many decision trees that are not related to each other.

Table 2. Processing activities

Activities	Descriptions	Tasks
One-Hot encoding	Convert categorical variables into a format that can be provided to machine learning algorithms for better prediction.	Identification of distinct categorical columns in the form of ranges.
		Conversion to multiple binary columns ("True" or "False").
		Identification of categorical columns that need to be transformed into numeric values.
Coding of categorical variables.	We transformed categorical columns into numeric values using "LabelEncoder".	Identification of categorical columns that need to be transformed into numerical values.
		Conversion of columns to numeric values.
Missing values management with special label.	Handle missing values by assigning special labels to null values in specific columns.	The special label is assigned to missing values.
		Counting Values with Special Label.
		Replacing a Special Label with Null Values
Determination of the best value of k for KNN by cross-validation.	The performance of the KNN model with different values of k was evaluated using the average score obtained via cross validation.	Feature Selection and Missing Value Management.
		Data Scaling.
		Target Variable Definition.
		Definition of target variable and k values to test.
		Evaluation of the KNN classifier.
Missing value imputation with KNN.	Imputation of missing values in node caps and breast quad columns using the KNN algorithm.	Selection of Characteristics.
		Missing Value Allocation with KNN.
Data balancing with SMOTE.	Balancing the dataset using the Synthetic Minority Over-sampling Technique (SMOTE) method.	Selection of Characteristics and Target Variable.
		Application of SMOTE for oversampling.
Feature Engineering	Create new features using the Feature Engineering principle.	Feature selection based on clinical studies.
		Implementation of the new features in the dataframe.

During the classification process, new input samples are added, and each decision tree in the forest is evaluated and categorized independently [5].

$$P(y = 1 \mid x) = \frac{1}{1 + e^{-(w^T x + b)}} \tag{3}$$

- **LR:** The LR algorithm is used to obtain odds ratio in the presence of more than one explanatory variable. The procedure is quite like multiple linear regressions except that the response variable is binomial. The result shows the impact of each variable on the odds ratio of the observed event of interest [21].

Table 3 Mean k-scores

k	Average score	k	Average score	k	Average score
1	56.23%	6	73.42%	11	73.07%
2	72.37%	7	72.37%	12	**74.48%**
3	55.88%	8	72.72%	13	**74.48%**
4	72.02%	9	73.07%	14	73.42%
5	68.51%	10	73.07%	15	**74.48%**

- **XGBoost:** The XGBoost algorithm is an optimized distributed gradient boosting framework that is designed to have high effectiveness, adaptability, and portability. Its goal is to leverage the extreme computational capabilities of modern hardware to offer a scalable and accurate end-to-end boosting system [22] (Table 5).

$$L(\theta) = \sum_{i=1}^{n} l(y_i, f(x_i)) + \sum_{k=1}^{K} \Omega(f_k) \tag{4}$$

3.4 Evaluation

The following metrics were used to measure model performance: Accuracy, Recall, Precision, Specificity, F1-Score and ROC.

- **Accuracy:** It helps to know how accurate or close the result is to the true value and provides information about possible errors that can be found in the classification [23].

$$Accuracy = \frac{TP + TN}{TP + TN + FP + FN} \tag{5}$$

- **Recall:** Recall and sensitivity measures allow us to determine the proportion of positive cases that were classified correctly. In a perfect model, recall is equal to 1 for each class [24].

$$Recall = \frac{TP}{TP + FN} \tag{6}$$

- **Precision:** This ensures that the classifier does not label a negative class sample as a positive class sample [23].

$$Precision = \frac{TP}{TP + FP} \tag{7}$$

- **Specificity:** This section discusses the negative cases where the algorithm was classified correctly. It expresses how well the model detects that class [25].

$$Specificity = \frac{TN}{TN + FP} \tag{8}$$

where:

Table 4. Implementation activities

Activities	Descriptions	Tasks
SVM	Implementation of a classification model using the SVM algorithm, which attempts to construct an optimal hyperplane in a feature space to effectively separate classes	Hyperparameter tuning
		Model training using the training set
		Evaluation of models using the test set
KNN	Implementation of the KNN algorithm, which classifies observations based on the classes of nearest neighboring samples in the feature space	Select an appropriate k value using techniques such as cross validation
		Model training using the training set
		Evaluation of the model using the test set
DT	Implementation of the DT algorithm, which divides the dataset into smaller subsets based on features for observation classification	Training the Decision Tree model using the training set
		Tuning of hyperparameters, such as the maximum depth of the tree by cross-validation
		Model training using the training set
		Evaluation of the model using the test set
RF	Implementation of the RF algorithm, which is an ensemble technique that combines multiple decision trees to improve model accuracy and generalization	Training of the RF model using the training set
		Tuning of hyperparameters, such as number of trees and maximum depth, by cross-validation
		Training of the model using the training set
		Evaluation of the model using the test set
LR	Implementation of the LR algorithm, which models the probability that an observation belongs to a particular class using a logistic function	Training of the RL model using the training set
		Adjustment of hyperparameters, such as regularization. by cross-validation.
		Training of the model using the test set
		Evaluation of the model using the test set
XGBoost	Implementation of the XGBoost algorithm, which uses a set of decision trees to improve accuracy. This algorithm builds the trees sequentially, each correcting for errors in the previous trees.	Training the XGBoost model using the training set
		Adjustment of hyperparameters, such as regularization, by cross-validation
		Training of the model using the test set
		Evaluation of the model using the test set

- TP = True Positives
- TN = True Negatives
- FP = False Positives
- FN = False Negatives

- **F1-Score:** It merges the metrics accuracy with recall, presenting differences in the performance of a classifier that are not revealed by accuracy alone [23].

$$F1 = \frac{2\ x\ Precision\ x\ Recall}{Precision + Recall} \tag{9}$$

- **ROC:** A two-dimensional plot of sensitivity versus (1 - specificity) for each class. The measure of comparison in this graph is the AUC, which corresponds to the area under the ROC curve, with values between 0 and 1 [24]. A wide range of diagnostic tests report their results quantitatively using continuous scales. The analysis of ROC curves

Table 5. Hyperparameters of the algorithms

SVM	C = [0.1, 1]
	gamma = [scale]
	kernel = [linear, rbf]
KNN	n neighbors = [3, 5, 7]
	weights = [uniform, distance]
	metric = [euclidean, manhattan]]
DT	max depth = [None, 10, 20]
	min samples split = [2, 5]
	min samples leaf = [1, 2, 4]
RF	n estimators = [50, 100]
	max depth = [None, 10, 20]
	min samples split = [2, 5]
	min samples leaf = [1, 2]
LR	C = [0.1, 1, 10]
	solver = [liblinear]
XGBoost	learning_rate = [0.01]

(receiver operating characteristic curve) constitutes a statistical method to determine the diagnostic accuracy of these tests, being used for three specific purposes: to determine the cut-off point of a continuous scale at which the highest sensitivity and specificity is reached, to evaluate the discriminative capacity of the diagnostic test, that is, its ability to differentiate healthy versus diseased subjects, and to compare the discriminative capacity of two or more diagnostic tests that express their results as continuous scales [26].

3.5 Validation

The validation process allows us to identify and confirm the best performing model. The results were generally very similar, with the SVM model exhibiting the highest accuracy of 82.72% (Table 6).

Table 6. Model results

Model	Parameters	Accuracy	Recall	Precision	Specificity	F1-Score
KNN	metric: euclidean	79.01%	76.47%	74.29%	80.85%	75.36%
	n_neighbors: 3					
	weights: uniform					
SVM	C: 0.1	**82.72%**	79.41%	79.41%	**85.11%**	**79.41%**
	gamma: scale					
	kernel: rbf					
DT	max_depth: 10	75.31%	70.59%	70.59%	78.72%	70.59%
	min_samples_leaf: 4					
	min_samples_split: 2					
RF	max_depth: None	79.01%	79.41%	72.97%	78.72%	76.06%
	min_samples_leaf: 2					
	min_samples_split: 2					
	n_estimator: 100					
LR	C: 0.1	77.78%	**82.35%**	70.00%	74.47%	75.68%
	solver: liblinear					
XGB	learning rate: 0.01	81.48%	78.79%	**81.48%**	**85.11%**	77.61%

4 Results

Figure 2 shows the ROC (Receiver Operating Characteristic) curve of the models, plotted as a function of their AUC (Area Under the Curve) values. The Random Forests (RF) model proved to be the most accurate, with an AUC of 0.8601, followed closely by SVM with 0.8533 and XGBoost with 0.8579. These models stand out for their ability to correctly classify positive and negative breast cancer cases. The Logistic Regression (LR) and K-Nearest Neighbors (K-NNN) models also demonstrated good performance, with scores of 0.8388 and 0.8548, respectively. In contrast, the Decision Tree (DT) model obtained a more moderate AUC of 0.7981. All models evaluated in this study showed promising performance, with AUCs above 0.79, indicating a reasonable ability to predict breast cancer. However, the RF and XGBoost models stood out as the most accurate for this purpose. These results underscore the utility of machine learning models for early detection and accurate diagnosis of breast cancer, which could significantly improve patient care and treatment.

Figure 3 shows the confusion matrix of the SVM model:

- **TN:** Identified 40 cases where the model classified the absence of breast cancer.
- **FP:** Identified 7 cases where the model incorrectly classified the presence of breast cancer.
- **FN:** Identified 7 cases where the model did not classify the presence of breast cancer. But there were.

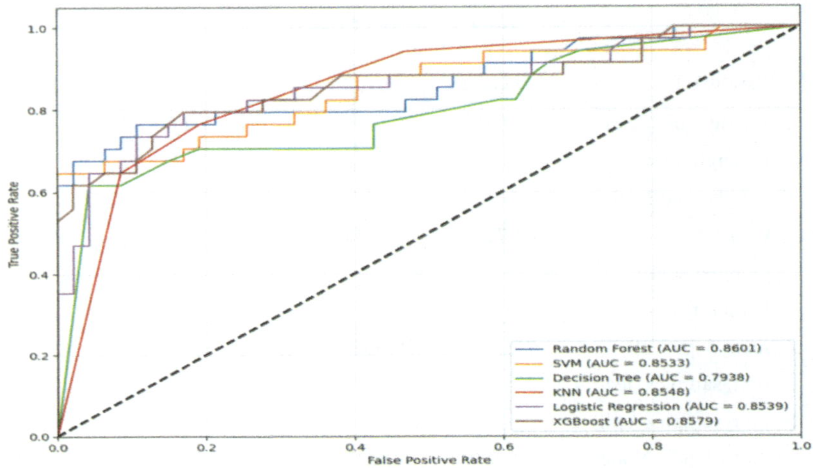

Fig. 2. ROC curve of the classification models.

- **TP:** Correctly identified 27 positive cases. These are the cases in which the model correctly classified breast cancer.

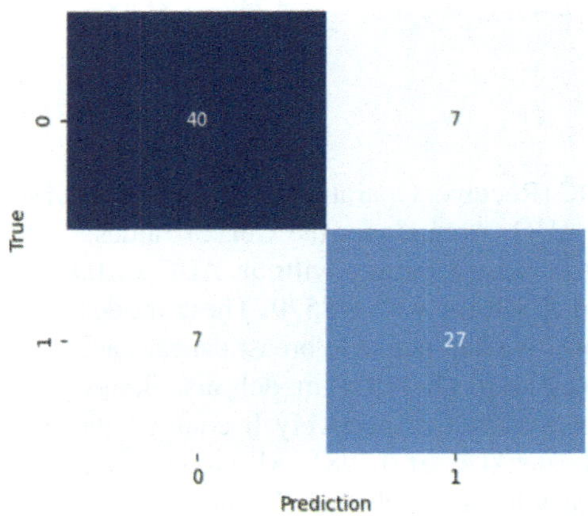

Fig. 3. SVM model confusion matrix

5 Conclusions

Breast cancer is one of the most prevalent oncological diseases worldwide, with a significant impact on public health and the quality of life of those affected. According to the World Health Organization [1], it is estimated that by 2024 there will be 310,720 new cases of breast cancer in women and 2,790 in men. In addition, the mortality rate will reach 42,250 women and 530 men who will lose their lives to this disease. However, despite these challenges, artificial intelligence and machine learning are emerging as promising tools in the fight against breast cancer. Advances in machine learning are revolutionizing disease diagnosis and treatment, enabling early and accurate detection of abnormalities in medical images such as mammograms and ultrasounds. In this study, machine learning models were used to improve early detection of breast cancer, with the goal of improving clinical outcomes, reducing mortality and offering hope to millions of people worldwide. The results obtained in breast cancer detection demonstrate the effectiveness of several machine learning algorithms, including SVM (Support Vector Machines), RF (Random Forests), DT (Decision Trees), KNN (K-Nearest Neighbors), RL (Logistic Regression) and XGBoost. The most outstanding results were obtained with the SVM model, which achieved an accuracy of 82.72%, a specificity of 85.11% and the following hyperparameters: C (1), gamma (scale) and kernel (linear). These findings underscore the potential of machine learning models to effectively identify breast cancer stages.

References

1. Organización Mundial de la Salud (OMS): Breast Cancer: Prevention and Control. (2023). Recuperado de: https://www.who.int/es/news-room/fact-sheets/detail/breast-cancer
2. Organización Panamericana de la Salud (OPS): Cáncer en las Américas: Datos y cifras. (2023). Recuperado de: https://www.paho.org/es/temas/cancer-mama
3. Gobierno de Perú. Ministerio de Salud: Plan Nacional para el Control del Cáncer 2020-2024. (2020). Recuperado de: https://www.gob.pe/institucion/minsa/noticias/662077-cancer-de-mama-si-se-detecta-a-tiempo-tiene-un-90-de-probabilidades-de-curacion
4. Peng, X.: Research on software defect prediction and analysis based on machine learning. J. Phys. Conf. Series. **2173**(1) (2022). https://doi.org/10.1088/1742-6596/2173/1/012043
5. Li, S.: A study on the crucial indicators for breast cancer detection using machine learning algorithm. J. Phys. Conf. Series. **2646**(1) (2023). https://doi.org/10.1088/1742-6596/2646/1/012042
6. Lou, J.: Prediction of breast cancer based on RF, SVM and PCA. J. Phys. Conf. Series. **2646**(1) (2023). https://doi.org/10.1088/1742-6596/2646/1/012036
7. Pérez Núñez, J.R.: Investigación sobre el método de predicción del cáncer de mama basado en aprendizaje automático. Revista de Investigación de Sistemas e Informática. **15**(2), 5–12 (2022). https://doi.org/10.15381/risi.v15i2.23402
8. Chen, H., Wang, N., Du, X., Mei, K., Zhou, Y., Cai, G.: Classification prediction of breast cancer based on machine learning. Comput. Intell. Neurosci. **2023**, 1–9 (2023). https://doi.org/10.1155/2023/6530719
9. Meza, A., Paucar, K., Morales, E., & Ticona, W.: Predicting anemia in patients through clinical and hematological data using machine learning algorithms (n.d.)
10. Torres, C., Torres, P., & Ticona, W.: Proposed model for the detection of diabetic retinopathy using convolutional neural networks (n.d.)
11. Ildefonso, A., & Ticona, W.: Proposal of a computational vision model for the pre-diagnosis of anemia based on the image of the ocular conjunctiva (n.d.)
12. Breast Cancer Dataset. (2023, 14 agosto). Kaggle. https://www.kaggle.com/datasets/faysalmiah1721758/breast-cancer-data
13. One hot encoding | Interactive Chaos. (s. f.). https://interactivechaos.com/es/manual/tutorial-de-machine-learning/one-hot-encoding

14. LabelEncoder | Interactive Chaos. (s. f.). https://interactivechaos.com/es/python/function/labelencoder
15. Laura-Ochoa, L.: Evaluation of classification algorithms using cross validation. In: Proceedings of the LACCEI International Multi-Conference for Engineering, Education and Technology (2019). https://doi.org/10.18687/LACCEI2019.1.1.471
16. Bello, A.M., Cuta, J.A., García, E.K.: Técnicas de imputación para datos de precipitación máxima mensual en la zona central de boyacá. Ingeniería Investigación y Desarrollo. **19**(1), 64–79 (2020). https://doi.org/10.19053/1900771X.v19.n1.2019.12209
17. Na, & Na.: Algoritmo K-Nearest Neighbor | Aprende Machine Learning (2020, 15 julio). https://www.aprendemachinelearning.com/clasificar-con-k-nearest-neighbor-ejemplo-en-python/
18. Spositto, O., Blanco, G., Matteo, L., Levi, M., & Bossero, J.: SMOTE, Algoritmo para balanceo de clases en un estudio aplicado a la ganadería (2020). www.cs.waikato.ac.nz/~ml/weka/
19. Zheng, A., & Casari, A.: Feature engineering for machine learning principles and techniques for data scientists (n.d.).
20. Narod, S. A. (2012). Tumour size predicts long-term survival among women with lymph node-positive breast cancer. Curr. Oncol. (19, 5, 249–253). Multimed Inc. https://doi.org/10.3747/co.19.1043
21. Kumar Ghosh, D.: Perspective chapter: linear regression and logistic regression models. In: Recent Advances in Biostatistics. IntechOpen (2024). https://doi.org/10.5772/intechopen.1003183
22. Hakkal, S., Lahcen, A.A.: XGBoost to enhance learner performance prediction. Comput. Educ. Artif. Intell. **7** (2024). https://doi.org/10.1016/j.caeai.2024.100254
23. Zapeta Hernández, A., Galindo Rosales, G.A., Juan Santiago, H.J., Martínez Lee, M.: Métricas de rendimiento para evaluar el aprendizaje automático en la clasificación de imágenes petroleras utilizando redes neuronales convolucionales. Ciencia Latina Revista Científica Multidisciplinar. **6**(5), 4624–4637 (2022). https://doi.org/10.37811/cl_rcm.v6i5.3420
24. Borja-Robalino, R., Monleon-Getino, A., Monleón-Getino, A., & Rodellar, J.: Estandarización de Métricas de Rendimiento para Clasificadores Machine y Deep Learning (2020). https://www.researchgate.net/publication/339943922
25. Arce, J. I. B.: La matriz de confusión y sus métricas. Juan Barrios (2023, 31 agosto). https://www.juanbarrios.com/la-matriz-de-confusion-y-sus-metricas/
26. Cerda, J., Cifuentes, L.: Uso de curvas ROC en investigación clínica: Aspectos teórico-prácticos. Revista Chilena de InfectologíA. **29**(2), 138–141 (2012). https://doi.org/10.4067/s0716-10182012000200003

RETRACTED CHAPTER: Computer Vision Model for the Pre-diagnosis of Autism from Facial Images

Franklin Arellano, Antony Huaman, Pedro Molina, and Wilfredo Ticona

The Editor has retracted and removed this publication. Concerns were raised regarding the use of images of minors from a non-curated dataset. The dataset is reported to contain images of children with ASD. However, the images appear to have been collected from the internet without any documented clinical history or confirmation of an actual ASD diagnosis. Additionally, there is no documented ethical oversight, or consent of the children included in the dataset or their parents and legal guardians.

In light of these ethical concerns and the reliance on unverifiable data, the Editor and Publisher no longer have confidence in the reliability of this article.

The authors have not responded to correspondence regarding this retraction.

R. Silhavy and P. Silhavy (Eds.): CoMeSySo 2024, LNNS 1489, pp. 505-516, 2025.
https://doi.org/10.1007/978-3-031-96798-6_38

RETRACTED CHAPTER

Exploring Novel Image Generation via Script-Directed Scene Formation

Kaveesh Khattar, Tejas Goyal(✉), K. P. Dhruv, Aditya Hombal, and H. R. Mamatha

Department of Computer Science and Engineering, PES University, Bengaluru, Karnataka, India
jaz.goyal@gmail.com, mamathahr@pes.edu

Abstract. Deep Fusion Generative Adversarial Networks (DF GANs) have shown to be effective tools for producing realistic pictures. In this research, we demonstrate our implementation that creates unique representations of objects and human beings using a pretrained DF GAN. With the help of labelled input photos and identification tokens, our model was able to create new representations of objects and show people doing interesting things. By using script-based prompts, our system was able to generate scenarios with many pictures that made sense. We also made accessibility easier with a Gradle-implemented user-friendly interface. We illustrate the efficiency and adaptability of our method for producing a wide range of visual material through experimental validation and qualitative evaluation.

Keywords: Image Models · Text-to-Image · Generative AI

1 Introduction

The creation of high-fidelity pictures from massive datasets has been revolutionised by Generative Adversarial Networks (GANs), especially Deep Fusion GANs. In this work, we explore new picture creation using a pretrained DF GAN. Our primary objectives are to: increase the model's capacity to generate new object representations and human actions; and secondlyallow the creation of scenes that are prompted by text. We challenge the model to produce human movements and unseen object descriptions by providing labelled input pictures and tokens, therefore expanding the limits of its generalisation abilities. We also include scripted building cues for cohesive scene development, with the goal of achieving dynamic visual narrative. To guarantee usability, we employ Gradle to create an interface that is easy to use and accessible to a wide range of user demographics. By advancing pretrained DF GANs, we contribute to the evolution of image synthesis techniques, fostering innovation in artificial intelligence and computer vision realms.

R. Silhavy and P. Silhavy (Eds.): CoMeSySo 2024, LNNS 1489, pp. 517–526, 2025.
https://doi.org/10.1007/978-3-031-96798-6_39

2 Dataset

MS-COCO (Microsoft Common Objects in Context): The MS-COCO benchmark dataset is highly advantageous for tasks in object identification, segmentation, and captioning within the field of computer vision. Since it has over 200,000 photos, it provides data that can help in building robust models. Thorough annotations of images in the dataset such as segmentation masks give the dataset its advantage. With a combination of large number of images and high quality annotation it gives the model what it needs to provide more accurate results.

Furthermore, MS-COCO is used by state of the art models in the field of computer vision, which further solidifies our stance in choosing it as our primary dataset. Also the dataset's accessibility makes research contributions to it more seamless and helps it stay relevant for further use cases.

Models trained on MS-COCO perform extraordinarily well. Improved performance can be achieved by efficiently fine-tuning on smaller, task-specific datasets by pre-training on this dataset.

Finally, the photos from MS-COCO's real-world relevance, which show commonplace scenarios, improve the applicability of models trained on this dataset. These include anything from robots and self-driving cars to general visual comprehension. For this reason, the MS-COCO dataset remains a fundamental tool that propels progress and creativity in computer vision research.

3 Architecture

3.1 Sourcing Images and Performing Transformations

We obtained a variety of datasets from publicly accessible sources that included photographs of both human individuals and objects. To ensure a broad range of visual representations, we carefully selected a collection of high-resolution photographs from different categories for items. To provide more context about humans to the model, we chose pictures that had people in multiple settings performing different activities.

We pre-processed the photos by converting them to a specified resolution the model could take and cropping them to size for compatibility. Then we give these images to the pre-trained Deep Fusion Generative Adversarial Network (DF-GAN). The process involved normalization techniques to produce pixel values in-line with all the images in the dataset and re-scaling the photos to 512×512 pixels (Fig. 1).

To give the model context and direction throughout the creation process, the photos were annotated with relevant labels and tokens. This step aims to enable the model to produce output that is more coherent and contextually relevant.

Finding and preparing the photographs was only one of the many tasks involved in getting the dataset ready for training and evaluation. These procedures prepared the way for the pre-trained DF-GAN to be applied with success in trials involving picture generation (Fig. 2).

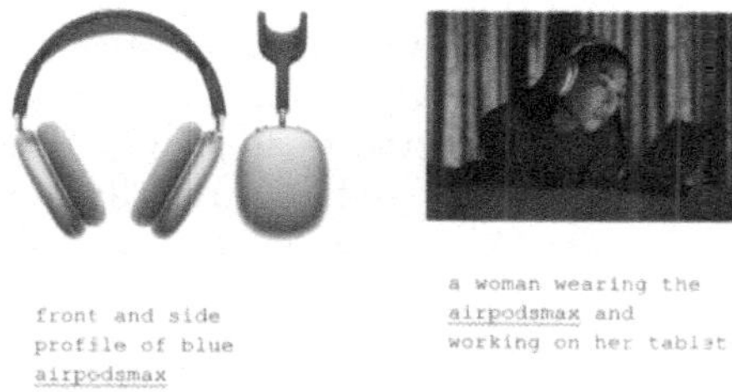

Fig. 1. Providing Captions for Images

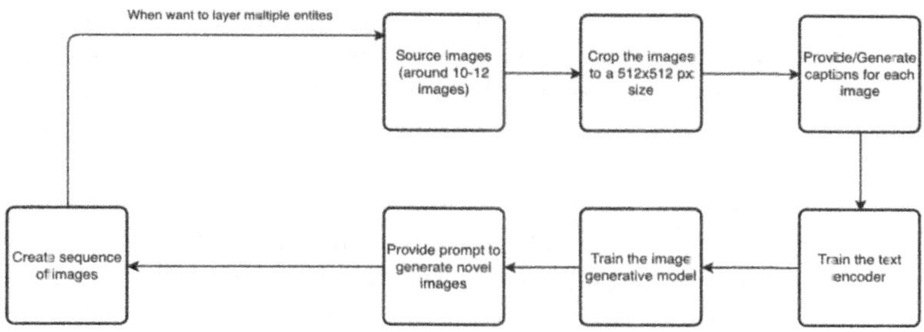

Fig. 2. Simplified flow of the model

3.2 Train Text Encoder

Interpreting and using the provided labels, and tokens is the most important phase in the training process for the text encoder. Textual inputs, such as labels indicating item categories, tokens for picture identification, and prompts guiding scene development, must be transformed into a format that can be combined with the image data by the text encoder. Through iterative training, the text encoder learns to precisely extract relevant characteristics and context from the textual input, which makes it easier for the model to understand and comprehend the intended instructions. By improving the text encoder in combination with the picture production model, we ensure that the generated outputs are consistent with the supplied textual direction. This results in representations that are both contextually relevant and visually appealing. The synergy between textual and visual modalities is increased and the model's ability to generate a wide range of outstanding pictures that match preset requirements is improved through this iterative training process.

3.3 Train Image Encoder

Training the image encoder is one of the many steps involved in developing dependable and effective picture production models.

Once the text encoder is trained on the captions and tokens assigned to an image, the image encoder is then trained on the corresponding image. Before this step, as mentioned before, the image is pre-processed and transformed to meet the model's compatibility needs.

The model's primary task is to extract salient visual features from the raw input photographs, identifying details and patterns that are required to remember the subject of the image. Then later on, the model's primary task is to produce coherent and visually content that is realistic and contextually appropriate.

Throughout the training phase, the image encoder is optimised to extract relevant visual information from the input photos through a series of runs. The parameters of the image encoder were tuned in tandem with those of other components of the DF-GAN framework, such as the text encoder so that it only learns the tokens associated with that image and binds that learning to an image. Continuous enhancement of the image encoder enables the model to better learn how to integrate visual cues with written instructions to generate precise representations of the input data.

The image encoder and the text encoder work hand in hand to achieve high consistency results between the textual and visual modalities. Through collaborative training, the model gains an understanding of the input data. This enables it to create realistic images that match the provided written descriptions. The iterative approach used in this paper strengthens the model's capacity to provide high-quality visual information and enhances contextual coherence in the generated pictures.

3.4 Prompt and Epoch Configurations for Optimum Outputs

To optimize for image generation, epoch parameters, and text prompt settings must be given throughout the training phase. To make impactful prompts we write brief textual descriptions that instruct the model to generate desired visual outcomes. These prompts, which act as contextual signals, help assist the model in focusing its attention on certain distinctive aspects of images or scenes. Correct prompt formulation and selection help ensure that the photos generated reflect the intended creative concept.

The extent and rigor of model training are also heavily controlled by the epoch configurations used. The training dataset is separated into epochs, or iterations, with each epoch allowing the model to improve performance and make parameter adjustments as needed. Training efficiency and model convergence are balanced by selecting appropriate period settings. By modifying these parameters, we would improve the model's performance and ensure that the pictures created contain the desired properties, such as realism, relevance, and contextual importance.

Prompt and epoch configurations must be adjusted by ongoing testing and validation. We iteratively tweak these parameters to assist the model's ability to generate high-quality photographs that match preset requirements. We can fully utilize the model's capabilities and adjust its behavior via an iterative method to satisfy specific picture-producing criteria and creative ambitions.

3.5 Scene Generation for Prompts with Multiple Sentences

During setting up scenes for multiple sentence prompts while training, it is vital to set up scene-creation configuration with multi-sentence prompts to allow for the model's capability to generate cohesive contextually rich, and correct narratives. When we give the model multi-sentence prompts, we need to provide a sequence of textual or verbal descriptions that explain complex and multi-faceted scenarios or stories. This instructs the model to generate multiple pictures that collectively depict the intended scene and sequence. These prompts serve as a sort of scaffolding for the story, guiding the model while it creates the scene and ensuring coherence and consistency in the images that are generated.

To learn how to employ multi-sentence prompts for scene generation most effectively, let's examine the following examples: In one scene, we see two men drinking orange juice in a cosy house, farmers gathering oranges in a grove before dawn, and orange boxes on an old truck parked on a road. These lines serve as the model's guide, creating a sequence of images that depict the process from harvest to consumption.

Another use case features a close-up shot of a burger smeared in ketchup. The setting of a dinner table with the ketchup bottle. As the narrative progresses, a man is seen eating this burger in a rather crowded restaurant. These few words effectively set the tone for a series of pictures that portray the joy and enthusiasm of a hearty meal which can be used in many marketing placeholders.

Every line that is produced using multi-sentence prompts gives a different facet to the larger plot, such as a place, a character, an event, or a dialogue. These sentences are deliberately placed differently to allow the scene to unfold gradually. Subsequently, the model would build upon these signals to generate coherent images that keep the storyline consistent. Utilizing this method, the model can create a range of dynamic situations by generating rich and well-structured visual tales from multi-sentence prompts.

4 Results

We show the experimental findings demonstrating the effectiveness of our work in making a diverse range of visual media using pre-trained Deep Fusion Generative Adversarial Networks (DF-GANs). We measured the model's capability to generate unique object representations by giving it labeled photographs as input and distinguishing tokens. The results successfully demonstrate the pre-trained DF GAN's capacity to extrapolate and generalize using existing data by generating realistic representations of items/objects that had not yet been seen in the dataset. We also noticed how effectively the model would use script-based prompts to build logically consistent scenarios. Short descriptions of diverse scenarios were given to the pre-trained DF-GAN, requesting it to generate many pictures that correctly matched the given context.

5 Performance Metrics

5.1 Inception Score (IS)

The inception score (IS) is a mathematical algorithm used to measure or determine the quality of images created by a generative adversarial network (GAN). The inception score offers objective and consistent measures of generated images; and by extension, the quality and capability of the underlying generative model.

5.2 Fréchet Inception Distance (FID)

This metric measures the distribution similarity of feature representations from a pre-trained Inceptionv3 model for real and produced images by combining Inception Distance and Frechet Distance. Higher quality image production is indicated by a smaller Frenchet Inception Distance.

5.3 Mean Opinion Score (MOS)

Utilize the Mean Opinion Score (MOS), which is derived from human participants rating images according to perceived fidelity or quality, to assess the quality of generated photos. The MOS is obtained by averaging these assessments; higher numbers denote more visually appealing or realistic images, while lower values denote inferior quality or fidelity.

6 Future Research Directions

Text to image generation models have come a long way, but there are still a number of areas that need more research and development. This section highlights potential future research directions based on the current state of models and identified areas for improvement (Figs. 3, 4, 5 and 6)

a woman with a hat and black ((airpodsmax)) around her neck

pair of green airpods max

Fig. 3. Novel representation

Fig. 4. Before providing images for training v/s after providing images for training

Fig. 5. Scene visualisation

Improved Semantic Understanding: Subsequent studies on text-to-image generation ought to improve semantic comprehension by utilizing sophisticated natural language processing methods such as knowledge graphs or pre-trained language models. By using knowledge graphs created from text embeddings, models may be able to produce visuals that are more closely matched with the intended meaning of the input text, potentially improving positional and contextual understanding.

Increased Resolution and Realism: Though current models have come a long way in producing high-quality photographs, resolution and photo-realism may still use some work. Future research could focus on developing techniques

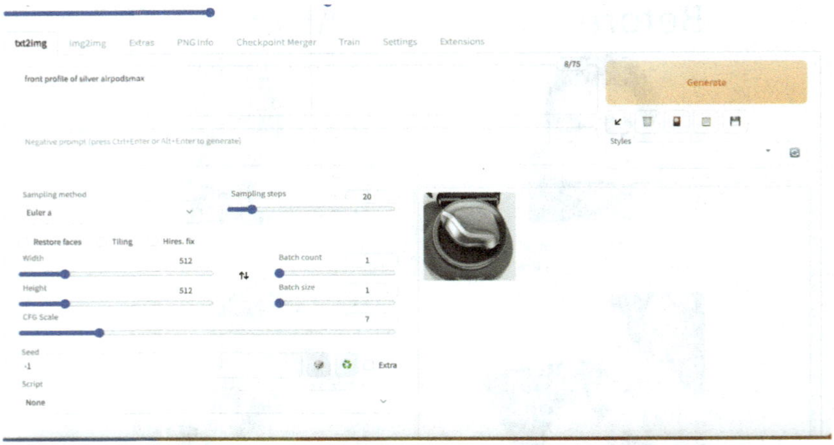

Fig. 6. Dashboard for Fine-Tuning

to generate images at higher resolutions, allowing for more detailed and visually appealing results. Additionally, exploring advanced loss functions or perceptual similarity metrics could further enhance the realism of generated images, making them indistinguishable from real photographs.

Fine-Grained Control and Manipulation: Current text-to-image models often lack fine-grained control over generated images. Future research could investigate methods to enable precise control and manipulation of image attributes, such as object positions, colors, and styles, based on textual input. This could involve exploring novel conditioning techniques or incorporating additional information during the generation process to produce images that align with specific user requirements.

Handling Ambiguity and Multi-modal Outputs: In order to achieve a variety of outputs that might capture many meanings, future research in text-to-image generation should investigate strategies for handling ambiguity in textual descriptions. One could use strategies like adversarial learning, variational techniques, and uncertainty estimation. To potentially reduce ambiguity, one way is to train the image generator using the prompt and the parsed scene graph output.

Incorporating User Feedback and Interactive Generation: Interactive text-to-image generation systems that incorporate user feedback and preferences hold great potential for enhancing user satisfaction and enabling personalized image generation. Subsequent investigations may concentrate on creating models that can adjust and enhance their creation procedure in response to user interactions, enabling users to offer input and direct the picture synthesis procedure in real-time.

7 Conclusion

Through this effort, we successfully show the ability of pre-trained Deep Fusion Generative Adversarial Networks (DF-GANs) that help expand the possibilities of image generation. We show significant work in the generation of unseen novel objects and human actions, as well as in the development of coherent text-prompted situations, by using these advanced models. With the help of comprehensive testing and qualitative research, our method has shown its value and versatility in creating novel and contextually relevant visual media.

Our findings portray the effectiveness of pre-trained DF GANs as efficient tools for creative picture synthesis. These models contain large potential for a diverse range of applications, including but not limited to content creation, building virtual worlds and narratives, and the creation of human behaviors and novel object representations. They also facilitate the creation of scenario-driven narratives.

Additionally, we consider accessibility and usability important, as seen by the Gradle user interface, demonstrating our commitment to helping a variety of academics, practitioners, and artists have access to cutting-edge AI technology.

In the future, deep learning and artificial intelligence research might lead to the creation of videos in addition to further advancements in picture production techniques. By building on the foundation created in this study, we want to further enhance pre-trained DF GANs and encourage additional computer vision research.

References

1. Thomee, B., et al.: YFCC100M: The New Data in Multimedia Research. https://dl.acm.org/doi/10.1145/2812802
2. Lin: Microsoft COCO: Common Objects in Context. https://arxiv.org/abs/1405.0312
3. Wah: The Caltech-UCSD Birds-200-2011 Dataset. http://www.vision.caltech.edu/visipedia/CUB-200-2011.html
4. Nilsback: Oxford 102 Dataset; Automated flower classification over a large number of classes. http://www.robots.ox.ac.uk/~vgg/data/flowers/102/
5. Schuldt: KTH Action Recognition; Recognizing human actions: a local SVM approach. http://www.nada.kth.se/cvap/actions/
6. Rodriguez: UCF Sports; Action mach a: a new representation for human action recognition. https://www.crcv.ucf.edu/data/ucf_sports_actions/
7. Ding, M.: CogView: Mastering Text-to-Image Generation via Transformers. arXiv preprint arXiv:2106.13700 (2021)
8. Biswas, B., Ghosh, S.K., Ghosh, A.: DVAE: Deep Variational Auto-Encoders for Denoising Retinal Fundus Image (2020)

9. Xu, T., et al.: AttnGAN: Fine-Grained Text to Image Generation With Attentional Generative Adversarial Networks (2018)
10. Tsue, T., Sen, S., Li, J.: Cycle text-to-image GAN with BERT. arXiv preprint arXiv:2003.12137 (2020)
11. Qiao, T., Zhang, J., Xu, D., Tao, D.: MirrorGAN: learning text-to-image generation by redescription. arXiv preprint arXiv:1903.05854 (2019)
12. Tao, M., Tang, H., Wu, F., Jing, X., Bao, B., Xu, C.: DF-GAN: a simple and effective baseline for text-to-image synthesis. arXiv preprint arXiv:2008.05865 (2022)
13. Srivastava, N., Mansimov, E., Salakhudinov, R.: Unsupervised learning of video representations using LSTMs. In: Proceedings of the 32nd International Conference on Machine Learning (2015)
14. Zhang, H., et al.: StackGAN: Text to Photo-Realistic Image Synthesis With Stacked Generative Adversarial Networks (2017)
15. Ramesh, A., Chu, C., Dhariwal, P., Nichol, A., Chen, M.: Hierarchical text-conditional image generation with CLIP latents. arXiv preprint arXiv:2204.06125 (2022)
16. Gafni, O., Polyak, A., Ashual, O., Sheynin, S., Parikh, D., Taigman, Y.: Make-A-Scene: Scene-Based Text-to-Image Generation with Human Priors (2022)

Approximate Estimation of the Volume of Educational Material Mastered by Students in the Process of E-Learning

Roman Tsarev[1,2](✉), Biswaranjan Senapati[3], Darío Salguero García[4], Ivana Roncevic[5], and Natalia Bystrova[6]

[1] MIREA - Russian Technological University (RTU MIREA), Moscow, Russia
tsarev.sfu@mail.ru
[2] Bauman Moscow State Technical University, Moscow, Russia
[3] University of Arkansas at Little Rock, Arkansas, IL, USA
[4] Almería University, Almería, Spain
[5] Prince Sultan University, Riyadh, Saudi Arabia
[6] Minin State Pedagogical University of Nizhny Novgorod, Nizhny Novgorod, Russia

Abstract. One of the most crucial factors in estimating students' academic performance is the volume of educational material studied by them. The article proposes a method of calculating the volume of educational material corresponding to the studied topics. Besides, the method of calculating the volume of actually studied by the student educational material is proposed. When estimating the volume of educational material, the textual and graphic information presented in different topics of the course is taken into account, as well as its semantic load, its complexity and labor-intensiveness in studying. The proposed approach is based on the expert knowledge of the teacher, so subjectivity takes place. However, the estimation of the expert teacher correlates with the volume and complexity of the educational material and can be sufficiently reliable information for estimating the current situation with the mastering of the educational material by students. The approach proposed for estimating the volume of the studied material allows tracking the progress of both individual students and groups for the purpose of effective management of the educational process. In particular, this approach allows early identification of students who fall behind in order to motivate and support them in their studies.

Keywords: E-Learning · Moodle · Teaching-Learning Material · Text-Graphic Information · Electronic Testing · Practical Assignment

1 Introduction

Estimation of learning outcomes is an essential aspect of the educational process, as it is inextricably related to its quality [1–8]. The problem of estimating the volume of the educational material studied by students remains one of the most urgent issues, because the teacher is facing multiple difficulties: the objectivity of the assessment, its "visibility" and the labor intensity of the assessment process [9–11].

R. Silhavy and P. Silhavy (Eds.): CoMeSySo 2024, LNNS 1489, pp. 527–536, 2025.
https://doi.org/10.1007/978-3-031-96798-6_40

It is necessary to take into account not only the results of assessment (students' marks: the majority of excellent and good marks in successful students-excellent students and satisfactory and unsatisfactory marks during the semester in weak, unsuccessful students), but also the process of learning the educational material by students, which could be presented visually with the help of tools that optimize and automate such assessment activities. This can be provided by e-learning environments that are developing worldwide [12].

This is caused by the need of the labor market for highly qualified personnel who are ready to solve problems of any complexity, including those involving a good scientific background [13–17]. The challenges of today require that a modern specialist has a systemic thinking, skills in handling information and communication technologies, has outstanding knowledge in his narrower professional area [18–21]. This applies to industry, scientific research, agricultural sector, issues related to the development of society and the state [22–27]. The demand to estimate the volume of studied material and related issues of assessment of practical assignments and tests makes even more relevant the use of e-learning tools, which significantly facilitate the activities of the teacher in terms of providing students with educational material and estimating their knowledge.

Thus, by means of LMS Moodle a teacher can create e-learning courses and track their attendance by students, monitor completed assignments, easily and quickly collect and analyze the results of electronic tests [28–33]. Focusing on electronic testing, we note that this form of assessment is characterized by a number of advantages in relation to the classical method of assessment using simple pieces of paper, when the teacher, in consequence, takes a lot of time to check and may not be objective [34, 35].

The electronic educational environment Moodle has a great functionality that facilitates the teacher's educational activities [36–40]. Thanks to the creation of electronic courses, its individual modules and topics with electronic tests, the teacher can get a clear picture of students' performance indicators, track the dynamics of their results [38, 41]. The advantages of electronic testing are automation, storing test tasks in the database, saving time and eliminating the subjective factor when assesing the test results [42–46].

Electronic test, developed on the basis of Moodle, stores test questions in the test bank, provides an opportunity for automated checking and assessing tasks, which saves the teacher's time, has the least psychological burden on students and most reflects the real knowledge of students [35, 43]. Electronic test in Moodle allows to create test questions of different types (open and closed, with one correct answer or several, true/false, matching and so on) for different purposes in accordance with the topics: introductory testing, self-assessment, intermediate or final testing.

Electronic testing does not require additional software and can be conducted both in face-to-face classes in the classroom and remotely, it can be configured in Moodle based on the organization of the course.

The test has various settings, such as time of test passing, number of attempts, which questions from the test bank to use, availability of feedback, etc. One of the main advantages is the saving of information about test results, which makes it possible to analyze test results in a multilateral way. The estimation of the volume of the studied educational material by students is of particular interest for us within the framework of our study, as it allows us to understand more deeply the level of students' knowledge quality and visualize their indicators.

The volume of the studied educational material affects the degree of its mastering and memorization. The greater the volume of the educational material to study, the more time is required to master and memorize it.

For a more accurate determination of the volume of the educational material being studied and its influence on memorization, it is necessary that the semantic units of the material allocated by the teacher coincide with those allocated by the student.

In addition, it is important to take into account the way in which the educational material is presented: theoretical information, graphs, charts and tables, examples, exercises, tasks, etc.

These forms serve to compress information to varying degrees. Any data compression aims to reduce the amount of information by transforming it. For example, text has the most voluminous representation, while graphs and tables, on the other hand, are more concentrated information. Thus, it is believed that visualized information has many advantages: visibility, attractiveness, compression, etc. There is also an opinion that even arrangement of schemes or graphs on the page influences the quality of perception, assimilation and reproduction in the future.

The number of units in the text, the type of information and its content influence the learning of a certain amount of educational material. So, on the one hand, certainly, it is much faster to study one page than fifteen, and, on the other hand, the same page in terms of its content can be very sophisticated, which will require deep understanding, analysis and processing. As an example, a theorem with its proof, where the student needs to be attentive and concentrated at every stage of reading in order to further reproduce it in a test or exam. A fifteen-page educational material may, on the contrary, not require mental exertion and not be overloaded with mathematical concepts or formulas. At present, there is no definitive evidence as to what presents the greatest challenge to students: volume or content of the educational material.

2 Methods

There are two factors to consider when estimating the volume of educational material. The first one is related to the volume of text, the number of graphs and tables, as well as formulas. The estimation of the first factor can be done mechanically. So we can count the number of symbols, formulas, graphs, tables.

The second factor is related to the semantic load of the studied material. In particular, two texts that contain the same number of symbols may require different efforts to perceive and realize their semantic content. In terms of the depth and complexity of the educational material, its comprehension and assimilation, there are also difficulties in estimation, since this factor can only be evaluated by experts. Thus, we can talk about an approximate estimation of the volume of educational material provided to students in e-learning to study. Despite the fact that this estimation will be quite subjective, nevertheless, it gives an opportunity to estimate the volume of educational material studied by an individual student or a whole group. There is an opportunity to monitor and manage the educational process. The teacher can identify students who are lagging behind at an early stage and pay special attention to them in order to increase their motivation for learning, provide them with more accessible presentation of educational material and to get higher learning results in the future.

It is also worth noting that despite the fact that the estimation of the semantic load is carried out by an expert teacher, as a rule, the educational material is well known to the teacher, he/she is able to evaluate the complexity and labor intensity of individual topics or modules of educational information. In addition, the evaluation is carried out by one expert teacher. Thus, his/her estimation significantly correlates with the volume and semantic load of the educational material.

The volume of educational material of one topic can be calculated as follows:

$$V_{em} = w_1 V \cdot w_2 S,$$

where V is the volume of textual and graphic information (for example, the number of lines, formulas, figures, tables, etc.); S is the semantic load of the educational material of a given topic; w_1 and w_2 are weighting coefficients that allow changing the importance of one or the other multiplier depending on the teacher's preferences.

Obviously, the topics that make up the whole volume of the educational material will be differentiated due to the difference in their actual volume and complexity.

Figure 1 shows the entire course divided into the topics to be studied.

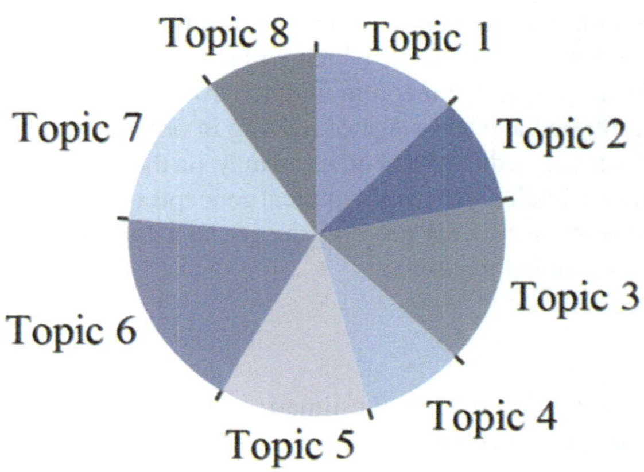

Fig. 1. Volume of course topics.

As we can see in Fig. 1, each topic corresponds to a separate sector. The size of the sectors is different. This is due to the fact that one topic is easier, the other is more complicated, and besides, the textual and graphic material of each topic is different. The volume and labor intensity of each topic depends on the content. So, for example, it can be a brief information about some theorem, but if we consider the proof of this theorem, and in addition, its application, it, of course, increases the complexity and, accordingly, the volume of the topic [47].

The sum of the volume of educational material of all topics makes up the full volume of the course (100%) and corresponds to the maximum possible volume that can be mastered by the student during its study.

3 Results

The course was divided into eight topics within the framework of the conducted research. The ratio of the volume of topics corresponds to the one presented in Fig. 1. In the process of studying each individual topic, the student performs a practical assignment, after which a test is available to him/her in the Moodle e-learning system. Thus, the evaluation activity was carried out according to two parameters - evaluation for the practical assignment and evaluation for the passed test.

It should be noted that the quality of an electronic test is of great importance for assessment activities. The better the bank of test questions is prepared, the more adequate the testing will be performed and the more objective the evaluation will be [33, 43]. When preparing test questions, we paid special attention to the fact that the test should not be contradictory and cover all the educational material. In addition, the tasks were diverse, of different types, the questions were not ambiguous.

Each topic included a practical assignment. In fact, practical assignments can be practical exercises, seminars, laboratory works. This is the second, no less important part, which allows to estimate the studied volume of educational material. The better it covers the topic, the more objective the estimation will be. However, it is not always possible to fully cover the theoretical material of the topic when performing a practical assignment. Therefore, coefficients are used to calculate the volume of educational material actually studied by the student, and it is possible to allocate more weight to the test than to the practical assignment. This will reflect the greater importance of the test grade. Otherwise, it is possible to assign more weight to practical assignment if it is more important. In this study, the weights were equal.

The volume of educational material actually studied by the student on the current topic was calculated using the following formula:

$$v_i = V_{em} \cdot (c_1 a_i + c_2 t_i)/(c_1 A + c_2 T),$$

where a_i is the grade for the practical assignment of the i-th student; t_i is the grade for testing of the i-th student; A is the maximum possible grade for the practical assignment on the given topic; T is the maximum number of points that can be obtained for the test on the given topic; c_1 and c_2 are weighting coefficients expressing the importance of the grade for the practical assignment and testing, respectively.

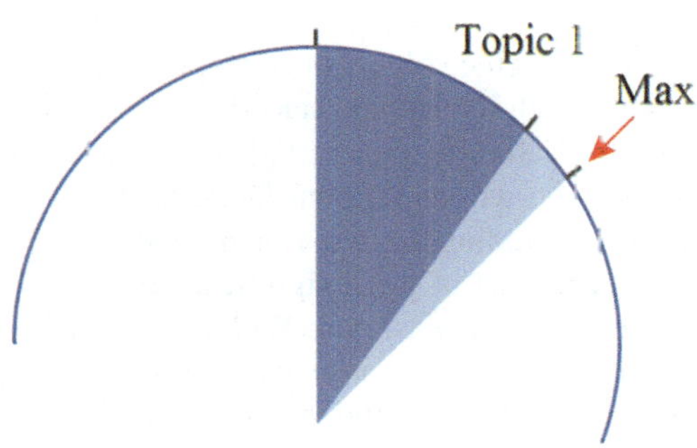

Fig. 2. Volume of educational material studied by the student on the first topic.

As a result of the calculations, we can determine that a particular student has studied the educational material on the first topic by 85% (Fig. 2).

Then, during the semester, at a certain point in time, when, for example, four topics of the course have already been studied, we can sum up the volume of material studied by the student and see what proportion of the maximum possible material the student has studied (Fig. 3).

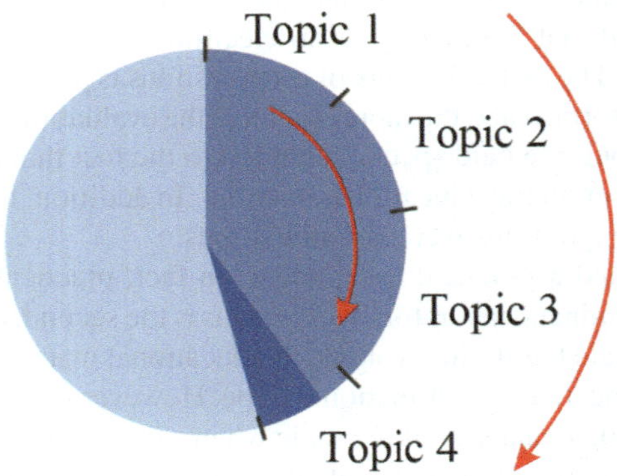

Fig. 3. The volume of educational material studied by the student on several topics.

At the time of the research, the volume of material mastered by students was monitored by means of a developed computer program. The initial data for the program were the data on grades downloaded from the Moodle system. At the moment the preparation for the realization of this functionality in the form of a module directly in the Moodle system is underway.

The wide functionality of Moodle gives the opportunity to view statistics on academic performance and the volume of studied educational material both for an individual student and for a group. Moreover, it is possible to compare the results of different groups or different students from the same group.

4 Conclusion

Thus, the presented estimation of the volume of educational material studied by students is universal and can be used both in classical (face-to-face, in classrooms) learning and in the process of e-learning. Naturally, e-learning environments have become a part of educational system all over the world. Without them, teachers can no longer imagine effective learning activities, including the assessment of students' results. These results can depend on various factors, but of great significance is the volume of educational material that has been studied and mastered. Within our paper, we have considered various conditions that influence the extent of material learned. In estimating the volume of educational material studied by students, their grades for testing and practical assignments were taken into account.

LMS Moodle opens us a huge amount of opportunities and facilities for estimating the volume of educational material studied by students. First of all, teachers create electronic courses, develop modules for them (the teacher himself can evaluate the complexity of each topic) and electronic tests, which, obviously, should be of high quality, valid and reliable.

LMS Moodle makes it possible to visualize the data obtained, to explain the relationship of information in a concrete and understandable form. In addition, the learning situation is tracked step-by-step, in the current moment at learning checkpoints or upon completion of the entire course. The electronic environment provides the possibility of downloading student assessment data, for example, for further visualization by means of Excel and self-developed computer programs.

The visual form of presentation of the results of the estimation of the volume of the studied educational material allows to demonstrate an understandable picture of the results of academic performance of every student. For teachers, it is an opportunity to evaluate the effectiveness of educational programs and teaching methods used and, if necessary, to make changes in the course, improve it in order to improve the performance of students and increase the quality of learning in general. In addition, thanks to the fact that we can track the situation for each student, it is possible to carry out more effective individual and differentiated teaching of students.

References

1. Dol, S.M., Jawandhiya, P.M.: ECSYAPPS – a framework for analyzing the effectiveness of classification techniques for early prediction of students academic performance in education sector. Eng. Appl. Artif. Intell. **134**, 108688 (2024). https://doi.org/10.1016/j.engappai.2024.108688
2. Hernández-García, A., Cuenca-Enrique, C., Del-Río-Carazo, L., Iglesias-Pradas, S.: Exploring the relationship between LMS interactions and academic performance: a learning cycle approach. Comput. Hum. Behav. **155**, 108183 (2024). https://doi.org/10.1016/j.chb.2024.108183
3. Mamatova, G.D., Kuchkarov, T.S.: Relevance of the concept of "Digital university": literary review of domestic and foreign research. Inform. Econ. Manag. **3**(1), 101–158 (2024). https://doi.org/10.47813/2782-5280-2024-3-1-0101-0158
4. Martínez-Serna, M.I., Baixauli-Soler, J.S., Belda-Ruiz, M., Yagüe, J.: The effect of online class attendance on academic performance in finance education. Int. J. Manag. Educ. **22**(3), 101023 (2024). https://doi.org/10.1016/j.ijme.2024.101023
5. Mousse, M.A., Almufti, S., García, D.S., Jebbor, I., Aljarbouh, A., Tsarev, R.: Application of fuzzy logic for evaluating student learning outcomes in e-learning. Lect. Notes Netw. Syst. **935**, 175–183 (2024). https://doi.org/10.1007/978-3-031-54820-8_15
6. Savchenkov, A.V., Uvarina, N.V., Shchagina, G.V.: Implementation in the educational process of the university of modern interactive educational technologies aimed at the development of flexible skills of future teachers. Vestnik of Minin Univ. **11**(2), 1455 (2023). https://doi.org/10.26795/2307-1281-2023-11-2-5
7. Vimarsha, K., Prakash, S.P.S., Krinkin, K., Shichkina, Y.A.: Student performance prediction: a co-evolutionary hybrid intelligence model. Proc. Comput. Sci. **235**, 436–446 (2024). https://doi.org/10.1016/j.procs.2024.04.043

8. Wang, Y., Xue, L.: Using AI-driven chatbots to foster Chinese EFL students' academic engagement: an intervention study. Comput. Hum. Behav. **159**, 108353 (2024). https://doi.org/10.1016/j.chb.2024.108353
9. Doz, D.: Factors influencing teachers' grading standards in mathematics. Oxf. Rev. Educ. **49**(6), 819–837 (2023). https://doi.org/10.1080/03054985.2023.2185217
10. Kovalev, I.V., Losev, V.V., Kalinin, A.O.: Formalized approach to the design of microprocessor systems with elements of human-machine interaction. Mod. Innov. Syst. Technol. **3**(2), 243–253 (2023). https://doi.org/10.47813/2782-2818-2023-3-2-0243-0253
11. Lu, O.H.T., Huang, A.Y.Q., Yang, S.J.H.: Impact of teachers' grading policy on the identification of at-risk students in learning analytics. Comput. Educ. **163**, 104109 (2021). https://doi.org/10.1016/j.compedu.2020.104109
12. Voroshilova, A.A., Kovalev, I.V., Pivovarov, G.O.: Innovative approaches to distance learning for volunteers in the field of museum activities and cultural initiatives: opportunities and prospects. Inform. Econ. Manag. **2**(4), 301–316 (2023). https://doi.org/10.47813/2782-5280-2023-2-4-0301-0316
13. Baryshnikova, K.V., Kadochkin, A.S., Shalin, A.S.: Nanostructural antireflecting coatings: classification analysis (a review). Opt. Spectrosc. **119**, 343–355 (2015). https://doi.org/10.1134/S0030400X15090040
14. Canós Valero, A., Borovkov, D., Kalganov, A., Dudnikova, A., Sidorenko, M., Dergachev, P., Gurvitz, E., Gao, L., Bobrovs, V., Miroshnichenko, A., Shalin, A.S.: On the existence of pure, broadband toroidal sources in electrodynamics. Laser Photonics Rev. **18**(4), 2200740 (2024). https://doi.org/10.1002/lpor.202200740
15. Kuznetsov, A.V., Canós Valero, A., Shamkhi, H.K., Terekhov, P., Ni, X., Bobrovs, V., Rybin, M.V., Shalin, A.S.: Special scattering regimes for conical all-dielectric nanoparticles. Sci. Rep. **12**(1), 21904. https://doi.org/10.1038/s41598-022-25542-2
16. Malozyomov, B.V., Tynchenko, V.S., Kukartsev, V.A., Bashmur, K.A., Panfilova, T.A.: Investigation of properties of laminar antiferromagnetic nanostructures. CIS Iron Steel Rev. **27**, 84–90 (2024). https://doi.org/10.17580/cisisr.2024.01.13
17. Novitsky, D.V., Valero, A.C., Krotov, A., Salgals, T., Shalin, A.S., Novitsky, A.V.: CPA-lasing associated with the Quasibound states in the continuum in asymmetric non-Hermitian structures. ACS Photon. **9**(9), 3035–3042 (2022). https://doi.org/10.1021/acsphotonics.2c00790
18. Golubev, I.M., Tsarev, R.J., Semenko, T.I.: N-version software systems design. In: Proceedings of the International Scientific and Practical Conference of Students, Post-graduates and Young Scientists - Modern Technique and Technologies, pp. 147–149. IEEE, Tomsk (2005). https://doi.org/10.1109/spcmtt.2005.4493232
19. Kucherik, A., Kutrovskaya, S., Osipov, A., Gerke, M., Chestnov, I., Arakelian, S., Shalin, A.S., Evlyukhin, A.B., Kavokin, A.V.: Nano-antennas based on silicon-gold nanostructures. Sci. Rep. **9**(1), 338. https://doi.org/10.1038/s41598-018-36851-w
20. Kukartsev, V., Kravtsov, K., Stefanenko, O., Podanyov, N., Bezvorotnykh, A.: Using machine learning techniques to simulate network intrusion detection. In: Proceedings of the International Conference on Intelligent Systems for Cybersecurity (ISCS), pp. 1–4. IEEE, Gurugram (2024). https://doi.org/10.1109/iscs61804.2024.10581097
21. Novitsky, D.V., Shalin, A.S., Novitsky, A.: Nonlocal homogenization of PT-symmetric multilayered structures. Phys. Rev. A. **99**(4), 043812 (2019). https://doi.org/10.1103/physreva.99.043812
22. Ang, A.S., Sukhov, S.V., Dogariu, A., Shalin, A.S.: Scattering forces within a left-handed photonic crystal. Sci. Rep. **7**(1), 41014 (2017). https://doi.org/10.1038/srep41014
23. Baryshnikova, K.V., Novitsky, A., Evlyukhin, A.B., Shalin, A.S.: Magnetic field concentration with coaxial silicon nanocylinders in the optical spectral range. J. Opt. Soc. Am. B. **34**(7), D36–D41 (2017). https://doi.org/10.1364/josab.34.000d36

24. Kovrov, A., Novitsky, A., Karabchevsky, A., Shalin, A.S.: A photonic Nanojet as tunable and polarization-sensitive optical tweezers. Ann. Phys. **530**(9), 1800129 (2018). https://doi.org/10.1002/andp.201800129
25. Kozlov, V., Filonov, D., Shalin, A.S., Steinberg, B.Z., Ginzburg, P.: Asymmetric backscattering from the hybrid magneto-electric meta particle. Appl. Phys. Lett. **109**(20), 203503 (2016). https://doi.org/10.1063/1.4967238
26. Rozhkova, A., Kukartsev, V., Kvesko, M., Suprun, E., Andreev, V.: Improving the efficiency of water use in agriculture by modelling the classification of groundwater quality. BIO Web Conf. **116**, 03020 (2024). https://doi.org/10.1051/bioconf/202411603020
27. Tynchenko, V., Kravtsov, K., Podanyov, N., Fedorova, A., Bezvorotnykh, A.: Automation for the sustainable development of agriculture. BIO Web Conf. **113**, 05010 (2024). https://doi.org/10.1051/bioconf/202411305010
28. Chen, C.-J., Tsai, H.-J., Lee, M.-Y., Chen, Y.-C., Huang, S.-M.: Effects of a Moodle-based E-learning environment on E-collaborative learning, perceived satisfaction, and study achievement among nursing students: a cross-sectional study. Nurse Educ. Today. **130**, 105921 (2023). https://doi.org/10.1016/j.nedt.2023.105921
29. Dobashi, K., Ho, C.P., Fulford, C.P., Lin, M.-F.G., Higa, C.: Learning pattern classification using moodle logs and the visualization of browsing processes by time-series cross-section. Comput. Educ. Artif. Intell. **3**, 100105 (2022). https://doi.org/10.1016/j.caeai.2022.100105
30. Gao, T.: Research on the design of online gamified tourism education activities based on Moodle platform. Entertain. Comput. **2024**, 100823 (2024). https://doi.org/10.1016/j.entcom.2024.100823
31. Marticorena-Sánchez, R., López-Nozal, C., Serrano-Mamolar, A., Olivares-Gil, A.: UBU-Monitor: desktop application for visual e-learning student clustering with Moodle. SoftwareX. **26**, 101727 (2024). https://doi.org/10.1016/j.softx.2024.101727
32. Merzlyakova, E.Y., Skotnikov, D.S., Eltyshev, Y.V.: Extension of TRIZ with a digital data field through an intelligent technical agent of the human-machine interface. Modern Innov. Syst. Technol. **4**(2), 242–256 (2024). https://doi.org/10.47813/2782-2818-2024-4-2-0242-0256
33. Yamaguchi, S., Kondo, H., Ohnishi, Y., Nishino, K.: Design of Question-and-Answer Interface using Moodle DATABASE Function. Proc. Comput. Sci. **207**, 976–986 (2022). https://doi.org/10.1016/j.procs.2022.09.153
34. Mustakerov, I., Borissova, D.: A conceptual approach for development of educational Web-based e-testing system. Expert Syst. Appl. **38**(11), 14060–14064 (2011). https://doi.org/10.1016/j.eswa.2011.04.214
35. Tsarev, R., Bhuva, A., Bhuva, D., Gogoleva, I., Nikolaeva, I., Bystrova, N., Varyan, I., Shamina, S.: Improving test quality in e-learning systems. Lect. Notes Netw. Syst. **723**, 62–68 (2023). https://doi.org/10.1007/978-3-031-35317-8_6
36. Arce, E., Suárez-García, A., López-Vázquez, J.A., Fernández-Ibáñez, M.I.: Design Sprint: enhancing STEAM and engineering education through agile prototyping and testing ideas. Think. Skills Creat. **44**, 101039 (2022). https://doi.org/10.1016/j.tsc.2022.101039
37. Bensalah, F., Daniel, M.P., Patra, I., Salguero García, D., Irgasheva, S., Tsarev, R.: Implementation of individual learning trajectories in LMS Moodle. Lect. Notes Netw. Syst. **935**, 159–174 (2024). https://doi.org/10.1007/978-3-031-54820-8_14
38. Bengueddach, A., Boudia, C., Bouamrane, K.: Interpretive analysis of online teaching labs constructed using moodle during the pandemic period. Heliyon. **9**(5), e16007 (2023). https://doi.org/10.1016/j.heliyon.2023.e16007
39. Hamamoto, N., Yokoyama, S., Takefusa, A., Aida, K.: Implemention of secured log analysis environment for Moodle using virtual cloud provider service. Proc. Comput. Sci. **192**, 3154–3164 (2021). https://doi.org/10.1016/j.procs.2021.09.088

40. Yorulmaz, M., Can, G.F.: Task and user-based entropy-rank sum-TPOP integration proposal for usability evaluation of web applications. J. King Saud Univ. Comput. Inform. Sci. **34**(8), 6466–6480 (2022). https://doi.org/10.1016/j.jksuci.2022.01.011
41. Bojović, Z., Bojović, P.D., Vujošević, D., Šuh, J.: Education in times of crisis: rapid transition to distance learning. Comput. Appl. Eng. Educ. **28**, 1467–1489 (2020). https://doi.org/10.1002/cae.22318
42. Castaño, M., Noeller, C., Sharma, R.: Implementing remotely proctored testing in nursing education. Teach. Learn. Nurs. **16**(2), 156–161 (2021). https://doi.org/10.1016/j.teln.2020.10.008
43. Deetjen-Ruiz, R., Esponda-Pérez, J.A., Haris, I., García, D.S., Osorio, J.L.Q., Tsarev, R.: Evaluating the reliability of tests used in LMS Moodle for e-learning. Lect. Notes Netw. Syst. **935**, 1–8 (2023). https://doi.org/10.1007/978-3-031-54820-8_1
44. Shilova, T.V., Artamonova, L.V., Averina, S.Y.: Computer-based tests as an integral component of an EFL course in Moodle for non-linguistic students. Procedia. Soc. Behav. Sci. **154**, 434–436 (2014). https://doi.org/10.1016/j.sbspro.2014.10.187
45. Smoline, D.V.: Some problems of computer-aided testing and "interview-like tests". Comput. Educ. **51**(2), 743–756 (2008). https://doi.org/10.1016/j.compedu.2007.07.008
46. Urbanovich, M.V., Kovaleva, K.A.: Principles of user interface design. Mod. Innov. Syst. Technol. **3**(4), 363–374 (2023). https://doi.org/10.47813/2782-2818-2023-3-4-0363-0374
47. Tsarev, R., Kuzmich, R., Anisimova, T., Senapati, B., Ikonnikov, O., Shestakov, V., Pupkov, A., Kapustina, S.: Automatic generation of an algebraic expression for a Boolean function in the basis $\{\wedge, \vee, \neg\}$. Lect. Notes Netw. Syst. **910**, 128–136 (2024). https://doi.org/10.1007/978-3-031-53552-9_12

Comparative Analysis of Convolutional Neural Networks and Long Short-Term Memory for Predicting Speeding Violations on Roads

Renato Arias[1(✉)], Ricardo Arias[2], and Kelly Ochoa[3]

[1] Universidad Autónoma del Perú, Lima, Peru
renato.arias@me.com
[2] Universidad Tecnológica del Perú, Lima, Peru
[3] Universidad Pompeu Fabra, Barcelona, Spain
https://www.autonoma.pe/

Abstract. Traffic accidents result in around 1.35 million deaths annually, and it is projected that they will rank as the fifth leading cause of mortality by the year 2030. This study compared two neural network architectures CNN and LSTM, to predict speeding violations on highways using 2022 speeding ticket data from speed cameras, obtained from the Peruvian government's open data platform. The data were preprocessed and adapted for the models, which were then trained and evaluated with regularization techniques and hyperparameter tuning to optimize performance. The results revealed that the CNN model achieved a loss of 2.96 and an error rate of 0.57%, whereas the LSTM model had a significantly higher loss of 195.10 and an error rate of 8.97%. These findings suggest that the CNN architecture is more effective for predicting speeding violations on highways. Future research should consider incorporating additional factors and exploring hybrid architectures that combine CNN and LSTM to enhance prediction accuracy.

Keywords: Speeding Prediction · Neural Networks · Traffic Safety · Machine Learning

1 Introduction

The law has traditionally served as a policy instrument to achieve social goals like safety, penalize risky behaviors, and regulate technological advancements [1]. However, with increasing economic growth and human activities, governments need more effective strategies to tackle pressing urban planning issues. In this context, recent advancements in AI and ML have positioned neural networks as the optimal tools for accident prediction and traffic management. Additionally, deep learning models have been utilized across various transportation-related scientific fields, particularly in promoting safe driving [2].

R. Silhavy and P. Silhavy (Eds.): CoMeSySo 2024, LNNS 1489, pp. 537–549, 2025.
https://doi.org/10.1007/978-3-031-96798-6_41

This study focuses on Peru, a developing country with a GDP per capita lower than neighboring countries such as Brazil, Chile, and Argentina. Nevertheless, the purchase of land transport vehicles has increased in recent years, despite road safety remaining a constant issue, with more than 2000 people killed in road accidents in 2023. Given that speeding is the main cause of traffic accidents, policymakers must urgently address this growing problem and propose feasible solutions supported by AI and ML.

It is noteworthy that conventional traffic monitoring systems in Peru are insufficient due to three main reasons: (1) poor urban planning; (2) problems with continuous power supply; and (3) irregular traffic patterns. Therefore, since 2018, the Peruvian government has promoted the implementation of speed cameras on various sections of its main roads with the highest concentration of vehicle accidents. These devices are designed to measure in real-time the speed of a moving vehicle through infrared light pulses that bounce off the unit and return to the speed camera, capturing the vehicle's image, speed, date, and location to generate a ticket if the speed limit is exceeded.

The results of the sections monitored by speed cameras have been published on the Peruvian government's open data platform. For this research, we used the database corresponding to 2022, considering that during this period, the Peruvian government lifted all measures implemented due to the Covid-19 pandemic. It is worth noting that in 2020, 176 traffic accidents were recorded per 100,000 inhabitants, the lowest figure in the last 30 years, which increased by 28% in 2021, repeating the same situation in 2022 when the state of emergency restrictions were officially lifted due to the pandemic's evolution and vaccination progress.

In this context, the aim of this research was to utilize data from the aforementioned speed cameras during the post-pandemic period, specifically in 2022, to apply deep learning models for predicting speeding violations, thus creating a tool to prevent road traffic accidents. To this end, two neural network architectures, CNN and LSTM, were compared, as both have demonstrated high effectiveness in time series applications and pattern recognition. The initial hypothesis posited that while both architectures are effective in predicting traffic accidents due to speeding, CNN might offer superior performance owing to its capability to identify pertinent features and its lower computational complexity compared to LSTM.

Regarding the structure of this research, Sect. 2 describes the methodology employed, detailing the packages, variables, and Python code used. Section 3 presents the results obtained from the predictive models. Section 4 discusses trends in the implementation of AI and ML in preventing traffic accidents due to speeding, and finally, Sect. 5 presents our conclusions and recommendations.

2 Methodology

2.1 Database Details

The present research utilized an open data database from the Peruvian government on speeding violations detected by speed cameras during 2022, which was preprocessed using the following packages (Table 1):

Table 1. Packages used

Package	Version	Purpose
Pandas	2.2.2	Data manipulation and analysis
Scikit-learn	1.4.2	Data preprocessing and splitting
Tensorflow	2.17.0	Implementation of neural networks
Matplotlib	3.8.4	Visualization of results

2.2 Predictive Modeling

The development of the predictive model involved several stages, beginning with data preprocessing and concluding with the implementation and assessment of two neural network architectures: CNN and LSTM.

The database used was obtained from the Peruvian government's open data platform, which includes detailed records of speeding violations detected from January to December 2022. The variables analyzed were department/region, enforced speed limit, detected speed, road, and vehicle type. As the first step in this methodology, categorical data were preprocessed into numerical data using the 'pd.get_dummies' function [3], which allowed the data to be interpreted and processed by machine learning models. Subsequently, data normalization was performed using 'StandardScaler' from the Scikit-learn package [4], which is essential for ensuring that all features are on a comparable scale, enhancing the model's prediction efficiency and accuracy. Subsequently, the data were divided into 80% for training and 20% for evaluation, enabling the model to learn from the majority of the data and be evaluated on a separate subset that was not used during training.

The CNN model architecture consisted of multiple Conv1D layers each containing 128 filters and a kernel size of 3 [5], activated by the 'relu' function and with L2 regularization, a first BatchNormalization layer to stabilize and accelerate the learning process, and a Dropout layer set at 0.5 to minimize the risk of overfitting. This was followed by a second Conv1D layer with 64 filters, repeating the process with another BatchNormalization and Dropout layer, flattening the output of these layers with a Flatten layer for processing in dense layers, featuring a Dense layer with 64 units, using 'relu' activation and L2 regularization, followed by another round of BatchNormalization and Dropout, concluding with an output Dense layer with one unit and linear activation to perform continuous prediction. Finally, the CNN model was trained using the Adam optimizer with a learning rate of 0.00005 and a maximum of 100 epochs and Early Stopping was employed to halt training if validation loss failed to improve, with a batch size of 64 [6].

On the other hand, the second model implemented was an LSTM network designed to optimize performance while minimizing overfitting. The architecture started with an LSTM layer containing 256 units, activated by 'relu', using L2 regularization, and configured to return sequences (return_sequences=true), followed by an initial Dropout layer with a 0.3 rate to prevent overfitting. Subsequently, a second LSTM layer was added with 128 units, activated by 'relu', incorporating L2 regularization, without sequence return (return_sequences=False), and another Dropout layer at a 0.3 rate to further reduce overfitting. The output from these layers was then passed through a Dense layer with 64 units, 'relu' activation, and L2 regularization, followed by a final Dropout layer before the output layer. The final layer was a Dense output layer with a single unit and linear activation for predictions. The model was trained using the Adam optimizer with a learning rate of 0.00005, for up to 100 epochs, with Early Stopping, and a batch size of 64 [7]. To evaluate the models' performance, a test dataset was used, and key metrics such as loss and MAE were calculated to assess each model's prediction accuracy.

```
# Importing libraries
import pandas as pd
from sklearn.model_selection import train_test_split
from sklearn.preprocessing import StandardScaler
from sklearn.metrics import mean_absolute_error
import tensorflow as tf
from tensorflow.keras.models import Sequential
from tensorflow.keras.layers import Dense, Conv1D, Flatten
    , Dropout, BatchNormalization, LSTM
from tensorflow.keras.optimizers import Adam
from tensorflow.keras.callbacks import EarlyStopping
from tensorflow.keras.regularizers import l2
import matplotlib.pyplot as plt

# Load the data
csv_path = '~/Downloads/
    CINEMOMETROS_2022_ENERO_A_DICIEMBRE_2022.csv'
dataset = pd.read_csv(csv_path)

# Preprocess the dates
dataset['TICKET DATE'] = pd.to_datetime(dataset['TICKET
    DATE'], format='%Y-%m-%d')

# Select features and labels
features = dataset[['DEPARTMENT', 'DETECTED SPEED', 'ROAD'
    , 'VEHICLE TYPE']]
```

```
labels = dataset['DETECTED SPEED']

# Convert categorical data to numeric
features = pd.get_dummies(features)

# Normalize the data
scaler = StandardScaler()
scaled_features = scaler.fit_transform(features)

# Split the data into training and testing sets
X_train, X_test, y_train, y_test = train_test_split(
    scaled_features, labels, test_size=0.2, random_state
    =42)

# Reshape the data to be compatible with Conv1D and LSTM
X_train_cnn = X_train.reshape((X_train.shape[0], X_train.
    shape[1], 1))
X_test_cnn = X_test.reshape((X_test.shape[0], X_test.shape
    [1], 1))
X_train_lstm = X_train.reshape((X_train.shape[0], 1,
    X_train.shape[1]))
X_test_lstm = X_test.reshape((X_test.shape[0], 1, X_test.
    shape[1]))

# Early stopping callback
early_stopping = EarlyStopping(monitor='val_loss',
    patience=10, restore_best_weights=True)

# Define and train the CNN model
cnn_model = Sequential([
    Conv1D(filters=128, kernel_size=3, activation='relu',
        kernel_regularizer=l2(0.01), input_shape=(
        X_train_cnn.shape[1], 1)),
    BatchNormalization(),
    Dropout(0.3),
    Conv1D(filters=64, kernel_size=3, activation='relu',
        kernel_regularizer=l2(0.01)),
    BatchNormalization(),
    Dropout(0.3),
    Flatten(),
    Dense(64, activation='relu', kernel_regularizer=l2
        (0.01)),
    BatchNormalization(),
    Dropout(0.3),
    Dense(1, activation='linear')
])
cnn_model.compile(optimizer=Adam(learning_rate=0.00005),
    loss='mean_squared_error')

```

```
cnn_history = cnn_model.fit(X_train_cnn, y_train, epochs=100, batch_size=64, validation_data=(X_test_cnn, y_test), callbacks=[early_stopping])

# Evaluate the CNN model
best_cnn_loss = cnn_model.evaluate(X_test_cnn, y_test)
print(f'Best CNN Model Loss: {best_cnn_loss}')

# Define and train the LSTM model
lstm_model = Sequential([
    LSTM(256, activation='relu', kernel_regularizer=l2(0.01), input_shape=(X_train_lstm.shape[1], X_train_lstm.shape[2]), return_sequences=True),
    Dropout(0.3),
    LSTM(128, activation='relu', kernel_regularizer=l2(0.01), return_sequences=False),
    Dropout(0.3),
    Dense(64, activation='relu', kernel_regularizer=l2(0.01)),
    Dropout(0.3),
    Dense(1, activation='linear')
])
lstm_model.compile(optimizer=Adam(learning_rate=0.00005), loss='mean_squared_error')

lstm_history = lstm_model.fit(X_train_lstm, y_train, epochs=100, batch_size=64, validation_data=(X_test_lstm, y_test), callbacks=[early_stopping])

# Evaluate the LSTM model
best_lstm_loss = lstm_model.evaluate(X_test_lstm, y_test)
print(f'Best LSTM Model Loss: {best_lstm_loss}')

# Error comparison
cnn_mae = mean_absolute_error(y_test, cnn_model.predict(X_test_cnn))
cnn_error_percentage = (cnn_mae / y_test.mean()) * 100
```

3 Results

3.1 Descriptive Analysis of the Database

The database used included the following relevant features for the study:

- Department: Indicates the region where the violation was recorded.
- Detected Speed: Speed recorded by the speed camera.
- Road: The specific road where the violation was recorded.
- Vehicle Type: Classification of the vehicle as car or pickup truck, buses, trucks, and minibuses.

3.2 Data Preprocessing

As mentioned earlier, categorical data was converted to numerical data using the 'pd.get_dummies' function from the Pandas package to convert categorical columns into dummy variables. This process allows machine learning models to handle categorical data by converting them into a series of binary variables. Afterwards, the 'StandardScaler' method was utilized to standardize the data, ensuring that all features were on a comparable scale, which enhanced the accuracy of the predictions.

3.3 Initial Summary

From the review of speeding violation data on Peruvian highways, the following data was obtained:

- Total number of detections: 103,092
- Regions: 10
- Roads: 9
- Speed limit: Range from 30 to 100 km/h
- Detected speed: Range from 36 to 225 km/h

3.4 Descriptive Analysis

Peru is composed of 24 regions, also known as departments, with two additional provinces having special regimes. In Fig. 1, the speed camera data used comprises the 10 main regions with the highest incidence of speeding violations. However, it can be seen that 74,347 speeding tickets were issued solely in the Lima region.

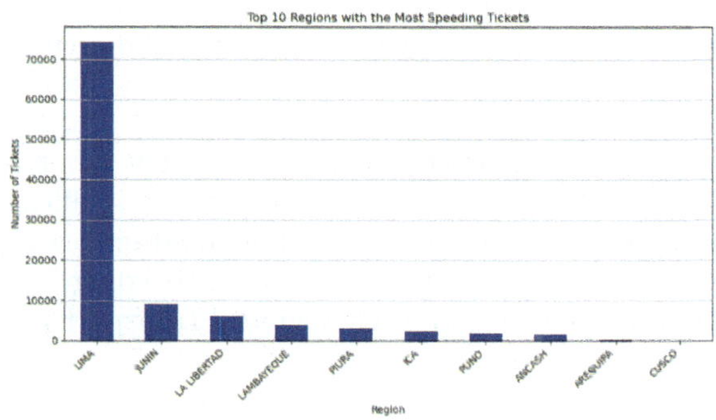

Fig. 1. Regions with the most speeding tickets

Similarly, Fig. 2 shows how speeding violations vary due to seasonal factors. It can be observed that during the summer and holiday season in Peru (December to March), there is a substantial increase in the number of tickets issued.

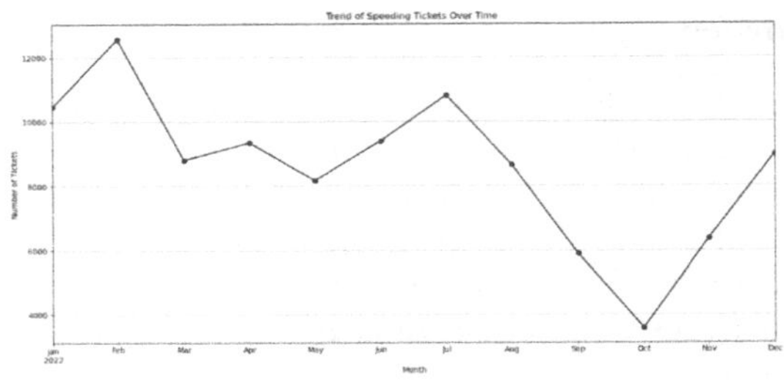

Fig. 2. Trend of speeding tickets over time

In contrast, during the winter (June to September) and spring (September to December) seasons, despite the peak during the national holidays in July, there is a reduction in the issuance of tickets, which could be due to both the natural decrease in traffic and adverse weather conditions.

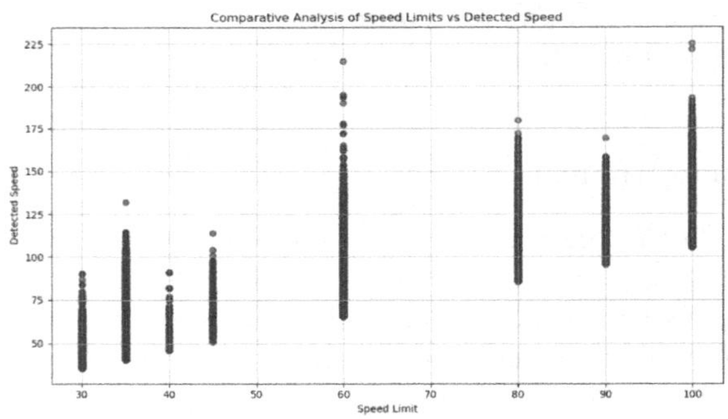

Fig. 3. Comparative analysis of speed limits vs. detected speed

On the other hand, as previously mentioned, there is a noticeable issue with road safety in Peru due to speeding on highways, even significantly as shown in Fig. 3. While the highest number of tickets is focused on private vehicles (cars and pickup trucks), there is also a percentage of speeding violations in cargo vehicles, with minimal incidence in interprovincial transport buses, as shown in Fig. 4.

3.5 Predictive Model

The goal of this study was to create and assess predictive models to detect speeding violations on highways. For this purpose, two neural network architectures, CNN and LSTM, were implemented.

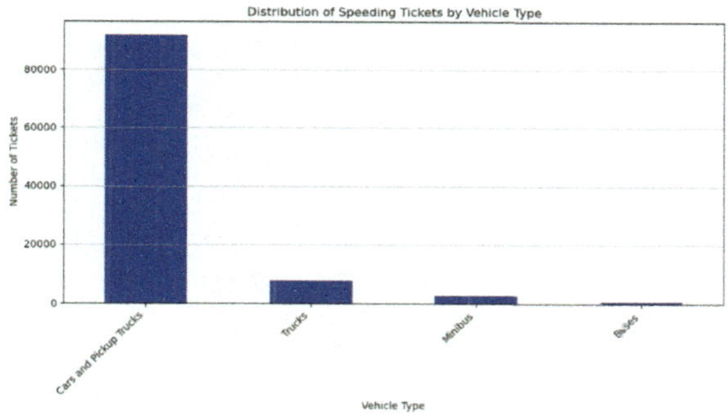

Fig. 4. Distribution of speeding tickets by vehicle type

Convolutional Neural Network. CNNs are primarily employed for analyzing images and videos but can also be used for time series data, such as speeding infractions. Consequently, a CNN comprises several convolutional layers designed to extract important features and patterns from the input data [8]. In this research, the CNN model was trained using normalized data, with specific configurations for both convolutional and dense layers. The evaluation metric utilized was the MSE, defined as follows:

$$MSE = \frac{1}{n}\sum_{i=1}^{n}(y_i - \hat{y}_i)^2 \quad (1)$$

where y_i represents the actual detected speed, $\hat{y}_i$ is the speed predicted by the model, and n denotes the number of samples.

The results showed a loss of 2.96 and an error rate of 0.57%, indicating high prediction accuracy, making the CNN model a viable option for detecting speeding violations. The error rate is calculated as follows:

$$\text{Error\%} = \left(\frac{1}{n}\sum_{i=1}^{n}\left|\frac{y_i - \hat{y}_i}{y_i}\right|\right) \times 100 \quad (2)$$

As depicted in Fig. 5, the training and validation loss of the CNN model steadily decreases over the epochs, indicating ongoing learning and improving accuracy with each iteration.

Long Short-Term Memory. LSTM is a well-known form of RNN capable of retaining significant information over extended time frames, often surpassing traditional statistical methods [9]. However, in this research, the LSTM model showed a higher loss of 195.10 and an error rate of 8.97%. Although LSTM performs well for other sequential data types, its performance in this scenario was inferior to that of the CNN model.

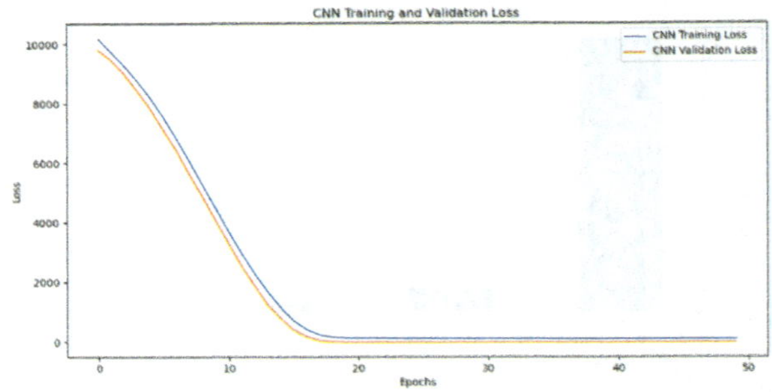

Fig. 5. CNN training and validation loss

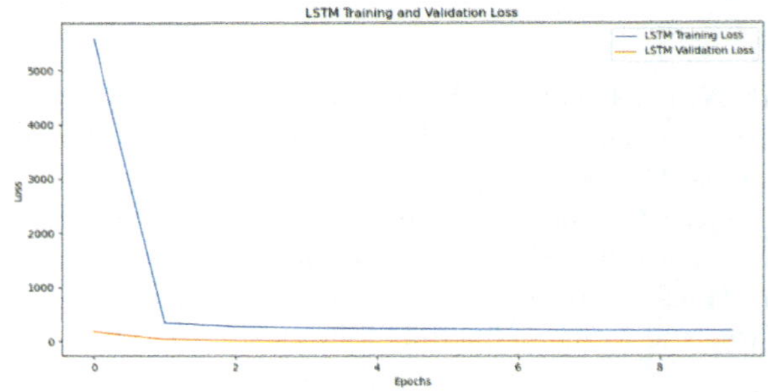

Fig. 6. LSTM training and validation loss

Figure 6 depicts the training and validation loss for the LSTM model. Despite the stabilization of the losses, a notable gap remains between the training and validation loss.

Therefore, to compare the performance of both models, the error rate of each was evaluated. As shown in Fig. 7, the CNN model has a significantly lower error rate of 0.57% compared to the LSTM model's 8.97%. This difference highlights the superiority of the CNN model for predicting speeding violations.

3.6 Bias Analysis

The bias analysis conducted in this study helped ensure that all obtained results were representative and reliable. Therefore, the mitigation of selection, performance, attrition, and reporting biases was prioritized, improving the applicability of the developed predictive models, as follows:

- Selection bias: Speed cameras were installed on roads with the highest incidence of speeding violations. Therefore, we ensured to include all regions and roads according to the 2022 database provided by the Peruvian government, ensuring that the sample was representative of the total population to evaluate speeding violations in Peru.

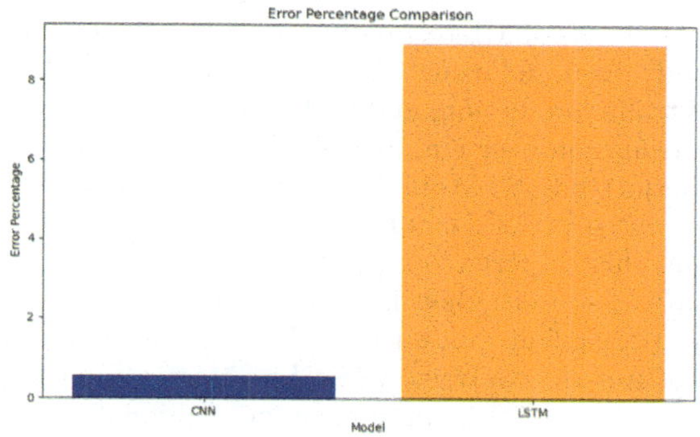

Fig. 7. Error percentage comparison

- Performance bias mitigation: The procedures for detecting violations were uniform across all regions, and the speed cameras had verification certificates issued by the National Institute of Quality (INACAL), ensuring standardized procedures for issuing speeding tickets.
- Attrition bias mitigation: The database was downloaded from the Peruvian government's open data platform. Therefore, all provided data were collected, ensuring that information from all regions and periods covered in the study was captured.
- Reporting bias mitigation: All obtained data were published, including those that did not support the initial hypothesis of this research, ensuring transparency and integrity of the results, providing all necessary information for a precise and comprehensive view of the predictive models' performance.

4 Discussion

The transportation industry today incurs significant costs due to the high number of accidents and injuries, which result in the loss of lives globally. The primary causes of these incidents include traffic violations such as speeding and reckless driving. Despite this issue being prevalent in many countries, the current detection systems used by patrol officers are inefficient [10]. Thus, accurately predicting the severity of traffic accidents can play a crucial role in enhancing road safety and preventing such incidents.

Historically, two types of models have been employed to analyze the causes and severity of traffic accidents. The first category includes statistical models that depend on specific probability distributions and data assumptions [8]. These models can produce inaccurate results if their assumptions are violated [11]. The second category consists of machine learning models, which frequently outperform statistical models as they do not rely on such assumptions. In this context, various models, including Random Forest [12] and Logistic Regression, have been applied to assess traffic accidents.

Given the effectiveness of ML techniques in addressing traffic accident-related problems, this study proposed using CNNs, which are among the most promising models for urban traffic prediction, and LSTM a popular RNN capable of retaining important information over long periods [8]. The results showed that CNNs significantly outperformed LSTMs in predicting speeding violations on highways, achieving an error rate of 0.57% compared to the LSTM model's 8.97%. Several factors can explain this disparity.

First, CNNs are created to identify specific patterns in the input data which can help identify specific features, such as the relationships between categorical and numerical features. In contrast, while LSTMs are more effective than traditional statistical methods, they may have faced difficulties due to the lack of temporal sequences in the provided data. Nonetheless, the use of regularization techniques like Dropout and L2 regularization enhanced the performance of both models, with the CNN model yielding superior results. It is important to highlight that normalizing the data and correctly converting categorical variables to numerical ones ensured that the models could learn effectively without being impacted by unequal feature scales.

The primary limitation of this study was the temporal scope of the provided database, which only covered nine main highways out of the 160 highways that constitute Peru's national road network. Therefore, to enhance the robustness and generalization of the models in future research, it would be advantageous to extend the temporal scope and geographical area analyzed, and to incorporate additional data such as weather conditions. This would allow for the exploration of hybrid models that combine CNN and LSTM architectures to capitalize on the strengths of both. Additionally, evaluating the integration of Raspberry Pi (RPi) modules and image processing algorithms [10] could improve the effectiveness of the speed cameras used in this study.

5 Conclusion

This research carried out a comparative analysis of CNN and LSTM models for predicting speeding violations on major highways in Peru. The findings underscored the superior effectiveness of CNNs compared to LSTMs, with CNNs achieving an error rate of 0.57% versus the significantly higher error rate of 8.97% for LSTMs. These results suggest that CNNs are better suited for this type of prediction, confirming the initial hypothesis. However, we consider that LSTMs would be more suitable for other contexts, such as predicting traffic accidents, where each accident represents a sequence of features. Nevertheless, the implementation of regularization methods like Dropout and L2 significantly improved the performance of both models, although CNNs demonstrated a greater ability to benefit from these techniques, further validating their suitability for this type of predictive model.

Finally, it is crucial to emphasize the study's limitations, particularly concerning the temporal scope of the database and the geographical coverage included. To enhance the robustness and generalizability of the models, it is recommended to integrate additional factors such as weather conditions and explore the use of hybrid architectures that combine CNN and LSTM to enhance prediction accuracy, as well as investigate the incorporation of RPi modules and image processing algorithms in speed cameras, which could improve detection systems and reduce traffic violations and accidents.

Acknowledgments. Universidad Autónoma del Perú.

References

1. Leiman, T.: Law and tech collide: foreseeability, reasonableness and advanced driver assistance systems. Policy Soc. **40**(2), 250–271 (2021). https://doi.org/10.1080/14494035.2020.1787696. ISSN 18393373
2. Cho, S., Shrestha, B., Salah, B., Ullah, I., Salem, N.M.: A proposed waiting time algorithm for a prediction and prevention system of traffic accidents using smart sensors. Electronics (Switzerland) **11**(11) (2022). https://doi.org/10.3390/electronics11111765. ISSN 20799292
3. Nandanwar, H., Katarya, R.: Deep learning enabled intrusion detection system for Industrial IOT environment. Expert Syst. Appl. **249** (2024). https://doi.org/10.1016/j.eswa.2024.123808. ISSN 09574174
4. Leng, L., et al.: Machine learning predicting wastewater properties of the aqueous phase derived from hydrothermal treatment of biomass. Bioresour. Technol. **358** (2022). https://doi.org/10.1016/j.biortech.2022.127348. ISSN 18732976
5. Abbasi, M., et al.: Multi-temporal change detection of asbestos roofing: a hybrid object-based deep learning framework with post-classification structure. Remote Sens. Appl. Soc. Environ. **34** (2024). https://doi.org/10.1016/j.rsase.2024.101167. ISSN 23529385
6. Farrukh, Y.A., Wali, S., Khan, I., Bastian, N.D.: SeNet-I: an approach for detecting network intrusions through serialized network traffic images. Eng. Appl. Artif. Intell. **126** (2023). https://doi.org/10.1016/j.engappai.2023.107169. ISSN 09521976
7. Tukymbekov, D., Saymbetov, A., Nurgaliyev, M., Kuttybay, N., Dosymbetova, G., Svanbayev, Y.: Intelligent autonomous street lighting system based on weather forecast using LSTM. Energy **231** (2021). https://doi.org/10.1016/j.energy.2021.120902. ISSN 03605442
8. Alhaek, F., Liang, W., Rajeh, T.M., Javed, M.H., Li, T.: Learning spatial patterns and temporal dependencies for traffic accident severity prediction: a deep learning approach. Knowl.-Based Syst. **286** (2024). https://doi.org/10.1016/j.knosys.2024.111406. ISSN 09507051
9. Alam, W., et al.: Analysis and prediction of risky driving behaviors using fuzzy analytical hierarchy process and machine learning techniques. Sustainability (Switzerland) **16**(11) (2024). https://doi.org/10.3390/su16114642. ISSN 20711050
10. Khan, S.U., Alam, N., Jan, S.U., Koo, I.S.: IoT-enabled vehicle speed monitoring system. Electronics (Switzerland) **11**(4) (2022) https://doi.org/10.3390/electronics11040614. ISSN 20799292
11. Abrari Vajari, M., Aghabayk, K., Sadeghian, M., Shiwakoti, N.: A multinomial logit model of motorcycle crash severity at Australian intersections. J. Safety Res. **73**, 17–24 (2020). https://doi.org/10.1016/j.jsr.2020.02.008. ISSN 00224375
12. Abellán, J., López, G., De Oña, J.: Analysis of traffic accident severity using Decision Rules via Decision Trees. Expert Syst. Appl. **40**(15), 6047–6054 (2013). https://doi.org/10.1016/j.eswa.2013.05.027. ISSN 09574174

Robust Model for Melanoma Classification Using Deep Learning and Machine Learning Techniques

José Cárdenas[1], Mariano Calderón[1], Junior Fabian[1], and Wilfredo Ticona[1,2](✉)

[1] Universidad ESAN, Lima, Peru
{21100885,21100602}@ue.edu.pe, {jfabian,wmamani}@esan.edu.pe
[2] Universidad Tecnológica del Perú, Lima, Peru

Abstract. A major problem facing society today is skin cancer, particularly melanoma. This disease causes millions of deaths annually. Therefore, the aim of this research was to classify melanoma, also known as "malignant mole", to distinguish it and eventually detect it in time. The methodology employed con-tained six (06) phases: dataset acquisition, preprocessing (normalization, resizing), feature vector extraction (SIFT, HOG), Machine Learning models (RF, K-NN, SVM), Deep Learning models (EffientNetB7, NASNet, Vision Transformer (ViT), VGG-19), and evaluation (Precision, Accuracy, Recall and F1-Score). The results indicated that the ViT model performed best with an accuracy of 85.34%, followed by VGG-19 (83.78%) and NASNet (82.19%). In addition, the ViT model demonstrated high sensitivity with an AUC of 0.95. In conclusion, the implementation of artificial intelligence techniques can significantly improve the classification of melanoma, providing an objective and accurate tool compared to traditional visual inspection methods.

Keywords: Artificial Intelligence · Melanoma · Deep Learning · Machine Learning · Hyperparameters · Vision Transformer · NASNet

1 Introduction

Melanoma is a type of skin cancer that originates in melanocytes, the cells that produce melanin, the pigment that gives skin its color. It is one of the most dangerous forms of skin cancer because it can spread rapidly to other parts of the body if not detected and treated early. Early detection is crucial, as treatment in the early stages can be very effective and significantly improve survival rates [1].

The classification of melanoma using Deep Learning and Machine Learning techniques has become a key research topic in the health and medical sectors due to the increase in skin cancer caused by global warming. Early identification of individuals at risk of developing a malignant mole can significantly impact the prevention and treatment of this disease. Numerous studies have explored the relationship between medical variables and the likelihood of a mole being malignant or benign, addressing the issue from a correlation perspective.

R. Silhavy and P. Silhavy (Eds.): CoMeSySo 2024, LNNS 1489, pp. 550–566, 2025.
https://doi.org/10.1007/978-3-031-96798-6_42

According to the American Cancer Society [2], skin cancer, particularly melanoma, is one of the most prevalent and dangerous malignancies worldwide. This type of cancer is notable not only for its high incidence but also for its lethal potential, especially when not detected early. The American Cancer Society also states that early detection is crucial because it significantly increases survival rates. Traditionally, melanoma evaluation relied on visual inspection by dermatologists and, when necessary, biopsy. However, the subjectivity of these methods and variability in specialist experience can compromise diagnostic efficacy. In this context, artificial intelligence (AI) has emerged as a revolutionary tool for skin cancer diagnosis, offering the ability to analyze dermatological images with unprecedented accuracy. In the United States, approximately 97,610 patients were diagnosed: 57,120 in men and 39,490 women. The lifetime risk of developing melanoma was approximately 2.6% (1 in 38) in whites, 0.1% (1 in 1,000) in blacks, and 0.6% (1 in 167) in Hispanics and Latinos.

Recent data from EsSalud [3] in Peru indicated 1,877 new cases of melanoma and other skin tumors detected in 2022, with approximately 1,300 melanoma cases diagnosed annually. This highlights the importance of more effective and accessible detection strategies.

Cabello et al. [4] reported that cutaneous melanoma represented 4% of all new cancer diagnoses in EU27 countries in 2020 and 1.3% of all cancer-related deaths. According to the European Cancer Information System [5], the countries with the highest incidence rates of melanoma in the European Union for both men and women are Denmark, the Netherlands, Sweden, Finland, and Germany.

According to Beyond Suncare [6], skin cancer is the leading cause of death among people with albinism in Africa, dramatically shortening the lives of those affected: 90% die before age 30 due to skin cancer, and 40% of children have precancerous lesions.

Additionally, Diario Libre [7] reported that in Latin America, the countries with the highest skin cancer rates were Argentina, Brazil, and Mexico. In Mexico, melanoma statistics are lower, between 15% and 20%, but it is the deadliest, claiming the lives of 40% of those affected. A study by the British Medical Jour-nal Oncology [8] reported an 80% increase in cancer incidence among people aged 50 years worldwide over the last three decades (1990–2019).

This demographic shift in cancer suggests the need to continuously adapt and improve diagnostic tools, including the integration of AI technologies. The methodological details of how these AI algorithms were implemented and compared will be discussed throughout the article, along with a detailed evaluation of their results. Our goal is to contribute to the advancement of diagnostic practices in dermatology, enhancing classification possibilities for timely melanoma detection, and providing a foundation for future research in this vital field.

2 Related Work

In the work of Dahdouh Y., Anouar A., and Ahmed M. [9], the authors present-ed a model based on the application of convolutional neural networks (CNN) for the detection and classification of melanoma. This study used a dataset compiled by the International Skin Imaging Collaboration (ISIC), which contains dermatoscopic images of benign and malignant cancerous moles. The model was trained with a deep CNN and compared with other CNN models, such as a 9-layer CNN, MobileNetV2-LSTM, DenseNet-EfficientNet, modified GoogleNet, and CNN + SVM, to classify images and detect malignant cancer. The model was evaluated using performance metrics, such as accuracy, precision, recall/sensitivity, and the F1 score. The results demonstrate that the proposed deep CNN model obtained the highest accuracy (92%), whereas MobileNetV2-LSTM achieved 90.72%. In contrast, the model with the lowest accuracy was the 9-layer CNN (80.52%). The results demonstrate that the proposed model is more efficient for detecting skin cancer samples.

The study conducted by Llaguno J. et al. [10] focuses on evaluating a super-vised learning K-NN algorithm using images. It uses the DTW metric to measure the similarity distance between time series to improve accuracy in data classification. For this research, two scenarios were considered: one with three time series (healthy, benign breast pathology, breast cancer) and the other with two classes (healthy, breast cancer). These time series were used in two ways: 1) with raw values, and 2) with Gaussian smoothing in the spatial domain. The study utilized 149 images from the area of Biology and Integral Health of the Institute of Bio-logical Research of the Universidad Veracruzana: 50 images of patients with breast cancer, 50 of patients with benign breast pathology, and 49 of healthy patients. The time series were obtained through banding patterns of images. Additionally, for the proposed experiments, the value of K was 1, 3, 5, 7, 10. The results show that the best classification accuracy was 66.88% in the time series with raw data (RAW), with K = 3 in the K-NN classifier and the similarity distance without a restriction window, and for the experiment with two classes, it was 88.77% with the same delimitations.

The research conducted by Islam, T. et al. [11] classified breast cancer using an explainable AI machine learning approach. The model evaluates and compares the classification accuracy, precision, recall, and F1 scores of five machine learning methods on a primary dataset. The dataset was obtained from 500 patients at Dhaka Medical College Hospital. The model was trained using a 9-layer convolutional neural network (CNN) to identify breast tissue with a sensitivity of 93.4% and specificity of 94.6%. Among the five machine learning methods used on this dataset, XGBoost and RF demonstrated exceptional performance, achieving ac-curacy values of 97% and 96%. However, the article concludes the following: "It is necessary to recognize some limitations. More extensive real-world data are needed to confirm the model's generalization capability in larger populations. Additionally, the dataset only included derived image attributes rather than raw scans, limiting the possibility of direct image-based analyses. The proposed method can be improved using the most recent version of XGBoost because the version used is from a 2017 repository.

Bae K. et al. [12] analyzed age-related macular degeneration (AMD). This study focused on developing a deep learning model using the DenseNet201 architecture to predict the recurrence of AMD within 3 months after verifying the absence of disease activity 1 month following the administration of three loading injections to treatment-naïve patients. The model was trained using a total of 1076 spectral domain optical coherence tomography (OCT) images from 269 patients diagnosed with NAMD. To evaluate the model's effectiveness, its performance was compared with that of six experienced ophthalmologists using a set of 100 randomly selected images. The DenseNet201-based model outperformed the experienced ophthalmologists in predicting AMD recurrence. Using only one image before the first injection, the model achieved an accuracy of 53.0%, whereas the ophthalmologists achieved an average accuracy of 52.17%. After analyzing images after the first three injections, the model accuracy improved to 60.2%, compared to 53.3% for the ophthalmologists after reviewing four images. However, to improve the results obtained in this study, a more current version of the DenseNet201 algorithm, such as the more powerful DenseNet264, can be implemented. One drawback of this update is that having more training and work-ing layers increases memory usage and processing time.

In the study by Yolcu, G. [13] a quantitative and objective classification tool has been developed to assist dermatologists and scientists. The proposed system classifies seven skin lesions including monkeypox. A popular Vision Transformer approach and some popular deep learning convolutional networks have been trained with the transfer learning approach and all the results have been compared. The dataset was obtained from 2298 PAD-UFES-20 images and 102 MSLD images. Then, the models that demonstrated the best accuracy point were combined to perform the final prediction using bagging-ensemble learning. The ensemble-based proposal system yielded values of 81.91% accuracy, 65.94% Jaccard, 87.16% precision, 74.12% recall, and 78.16% Fscore. The system results were competitive, if not better, in terms of the various metrics reported in the literature.

Baykal, E. & Ayas, S., [14] developed a novel approach based on conjuntos was developed for the classification of cutaneous lesions, utilizing the various information captured by various ConvNeXt model architectures that have been shown to achieve performance comparable to most visual transformators using a CNN. More specifically, several versions of the preentrenado and adjusted ConvNeXt models Tiny, Small, Base, and Large were used for the classification of cutaneous lesion images to analyze and compare the classification results in the publicly available ISIC 2019 data set. ConvNeXt-Large achieved the highest precision of 97.2% among the separate models, thereby making it the best-performing model. The four ConvNeXt models were combined using confidence levels to increase the classification accuracy. The combined focus achieved a global classification accuracy of 97.7%, which exceeded both the performance of individual models and more sophisticated methods. Furthermore, 84.2% sensitivity and 97.9% specificity were obtained. The findings showed that the proposed method accurately and efficiently classified skin lesions based on dermatoscopy images.

Finally, the research conducted by Zhang, C. et al. [15] improved lung cancer diagnosis based on data fusion using DenseNet and CNN. This study introduces an innovative method that integrates a convolutional neural network (CNN) with DenseNet to enhance the identification and classification of lung cancer. It employs data fusion techniques to combine information from various sources, thereby increasing the model's accuracy. Additionally, mobile edge computing accelerates the processing of computed tomography images, thereby allowing real-time analysis near the data source. The proposed method enables excellent feature extraction and learning to differentiate normal, benign, and malignant lung tissue samples with high accuracy. In tests, this model outperformed other advanced methods, achieving an accuracy of 99%, highlighting its potential use for the early diagnosis of lung cancer. DenseNet121, a variant of dense convolutional networks (DenseNet), offers a unique architecture that facilitates more efficient training of neural networks with fewer parameters than traditional CNNs. In this case, as in the previous case, the device uses technology that can be improved. For DenseNet121, its performance can be better exploited if it is upgraded to one of its two subsequent versions: DenseNet201 or DenseNet264.

3 Methodology

The proposed methodology is based on [9, 10]. The model comprises six phases: Data Acquisition, Preprocessing, Features Vector Extraction, Application of Ma-chine Learning Models, Application of Deep Learning Models and Evaluation. Subsequently, each stage includes general activities crucial to research development. The proposed methodology is illustrated in Fig. 1.

3.1 Dataset Acquisition

As a first phase, the first step was to identify the most appropriate, reliable and relevant data sources. From previous studies, it was found that the Claudio Fan-conic dataset is robust and reliable for making melanoma-related predictions, as these are images from the ISIC (The International Skin Imaging Collaboration) database. This data contains organized information on patients who have been evaluated for skin cancers such as melanoma. It is worth mentioning that ISIC, according to the University of Queensland [16], is an international collaboration between academia and industry whose goal is to develop and promote standards for digital skin imaging to help reduce melanoma mortality.

The dataset was obtained from the URL https://www.kaggle.com/datasets/fanconic/skin-cancer-malignant-vs-benign/data. Additionally, Claudio Fanconic's dataset is divided into two folders categorized into benign and malignant types of moles: benign and malignant.

This dataset contains 1800 images (224x224) of each mole type. Table 1 lists the dataset used for model training and testing. The number of images of the two types of lesions (benign and malignant) and their dimensions are presented (Fig. 2).

3.2 Preprocessing

In this phase, data preprocessing is applied to develop the best strategy for distinguishing between benign and malignant tumors. Therefore, the proposed method is applied to the background images of a raw mole prior to feature extraction. Table 2 lists the activities.

- ***Normalize the images in the dataset:*** According to FasterCapital [17], data normalization scales data to a common range. This is important because deep learning models are sensitive to the scale of the input features. If features are not normalized, some may dominate other features, which leads to biased results. By normalizing the pixel values, we can ensure that the model treats all images equally, which improves the model training efficiency and stability.

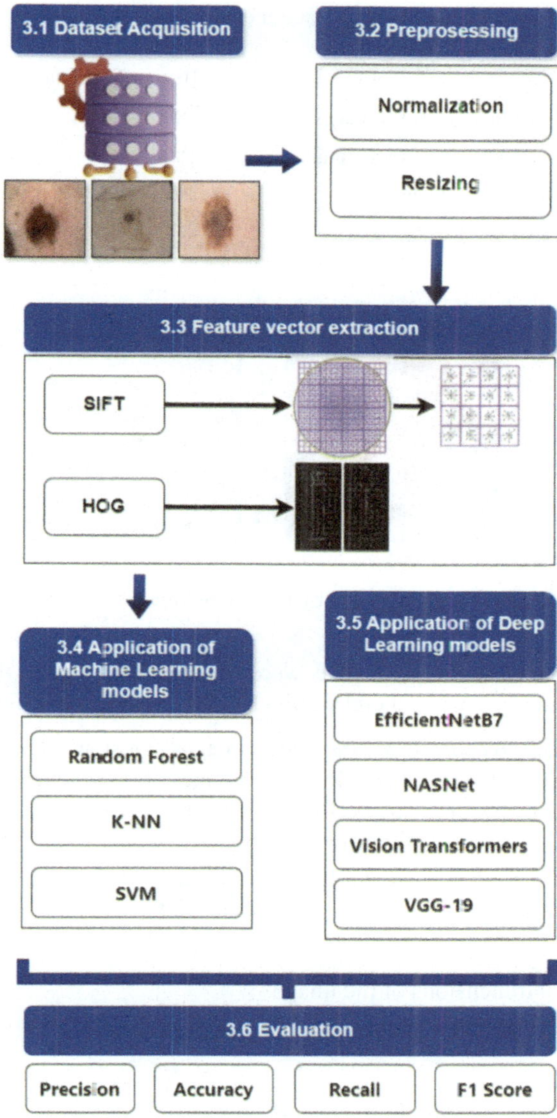

Fig. 1. PDCA Methodology with ISO 27001. Source: Own elaboration

$$Z = \frac{X}{255} \tag{1}$$

Where:

- Z: Normalized pixel value in the range [0, 1].
- X: Original pixel value in the range [0, 255].

Table 1. Dataset description

Type	Benign/Malignant
Number of benign moles images	1800
Number of malignant mole images	1800
Total number of images	3600
Image dimensions	224x224 pixels

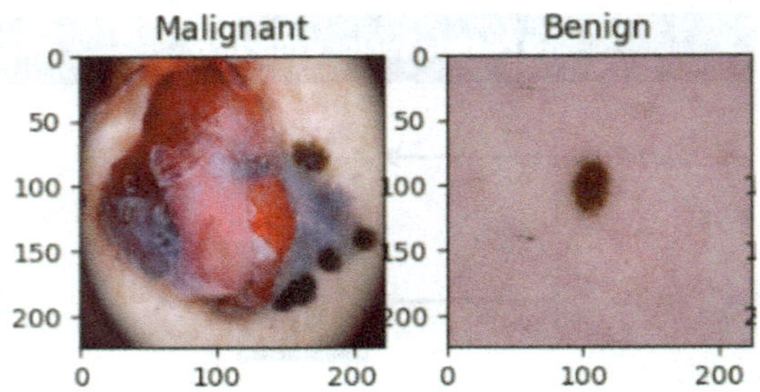

Fig. 2. Deliverables Model by Phases. Source: Own elaboration

Using range scale normalization, the data were adjusted according to the minimum and maximum values in the dataset, thereby allowing features to be within the range [0,1].

- ***Resize the dataset to a specif-ic size:*** Large images can slow webpage loading, negatively impacting user experience. Resizing images allows the images to load faster, which improves overall website performance [18].

$$new\_hight = \frac{hight}{aspect\_ratio} \tag{2}$$

Where:

- Height: Original dimensions of the image.
- aspect_ratio: Desired aspect ratio.

Table 2. Preprocessing activities

Activities	Description	Tasks
Normalize the images in the dataset.	Adjust the data according to the minimum and maximum values in the dataset.	Apply a formula to fit an algorithm.
Resize the dataset to a specific size.	Adjust the image size to obtain uniform images.	Resize some images to the specified size.

3.3 Features Vector Extraction

In this phase, two feature extraction techniques were applied: SIFT and HOG (see Table 3).

- ***Feature extraction with SIFT:*** The feature extraction procedure using the scale-invariant feature transform (SIFT) method is divided into a series of steps that ensure effective identification and description of key points in an image. Initially, the image is transformed into a grayscale version to reduce its complexity and prepared for texture and structure analysis by SIFT. Subsequently, the construction of the Gaussian scale pyramid is performed. This process generates different levels of "blurring" of the original image, each of which captures details of the image structure at a specific scale. The difference in Gaussian scale space is obtained by differentiating adjacent Gaussian blurred images, thereby capturing key details across scales. From this difference in Gaussian scale space, key points are detected by identifying local maxima and minima in the scale space and orientation [19].
- ***Feature extraction with HOG:*** The feature extraction process using the histogram of oriented gradients (HOG) method is divided into a series of steps that effectively capture the shape and appearance of objects in an image. The procedure begins by transforming the original image into a grayscale version. This conversion leads to a simplified representation of the image, which is suitable for the detailed gradient analysis performed by HOG. Essentially, the HOG algorithm evaluates the direction and magnitude of the gradient for each pixel in the image. This gradient calculation can identify rapid changes in pixel intensity and provides critical information about the boundaries and textures of objects in an image. After determining the gradients, oriented gradient histograms are constructed. This process is performed within specific regions of the image called cells. The histograms count the frequency of gradients in different orientations within each cell. The next step in the HOG procedure was block normalization. This step enhances the robustness of the HOG descriptor by adjusting the histograms of each cell with information from adjacent cells, thereby increasing the invariance to changes in lighting and shadows [20].

Table 3. Modeling and classification.

Activities	Description	Tasks
Feature extraction with SIFT	Feature extraction for each image using the SIFT method.	Image preprocessing
		Apply SIFT
		Store and validate results
Feature extraction with HOG	Feature extraction for each image using the HOG method.	Image preprocessing
		Apply HOG
		Store and validate results

3.4 Application of Machine Learning Models

- ***RF:*** A machine learning algorithm proposed by Breiman and Cutler that combines the outputs of multiple decision trees to obtain the result. The ease of use and flexibility have driven its adoption because it handles both classification and regression problems [21].

$$\hat{y} = mode\{T1(x), T2(x), \ldots, TN(x)\} \tag{3}$$

- ***KNN:*** A nonparametric supervised learning classifier that uses proximity to classify or make predictions about the grouping of an individual data point. It is currently one of the most popular and simple classification and regression classifiers used in machine learning. The KNN algorithm can be used for regression or classification problems, and it is typically employed as a classification algorithm based on the assumption that similar points are close to each other [22].

$$\hat{y} = \frac{1}{k}\sum_{j=1}^{k} y_{i_j} \tag{4}$$

- ***SVM:*** Supervised learning algorithms are used in many classification and regression problems, including medical signal processing, natural language processing, and image and voice recognition. The goal of the SVM algorithm is to find a hyperplane that best separates two different classes of data points. "Best separation" denotes the hyperplane with the widest margin between the two classes, represented by the plus and minus signs in the following figure. The margin is defined as the maximum width of the region parallel to the hyperplane that does not contain any data points. The algorithm can only find this hyperplane in problems that allow linear separation; in most practical problems, the algorithm maximizes the flexible margin by allowing some misclassifications [23].

$$w \cdot x + b = 0 \tag{5}$$

Parameters for Machine Learning models
Furthermore, the GRID hyperparameter configuration was employed with the following parameters for the machine learning algorithms.

- ***Random Forest:***
  - n_estimators = [10, 100, 200]
  - maximum features = ['sqrt', 'auto']
  - criterion = ['gini']
  - rf_grid = dict(n estimators=n estimators, max_features=max_features)
- ***KNN:***

- n_ neighbors = rango (1, 21, 2)
- weight = ['uniforme', 'distancia']
- metric = ['euclidiana', 'manhattan', 'minkowski']
- knn_grid = dict(n vecinos=n neighbord, weigth=weight, metric=metric)

- *SVM:*

  - core = ['rbf', 'sigmoide']
  - C = [10, 1,0, 0,1, 0,01]
  - gamma = [0.1, 0.01, 0.001]
  - param_grid_svm == dict(core=core, C=C, gamma=gamma)

3.5 Application of Deep Learning Models

In the training phase, using the prepared and preprocessed data, training can be effectively executed for the deep learning model. This involves associating the training features (referred to as "X_train") with the training labels (referred to as "y_train"). In other words, the respective model uses the images (stored in X_train) and their corresponding labels (stored in y_train) to learn to recognize patterns and features that distinguish between the two classes of moles (benign and malignant), thereby enabling melanoma detection.

Many training algorithms are currently available to build prediction models. For this study, the following algorithms are used:

- ***EfficientNetB7:*** A CNN algorithm that employs a scaling method uniformly scaling all depth, width, and resolution dimensions using a compound coefficient. Unlike conventional practices that arbitrarily scale these factors, EfficientNet's scaling method uniformly scales the width, depth, and resolution of the network with a set of fixed scale coefficients [24].

$$Escalado\ Compuesto : d = \alpha^p, w = \beta^p, r = \gamma^p \tag{5}$$

- ***NASNet:*** It is a type of convolutional neural network discovered through neural architecture search. The building blocks consist of normal and reduction cells [25].

$$\hat{y} = softmax(fdense(fGAP(Creduction(\ldots Cnormal(x)\ldots)))) \tag{6}$$

- ***Vision Transformer (ViT):*** The image classification model employs a Transformer-like architecture over image patches. An image is split into fixed-size patches, each of which is linearly embedded. Positional embeddings are added, and the resulting vector sequence is passed through a standard Transformer encoder. For classification, the standard approach of adding an additional "classification token" that can be learned to the sequence is used [26].

$$\hat{y} = softmax\Big(W_{cls} \cdot Transformer(PatchEmbedding(x))^{[CLS]} \tag{7}$$

- ***VGG-19:*** Convolutional neural network with 19 layers of depth. The proposed model can load a pretrained version of the network trained on more than one million images from the ImageNet database. The pretrained network can classify images into 1000 object categories (e.g., keyboard, mouse, pencil, and many animals). As a result, the network can learn feature-rich representations of various images. The input image size of the network is 224 by 224 [27].

$$\hat{y} = softmax\left(W_{fc3} \cdot ReLU\left(W_{fc2} \cdot ReLU\left(W_{fc1} \cdot Flatten(ConvPool(x)) + b_{fc1}\right) + b_{fc2}\right) + 3b_{fc3}\right) \quad (8)$$

Parameters for Deep Learning Models
The following hyperparameters were also used by the deep learning models:

(a) ***EfficientNetB7:***

- Input shape: (200, 200, 3)
- Learning Rate: 0.001
- Epochs: 10
- Batch Size: 32
- Optimizer: Adam

(b) ***NASNet:***

- Input shape: (200, 200, 3)
- Learning Rate: 0.001
- Epochs: 10
- Batch Size: 32
- Optimizer: Adam

(c) ***ViT:***

- Input shape: (200, 200, 3)
- Learning Rate: 0.001
- Epochs: 10
- Batch Size: 32
- Optimizer: Adam

(d) ***VGG-19:***

- Input shape: (200, 200, 3)
- Learning Rate: 0.001
- Epochs: 10
- Batch size: 32
- Optimizer: Adam

3.6 Evaluation

After training the models using the training dataset, it is essential to determine how well the models can predict data that they have not previously seen. This data is stored in a test image set. This step is crucial in the model evaluation phase of deep and automatic learning. To evaluate model performance, the fol-lowing metrics were used: precision, accuracy, recall, and Score-F1.

4 Results

For each trained model, the parameters that maximized performance were de-termined. These parameters, selected after rigorous trial-and-error testing, repre-sent the optimal conditions under which each model provides the most accurate results. Table 4 presents the training results of the machine learning models, and Table 5 presents the results of the deep learning models.

Table 4. Results obtained using optimal model parameters: RF, KNN and SVM models

	Feature Extractor	Parameters	Train	Test
RF	SIFT	{'n_estimators': 200, 'max_features': 'auto', 'max_depth': 20, 'criterion': 'gini'}	1.0	0.6888
	HOG	{'n_estimators': 100, 'max_features': 'auto', 'max_depth': None, 'criterion': 'gini'}	1.0	0.7277
KNN	SIFT	{'weights': 'uniform', 'n_neighbors': 1, 'metric': 'euclidean'}	1.0	0.6259
	HOG	{'weights': 'distance', 'n_neighbors': 2, 'metric': 'manhattan'}	0.9626	0.6092
SVM	SIFT	{'kernel': 'rbf', 'gamma': 0.001, 'C': 10}	1.0	0.5129
	HOG	**{'kernel': 'rbf', 'gamma': 0.01, 'C': 1}**	**1.0**	**0.7685**

Table 5. Result of Deep Learning models using specific parameters

Model	LR	Epochs	Optimizer	Loss	Accuracy
EfficientNetB7	0.001	10	Adam	0.6931	0.5
NASNet	0.001	10	Adam	0.4846	0.8750
ViT	0.001	10	Adam	0.4326	0.7917
VGG-19	0.001	10	Adam	0.4783	0.8050

4.1 Results of Machine Learning

Table 4 presents the top results obtained with the SVM-HOG model with the following hypotheses: kernel: 'rbf', 'gamma': 0.01, 'C': 1.

Table 4 shows that the SVM model combined with the HOG feature extractor has the best performance in terms of generalization, with a test accuracy of 0.7685. Random Forest (RF) models exhibit overfitting, achieving perfect accuracy in training but lower in testing. Furthermore, HOG consistently outperforms SIFT as a feature extractor, especially in combination with KNN, indicating that HOG is more effective for this dataset.

4.2 Results of Deep Learning

Table 5 lists the results obtained from the deep learning models and their hyperparameters.

In Fig. 3, they show the precision and loss curves during training of the ViT model. The precision of the ViT method increased, and the loss was reduced.

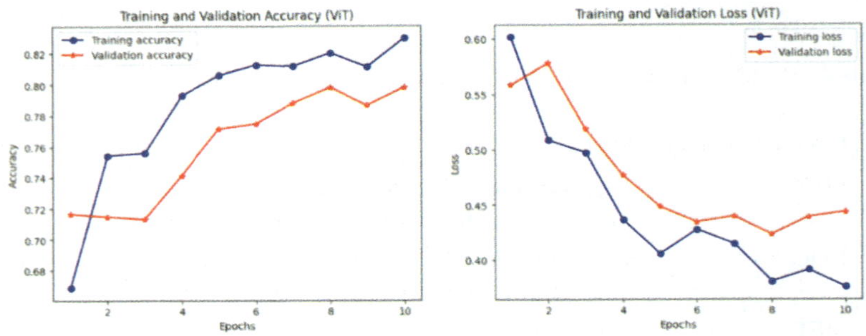

Fig. 3. (a) Accuracy of the ViT model in the training and validation sets. (b) Loss of the ViT model in the training and validation sets.

Figure 4 (a) and (b) show the differences between the training and validation accuracy and loss for the ViT model, respectively.

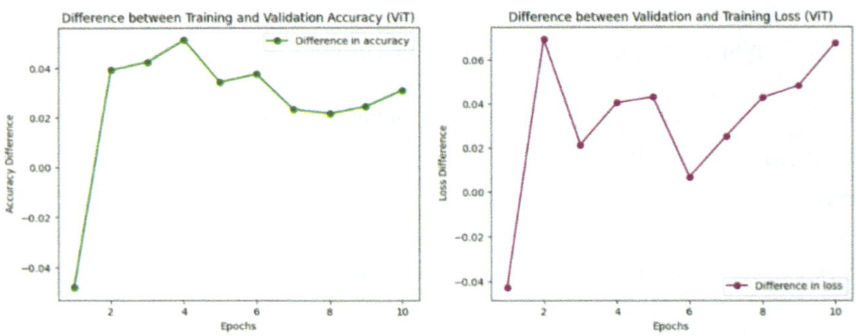

Fig. 4. (a) Difference between training accuracy and validation for ViT. (b) Difference between training loss and validation for the ViT model.

The table shows the results of different deep learning models with their specific hyperparameters. The NASNet model achieves the highest accuracy with 87.5%, followed by VGG-19 with 80.5%. The ViT model has a validation accuracy of 79.17% but shows a large discrepancy between accuracy and loss during training and validation, which could suggest generalization problems. Although the ViT accuracy improves during training, the difference between the training and validation curves suggests that there might be a slight overfitting. EfficientNetB7 has the worst performance, with an accuracy of only 50%.

4.3 Model Performance

The efficacy of the models, evaluated in terms of precision, accuracy, recall, and F1 score, is presented in Tables 6 and 7. These results reflect the ability of each model to make predictions, particularly regarding melanoma detection.

Table 6. Performance results of Machine Learning models

Model	Feature Extractor	Precision	Accuracy	Recall	Score F1
RF	HOG	0.7220	0.7278	0.7407	0.7313
KNN	SIFT	0.6096	0.6259	0.7	0.6517
SVM	HOG	0.7695	0.7685	0.7667	0.7681

Table 7. Model performance results of Deep Learning models.

Model	Precision	Accuracy	Recall	Score F1
EfficientNetB7	0.25	0.5	0.5	0.3333
NASNe	0.8219	0.8217	0.8217	0.8216
ViT	**0.8534**	**0.808**	**0.805**	**0.7981**
VGG-19	0.8378	0.81	0.81	0.8060

Figure 5 shows the results of the confusion matrix of the ViT model.

Figure 6 the Receiver Operating Characteristic (ROC) curve of a ViT (Vision Transformer) model, with an area under the curve (AUC) of 0.89. The ROC curve illustrates the relationship between the true-positive and false-positive rates, with the dotted blue diagonal being the non-discrimination line (a random prediction). An AUC of 0.89 indicates that the model performed well in classification, achieving a high hit rate and low error rate.

5 Conclusions

This study demonstrated that the use of advanced Deep Learning and Machine Learning techniques is highly effective for the classification and early detection of melanoma. CNNs demonstrated high precision and accuracy. In the introduction, the problem of melanoma was raised as a serious and growing disease due to global warming and the importance of early detection to improve the chances of survival was highlighted. It was mentioned that artificial intelligence (AI) offers an accurate and objective tool than traditional methods based on visual inspection. In this study, the ISIC database was used to collect skin cancer images and preprocessing techniques such as image resizing, and feature normalization were applied. The methodology was based on five phases: data acquisition, preprocessing, feature extraction, in-training, and evaluation and validation.

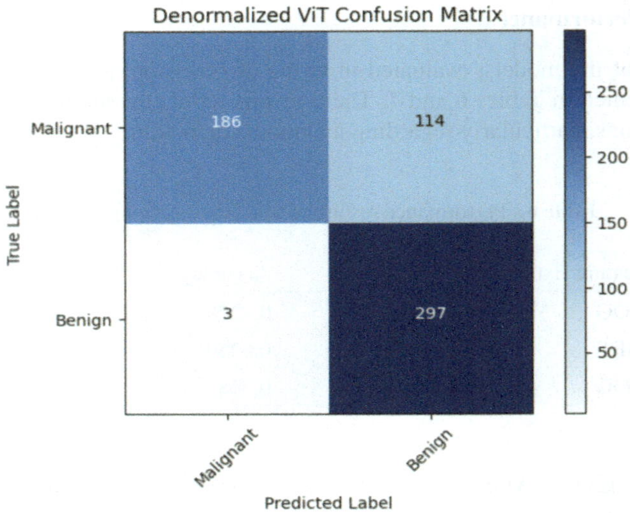

Fig. 5. Denormalized and normalized confusion matrix of the ViT model.

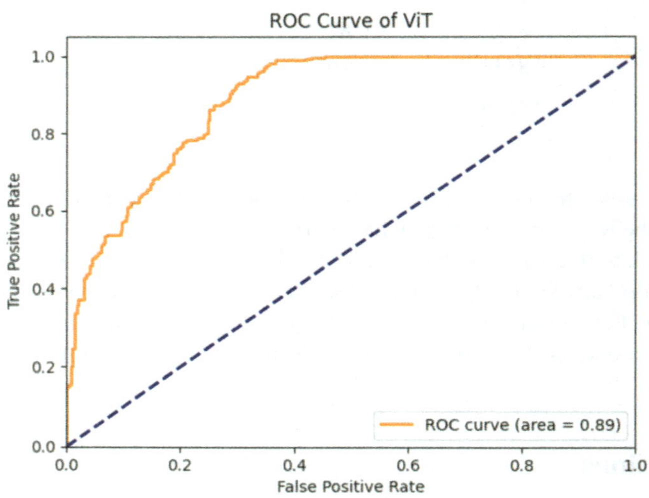

Fig. 6. ROC curve for ViT model

Deep Learning and Machine Learning models were used to classify melanoma cases and compare their results. The results demonstrate that deep learning models, in particular ViT and NASNet, show superior performance compared to traditional Machine Learning models. The ViT model obtained the highest accuracy (85.34%), followed by VGG-19 (83.78%) and NASNet (82.19%). The accuracy of the Deep Learning models was also remarkable, with ViT exhibiting 80.8% accuracy. NASNet and VGG-19 proved to be robust models with F1 scores of 82.16% and 80.6%, respectively, indicating a good balance between accuracy and recall. However, EfficientNetB7 demonstrated significantly lower performance on all metrics, suggesting that it is not suitable for this specific task. In summary, our findings underline the effectiveness of deep learning models for melanoma classification. The high accuracy and F1-score of the ViT and NASNet models highlight their robustness and ability to correctly classify melanoma, making these models valuable tools for early diagnosis of this disease. These results reinforce the potential of deep learning in medical applications by providing models that significantly improve the accuracy and efficiency in melanoma diagnosis. In conclusion, the proposed methodology, which is based on advanced Deep Learning and Machine Learning techniques, has proven effective in melanoma classification. This approach not only improves diagnostic practices in dermatology and provides a solid foundation for future research, with the potential to further integrate emerging technologies for detecting and treating melanoma.

References

1. ¿Qué es el cáncer de piel tipo melanoma? In: Cancer.org. https://www.cancer.org/es/cancer/tipos/cancer-de-piel-tipo-melanoma/acerca/que-es-melanoma.html
2. American Cancer Society. "Estadísticas importantes sobre el cáncer de piel tipo melanoma." Accessed: Apr. 24, 2024. [Online]. Available: https://www.cancer.org/es/cancer/tipos/cancer-de-piel-tipo-melanoma/acerca/estadisticas-clave.html
3. "Día del Lunar: EsSalud reporta 1877 casos de melanoma y otros tumores de piel durante el 2022 - Essalud," Essalud, 2022. http://noticias.essalud.gob.pe/?inno-noticia=dia-del-lunar-essalud-reporta-1877-casos-de-melanoma-y-otros-tumores-de-piel-durante-el-2022 (accessed Jun. 05, 2024)
4. Cabello, C.: "Cáncer de piel: ¿Qué países de la UE tienen las mayores tasas de incidencia por melanoma? | Euronews," Eurnonews. Accessed: Apr. 26, 2024. [Online]. Available: https://es.euronews.com/salud/2023/07/19/cancer-de-piel-que-paises-de-la-ue-tienen-las-mayores-tasas-de-incidencia-por-melanoma
5. European Cancer Information System, "Melanoma skin cancer factsheet in 2020 for EU-27 countries," Aug. 2021. [Online]. Available: https://cancer-code-europe.iarc.fr
6. Beyond Suncare, "Cáncer de piel, una epidemia causada por el sol," Beyond Suncare. [Online]. Available: https://beyondsuncare.org/cancer-de-piel/. Accessed: Apr. 26, 2024
7. EFE, "El melanoma que acecha a la población latinoamericana - Diario Libre," Diario Libre. [Online]. Available: https://www.diariolibre.com/actualidad/salud/el-melanoma-que-acecha-a-la-poblacion-latinoamericana-ME9107907. Accessed: Apr. 26, 2024.
8. Zhao, J., et al.: Global trends in incidence, death, burden and risk factors of early-onset cancer from 1990 to 2019. BMJ Oncol. **2**(1), e000049–e000049 (2023). https://doi.org/10.1136/bmjonc-2023-000049
9. Dahdouh, Y., Anouar, A.B., Ben Ahmed, M.: Embedded artificial intelligence system using deep learning and raspberrypi for the detection and classification of melanoma. IAES Int. J. Artif. Intell. **13**(1), 1104–1111 (2024). https://doi.org/10.11591/ijai.v13.i1.pp1104-1111
10. Roque, J.L.L., Martínez, R.E.B., González, T.R., Mesa, H.G.A.: Usos de la Inteligencia Artificial para el diagnóstico temprano del cáncer de la mama. Komputer Sapiens. **1**, 52–55 (2023). [Online]. Available: http://www.komputersapiens.org
11. Islam, T., Sheakh, M.A., Tahosin, M.S. et al. Predictive modeling for breast cancer classification in the context of Bangladeshi patients by use of machine learning

12. Jung, J., Han, J., Han, J.M., et al.: Predicción de la recurrencia de la degeneración macular neovascular relacionada con la edad mediante imágenes de tomografía de coherencia óptica con una red neuronal profunda. Representante científico. **14**, 5854 (2024). https://doi.org/10.1038/s41598-024-56309-6
13. Yolcu Oztel, G.: Vision transformer and CNN-based skin lesion analysis: classification of monkeypox. Multimed. Tools Appl. **83**, 71909–71923 (2024). https://doi.org/10.1007/s11042-024-19757-w
14. Baykal Kablan, E., Ayas, S.: Skin lesion classification from dermoscopy images using ensemble learning of ConvNeXt models. Sign. Image Video Proc. **18**, 6353–6361 (2024). https://doi.org/10.1007/s11760-024-03321-y
15. Zhang, C., Aamir, M., Guan, Y., et al.: Mejora del diagnóstico de cáncer de pulmón con fusión de datos y computación de vanguardia móvil utilizando DenseNet y CNN. J. Cloud. Comp. **13**, 91 (2024). https://doi.org/10.1186/s13677-024-00597-w
16. "International skin imaging collaboration," Uq.edu.au, Oct. 03, 2019. https://coh.centre.uq.edu.au/project/international-skin-imaging-collaboration (accessed Jun. 05, 2024)
17. Técnicas para la normalización de datos, *FasterCapital*. [Online]. Available: https://fastercapital.com/es/tema/t%C3%A9cnicas-para-la-normalizaci%C3%B3n-de-datos.html. (Accessed: Jun. 11, 2024)
18. Cyberstream (2023) Guía completa para redimensionar imágenes en diseño web. In: Byron Vargas ®. https://www.byronvargas.com/web/como-cambiar-el-tamano-de-las-imagenes/. Accessed 6 jul 2024
19. Lowe, D.G.: Distinctive image features from scale-invariant keypoints. Int. J. Comput. Vis. **60**(2), 91–110 (2004)
20. Dalal, N., Triggs, B.: Histograms of oriented gradients for human detection. In: Proceedings of the IEEE Computer Society Conference on Computer Vision and Pattern Recognition, vol. 1, pp. 886–893. IEEE (2005)
21. Random Forest, IBM. [Online]. Available: https://www.ibm.com/topics/random-forest#:~:text=Random%20forest%20is%20a%20commonly,Decision%20trees. [Accessed: Jun. 1, 2024]
22. K-Nearest Neighbors (KNN), IBM. [Online]. Available: https://www.ibm.com/topics/knn#:~:text=The%20k%2Dnearest%20neighbors%20(KNN,used%20in%20machine%20learning%20today. [Accessed: Jun. 1, 2024]
23. Support Vector Machine, MathWorks. [Online]. Available: https://la.mathworks.com/discovery/support-vector-machine.html. [Accessed: Jun. 1, 2024]
24. Tan, M., Le, Q.V.: EfficientNet: Rethinking Model Scaling for Convolutional Neural Networks. In: Proceedings of the 36th International Conference on Machine Learning, 2019. [Online]. Available: https://paperswithcode.com/method/efficientnet. [Accessed: Jun. 1, 2024]
25. Papers with code - NASNet. In: Paperswithcode.com. https://paperswithcode.com/model/nasnet?variant=nasnetalarge. Accessed 26 ago 2024d
26. Vision Transformer. In: Paperswithcode.com. https://paperswithcode.com/method/vision-transformer. Accessed 26 ago 2024e
27. VGG-19, MathWorks. [Online]. Available: https://es.mathworks.com/help/deeplearning/ref/vgg19.html. [Accessed: Jun. 1, 2024]

Application of Polynomial Regression in Assessing Students' Academic Performance in Blended Learning

Roman Tsarev[1,2(✉)], Darío Salguero García[3], Biswaranjan Senapati[4], Elena Denisova[5], Irina Nikolaeva[6], and Igor Kukhar[7]

[1] MIREA - Russian Technological University (RTU MIREA), Moscow, Russia
tsarev.sfu@mail.ru
[2] Bauman Moscow State Technical University, Moscow, Russia
[3] Almería University, Almería, Spain
[4] University of Arkansas at Little Rock, Arkansas, IL, USA
[5] Kaluga Branch of the Financial University under the Government of the Russian Federation, Kaluga, Russia
[6] North-Eastern Federal University, Yakutsk, Russia
[7] Reshetnev Siberian State University of Science and Technology, Krasnoyarsk, Russia

Abstract. The paper focuses on the assessment of student performance in blended learning. Blended learning is a combination of classroom and online formats, combining all the advantages of these formats and excluding their disadvantages. The advantages of blended learning include flexibility, variety of pedagogical approaches used, reduction of time costs, no loss of social interaction of students with each other and the teacher, independence from time and place, individual control over learning, giving students the opportunity to organize their own learning activities, control and analyze their results. In the conducted research, part of the students' out-of-class work in the framework of blended learning took place in an electronic course created in LMS Moodle. Students mastered theoretical material, performed practical tasks, and passed electronic tests. The Moodle provides information about both student learning outcomes and course attendance, student's time in the e-course and their activity. This paper investigates the relationship between student's time spent in e-course and student's academic performance using polynomial regression. The conducted study showed that students who spent more time in the e-course tend to have higher scores and better academic performance. #COMESYSO1120

Keywords: E-Learning · E-Course · Blended Learning · LMS Moodle · Polynomial Regression · Learning Outcome · Academic Performance

1 Introduction

E-learning and blended learning are among the modern pedagogical forms that have changed the approach to global education and transformed it. E-learning and blended learning are united, first of all, by the incorporation of information and communication

R. Silhavy and P. Silhavy (Eds.): CoMeSySo 2024, LNNS 1489, pp. 567–577, 2025.
https://doi.org/10.1007/978-3-031-96798-6_43

technologies (ICT) into the educational process [1–3]. However, it is necessary to distinguish between these forms and analyze the advantages and disadvantages of each of them.

E-learning is certainly characterized by such characteristics as flexibility (independence of the student and teacher from place and time), access to electronic resources, the ability to study learning materials in the LMS Moodle system, saving time and money of all participants of the educational process [4–8].

Despite this, e-learning limits students in personal communication with teachers, which is often necessary for students to better understand and master the material. Lack of direct contact with the teacher can also lead to isolation, lack of socialization, violation of self-discipline [9].

In addition, technical difficulties that teachers and students may encounter in e-learning cannot be avoided. A disadvantage of e-learning that needs to be emphasized is the limitation of learners' practical skills, as the virtual environment does not always provide in-depth practical training.

Blended learning is usually a combination of traditional learning methods (face-to-face classes) and e-learning [10–12]. This means that part of the learning process takes place in the classroom, while the other part takes place in an online format. This form is recognized by most educational researchers as the most effective, as it offers the same advantages as e-learning and eliminates all the disadvantages [13, 14].

Indeed, blended learning is characterized by the fact that students can get everything they need from live interaction with an educator and also be flexible, thanks to online classes [12, 15–18]. This combination leads to a more in-depth understanding of the subject and effective development of practical skills.

One way or another blended learning is noticeably replacing traditional learning around the world, the role of the learner is changing: in the traditional approach the teacher is in charge (teacher-centered), while in the blended learning model the focus shifts to the student (student-centered).

This means that in the framework of blended learning the teacher becomes an observer and assistant who only helps to learn, the independent work of the learner is given the main role: the student, mastering online learning material, organizes himself and his time.

There are several blended learning model groups. "Face-to-Face Driver" refers to the delivery of predominantly classroom-based instruction with the addition of e-learning technologies to the core curriculum [10, 19, 20]. This also includes the so-called extended face-to-face model, where students use digital learning resources in the classroom or complete an electronic test in a computer lab after lecture material.

Rotation model includes a more extensive list and is characterized by alternating modes of learning activities (classroom learning and online learning with LMS Moodle) [21]. This group includes classroom, laboratory (online), individual rotation, as well as the method of flipped classroom, when students study the theoretical part independently in the electronic environment and come to the practical part of the course face-to-face, namely classroom and online work change places [22–24]. The process of knowledge transfer is fully or partially transferred to independent study. This option is the most common and effective, because the study of theory by students independently online allows to better motivate them: students become more interested in obtaining or updating theoretical knowledge, without which they will not be able to perform practical tasks, because this will no longer be allocated time in the classroom in the framework of practice. The efficiency of flipped classes is investigated in our previous work [25].

There is also a flexible model (Flex) [20]. According to this model, most of the training is online, students work individually and in groups.

In any case, the student always refers to Moodle, which, in turn, having a large functionality, allows to place for students learning materials in different forms, create interactive tasks, electronic tests and automatically receive results on them (the ability to download reports) [26–31]. In addition, the teacher can track the attendance (and time spent) of the student in the course, build an individual learning trajectory for him/her, analyze the volume and quality of the studied educational material [32, 33].

Students can also contact with the teacher through chat or forum in the e-course. Thus, the main elements of the blended model, which allows using Moodle among others, are online course lectures with presentations, files with textbooks and manuals, online projects, audio and video files, virtual classrooms for feedback from teachers [27, 28, 32, 34].

It is particularly important to take into account the peculiarities of students and the specifics of the subject studied in the training of specialists whose professional activities are associated with technical solutions and research at the cutting edge of science [35–40]. Having actually the same time for learning as students of other specialties, future professionals of the fields related to high-tech developments should master the material at a level sufficient not only to fulfill their professional duties, but also to create new intellectual solutions [41–45]. Such fields that are leading today in terms of innovation, such as, for example, information technology or nanotechnology, require particularly high attention to the training of specialists [46–52]. This, in turn, is expressed in high requirements to theoretical teaching material, practical tasks, practical work in laboratories and production, as well as to the testing of students' knowledge and skills.

Naturally, it is impossible not to mention the difficulties that teachers may encounter in the process of implementing blended learning. Firstly, the teaching materials, which in this case should be adapted in such a way that they are suitable for both formats of lessons or differentiated. In addition, the teaching materials should be organized in such a way that the student can apply the acquired knowledge in practice both in the classroom and online. Secondly, it is necessary to take into account all possible technical difficulties, especially those related to online learning.

Finally, both students and instructors should have a sufficient level of information competence to work in an e-learning environment.

Student academic performance is a key aspect of the educational process, reflecting the level of mastering the educational material, as well as the ability to apply the acquired knowledge in practice. There are some groups of parameters of academic performance of students of higher education institution. The first group includes academic achievements directly: grades (for control and practical works, tests, examinations, attestations, coursework, etc.) or the average score of the student, which is calculated on the basis of all the grades received by the student. The second group includes student participation in classes, i.e. attendance in face-to-face classes and time spent and activity of the student in the e-course in Moodle.

In addition, performance is not limited to academic knowledge only. Communication and group interaction skills, critical thinking, personal qualities of the student are also important aspects. Assessment can be formative (through feedback from the instructor during the course to identify the student's own strengths and weaknesses), reciprocal (where students assess each other), or summative (assessment at the end of the course based on final works and exams).

Assessment of students' academic performance is a complicated process that requires taking into account many factors. It must be based on clear and known to the student criteria, be regular and aimed not only at identifying knowledge, but also at stimulating the student's motivation to learn, it is necessary to encourage successes, to consider mistakes as an opportunity for personal growth [8, 53–57].

Blended learning adapts assessment methods to accommodate the hybrid nature of classes. Traditional methods (performing and checking tasks and tests on worksheets with counting the correct answer options) are replaced by innovative methods - electronic testing [28, 30].

The organization of electronic testing in Moodle includes access settings (opening it only for the time of testing, disabling visibility on the page, limiting access to certain dates), as well as time limits, the requirement to complete the test and send the results [28, 30, 34]. Moreover, the teacher can customize the electronic test assessment settings: for example, the ability to take the test twice, and select the order of questions (including random). All these features and rich customization options cannot be compared to traditional printed tests in the classroom.

In addition, the teacher can create test questions of different types. Test tasks can be the same for all users or unique for each student, which will indicate compliance with the principle of differentiated individual approach to learning.

The main advantage of electronic testing is, as it was mentioned earlier, the possibility of automatic checking and obtaining the results in a visual form for their further analysis.

So, a special interest for us in this paper is the study of the dependence of student performance on the time of being in the e-course.

Certainly, the attendance of an e-course, which in one way or another is accessed by students in the framework of blended learning, can depend on many factors: the level of student engagement, teacher feedback and support, and course structuring. Let us focus more on the last factor: if an e-course is structured in such a way that assignments and materials are available within strictly defined timeframes, students who regularly attend the course are more likely to keep up with and understand the material at a deeper level.

LMS Moodle allows to record the time a student has spent studying the material. The teacher can access a report for each user (student) about the dates, time and duration of the course attendance, as well as what actions were performed by the student. In the Statistics section, students and the teacher can get a graph of course attendance, analyze these results, and find ways to adjust them if necessary.

It is important to note that the level of course attendance is influenced by such qualities of a student as self-discipline, self-organization, self-regulation and self-control.

Failure to develop these personal qualities can lead to academic dishonesty and, as a consequence, poor academic performance.

2 Methods

In our research, polynomial regression was used in order to identify a trend based on the values of the indicators of the time spent by the student in the e-course and the results of his/her academic performance.

In fact, a trend is a function of a given kind that can be used to approximate a graph constructed from the table data. It can be used both to identify trends in the development of the process, as well as perform a forecast, if the task at hand suggests it.

When establishing the most appropriate type of regression dependence to describe the process of change in the indicators of any value, the reliability indicator of the description of the function is used. The value of the approximation reliability index R^2 lies in the range from 0 to 1. The closer this value is to 1, the better the approximation is performed. If the value of the approximation reliability index R^2 is less than 0.6, it is considered that the selected type of dependence to describe the process of indicator change is not suitable.

The general form of the polynomial used in polynomial regression can be presented as follows:

$$y = \sum_{i=0}^{k} c_i \cdot x,$$

where c_i are the coefficients of the polynomial, among them c_0 is the free term, k is the degree of the polynomial.

The c_i coefficients can be found using the well-known least squares method [25, 58]:

$$y = \sum_{j=1}^{n} \left(\sum_{i=0}^{k} c_i \cdot x_j^i - y_j \right)^2 \rightarrow \min,$$

where y_j is the actual value of the dependent variable, the total number of observations is n, so $j = 1..n$.

3 Results

The research was conducted on a group of 30 students. Statistical results about the time spent in the e-course were obtained from the LMS Moodle. Figure 1 shows the dependence of the students' academic performance results on the time spent in the course. The maximum number of points a student could score during a semester is 100, the time in Fig. 1 is given in minutes.

Figure 1 shows linear and polynomial regression lines in addition to the indicated values. It can be seen that polynomial regression shows better approximation of the data and more clearly reflects the existing trend.

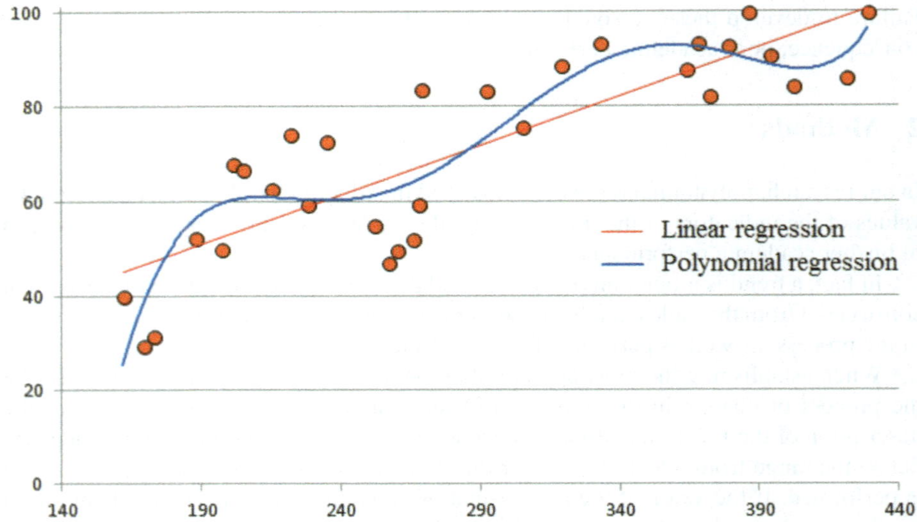

Fig. 1. Dependence of students' academic performance on the time spent in the e-course.

The equation of linear regression (actually it is a first order polynomial):

$$y = 0.21 \cdot x + 11.69.$$

The value of the approximation reliability index for linear regression is equal to:

$$R^2 = 0.71.$$

Polynomial regression equation (fifth order polynomial was used) is as follows:

$$y = -4798.2 + 88.3 \cdot x - 0.6 \cdot x^2 + 2.2E - 3 \cdot x^3 - 4E - 6 \cdot x^4 + 2E - 9 \cdot x^5.$$

The value of the approximation reliability index for polynomial regression is:

$$R^2 = 0.80.$$

The value of approximation reliability index for polynomial regression confirms that it is more appropriate to use polynomial regression for approximation of the values presented in Fig. 1.

4 Conclusion

The emergence of e-learning and blended learning was facilitated by the development of the info-communication paradigm of learning: the spread of ICT in education, increased information flows, constant updating of information, active development of means and methods of communication and the Internet.

Blended learning is characterized by a combination of learning formats: face-to-face classroom classes (classroom-lesson system) and e-learning (online work in an e-course created with Moodle).

Blended learning is the integration of traditional and e-learning, which opens up great opportunities and has many advantages: effective learning management tools, complementarity of ICT and traditional classroom teaching methods, which provides a variety of pedagogical approaches, reduction of time and financial costs, no loss of social interaction of students with each other and the teacher, independence from time and place, individual control over learning thanks to the functional capabilities of Moodle, providing students with the opportunity to independently organize their learning activities, control and analyze their results, and the teacher, in turn, to organize individual support for the learning activities of each student.

In this paper, polynomial regression was used to establish the relationship between student performance and time spent in the e-course. The study showed that students who spent more time in the e-course have higher academic performance, higher grades on electronic tests and completed assignments. The studies also confirm that the use of polynomial regression provides a more accurate approximation.

References

1. Mamatova, G.D., Kuchkarov, T.S.: Relevance of the concept of "Digital university": literary review of domestic and foreign research. Inform. Econo. Manag. **3**(1), 0101–0158 (2024). https://doi.org/10.47813/2782-5280-2024-3-1-0101-0158
2. Zenyutkin, N., Kovalev, D., Tuev, E., Tueva, E.: On the ways of forming information structures for modeling objects, environments and processes. Modern Innov. Syst. Technol. **1**(1), 10–22 (2021). https://doi.org/10.47813/2782-2818-2021-1-1-10-22
3. Begimkulov, U.S., Olimov, S.A.: Information and communication technologies of organization and management of innovative methodological support in preschool educational organizations. Inform. Econo. Manag. **3**(1), 0159–0168 (2024). https://doi.org/10.47813/2782-5280-2024-3-1-0159-0168
4. Al-smadi, A.M., Abugabah, A., Al Smadi, A.: Evaluation of e-learning experience in the light of the covid-19 in higher education. Proc. Comput. Sci. **201**, 383–389 (2022). https://doi.org/10.1016/j.procs.2022.03.051
5. Maulana, F.I., Febriantono, M.A., Raharja, D.R.B., Khaeruddin, Herasmara, R.: Twenty years of e-learning in health science: A bibliometric. Proc. Comput. Sci. **216**, 604–612 (2023). https://doi.org/10.1016/j.procs.2022.12.175
6. Shi, D., Wang, T., Xing, H., Xu, H.: A learning path recommendation model based on a multidimensional knowledge graph framework for e-learning. Knowl.-Based Syst. **195**, 105618 (2020). https://doi.org/10.1016/j.knosys.2020.105618
7. Veeramanickam, M.R.M., Ramesh, P.: Analysis on quality of learning in e-learning platforms. Adv. Eng. Softw. **172**, 103168 (2022). https://doi.org/10.1016/j.advengsoft.2022.103168

8. Zamecnik, A., Kovanović, V., Joksimović, S., Liu, L.: Exploring non-traditional learner motivations and characteristics in online learning: a learner profile study. Comput. Educ. Artif. Intell. **3**, 100051 (2022). https://doi.org/10.1016/j.caeai.2022.100051
9. Merzlyakova, E.Y., Skotnikov, D.S., Eltyshev, Y.V.: Extension of TRIZ with a digital data field through an intelligent technical agent of the human-machine interface. Mod. Innov. Syst. Technol. **4**(2), 0242–0256 (2024). https://doi.org/10.47813/2782-2818-2024-4-2-0242-0256
10. Alsalhi, N.R., Al-Qatawneh, S., Eltahir, M., Aqel, K.: Does blended learning improve the academic achievement of undergraduate students in the mathematics course?: A case study in higher education. Eurasia J. Math. Sci. Technol. Educ. **17**(4), em1951 (2021). https://doi.org/10.29333/ejmste/10781
11. Ustun, A., Tracey, M.: An effective way of designing blended learning: a three phase design-based research approach. Educ. Inf. Technol. **25**(1), 1529–1552 (2020). https://doi.org/10.1007/s10639-019-09999-9
12. Adiguzel, T., Kamit, T., Ertas, B.: Teaching and learning experiences with enhanced books in engineering math and science courses. Contemporary. Educ. Technol. **11**(2), 143–158 (2020). https://doi.org/10.30935/cet.660725
13. Schmid, R.F., Borokhovski, E., Bernard, R.M., Pickup, D.I., Abrami, P.C.: A meta-analysis of online learning, blended learning, the flipped classroom and classroom instruction for pre-service and in-service teachers. Comput. Educ. Open. **5**, 100142 (2023). https://doi.org/10.1016/j.caeo.2023.100142
14. Inprasitha, M.: Blended learning classroom model: a new extended teaching approach for new normal. Int. J. Lesson Learn. Stud. **12**(4), 288–300 (2023). https://doi.org/10.1108/IJLLS-01-2023-0011
15. de Brito Lima, F., Lautert, S.L., Gomes, A.S.: Contrasting levels of student engagement in blended and non-blended learning scenarios. Comput. Educ. **172**, 104241 (2021). https://doi.org/10.1016/j.compedu.2021.104241
16. Chiu, T.K.F.: Digital support for student engagement in blended learning based on self-determination theory. Comput. Hum. Behav. **124**, 106909 (2021). https://doi.org/10.1016/j.chb.2021.106909
17. Sareen, S., Mandal, S.: Challenges of blended learning in higher education across global north-south: a systematic and integrative literature review. Soc. Sci. Human. Open. **10**, 101011 (2024). https://doi.org/10.1016/j.ssaho.2024.101011
18. Shakeel, S.I., Haolader, M.F.A., Sultana, M.S.: Exploring dimensions of blended learning readiness: validation of scale and assessing blended learning readiness in the context of TVET Bangladesh. Heliyon. **9**(1), e12766 (2023). https://doi.org/10.1016/j.heliyon.2022.e12766
19. Prokhorets, E.K., Plekhanova, M.V., Scherbinina, N.G.: Instructional design of foreign language blended courses. Procedia. Soc. Behav. Sci. **215**, 161–169 (2015). https://doi.org/10.1016/j.sbspro.2015.11.611
20. Tesch, A.: Implementing pre-post test designs in higher education evaluations. N. Dir. Eval. **151**, 85–96 (2016). https://doi.org/10.1002/ev.20195
21. Barros, A.P.R.M.D., Simmt, E., Maltempi, M.V.: Understanding a Brazilian high school blended learning environment from the perspective of complex systems. J. Online Learn. Res. **3**(1), 73–101 (2017)
22. Attard, C., Holmes, K.: An exploration of teacher and student perceptions of blended learning in four secondary mathematics classrooms. Math. Educ. Res. J. **34**(4), 719–740 (2022)
23. Cevikbas, M., Kaiser, G.: Can flipped classroom pedagogy offer promising perspectives for mathematics education on pandemic-related issues? A systematic literature review. ZDM–Math. Education. **55**(1), 177–191 (2023)
24. Cevikbas, M., Kaiser, G.: Promoting personalized learning in flipped classrooms: a systematic review study. Sustain. For. **14**(18), 11393 (2022)

25. Tsarev, R., Senapati, B., Alshahrani, S.H., Mirzagitova, A., Irgasheva, S.: Ascencio, J: evaluating the effectiveness of flipped classrooms using linear regression. Lect. Notes Netw. Syst. **910**, 413–427 (2024). https://doi.org/10.1007/978-3-031-53552-9_38
26. Arce, E., Suárez-García, A., López-Vázquez, J.A., Fernández-Ibáñez, M.I.: Design Sprint: enhancing STEAM and engineering education through agile prototyping and testing ideas. Think. Skills Creat. **44**, 101039 (2022). https://doi.org/10.1016/j.tsc.2022.101039
27. Bengueddach, A., Boudia, C., Bouamrane, K.: Interpretive analysis of online teaching labs constructed using Moodle during the pandemic period. Heliyon. **9**(5), e16007 (2023). https://doi.org/10.1016/j.heliyon.2023.e16007
28. Deetjen-Ruiz, R., Esponda-Pérez, J.A., Haris, I., Salguero García, D., Quispe Osorio, J.L., Tsarev, R.: Evaluating the reliability of tests used in LMS Moodle for e-learning. Lect. Notes Netw. Syst. **935**, 1–8 (2024). https://doi.org/10.1007/978-3-031-54820-8_1
29. Hamamoto, N., Yokoyama, S., Takefusa, A., Aida, K.: Implemention of secured log analysis environment for Moodle using virtual cloud provider service. Proc. Comput. Sci. **192**, 3154–3164 (2021). https://doi.org/10.1016/j.procs.2021.09.088
30. Tsarev, R., Bhuva, A., Bhuva, D., Gogoleva, I., Nikolaeva, I., Bystrova, N., Varyan, I., Shamina, S.: Improving test quality in e-learning systems. Lect. Notes Netw. Syst. **723**, 62–68 (2023). https://doi.org/10.1007/978-3-031-35317-8_6
31. Yorulmaz, M., Can, G.F.: Task and user-based entropy-rank sum-TPOP integration proposal for usability evaluation of web applications. J. King Saud Univ Comput. Inform. Sci. **34**(8), 6466–6480 (2022). https://doi.org/10.1016/j.jksuci.2022.01.011
32. Bensalah, F., Daniel, M.P., Patra, I., Salguero García, D., Irgasheva, S., Tsarev, R.: Implementation of individual learning trajectories in LMS Moodle. Lect. Notes Netw. Syst. **935**, 159–174 (2024). https://doi.org/10.1007/978-3-031-54820-8_14
33. Deetjen-Ruiz, R., Ikonnikov, O., Azizam, S.H., García, D.S., Gavilán, J.C.O., Otcheskiy, I., Tsarev. R.: Applying ant colony optimisation when choosing an individual learning trajectory. Lect. Notes Netw. Syst. **723**, 587–594 (2023). https://doi.org/10.1007/978-3-031-35317-8_53
34. Chango, W., Cerezo, R., Romero, C.: Multi-source and multimodal data fusion for predicting academic performance in blended learning university courses. Comput. Electr. Eng. **89**, 106908 (2021). https://doi.org/10.1016/j.compeleceng.2020.106908
35. Canós Valero, A., Borovkov, D., Kalganov, A., Dudnikova, A., Sidorenko, M., Dergachev, P., Gurvitz, E., Gao, L., Bobrovs, V., Miroshnichenko, A., Shalin, A.S.: On the existence of pure, broadband toroidal sources in electrodynamics. Laser Photonics Rev. **18**(4), 2200740 (2024). https://doi.org/10.1002/lpor.202200740
36. Chebykin, A.V., Orlov, A.A., Shalin, A.S., Poddubny, A.N., Belov, P.A.: Strong Purcell effect in anisotropic ε-near-zero metamaterials. Phys. Rev. B. **91**(20), 205126 (2015). https://doi.org/10.1103/physrevb.91.205126
37. Kozlov, V., Filonov, D., Shalin, A.S., Steinberg, B.Z., Ginzburg, P.: Asymmetric backscattering from the hybrid magneto-electric meta particle. Appl. Phys. Lett. **109**(20), 203503 (2016). https://doi.org/10.1063/1.4967238
38. Kucherik, A., Kutrovskaya, S., Osipov, A., Gerke, M., Chestnov, I., Arakelian, S., Shalin, A.S., Evlyukhin, A.B., Kavokin, A.V.: Nano-antennas based on silicon-gold nanostructures. Sci. Rep. **9**(1), 338. https://doi.org/10.1038/s41598-018-36851-w
39. Senapati, B., Rawal, B.S.: Adopting a deep learning Split-protocol based predictive maintenance management system for industrial manufacturing operations. Lect. Notes Comput. Sci. **13864**, 22–39 (2023). https://doi.org/10.1007/978-981-99-2233-8_2
40. Shalin, A.S., Ginzburg, P., Belov, P.A., Kivshar, Y.S., Zayats, A.V.: Nano-opto-mechanical effects in plasmonic waveguides. Laser Photonics Rev. **8**(1), 131–136 (2013). https://doi.org/10.1002/lpor.201300109

41. Kuznetsov, A.V., Canós Valero, A., Tarkhov, M., Bobrovs, V., Redka, D., Shalin, A.S.: Transparent hybrid anapole metasurfaces with negligible electromagnetic coupling for phase engineering. Nano. **10**(17), 4385–4398 (2021). https://doi.org/10.1515/nanoph-2021-0377
42. Markovich, D., Baryshnikova, K., Shalin, A., Samusev, A., Krasnok, A., Belov, P., Ginzburg, P.: Enhancement of artificial magnetism via resonant bianisotropy. Sci. Rep. **6**(1), 22546 (2016). https://doi.org/10.1038/srep22546
43. Naeem, A.B., Senapati, B., Islam Sudman, M.S., Bashir, K., Ahmed, A.E.M.: Intelligent road management system for autonomous, non-autonomous, and VIP vehicles. World Electr. Vehic. J. **14**(9), 238 (2023). https://doi.org/10.3390/wevj14090238
44. Novitsky, D.V., Valero, A.C., Krotov, A., Salgals, T., Shalin, A.S., Novitsky, A.V.: CPA-lasing associated with the Quasibound states in the continuum in asymmetric non-Hermitian structures. ACS Photon. **9**(9), 3035–3042 (2022). https://doi.org/10.1021/acsphotonics.2c00790
45. Terekhov, P.D., Evlyukhin, A.B., Shalin, A.S., Karabchevsky, A.: Polarization-dependent asymmetric light scattering by silicon nanopyramids and their multipoles resonances. J. Appl. Phys. **125**(17), 173108 (2019). https://doi.org/10.1063/1.5094162
46. Bogdanov, A.A., Shalin, A.S., Ginzburg, P.: Optical forces in nanorod metamaterial. Sci. Rep. **5**(1), 15846 (2015). https://doi.org/10.1038/srep15846
47. Kukartsev, V., Degtyareva, K., Dalisova, N., Mazurov, A., Bezvorotnykh, A.: Optimization of maintenance work by implementing an automated information system at a repair facility. E3S Web Conf. **549**, 09011 (2024). https://doi.org/10.1051/e3sconf/202454909011
48. Kukartsev, V., Kravtsov, K., Stefanenko, O., Podanyov, N., Bezvorotnykh, A.: Using machine learning techniques to simulate network intrusion detection. In: Proceedings of the International Conference on Intelligent Systems for Cybersecurity (ISCS), pp. 1–4. IEEE, Gurugram (2024). https://doi.org/10.1109/iscs61804.2024.10581097
49. Malozyomov, B.V., Tynchenko, V.S., Kukartsev, V.A., Bashmur, K.A., Panfilova, T.A.: Investigation of properties of laminar antiferromagnetic nanostructures. CIS Iron Steel Rev. **27**(84–90) (2024). https://doi.org/10.17580/cisisr.2024.01.13
50. Naeem, A.B., Senapati, B., Bhuva, D., Zaidi, A., Bhuva, A., Sudman, M.S.I., Ahmed, A.E.M.: Heart disease detection using feature extraction and artificial neural networks: a sensor-based approach. IEEE Access. **12**, 37349–37362 (2024). https://doi.org/10.1109/access.2024.3373646
51. Savchenko, M., Tynchenko, V.: Unsupervised production machinery data labeling method based on natural language processing. In: Proceedings of the International Russian Smart Industry Conference (SmartIndustryCon), pp. 416–421. IEEE, Sochi (2024). https://doi.org/10.1109/smartindustrycon61328.2024.10515763
52. Senapati, B., Rawal, B.S.: Quantum communication with RLP quantum resistant cryptography in industrial manufacturing. Cyber Sec. Appl. **1**, 100019 (2023). https://doi.org/10.1016/j.csa.2023.100019
53. Adam, M.S., Hamid, J.A., Khatibi, A., Azam, S.M.F.: Autonomous motivation in blended learning: effects of teaching presence and basic psychological need satisfaction. Learn. Motiv. **83**, 101908 (2023). https://doi.org/10.1016/j.lmot.2023.101908
54. De Moura, V.F., de Souza, C.A., Viana, A.B.N.: The use of massive open online courses (MOOCs) in blended learning courses and the functional value perceived by students. Comput. Educ. **161**, 104077 (2021). https://doi.org/10.1016/j.compedu.2020.104077
55. Tomas, N., Poroto, A.: The interplay between self-regulation, learning flow, academic stress and learning engagement as predictors for academic performance in a blended learning environment: a cross-sectional survey. Heliyon. **9**(11), e21321 (2023). https://doi.org/10.1016/j.heliyon.2023.e21321

56. Tong, D.H., Uyen, B.P., Ngan, L.K.: The effectiveness of blended learning on students' academic achievement, self-study skills and learning attitudes: a quasi-experiment study in teaching the conventions for coordinates in the plane. Heliyon. **8**(12), e12657 (2022). https://doi.org/10.1016/j.heliyon.2022.e12657
57. Tsarev, R., El Amrani, Y., Alshahrani, S.H., Al Momani, N.M., Ascencio, J., Losev, A., Zhigalov, K.: Motivation to learn in an e-learning environment with fading mark. Lect. Notes Netw. Syst. **910**, 91–99 (2024). https://doi.org/10.1007/978-3-031-53552-9_8
58. Esponda-Pérez, J.A., Mousse, M.A., Almufti, S.M., Haris, I., Erdanova, S.A., Tsarev, R.: Applying multiple regression to evaluate academic performance of students in e-learning. Lect. Notes Netw. Syst. (2024). https://doi.org/10.1007/978-3-031-70595-3_24

Author Index

R. Silhavy and P. Silhavy (Eds.): CoMeSySo 2024, LNNS 1489, pp. 579–581, 2025.
https://doi.org/10.1007/978-3-031-96798-6

The manufacturer's authorised representative in the EU is Springer Nature Customer Service Centre GmbH, Europaplatz 3, 69115 Heidelberg, Germany. If you have any concerns regarding our products, please contact ProductSafety@springernature.com

Printed and bound by CPI Group (UK) Ltd, Croydon, CR0 4YY
30/09/2026
02225561-0019